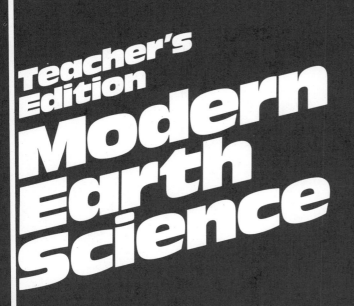

Teacher's Edition
Modern Earth Science

William L. Ramsey
Clifford R. Phillips
Frank M. Watenpaugh

Holt, Rinehart and Winston, Publishers
New York • Toronto • Mexico City • London • Sydney • Tokyo

William L. Ramsey
Head of the Science Department, Helix High School, La Mesa, California

Clifford R. Phillips
Science Department Chairman, Mount Miguel High School, Spring Valley, California

Frank M. Watenpaugh
Science Department, Helix High School, La Mesa, California

Copyright © 1983, 1979, 1973, 1969, 1965 by Holt, Rinehart and Winston, Publishers
All Rights Reserved
Printed in the United States of America

ISBN 0-03-059984-9

23456   071   9 8 7 6 5 4 3 2 1

contents

Introduction   Tiv
Reading Skills for Modern Earth Science   Tiv
Safety in the Science Classroom   Tvii
Alternative Approaches to Using Modern Earth Science   Tix
Materials for Chapter Activities   Txi
Equipment and Educational Material Suppliers   Txii
Learning Guide   Txiv

1 A Small Planet in Space   T1
2 The Sun   T4
3 Design of the Solar System   T7
4 The Moon   T11
5 Probing the Secrets of Space   T14
6 Models of the Planet Earth   T17
7 Earth Chemistry   T20
8 Materials of the Earth's Crust   T23
9 The Movement of Continents   T26
10 Movement of the Earth's Crust   T29
11 Weathering and Erosion   T32
12 Water and Rock   T35
13 Ice and Rock   T38
14 Shorelines   T41
15 The Unknown Sea   T45
16 Sea Water   T47
17 Motions of the Sea   T50
18 The Rock Record   T54
19 A View of the Earth's Past   T57
20 Air and Its Movements   T61
21 Water in the Atmosphere   T64
22 Weather   T67
23 Elements of Climate   T69
24 People and the Planet   T73

Masters of illustrations needed for text activities   T76

introduction

The *Modern Earth Science Teacher's Edition* contains the following items:

1. *Reading Skills for Modern Earth Science.* Suggestions are provided for practicing the skills needed by students to read and understand science content.

2. *Alternate Approaches to Using Modern Earth Science.* Suggestions are provided for using various topics in the text in different ways to meet different teaching requirements.

3. *Materials List.* This list can be helpful in obtaining equipment and materials for activities in the text.

4. *List of Equipment and Educational Materials Suppliers.* This list can be helpful in ordering supplemental materials.

5. *Learning Guide.* Teachers will find page references to subject matter, text activities, and questions that are keyed to each teacher objective.

In addition, the *Teacher's Edition* offers a chapter-by-chapter commentary. Each chapter commentary contains *Teacher Objectives*, a suggested *Approach* to the chapter, *Activity Hints*, *Other Activities*, *Answers* to *Vocabulary Review*, *Answers* to *Group A* and *Group B Questions*, *References* for students and teachers, and a list of *Audio-Visual Aids*.

Teachers who are interested in review activities and laboratory investigations will find *Activities and Investigations for Modern Earth Science* a useful supplement. This valuable aid follows the organization of the text; a *Teacher's Edition* is also available. In addition, *Tests for Modern Earth Science* following the sequence of material in the text is also available. Tests are in the form of duplicating masters with overprinted answers.

reading skills for modern earth science

readability of textbooks

There are several methods of determining the reading level of a textbook. Some methods are based on graded word lists, and others on readability scales, such as the Fry Scale. The Fry Scale is based on average sentence length and the total number of syllables. (*The Science Teacher*, March, 1974, pages 26-27, explains how to use the Fry Readability Scale.)

These methods focus on the mechanical aspects of reading rather than on the meaning of the words. Materials rated at a higher "reading level" according to mechanical tests might be easier for students to understand than materials rated at a lower "reading level."

diagnosis

A method of determining whether a student understands the text material is the Cloze test. The results of this test can help you decide if the student can work independently, requires instructions, or even with instructions will have reading difficulties.

To make a Cloze test, select at least a 250-word passage from the material you would like to assign. This material must be new to the student. Leave the first sentence complete. Starting with the second sentence, leave a uniform blank for every fifth word. Continue until there are 50 blanks. Then finish with a complete sentence. Do not start a sentence with a blank.

Give this or a similar test to your students with the instructions to fill in each blank with the word that makes the most sense. Score 2 points for each exact original word. Do not count synonyms. If the student scores below 40%, go back and score 2 points for each synonym or appropriate new word. Some students may even use better words than in the original sample.

Level of material for student:

60%-100% Student could study material independently.

40-59% Student will need teacher help to study material.

0-39% Student may experience frustration even with help.

following directions

The importance of following directions cannot be stressed too much in a science class. At times, even the safety of the experimenter is at stake if proper procedures or techniques are not followed. The exercise given here can be an interesting way to see immediate results and to correct errors in following directions. Duplicate the directions and ask students to carry them out one step at a time.

The answer to each question is yes. If you cannot answer yes, go back and redo the previous step or steps.
A. Obtain a piece of paper 21.5 cm square.
B. In pencil, mark one side of the paper "front" and the other side "back."
C. Place the paper front side up.
D. Fold the upper left corner diagonally to the lower right corner. Unfold.
   1. Do you now see a crease running from the upper right corner to the lower left corner?
E. Fold the upper right corner to the lower left corner. Undo the fold.
   2. Do you now see two diagonal creases forming an X on your paper?
F. Turn the paper over so that the front side faces the desk.
   3. Do you see the word "back" that

you wrote in pencil?
G. Fold the paper in half from left to right.
   4. Do you now have a rectangle that is taller than it is wide?
H. Undo the fold.
I. Fold the paper in half from top to bottom.
   5. Do you now have a rectangle that is wider than it is tall?
J. Undo the fold.
K. Turn the paper over so that the front side faces you.
L. Starting with the upper left corner, fold each corner to the center of the square (where all the creases cross).
   6. Is your paper now shaped like a diamond with four loose flaps shaped like triangles?
M. Fold the loose point of each triangle back to the center of the base of the triangle.
N. Cut out a 15-cm square picture from a magazine and place it inside your picture frame.

reading illustrations and graphs

The many photographs and graphic illustrations in *Modern Earth Science* are designed to help the students understand the concepts better or to help them follow directions in the activities more easily. Some students may need to improve their skill of reading illustrations.

An exercise that might help improve the skill of reading illustrations is having the students work with unit opener illustrations in the text. Direct the students to study the illustrations one at a time. Ask them to list what they see. After all of the facts are discussed or listed, ask students to interpret the illustrations. What is the main point? How does the illustration relate to the title of the unit? What do students think the unit will be about, based on a study of the illustration?

A graph is a means of organizing and illustrating data. Line graphs, bar graphs, and circle graphs are several types of graphs students should become familiar with in science. Often, students have problems reading graphs and interpreting the information they contain. You may want to refer to several graphs in the text to show how information is organized in a graph.

finding patterns in data

Scientists spend a lot of time collecting data and then looking for patterns in the data. In many of the text activities, students will collect the facts or data. To look for patterns, they will arrange the data in tables. In Chapter 13, in the Activity on page 279, the student is asked to prepare a table of the motion of particles on a simulated glacier. A given pattern emerges from the graph that is closely related to a graph given in Fig. 13-4, page 279, concerning a real glacier.

reading for cause and effect

Scientists are constantly relating cause and effect in their studies and experiments. Students should do the same in their reading and activities. They will learn facts faster if they look for cause/effect relationships among them.

Many of the geological changes of the earth can be related to cause and effect. These changes include erosion, earthquakes, volcanism, and weather.

Many exercises can be given to the students to test their understanding of cause and effect. For example, given two statements, have the students identify which is the cause and which is the effect.

reading to compare and contrast

Reading comprehension is improved if the reader has a definite purpose in mind. Scientists, for instance, might be thinking of comparing and contrasting information as they read. They may plan to analyze results of experiments by looking for similarities and differences. Such an analysis may be the key to a discovery. If the student knows how to look for similarities or differences in the reading or activity work, it will be easier to remember the facts, relate them to one another, and apply them to new situations. Examples in the text where these skills are used are:

In Chapter 1, the notion of constellations in the night sky was originally one of comparing groupings of heavenly bodies with shapes of common earth objects. In Chapter 4, the characteristics of the moon are given in terms of how it is like the earth and how it differs from the earth.

classification

In our everyday lives, we use classification constantly. Because of classification systems, we know our way around hardware stores and can find the book we need in the library.

In science, knowledge is classified so that it can be studied. Earth science is the study of geology, astronomy, meteorology, and oceanography. This vast body of information has already been grouped into these categories, although the boundaries overlap. Minerals or rocks are classified into groups according to their characteristics. The elements that make up a mineral are further grouped into families and arranged in the Periodic Table.

generalization

A generalization is a statement that encompasses common characteristics of a group of observations or statements. The observations could be made in an activity. The statement could be found in the reading material. Through analysis of their experiences, observations, and the facts given, the students should be able to form a generalization. As an example, you might refer to the activity concerning the rotation of the sun on its axis found in Chapter 2. Discuss with the class how the answers to the questions give data from which a general statement of the period of the sun's rotation can be deduced.

sequencing

Sequencing is an important part of science. The continuity of many scientific facts, concepts, and experiments relies on sequencing. Being able to follow things in a logical order is a valuable skill that will help students perform activities and recognize various concepts. An example of sequencing can be found in Chapter 12, page 256, where the students study the formation of a drainage pattern.

problem solving

Problem solving is defined as thinking of ways to combine observations with facts to find a solution. Scientists use the scientific method to solve problems. Students, too, should be encouraged to use the scientific method to solve problems in everyday matters as well as in science. It is important to realize, however, that every problem does not warrant a long, systematic approach for its solution. Some problems are so simple that the solution is easily discovered. The scientific method of problem solving should be employed only when there is a genuine problem to be solved.

Many of the Group B questions at the end of each chapter may be solved by using the scientific method. In addition, research exercises in *Activities and Investigations for Modern Earth Science* require the student to use the problem-solving technique.

safety in the science classroom

The responsibility for safety in the science classroom is that of the principal, teacher, and student. This means that proper attitudes toward safety should be developed from the outset. The information contained in the following pages is not intended to be all-inclusive, but it is designed to alert teachers to potential hazards and to provide suggestions for averting potential dangers.

Students should be required to read, understand, and sign a student safety contract, such as the one that follows, after they have been instructed in safety procedures.

student safety contract

I will:

- Follow all instructions given by the teacher
- Read directions thoroughly before starting an activity
- Protect eyes, face, hands, and body while conducting an activity
- Know the location of first-aid and fire-fighting equipment
- Conduct myself in a responsible manner at all times in a laboratory situation

I, _____, have read and agree to abide by the safety regulations as set forth above and also any additional printed instructions provided by the teacher and/or district. I further agree to follow all other written and verbal instructions given in class.

Date_____

_____
Student Signature

_____
Parent Signature

school safety audit

It is recommended that at the beginning of the school year and throughout the year as the need arises a safety audit of classroom and laboratory facilities be conducted by the teacher.

storing chemicals

It is not advisable to store large amounts of potentially hazardous chemicals over long periods of time. Contact your regional EPA solid waste office for information on proper disposal of surplus or dangerous chemicals. To aid in identification of chemicals, a standard labeling system should be used.

incident/accident report

In the science classroom, as well as in any other class, any accident, hazardous incident, or potentially hazardous incident should be officially recorded and reported as soon as possible. A standard report form can be used to facilitate the recording of such incidents and insure that all necessary information is recorded.

field trips

When taking students on field trips it is advisable for the teacher to visit the area beforehand to note any possible hazards. These potential dangers should then be discussed with the students; special clothing to be worn (such as hard hats) and rules for safe conduct should be established.

Adequate supervision by teachers and/or parents should be insured and school policies for obtaining proper consent prior to the field trip should be followed.

laboratory safety guidelines

The following is a list of some of the general guidelines that should be followed in the science laboratory at all times. Teachers should review these guidelines with their students and be sure that they are understood.

- All science laboratories should have and use the following safety items: fire extinguishers, first-aid kits, fire blankets, sand buckets, eyewash facilities, emergency shower facilities, safety goggles, laboratory aprons, heat-resistant gloves, tongs, respirators, and wire or ceramic-centered gauze.
- Good housekeeping is essential for maintaining safe laboratory conditions.
- Never have students conduct experiments alone in the laboratory.
- Every laboratory should have two exits.
- There should be an annual, verified safety check of each laboratory.
- Master cutoff valves or switches for gas, water, and electricity are advisable for laboratories.
- Emergency instructions or procedures for fires, explosions, chemical reactions, spillage, and first aid should be conspicuously posted near all storage areas.
- Always perform an experiment or demonstration prior to allowing students to do the activity. Look for possible hazards. Safety instructions should be given each time an experiment is begun.
- Horseplay or practical jokes of any kind are not to be tolerated.
- Never eat or drink in the laboratory or from laboratory equipment.
- Exercise great care in noting odors or fumes. Use a wafting motion of the hand.
- Never "force" glass tubing into rubber stoppers.
- A positive attitude toward safety is imperative. Students should not be afraid to do experiments or use reagents and equipment, but should respect them as potential hazards.
- Never allow the open end of a heated test tube to be pointed at anyone.

- Broken or chipped glassware should not be used.
- Breathing gases, especially in high concentrations, can be very dangerous.
- Never taste chemicals for identification.
- Do not pour water into acid. Always add acid to water.
- Students unable to safely perform certain activities should be excused from these activities.

bibliography

American National Safety Standard Practice for Occupational and Educational Eye and Face Protection. ANSI X87.1-1968. American National Standards Institute, New York, NY

Chemical and Biological Safety Guide. Stock No. 174000383. U.S. Department of Health, Education and Welfare, National Institutes of Health. Superintendent of Documents, U.S. Government Printing Office, Washington, DC

Fisher Safety Manual. Fisher Scientific Co., 711 Forbes Ave., Pittsburgh, PA 15219

Hazardous Materials Handbook. Beverly Hills, CA: Glencoe Press, 1972.

Laboratory Waste Disposal Manual. Manufacturing Chemists Association, 1825 Connecticut Ave., Washington, DC 20009

audio-visual aids on safety

Laboratory Safety—Part I
Motion picture, 16mm, sound, 20 minutes. This excellent film was produced by the Virginia State Department of Education. It is one of the very few films that directly addresses the school laboratory safety issue. NIOSH will supply one copy to each state. Contact your State Science Supervisor or another designated State Officer.

Safety Attitudes
Motion picture, 16mm, color, 10 minutes. Describes the causes and effects of unsafe attitudes. Illustrates that attitudes are learned, not inherited.
National Safety Council
425 N. Michigan Avenue
Chicago, IL 60611

suppliers of safety equipment

Syracuse Safety Service
1108 Spring Street
Syracuse, NY 13208

Sargent-Welch Scientific Co.
7300 N. Linder Avenue
Skokie, IL 60076

alternative approaches

module and "mini-course" approach

Each unit in *Modern Earth Science* has been planned as a complete treatment of a major segment of the earth sciences. While the sequence of these units used in the text represents a systematic approach, each unit stands by itself. This flexibility allows each unit to be used as a separate teaching "module" or complete abbreviated course dealing with a particular segment of the earth sciences. A teacher can select the modules and the sequence needed to assemble a complete earth science course to fit a desired emphasis or time span. This kind of flexibility is particularly useful with individualized contact and self-paced instruction.

Within each module the chapters can be "mini-courses" with their own laboratory investigation. An example follows.

Module: The Planet Earth

Purpose: The student should be able to give evidence that the earth is a large and complex object that gives evidence of internal processes that are constantly changing its surface.

Mini-course 1: Models of the Planet Earth (Chapter 6, page 116)
Laboratory Investigation: (Chapter 6, Topographic Maps)

If the nature of the substance of the earth is emphasized then:

Mini-course 2: Earth Chemistry (Chapter 7, page 138)
Laboratory Investigation: (Chapter 7, Chemical Analysis)

Mini-course 3: Materials of the Earth's Crust (Chapter 8, page 153)
Laboratory Investigation: (Chapter 8, Mineral Identification)

one-objective-a-day approach

A one-year plan for studying *Modern Earth Science* can be made by using the Learning Guide in the *Teacher's Edition.*

individual curriculum approach

The following guide may be helpful in developing effective curriculum, with considerations given to: (1) Minimal Course (2) Average Course (3) Enriched Course.

Part of Course	Minimal	Average	Enriched
Topics and Chapters	Geology 6, 9–13 Oceanography 14–17 Meteorology 20–22	All minimal plus Astronomy 1–5 Earth Chemistry 7 Rocks and Minerals 8	All average plus Earth's History 18–19 Climate 23 Resources and Energy 24
Approach	Class time for reading instruction, text activities, and demonstrations	Class discussions, text activities, demonstrations, field trips, laboratories	Text reading as homework to allow more challenging laboratory and activities in class
Review	Review Activities from workbook done in class	Review Activities from workbook done in class	Review Activities assigned as homework
Questions	Vocabulary and Group A	Vocabulary, Group A, and selected Group B	Vocabulary, Group A, and more Group B
Further Reading		Selected Research in workbook	Workbook research as class reports
Text Activities	As time and ability permit	Most activities except more challenging	All activities; many can be done at home
Laboratory Investigations		Most Investigations	All Investigations

materials for chapter activities

Chapter	Page	Materials	Quantity for 30 students	Chapter	Page	Materials	Quantity for 30 students
1	15	drawing compass	30			plastic straw; large test tube	each
		pins	60			tape (masking)	1 roll
		string	20 m			clay	1 pkg
		cardboard, 8½ × 11	30			salt water solution*	20 mL/student
2	28	none				test tube rack	15
3	62	metric ruler	30			Part II—food coloring (dark)	1 box
4	77	meter stick	30			saltwater solution**	100 mL/student
		index cards (3 × 5)	50			plastic shoe box	30
5	102	NASA equipment list (see Sources of Information, Tx)	30	17	354	beaker; thermometer	30 of each
						rubber bands	1 box
6,7,8		none				250 watt lamp	4
9	186	map (see page T76)	30	17	358	beaker; plastic shoebox	30 of each
10	222	map (see page T76)	30			ice cubes	2/student
11	235	paper cup	30			food coloring (dark)	1 box
		tissue paper	1 box	18	381	coins or discs	100/student
		sand	5 lbs			shoebox	30
		table salt	1 box	19	409	cross-section of rock layers (see page T77)	30
12	267	glass or plastic tubing (1 cm diameter, 15 cm long)	60 pieces	20	434	candle (birthday); index cards (4×6); metric ruler; 16d finishing nails	30 of each
		cotton; gauze; rubber bands	1 box each	21	442	shiny tin can or plastic glass	30
		dry, fine sand; dry, coarse sand	5 lbs of each			ice cubes	5 lbs
		metric ruler; plastic or glass container (about one quart)	30 of each			thermometer	30
13	279	silicone putty	15	22	471	weather map (see page T78)	30
		4×6 index cards	60	23	480	almanac or magazine articles	
		tape (masking)	1 roll	24	501	10-cm test tube; ring stand; glass tubing; file***	30 of each
		small beads (ball bearings)	150			Bunsen burner	15
14		none				wood splints or saw dust	one bundle or can
15	328	large test tube; magnifier	30 of each				
		sandy soil	5 lbs				
		aluminum foil	1 roll				
		test tube rack	15				
16	340	Part I—metric ruler;	30 of				

*Make salt solution by adding 3.6 grams of table salt to 10 mL of distilled water.
**Same as previous procedure.

***Caution: Show students how to cut and file the glass tubing. Point out the dangers of this procedure.

equipment and educational material suppliers

equipment suppliers

Calbiochem, 3625 Medford Street, Los Angeles, CA 90054

Carolina Biological Supply Co., Burlington, NC 27215

Eduquip-Macalaster Corp., 1085 Commonwealth Avenue, Boston, MA 02215

Fisher Scientific Co., 633 Greenwich Street, New York, NY 10014

LaPine Scientific Co., 6001 S. Knox Avenue, Chicago, IL 60629

Macmillan Science Co., 8200 S. Hoyne Avenue, Chicago, IL 60620

Sargent-Welch Scientific Co., 7300 N. Linder Avenue, Skokie, IL 60076

Wards Natural Science Establishment, Inc., P.O. Box 1712, Rochester, NY 14603

The following is a list of various agencies and publications that offer suggestions that you can use to help teach handicapped students.

American Foundation for the Blind, 15 West 16th Street, New York, NY 10011 (aids and appliances for the blind and visually impaired)

Association for Education of the Visually Handicapped, 919 Walnut, fourth floor, Philadelphia, PA 19107

Gearheart, B., and M. Weishahn. *The Handicapped Child in the Regular Classroom.* St. Louis: The C.V. Mosby Co., 1976.

National Easter Seal Society, Information Center, 2023 W. Ogden Avenue, Chicago, IL 60612

National Association for Gifted Children, 217 Gregory Drive, Hot Springs, AR 71901

National Association for Retarded Citizens, 2709 Ave. E East, Arlington, TX 76011

National Association of the Deaf, 814 Thayer Avenue, Silver Springs, MD 20910

National Information Center for Special Education Materials, University of Southern California, University Park, Los Angeles, CA 90007

sources of information

NASA, Public Relations, Washington, DC 20546

AEF American Education Films, 331 N. Maple Drive, Beverly Hills, CA 90210

AGS American Guidance Service, Inc., Publishing Building, Circle Pines, MN 55014

American Meteorological Society, 45 Beacon Street, Boston, MA 02100

BARR Arthur Barr Productions, 1029 North Allen Ave., Pasadena, CA 91104

Caterpillar Tractor Film Library, 160 E. Grand Ave., Chicago, IL 60611

Coronet Instructional Films, 65 E. South Water Street, Chicago, IL 60601

CF Churchill Films, 662 N. Robertson Blvd., Los Angeles, CA 90069

Creative Visuals, Box 1911-S3-A, Big Spring, TX 79720

DG Denoyer-Geppert, 5235 Ravenswood Ave., Chicago, IL 60640

Doubleday Multimedia, P.O. Box 11607, Santa Ana, CA 92705

Ealing Films, 2225 Massachusetts Avenue, Cambridge, MA 02140

EAV Educational Audio Visual Inc., Pleasantville, NY 10570

EBEC Encyclopedia Britannica Educ. Corp., 425 N. Michigan Ave., Chicago, IL 60611

EGH Eyegate House, 14601 Archer Avenue, Jamaica, NY 11435

FA Film Associates of California, Santa Monica Blvd., Los Angeles, CA 91014

FFE Films for Education, Audio Lane, New Haven, CT 06519

Filmstrip House, Inc., 432 Park Ave. South, New York, NY 10016

film and other audio-visual aids suppliers

ABP Arthur Barr Production, P.O. Box 7-C, Pasadena, CA 91104

HENNESSY John J. Hennessy, 1702 Marengo Ave., South Pasadena, CA 91030

Holt, Rinehart and Winston, Publishers, 383 Madison Ave., New York, NY 100l7

Hubbard Scientific Co., P.O. Box 105, Northbrook, IL 60062

IFB International Film Bureau, 332 South Michigan Ave., Chicago, IL 60604

Indiana University, Audio-Visual Center, Bloomington, IN 4740l

JFI Journal Films, Inc., 909 West Diversey Parkway, Chicago, IL 60614

Learning Arts, Box 917, Wichita, KA 67201

LIFE Life Education Program, Box 834 Radio City P.O., New York, NY 10020

Listening Library, 1 Park Avenue, Old Greenwich, CT 06870

McGraw-Hill Films, 1221 Ave. of the Americas, New York, NY 10020

Millikan Publishing Co., 1100 Research Blvd., St. Louis, MO 63132

MTPS Modern Talking Picture Service, 1212 Ave. of the Americas, New York, NY 10036

MIS Moody Institute of Science, 1200 E. Washington Blvd., Whittier, CA 90606

NASA Public Relations, Washington, DC 205‡6

NFBC National Film Board of Canada, 680 Fifth Ave., New York, NY 10016

NGS National Geographic Society, Educational Services, 17 and M Streets N.W., Washington, DC 20036

OSU Ohio State University, Motion Picture Division, 1885 Neil Ave., Columbus, OH 43210

PHOENIX Phoenix Films, 470 Park Avenue South, New York, NY 10016

PRENTICE-HALL Prentice-Hall, Inc., Englewood Cliffs, NJ 07632

Ramsgate Films, 704 Santa Monica Blvd., Santa Monica, CA 90401

Rand McNally and Co., School Dept., Box 7600, Chicago, IL 60680

Schloat Productions, 1500 White Plains Road, Tarrytown, NY 10591

Edwin Shapiro Co., 43-55 Kissena Blvd., Flushing, NY 11355

Shell Oil Co. Films, 450 North Meridian St., Indianapolis, IN 46204

SVE Society for Visual Educations, Inc., 1345 Diversey Parkway, Chicago, IL 60614

Sterling Educational Films, Inc., 241 East 34th Street, New York, NY 10016

Stuart Finley, Inc., 3428 Mansfield Road, Falls Church, VA 22041

Thorne Films, Scott Education Div., Holyoke, MA 01040

Time-Life Films, Inc., 43 West 16th St., New York, NY 10011

UEVA Universal Education & Visual Arts, 221 Park Ave. South, New York, NY 10003

University of Wisconsin, Film Library, Madison, WI 53706

USGA U.S. Geological Survey, Distribution Section, Washington, DC 20244

USSCS U.S. Soil Conservation Service, Motion Picture Service, U.S. Dept. of Agriculture, Washington, DC 20525

UWF United World Films, 1455 Park Ave., New York, NY 10029

Ward's Natural Science Est. Inc., P.O. Box 1712, Rochester, NY 14603

WORLD The World Publishing Co., 119 West 57th St., New York, NY 10019

modern earth science learning guide

Teachers will find page references to subject matter, activities, and questions that appear in the text. Each text reference is keyed to a teacher's and student's objective listed in the second and third columns and to corresponding questions. All the areas of study in this guide are intended for the average earth science student. Many of the Laboratory Investigation Thought Questions and Research Activities in the workbook (not listed in the column) are also particularly challenging. The questions in *italics* in this table are more challenging.

This Learning Guide may be used to help the teacher plan and prepare daily and weekly class activities, homework assignments, and materials for testing. Copies of the Learning Guide may also be made for students who are involved in individualized or contract learning.

chapter	objective teacher	objective student	text page	activity page	questions A	questions B
1	1	1	2-4		1,2,3	1
	2	1	5		4,5,6	2,3,4,5,6
	3	2	6-7		7,8	7,8
	4	3	7-8	15	9,10	9,10
	5	4	9-11		11,12,13,14,15	11
	6	5	11-15		16,17,18,19,20	12,13,14
	7	6	15-17		21,22,23	15,16
	8	7	17		24,25,26	17
2	1	1	21-23		1,2,3	1
	2	2	23-24		4	2
	3	3	24-26		5,6,7,8,9	3,4
	4	4	26-27	28	10,11,12	5
	5	4	27		13	6
	6	5	28-30		14,15	7,8
	7	6	30-34		16,17,18,19	9,10
	8	7	34-35		20,21,22	11,12
	9	8	35-38		23,24,25	13
	10	8	38-39		26,27	14,15
	11	9	39-40		28	16

chapter	objective teacher	objective student	text page	activity page	questions A	questions B
8	1	1,2	153-157		1-6	16,18,19
	2	3	158	163	22,23,28	3
	3	4	158-169		7-16	
	4	5	169-178		19-21,24-27,29,30	
9	1	1	182-183		1-3	1
	2	1	183		4,5	2,3
	3	2,3	183-188		6-17	4-7(6)
	4	4	188-189		18,19	8,9
	5	5	189-192	186	20-24	10-15(13)
	6	6	192-195		25-27	16
	7	7,8	196-203		28-30	17-20
10	1	1	207-208		1,2	1
	2	2,3	208-212		3-7	2-4
	3	4	212-213		8-11	5-6
	4	5	213-217		12-16	7-10(9)
	5	6	217-220	222-223	17-19	11-12
	6	7	220-221		20-22	13-15(14)

Sections 11–16

Sec			Page			
11	1	1	232-235	235	2-8	1-7(2,3,5,6)
	2	2	235-238		9-11	8,17
	3	3,4	238-239		1,12-19	9,10
	4	5	239-245		20-25	11-15(12)
	5	6	245-247		26-28	16,18
	6	7	247-250		29-30	19-21
12	1	1	253-255	267	1-5	1,8,17
	2	2	236		6-8	19
	3	3	256-260		9-16	3,5,11,13
	4	4	260-265		17-20	6,9,12,14,16,18
	5	5	265-268		21-23	2,7,15
	6	6	268-270		24-25	10
	7	7	270-271		26-27	4
13	1	1,2	275-281	279	1-10	1-5(2)
	2	3,4	281-286		11-20	6-11
	3	5,6	287-289		21-26	12-15
	4	7,8	289-292		27-30	16-18
14	1	1	296-297	304	1-7	1-5(2,3,4)
	2	2	297-301		8-17	6-9,14
	3	3,4	301-307		18-22	10-13
	4	5	307-309		23,24	16,17
	5	6	309-310		25,26	18,19
15	1	1	316	328	3	1-8(5,7)
	2	2,3,4	317-322		1,2,4-15	9-15(10-14)
	3	5	323-327		16-21	16-20(17)
	4	6	327-330		22-30	
16	1	1,2	335-339	340	1-8	1,2,4,6,10, 11,14-16,20,21
	2	3	340-343		9-14	7,8,12,13
	3	4	343-346		15-17	5,9,22,23,24
	4	5	346-348		18-21	3,17-19

Sections 3–7

Sec			Page			
3	1	1	44-47	62	1-9	1-8(5,6)
	2	2	47-49		10-16	9,12,15
	3	3	49-55		17-27	13
	4	3	56-60		28-30	10,11,14
	5	4	60-63		31,32	16,19,22
	6	5	63-65		33	17,20
	7	6	65-66		34,35	18,21
4	1	1,2	71-76	77	1-6	1-3
	2	3	76-77		7,8	4,5
	3	4	77-79		10,11	6,19,20,21
	4	4	80-81		12,13	7
	5	4	80-81		14-17	8
	6	5	82-83		18-21	9-12
	7	6	84-86		22-28	13-16(14,15)
	8	7	86-88		29-35	17,18,22
5	1	1	93-95	102	1,2,5,6,8	1,3
	2	1	95-98		3,4,9	2
	3	2	99-100		7,10-15	4-8,19
	4	3	100-102		16-18	9,10
	5	4	102-104		19-24	11-14,20,21
	6	5	104-107		25-32	15,16,22
	7	6			33-35	17,18
6	1	1	116-117	131-134	1,3	8
	2	2	117-118		4-5	1
	3	3	119-120		6-11	15,18,19
	4	4	120-122		12-21	2,5-7,9
	5	5	122-125		2,22-23	14
	6	6	125-127		26-27	11-13,16-17
	7	7	127-129		28-29	3,10
	8	7	129-131			
7	1	1,2	138-140	143	1-8	2,3
	2	3,4,5	140-143		9-23	1,5-12(6-10),15
	3	6	143-145		24-27	4,13,16,17
	4	7	145-149		28-30	18

modern earth science learning guide (continued)

Chapters 17–20

chapter	objective teacher	student	text page	activity page	questions A	questions B
17	1	1	352-354	354	1-5	1-2
	2	2	354-356		6-7,11,12	3-6,9,10
	3	3	357-359	358	8-10,14-17	8
	4	4	359		13	7
	5	5	359-361		18-22	11-13,16
	6	6	361-363		23-25	14,15
	7	7	363-367		17-19	
18	1	1,2	376-381	381	1-8	1-6,19,20
	2	3	381-383		9-16	7-10
	3	4	383-385		17-19	11
	4	5,6	385-391		20-30	12-18
19	1	1	396-397	409	1-4	1-3
	2	2	398-399		5-10,17,19	4-9(6,7)
	3	3	399-400		13-16	10,11
	4	4	400-402		11,12,20-22	12-14,18
	5	5	402-404		18,23-29	15-17,19,20
	6	6	404-414			
20	1	1	420-421		1-2	9
	2	2	421-424		3-7	4,7,11
	3	3	424-425		8-10	5,6
	4	4	426-427		11	8
	5	5	427-429		12-17	2,3,10,13
	6	6	429-433		18-21	1
	7	7,8	433-436	433		14-16

Chapters 21–24

chapter	objective teacher	student	text page	activity page	questions A	questions B
21	1	1	439-442	442	1-7	1-7
	2	2	442-444		8-12,14	8,9
	3	3	444-446		13,15-23	10
	4	4	446-448		24-26	11,12,18
	5	5	449-450		27,28	13,14
	6	6,7	450-452		29,30	15-17
22	1	1,2	458-459	471	1-7	3
	2	3	459-466		8-10	4,6,7
	3	4	466-469		11-14	1,2,8,9, 13,14
	4	5	470-471		15-21	5,10-12(11)
	5	6	471-476		22-23	
23	1	1,2	479-484	480	1-12	1-7(3-6)
	2	3	484		13,14,23,24	8,16
	3	4	485-489		15,16	9-11,14, 15,17
	4	5,6	489		17-22,25,27	12,18,19
	5	6	489-490		26,28-30	13,20
24	1	1	498-499	501	1,3-7,26	5
	2	2	499-503		2,8	6-8
	3	3	503-510		9-15,18-20	9-14(10,11),16,17
	4	4	510-512		16,21-23	15,18
	5	5	512-514		24-25,27	1-3
	6	6	514-516		28-29	4

1 a small planet in space

objectives

When finished with this chapter, the student will be able to:

1. Explain why the most important observation that can be made when looking at the sky is that there is an ever-changing pattern of movements by the celestial bodies.
2. Describe one of the constellations found in the night sky and explain why Polaris is considered the North Star.
3. Describe the ancient Greek model of the solar system as proposed by Ptolemy and explain why it was widely accepted for so long.
4. Use Kepler's Laws of Planetary Motion to describe how the planets of the solar system behave.
5. Given the mathematical statement of Newton's Law of Gravitation, explain how one object in the universe affects other objects, large or small, at close or far distances.
6. Identify two windows to the universe in the electromagnetic spectrum, and describe one device that is used to see the universe through each window.
7. Describe the Milky Way galaxy.
8. List the three basic types of galaxies and give their distinguishing characteristics.

approach

The study of astronomy presents an excellent opportunity to establish the concept that observation is the foundation for all science. It is easy to see how knowledge of the distant bodies in the universe comes from careful and systematic observation. It is also easy to see why instruments such as telescopes are needed. But it should be emphasized that all instruments used in science are usually only extensions of the observer's senses. The crucial step in the process of scientific thought is the synthesis of all available observations into a hypothesis or model that can be tested by experimentation. Appreciation for the experimental nature of astronomy can be developed if it is emphasized that the models, such as one that describes the solar system, are constantly being tested for predictability of the various moving parts.

Occasionally, a student may challenge information given in the text on the basis of current discoveries described in newspapers or magazines. Such conflicts can often be used to stimulate deeper research into a topic in an effort to resolve the problem. When researching questions, students should be guided toward the use of reliable current sources such as *Scientific American* rather than encyclopedias and similar general references.

Actual observation of the night sky by individual students is a desirable and worthwhile activity. This is possible if some instruction is given in locating the major constellations and brightest stars or planets. Information on the appearance of the sky at any season can be found in magazines such as *Natural History* and *Sky and Telescope*. Many simple but effective starfinders are also available.

activity hints

Drawing Elliptical Orbits (page 15)

21½ × 28 cm pieces of corrugated cardboard can be cut from cardboard boxes.

The following can be used to have students add orbits of Venus and Mercury to their drawings. Venus: pins l mm apart and a loop 74 mm around. Mercury: pins 8 mm apart and a loop 55 mm around. In drawing each orbit, the same pin, the "SUN," must remain untouched.

other activities

Geology and Earth Sciences Sourcebook (2nd edition, 1970) Holt, Rinehart and Winston.

This book is no longer in print, but has many worthwhile ideas. The school, college or public library or an experienced earth science teacher may have a copy you could refer to. See Chapter 11 for presentation ideas, problems and questions, unsolved problems, demonstrations, projects, and experiments.

Also see Chapter 1 in *Activities and Investigations for Modern Earth Science* (1983 edition) Holt, Rinehart and Winston.

answers—chapter 1

Vocabulary Review

1. g	4. b	7. i	10. f
2. c	5. h	8. j	
3. e	6. d	9. a	

Questions: Group A

1. a	8. b	15. b	22. a
2. c	9. c	16. c	23. d
3. b	10. a	17. d	24. a
4. d	11. c	18. c	25. c
5. a	12. b	19. a	26. b
6. c	13. c	20. d	
7. d	14. d	21. d	

Questions: Group B

1. After rising in the east, the sun makes an arc across the sky with a southward motion until it reaches its highest point in the southern sky. It then moves down to finally set in the west. In summer, the path shifts northward making less of an arc from sunrise to sunset.
2. Ursa Major, great bear; Leo, the lion; Scorpius, the scorpion; Draco, the dragon.
3. They are used to describe certain regions of the sky for locating sky objects such as individual stars.
4. Sirius, Vega, Capella, Arcturus, and Rigel.
5. They follow paths within a narrow belt across the sky, moving amongst the stars. Some follow odd, looping paths that carry them forward and backward amongst the stars.
6. A belief in which the exact positions of the heavenly bodies at the time a person is born is supposed to control that person's life.
7. The Ptolemaic model accurately predicted the positions of the heavenly bodies at all times. It also had the earth as the center of the universe, which proved more agreeable to people.
8. Copernicus removed the earth from the center of the universe and suggested that the earth and other planets moved in paths around the sun.

9. Kepler's Laws of Planetary Motion:
   (1) The planets move in elliptical paths.
   (2) A line joining a planet with the sun will sweep out equal areas in equal times.
   (3) The cube of the distance from the sun of a planet is equal to the square of the time of its revolution about the sun.
10. Kepler's Laws describe proven facts of nature while Copernicus' model tried to explain the make-up of the universe and was not a proven fact of nature.
11. F is proportional to $\dfrac{M_1 \times M_2}{d^2}$. As the distance "d" is doubled, d^2 becomes 4 times as great, which means the force is $\frac{1}{4}$ as great.
12. You can see the stars and other celestial bodies because the earth's atmosphere lets visible light come through as though it were a window.
13. Gamma rays, X-rays, ultraviolet, visible, infrared, radio waves.
14. It will be able to observe radiations that the atmosphere hides from earth-based telescopes and it will see 50 times farther.
15. Average diameter = 600,000 trillion miles 600,000,000,000,000,000 miles thickness—12,000 trillion miles 12,000,000,000,000,000 miles
16. It is shaped like a plate with spiral arms. It is 100,000 light years in diameter and 2000 light years thick. It slowly rotates like a wheel, one complete revolution in 200 million years. It is made up of about 100 million stars spread out fairly evenly with great clouds of gas that form the spiral arms.
17. Spiral, elliptical, barred spirals, and irregular.

references

Angrist, Stanley. *Other Worlds, Other Beings.* New York: Thomas Y. Crowell Co., 1973.

Bova, Ben. *Starlight and Other Improbabilities.* Philadelphia: Westminster Press, 1973.

Brandt, J. C., and S. P. Maran. *New Horizons in Astronomy.* San Francisco: W. H. Freeman & Co., 1972.

Cleminshaw, C. H. *The Beginner's Guide to the Skies.* New York: Thomas Y. Crowell Co., 1977.

Corliss, W. R. *Mysteries of the Universe.* New York: Thomas Y. Crowell Co., 1971.

Ebbighausen, E. G. *Astronomy.* Columbus, OH: Charles E. Merrill Publishing Co., 1971.

Friedman, H. *The Amazing Universe.* Washington, DC: National Geographic, 1975.

Hinkelbein, Albert. *The Origins of the Universe.* New York: Franklin Watts, Inc., 1973.

Moore, Patrick. *The Atlas of the Universe.* Chicago: Rand McNally & Co., 1970.

Press, F., and R. Siever. *Earth,* 2nd ed. San Francisco: W. H. Freeman & Co., 1978.

Rey, H. A. *Find the Constellations.* Boston: Houghton Mifflin, 1976.

audio-visual aids

16mm Sound Films

Galaxies. McGraw-Hill, 15 min., color
How Many Stars. Moody, 11 min., color
New View of Space. NASA, 28 min., color, HQ214
Space Science—Galaxies and the Universe. Coronet.
The Astronomer. IFB, 16 min., color
The Sky and Telescope. McGraw-Hill, 15 min., color

The Universe: Beyond the Solar System. EBEC, 18 min.

What Are Stars Made Of? IFB, 11 min., color

Filmstrips

Astronomy. Millikan, set of 6 with duplicating masters, color sound

The Stars. Creative Visuals, set of 6, color, MH 396020

The Universe. Creative Visuals, set of 6, color, MH 396027

The Universe. Edwin Shapiro Co., captions

8mm Film Loops

Using a Star Map. PHOENIX, super 8, color

35mm Slides

Galaxies—Nebulae—Star Clusters. Hubbard, set of 15, color, VAG 286

2 the sun

objectives

When finished with this chapter, the student will be able to:

1. Identify the reaction that produces the sun's energy and describe the role of hydrogen in this reaction.
2. Describe two ways solar energy can be used directly in homes.
3. Describe the role of each part of the sun in producing the energy it radiates.
4. Compare and contrast sunspots, prominences, and solar flares as disturbances found on the sun.
5. List and describe three effects solar flares can have on the earth.
6. Describe several ways that light is used to measure distances to stars and characteristics of these stars.
7. Identify the most intense kind of solar radiation.
8. Describe the pattern that occurs when star populations are classified according to temperature and brightness.
9. Describe the processes that are involved in the birth of a star.
10. Describe the processes that spell out the death of a star.
11. Give two observations that support the Big Bang theory.

approach

The sun is an important topic of discussion because of the enormous importance of its energy to life on earth and its value as a source of knowledge about the nature of all stars. When considering the nuclear reactions that are the source of all solar energy, it is best to avoid detailed study of the structure of atoms. Instead, stress the result of the nuclear reactions—the conversion of a part of the solar mass into energy. Some confusion may arise between the two types of nuclear reactions. Only *fusion reactions*, involving the grouping of lighter nuclei into heavier ones, are believed to occur in the sun and other stars.

activity hints

Sun's Rotation (page 28)

The student should answer the questions as follows:
1. One-fourth
2. 28 days
3. About halfway between. At the equator, the sun's rotation takes 25.33 days, at the poles 33 days,.and halfway between 28 days.

Bright Line Spectra (page 32)

The student should answer the questions as follows:

1. C
2. B
3. D

other activities

Geology and Earth Sciences Sourcebook (2nd edition, 1970) Holt, Rinehart and Winston.
Chapter 11, The Earth in Space and Time, pp. 243-283

Also see Chapter 2 in *Activities and Investigations for Modern Earth Science* (1983 edition) Holt, Rinehart and Winston.

answers—chapter 2

Vocabulary Review

1. e	**4.** j	**7.** c	**10.** i
2. h	**5.** a	**8.** l	
3. b	**6.** d	**9.** g	

Questions: Group A

1. b	**8.** c	**15.** d	**22.** a
2. a	**9.** b	**16.** b	**23.** b
3. c	**10.** c	**17.** a	**24.** b
4. d	**11.** c	**18.** c	**25.** c
5. b	**12.** a	**19.** b	**26.** a
6. a	**13.** d	**20.** d	**27.** b
7. b	**14.** c	**21.** b	**28.** d

Questions: Group B

1. All atoms are found as nuclei. The temperature range is in millions of degrees and the pressure more than a billion times greater than normal air pressure on earth. Nuclei of hydrogen are present. They undergo fusion very easily.
2. Passive systems for solar energy use the energy directly from the sun for heating. Active systems use some device to convert the solar energy to another form. An example is the solar cell, which produces electricity.
3. Energy produced in the sun's core bombards atoms in the zone of radiation, causing these atoms to give off X-rays and ultraviolet rays (short wavelength radiations). Most of the short wavelength radiations are converted to longer wavelengths, including visible light, in the photosphere. It is these short wavelength radiations that are responsible for the decreased amount of light given off by the chromosphere and the corona.
4. According to Einstein's formula, ($E = mc^2$), a very small amount of matter can be converted to a huge amount of energy. If reactions going on inside the sun were converting a small fraction of the sun's matter into energy, the amount of energy thus produced could be huge and at the same time allow the sun to remain an active energy source for an indefinite period.
5. Magnetic disturbances on the sun appear to block the movement of very hot gases on the sun's surface, thereby causing areas to form that are about 1500°C cooler than the surrounding surface. These cooler spots appear dark in relation to the rest of the surface. Usually occurring in pairs, sunspots are thought to mark the poles of a magnetic field.
6. Dark areas on the sun's photosphere are called sunspots and mark the locations of powerful magnetic fields. Prominences are giant clouds of glowing gases that arch from one sunspot to another and are controlled by the magnetic fields. Solar flares are more violent solar storms that occur near sunspots and throw charged particles into space.
7. Parallax is the apparent relative motion of two objects at different distances from a moving observer. It can be used to measure distances.
8. By observing color and variation in

brightness, the absolute magnitude of a star can be determined. By using knowledge of how this brightness appears at various distances and the apparent magnitude of the star, its distance can be determined.

9. The chemical elements of the sun have been identified by spectroscopic analysis of the sun's light. The kind of light given off by individual chemical elements when heated is characteristic of only those individual elements.

10. By analyzing the spectrum of starlight using a spectroscope, one can estimate a star's composition as well as determine its relative motion. This method will also help indicate the distance of a star from the earth.

11. Grouping of stars along a line will occur if the absolute magnitude is plotted against surface temperature. There are several points along the line where large groups of stars are found.

12. It is believed that the dominant groups of stars that appear in the main sequence represent varying stages in star development.

13. Multiple star systems could develop from the same vast cloud of dust and gases in space by several points shrinking separately, though at the same time. Each of these points would accumulate sufficient gases for a star to form.

14. The life history of a star involves many complex processes. They range from gravitational attraction of cosmic dust to the development of enough heat and pressure for nuclear reactions and the resulting emission of radiation to take place. More complete details on the development of a star from birth to death will be found on pages 37-41 in the text.

15. Pulsars are stars that give off radiation at regular intervals. These pulses of radiation may be the result of the star not being stable. These strange stars are not fully understood. It is known that neutron stars exhibit this pulsating effect.

16. All galaxies are moving away from us as determined by the red shift in their spectra. In addition, electromagnetic radiation of a type that would have accompanied the Big Bang should be present in space and in fact has been found to be present.

references

Asimov, Isaac. *What Makes the Sun Shine?* Boston: Little, Brown and Co., 1971.

Batler, S. T., and R. Raymond. *The Family of the Sun.* New York: Doubleday, 1977

Brauley, F. *Black Holes, White Dwarfs, and Superstars.* New York: Thomas Y. Crowell, 1976.

Fields, A. *The Sun.* New York: Franklin Watts, Inc., 1980.

Fisher, George. *The Creation of the Universe.* Indianapolis, IN: The Bobbs-Merrill Co., 1977.

audio-visual aids

16mm Sound Films

How to Make a Solar Heater. Handel Film Corp., 20 min., color

Our Nearest Star. McGraw-Hill, 29 min., color

The Sun and How It Affects Us. Coronet, 11 min., color

The Flaming Sky. McGraw-Hill, 29 min., color

The Solar Generation. Stuart Finley, 19 min., color

Filmstrips

Exploring the Sun. EBEC

Scanning the Universe. EBEC, set of 7, captioned, No. 9200

The Sun. FFE
The Sun's Awesome Impact. Time-Life
The Universe. Edwin Shapiro Co.

8mm Film Loops

Solar Activity I. Thorne, color, No. GC 420
Solar Activity II. Thorne, color, No. GC 421
Solar Flares. PHOENIX, super 8, color
Solar Prominences. PHOENIX, super 8, color
Solar Prominences and Sun Spots. Doubleday Multimedia, color, No. 10925
Use of Sun's Energy. Doubleday Multimedia, color, No. 10935

3 design of the solar system

objectives

When finished with this chapter, the student will be able to:

1. Trace the nebular theory of the origin of the solar system.
2. Describe asteroids, meteorites, and comets.
3. Compare similarities and differences of the inner planets.
4. Compare the outer planets with each other and with the inner planets.
5. Explain the summer and winter solstices and the vernal and autumnal equinoxes.
6. Compare apparent solar time and mean solar time.
7. Explain why we use standard time and have the International Date Line.

approach

The solar system is composed of an intricate arrangement of a multitude of varying objects. They are kept in an organized but complex system of orbits by the primary influence of the sun's gravity. The motions of all members of the solar system conform to certain physical laws that enable us to calculate their positions with considerable accuracy.

All theories concerning the origin of the solar system are highly speculative and are constantly being discarded or revised. The tentative nature of these theories should be strongly emphasized. The theories described in the text were chosen to provide a representative sample of the ideas currently accepted that explain the beginning of the solar system. To illustrate the great diversity of the proposed explanations, the following theory can also be mentioned, then compared with those already described in the text.

The Electromagnetic Theory: The sun was already in existence before it acquired a group of satellites. As it moved through space, the sun entered a nebula. A cloud of atoms from the gases of the nebula collected around it. The atoms that were attracted to the sun became ionized or electrically charged. Interaction between the sun's magnetic field and the cloud of charged particles drew the atoms out on the sun's equatorial plane and they eventually spread, condensing into planets and their satellites.

activity hints

Planet Orbits (pages 62-63)

You could have students use protractors to measure angles if you wish, but it isn't necessary for the main idea of the activity.

Answers to questions:

1. 4.0 cm
2. 0.4 A.U.
3. 60 million kilometers
4. 7.4 cm
5. 0.74 A.U.
6. 111 million kilometers

other activities

Geology and Earth Sciences Sourcebook (2nd edition, 1970) Holt, Rinehart and Winston.
Chapter 11, The Earth in Space and Time, pp. 243-283

Also see Chapter 3 in *Activities and Investigations for Modern Earth Science* (1983 edition) Holt, Rinehart and Winston.

answers—chapter 3

Vocabulary Review

1. f	5. b	9. e	13. p
2. h	6. i	10. k	14. d
3. g	7. l	11. q	15. o
4. a	8. j	12. r	

Questions: Group A

1. a	10. a	19. d	28. d
2. d	11. a	20. a	29. a
3. c	12. c	21. b	30. b
4. b	13. b	22. b	31. b
5. b	14. d	23. c	32. c
6. d	15. b	24. d	33. b
7. b	16. a	25. a	34. b
8. a	17. a	26. b	35. a
9. d	18. c	27. d	

Questions: Group B

1. The dust and gas cloud that existed in this part of the solar system was made to crowd together, possibly by a shock wave from a nearby supernova. Mutual gravity of the particles caused the cloud to shrink around one large lump of matter. As it shrunk, it began to slowly rotate and take on a disk shape of the solar nebula.

2. Protoplanets were the beginning stages of the formation of planets. They formed from other clumps of matter in the solar nebula at the same time the sun was forming.

3. Radioactive elements in the proto-planets heated the planets, melting them and allowing the heavier material, mostly iron, to settle to the center and form the core.

4. The explosions of the beginning sun stripped away the original atmosphere from the inner planets; the outer planets were struck by less force and therefore retained most of their atmosphere.

5. Mercury is so near the sun that it is invisible in daylight. Therefore, the sun must be below the horizon but not very far below for Mercury to be seen. This is the case just before sunrise and just after sunset.

6. Since it takes 687 days for Mars to go around the sun, the seasons on Mars are about twice as long as on the earth. The temperature extremes are thus greater than on the earth.

7. Large volcanoes, craters partly covered with sand and dust, an atmosphere, dust clouds, and windblown material.

8. Jupiter (15), Saturn (15), Neptune (2), Uranus (5), Earth (1), Venus (1), Mars (2), Pluto (unknown), Mercury (0).

9. This undeveloped planet is spoken of as the relatively large gap between Mars and Jupiter that contains thousands of asteroids. It has been suggested that this was the location and material of a planet that failed to develop.

10. A photographic edge-on view of Saturn and its rings occasionally fails to show visual evidence of the rings.

11. A transit of the sun is the name given to the passage of a planet across the sun's face. Only Mercury and Venus have orbits that bring them into a position between the earth and the sun.

12. The asteroids entering the earth's atmosphere glow brightly, usually breaking into pieces. As these pass

through the atmosphere, they are called meteors. Any surviving pieces that reach earth's surface are called meteorites. As the earth passes through a comet's path in space, it sweeps out the tiny pieces of the comet left in the path. As these pass through the atmosphere, they give off light streaks called meteors. Most meteors are from comets; meteorites are from asteroids.

13. They are all small, dense planets with a metallic core surrounded by lighter, rocky layers. They are all relatively close to the sun and therefore receive an appreciable amount of heat from the sun. They differ in their speed of rotation and the length of time to complete one orbit of the sun. Mercury has no atmosphere, Venus has a very dense atmosphere, and Mars has a very thin atmosphere compared to earth's.

14.
Inner Planets	**Outer Planets**
a. close to sun	very far from sun
b. appreciable sunlight	practically none
c. none to heavy gases such as N_2, O_2, H_2O, H_2SO_4	very thick atmosphere of light gases such as H_2, NH_3, CH_4, H_2O
d. dense, metallic, and rocky	liquid and solid light gases, mostly H_2
e. 88 to 687 days	12 years (Jupiter) to 165 years (Neptune)
f. slowly, 1 to 243 days for one rotation	rapid, about 10 hours for one rotation

15. a. Mostly in orbit about sun between Mars and Jupiter. Some orbits go inside Mars' and earth's orbits.

b. Three kinds: stony, metallic, and mixture of stone and metal.

c. Most have a mass of less than 1 kilogram; about 230 asteroids are greater than 100 kilometers diameter.

d. They are thought to be some of the original material from which the solar system formed. Study of them may answer questions about the early solar system.

e. Hundreds collide with earth each year as meteors. Those striking the surface usually cause little damage and are picked up as meteorites. The larger ones may be as destructive to make a crater large enough to destroy a city.

16. The time of sunrise and sunset, the apparent speed of the sun across the sky, and, therefore, the rate at which the shadow of a sundial moves across the clock, change with the motion of the earth around the sun. These effects are all due to the inclination of the earth's axis.

17. Apparent solar time and mean solar time are nearly the same at the time of the equinoxes. It is then that there are 12 hours of light and 12 hours of darkness.

18. The earth's surface is divided into 24 standard time zones. The standard time in each zone is the mean solar time for the center of that zone. The purpose of standard time zones is to avoid the confusion of thousands of different times, differing by only a few minutes.

19. Daily—the sun moves east to west. June 21 or 22—the sun reaches its northernmost position in the sky. Each day the sun is farther south. September 22 or 23—the sun is overhead at the equator. December 22 or 23—the sun is at its southernmost position in

the sky and starts to move north. March 20 or 21—the sun is overhead at the equator again but this time headed north. June 21 or 22—the sun has completed a cycle and a tropical year has passed.

20. (a) Apparent Solar Time—Time measured with a sundial. (b) Mean Solar Time—Time that is an average of sundial times for a period of one year. (c) Standard Time—Time measured by agreement that zones be established on the earth's surface. All cities in a zone will be on the same time.

21. If the earth's axis were perpendicular to the plane of its orbit, the only seasonal changes would be in response to the varying distance from the sun. The seasons as we now know them would disappear. In December, we are closest to the sun and would receive the greatest heat at that time. In June, we would receive the least heat since we are farthest from the sun.

22. Daylight Saving Time causes people to change their clock settings so that people are active during more hours of daylight when the days are longest. This is said to save energy. It also causes some people to go to work before the sun is up. Animals don't adjust to this special time.

references

Asimov, Isaac. *Jupiter, the Largest Planet*, revised ed. New York: Lothrop, Lee & Shepard Co., 1976.

Dermott, J. F. *The Origin of the Solar System*. New York: Wiley, 1978.

Guest, J., et al. *Planetary Geology*. New York: Halsted/Wiley, 1979.

Hopkins, Jeanne. *Glossary of Astronomy and Astrophysics*. Chicago: The University of Chicago Press, 1976.

Hoyle, Fred. *Highlights in Astronomy*. San Francisco: W. H. Freeman & Co., 1975.

Knight, David. *The Tiny Planets: Asteroids of Our Solar System*. New York: William Morrow & Co., 1973.

Mehlin, T. G. *Astronomy and the Origin of the Earth*, 2nd edition. Dubuque, IA: William C. Brown Co., 1973.

Nourse, A. F. *Venus and Mercury*. New York: Franklin Watts, Inc., 1972.

The Solar System: A Scientific American Book. San Francisco: W. H. Freeman & Co., 1975.

audio-visual aids

16mm Sound Films

New View of Space. NASA, 28 min., color, HQ214

The Solar System: Its Motions. McGraw-Hill, 9 min., color

Filmstrips

The Solar System. Creative Visuals, set of 6, color, MH396013

The Solar System. Doubleday Multimedia, set of 4, color

What Is Out There? Prentice-Hall, set of 7, color

Solar System, Edwin Shapiro Co.

Earth and Its Neighbors in Space, EBEC, set of 6, with cassettes

Earth, Rotation, and Orbit, Prentice-Hall, set of 2

8mm Film Loops

Comet Orbits. PHOENIX, color

Mars and Jupiter. PHOENIX, color

Scale Model of Solar System. PHOENIX, color

Solar System: Inner Planets. EBEC, color, No. S-81390

Solar System: Outer Planets. EBEC, color, No. S-81391

Earth Rotation and Orbit, P. H. Media

Day and Night, PHOENIX

Earth in Space, Hubbard

4 the moon

objectives

When finished with this chapter, the student will be able to:

1. Describe the surface of the moon and compare conditions with the earth.
2. Describe the stages in the theory of the origin of the moon.
3. Describe the orbit of the moon around the earth and around the sun.
4. Explain why the time between moon rises is greater than 24 hours.
5. Draw a diagram to show the sun, earth, and moon positions for the various phases of the moon.
6. Explain the reasons for solar and lunar eclipses.
7. Describe the moon's and sun's influence in producing tides in the ocean.
8. Compare the Julian, Gregorian, and World calendars.

approach

Direct observation of the moon by students should be encouraged. A telescope is desirable but not necessary since the moon's motions and appearance are apparent without optical aids. However, low-powered field glasses or binoculars are helpful. The full moon is too bright for detailed viewing. Therefore, the most satisfactory time to view it is in one of the partial phases. The moon appears larger when close to the horizon because of its comparative nearness to familiar objects. At a higher elevation it appears smaller because it is isolated. Students may challenge this explanation and should be encouraged to test it with appropriate experimentation.

Many students will probably find it difficult to understand why there is a 29½-day interval between two successive phases of the moon while it takes only 27½ days for the moon to complete one revolution around the earth. This problem can be illustrated with the aid of a globe (representing the earth), illuminated with a lamp (representing the sun). For the moon, an ordinary tennis ball would be useful. The earth and moon can be moved to show that the earth completes 1/12 of its orbit around the sun during the time the moon makes one revolution around the earth. This should make clear that some additional time is required for the earth and moon to return each month to the same positions relative to the sun. Thus, the additional two days following each complete revolution are required for the moon to return to the same phase.

Many students are likely to question the reasons for the appearance of the tidal bulge on the side of earth opposite the sun and moon. There is really no simple explanation for this. It is wise to postpone a detailed consideration of tides until other influences, described in Chapter 17, have been studied.

activity hints

Measuring the Size of the Moon (page 77)

Scientific, or power-of-ten, notation can be used here, but it is not essential. Changing units is also a useful process that can be shown here. The students will get results that are close to the correct value of 3476 km, but remember, the distance to the moon varies from 360,000 km to 405,000 km. Answers from 3200 to 3700 km are consistent with this variation. When the average of all measurements is calculated, you may want to speculate on whether the moon was near perigee or apogee.

other activities

Geology and Earth Sciences Sourcebook (2nd Edition, 1970) Holt, Rinehart and Winston.
Chapter 12, Lunar Resources, pp. 285-295

Also see Chapter 4 in *Activities and Investigations for Modern Earth Science* (1983 edition) Holt, Rinehart and Winston.

answers—chapter 4

Vocabulary Review

1. f	**4.** g	**7.** 1	**10.** j
2. c	**5.** i	**8.** k	
3. d	**6.** b	**9.** h	

Questions: Group A

1. a	**10.** b	**19.** a	**28.** c
2. c	**11.** a	**20.** c	**29.** b
3. a	**12.** c	**21.** b	**30.** c
4. d	**13.** d	**22.** c	**31.** b
5. c	**14.** a	**23.** d	**32.** b
6. b	**15.** c	**24.** c	**33.** d
7. b	**16.** b	**25.** d	**34.** a
8. d	**17.** a	**26.** a	**35.** c
9. c	**18.** b	**27.** d	

Questions: Group B

1. The impact of bodies with the moon's surface has pulverized the rocky surface to dust.

2. The slow rotation of the moon allows considerable time for its surface to heat and cool. The absence of an atmosphere accounts for the faster heating and cooling.

3. The maximum temperature would probably be somewhat lower and the minimum temperature higher. The average temperature would probably be the same.

4. (1) The moon may be a chunk of the earth that broke off at some time.
   (2) The moon may have developed in the same way and at the same time as the earth.
   (3) The moon may have been formed at some other place in the solar system and was later captured by the earth's gravity.

5. *First stage* is its origin. Little is known about the moon's origin except that it occurred about the same time and place as the earth's origin. *Second stage* is a molten stage in which the moon was covered by molten rock. In the *third stage*, the molten rock separated into layers. In the *fourth stage*, the crust cooled and became solid. There was a great deal of volcanic activity and mountain building in this stage. The *fifth stage* contains the period of time in which intense surface bombardment by other bodies formed the craters and dust surface. The *sixth stage* is marked by little bombardment, but great lava flows formed the maria. This last stage was followed by a quiet period with very little change.

6. The rotation of the moon does not affect moon phases. If, however, the moon rotated at a rate different from its speed around the earth, we would see different parts of the moon. We have only limited knowledge of the far side of the moon.

7. At full moon, the earth would be seen as "new earth." At new moon the earth would be seen as "full earth," and at the moon's first quarter, the earth would be at "last quarter."

8. Because the moon rotates on its axis at the same rate that it revolves around the earth, the same side of the moon is always toward the earth. We have some knowledge of the far side of the moon from space probes, but it is limited.

9. (1) Moon between earth and sun (new moon).
   (2) Moon at or near perigee.
   (3) Earth intercepts umbra of moon. All of these conditions must be met or a total eclipse will not occur.

10. Everyone on the dark side of the earth is within the earth's umbra. All people within the umbra will see a total eclipse of the sun. The sun will be completely covered by the moon.

11. At first quarter, the moon's umbra makes an angle of 90 degrees with the line of sight to the moon. Since the umbra doesn't come anywhere near the earth, there is no eclipse.

12. To see a total eclipse of the sun, a person must be in the umbra of the moon. The moon's umbra never casts a large shadow on the earth and only those people in this area will be able to see a total eclipse. Others, in the penumbra, will see a partial eclipse.

13. At full moon the sun, earth, and moon are all in a straight line and act together in the buildup of spring tides.

14. Opposite tides are the result of a reduction in the effect of gravity because of increased distance from the moon. This permits the earth-moon system to rotate around a common center, thus pushing water away from the earth.

15. Friction against the sea floor by the moving masses of water in the tidal bulges tends to slow the earth's rotation. The effect, however, is small. In 1000 years, a day will be about 0.2 second longer than it is at present.

16. Yes. The earth must rotate a little more than one turn before the moon is overhead on two consecutive occasions. A speedup is necessary because the moon completes $1/27.3$ of its orbit in one day, revolving around the earth in the same direction as the earth turns.

17. Basically, it is to keep the months more nearly in line with the seasons. The addition of an extra day every 4 years (leap year) puts about ¾ of a day more than was needed in the calendar every 100 years. Therefore, 3 out of 4 century-years are not leap years. The ¼ of a day that the calendar is short each century is brought back into line with the seasons by the addition of a leap year in century-years divisible by 400. Remember this still has a small error.

18. The Julian calendar made better provisions for keeping in step with the seasons by adding an extra day every 4 years. This made the length of the calendar year 365¼ days on the average.

19. The moon travels about 3665 km/hr in an elliptical path about the earth. At apogee it is 404,800 km from earth. At perigee it is 360,000 km from earth. It has an average distance from earth of 384,400 km.

20. (a) 20,400 km
    (b) 24,400 km

    (c) apogee $= \dfrac{20,400}{384,400} \times 100 = 5.3\%$

    perigee $= \dfrac{24,400}{384,400} \times 100 = 6.3\%$

21. It is not very elliptical. More like a circle.

22. The Julian calendar year was 11 minutes and 14 seconds short of one orbit of earth about the sun. This caused the seasons to get out of phase. The Gregorian Calendar corrected this.

references

Alter, Dinsmore. *Pictorial Guide to the Moon* (3rd ed.). New York: Thomas Y. Crowell Co., 1973.

Guest, J., et al. *Planetary Geology.* New York: Halsted/Wiley, 1979.

Lewis, Richard C. *The Voyage of Apollo: The Exploration of the Moon.* New York: Quadrangle/New York Times Book Co., 1974.

Musgrave, R., comp. *Lunar Photographs of Apollos 8, 10, and 11.* Washington, DC: NASA, SP-246, 1971.

Shuttlesworth, D. E., and L. A. Williams. *The Moon*. New York: Doubleday, 1977.

Whittingham, R. *Astronomy Fact Book*. Hubbard Scientific Co., 1971.

(For additional information, write to NASA: Scientific and Technical Information Division, NASA, Washington, DC 20546)

audio-visual aids

16mm Sound Films

Eclipses of the Sun and Moon. EBEC, 11 min., color

Man Looks at the Moon. EBEC, 15 min., color, No. 3000

The Moon: A Giant Step in Geology. EBEC, 24 min., color

Filmstrips

The Earth and Its Moons. Creative Visuals, set of 6, color, MH 396006

8mm Film Loops

Apollo 11. Holt, Rinehart and Winston, 4 loops, color, 89-3875/1

Lunar Eclipse. EBEC, color, No. S81150

The Moon. PHOENIX, color

5 probing the secrets of space

objectives

When finished with this chapter, the student will be able to:

1. Explain why the gravity barrier is a major obstacle to space study.
2. Describe in simple terms how a rocket works.
3. Explain the use of a minimum energy orbit in a flight to the inner planets.
4. Identify the main reasons for sending spacecraft to the other planets.
5. Describe the launching of an artificial satellite and its orbit.
6. Give examples of the ways satellites can be used to study the earth.
7. Describe the function of several artificial earth satellites and the *Space Shuttle*.

approach

Understanding the earth and discussion of management of the earth's resources are two points included in this chapter. The various missions of our satellite programs, including geological studies, natural resource studies, weather surveillance, communications, and possible industrial use of space are considered here. The material in the text is very fundamental and may cover ground already familiar to many students. In this case, the text can be supplemented easily with material from magazines and other sources of current information. This entire chapter can be presented with as much or as little emphasis as appears suitable for each class situation.

activity hints

Crash Landing (page 102)

The following list of items, salvaged from the moon crash landing, was ranked and supplied to NASA by astronauts. Their reasons for ranking these items as they did are also given. You could duplicate a similar list for student comparison. Collect them at the end of class.

Rank	Item
1	2 50 kg tanks of oxygen (required for breathing)
2	20 L water (replenish body loss)
3	Star chart (principal means to find directions)
4	Food concentrate (daily food requirements)
5	Solar powered receiver-transmitter (distress signaling)
6	20 m nylon rope (climbing, securing packs)
7	First aid kit (injury or sickness)

8 Large piece nylon fabric (shelter and carrying equipment)

9 Signal flares (location marker when within sight)

10 2 45-caliber pistols (self-propulsion devices)

11 1 case dehydrated milk (food when mixed with water)

12 Portable heater (useful on dark side only)

13 Flashlight (useful on dark side only)

14 Magnetic compass (useless, the moon probably has no magnetic field)

15 Matches (little or no use on moon)

The changes for a Mars crash might include lowering the importance of the pistols because gravity is greater on Mars. The heater would be important at all times because the temperatures never get above freezing. The flashlight would be useful since darkness occurs more often on Mars than on the moon.

other activities
See Chapter 5 in *Activities and Investigations for Modern Earth Science* (1983 edition) Holt, Rinehart and Winston.

answers—chapter 5

Vocabulary Review

1. d	4. b	7. e	10. f
2. g	5. k	8. l	
3. i	6. c	9. a	

Questions: Group A

1. b	10. c	19. a	28. d
2. c	11. d	20. c	29. b
3. a	12. b	21. a	30. b
4. c	13. c	22. c	31. a
5. c	14. b	23. d	32. d
6. c	15. b	24. b	33. c
7. b	16. c	25. a	34. d
8. a	17. a	26. c	35. d
9. b	18. d	27. d	

Questions: Group B

1. The force of gravity must be overcome in order to escape the earth. It is like climbing from a deep steel-walled pit. The force must be enough to first lift and then accelerate.

2. Gases given off by burning fuel in a closed container exert pressure in all directions. The rocket engine is provided with an exhaust nozzle that permits these hot gases to escape. Pressure on the front of the engine is no longer balanced toward the rear, and the rocket is pushed forward.

3. In going from earth into space, overcoming gravity can be compared to climbing out of a deep hole or pit. The walls would be very steep near the bottom, but less steep near the top until eventually almost no force is needed to keep the rocket from falling back into the pit.

4. A "minimum energy orbit" makes the spacecraft a temporary satellite of the sun and requires rocket propulsion only to correct its flight path.

5. To reach Venus with minimum energy, a spacecraft should be launched in a direction opposite to the earth's motion around the sun. For a Mars trip the launch should be in the direction of the earth's motion.

6. The main difficulty results from the tilt in Venus' orbital planes compared to the earth's tilt. A minimum energy orbit must account for this.

7. The periods when minimum energy orbits are possible are called "launch windows."

8. Jupiter's gravity can be used to change the path of a spacecraft and send it on toward Saturn.

9. (a) Discover how the earth developed.
(b) Attempt to find evidence of other life.
(c) Possible place for future colonies.

(d) For a better understanding of the earth.

10. Mars' huge dust storm resulted in a 20°C drop in temperature. On the earth, this could result in the start of an ice age set off by pollution of the atmosphere.

11. The term "fall" refers to falling out of a straight line. The face path results in a curvature that, if less than that of the earth, results in an orbit.

12. A synchronous orbit is one that keeps a satellite in step with the earth's rotation. It must be at an altitude of 36,100 km and would complete one revolution each 24 hours and move in the same direction the earth turns.

13. The higher a satellite orbits, the slower its speed will need to be.

14. In polar orbit, a satellite crosses the poles each time around, but because the earth is turning it will pass over the equator west of its previous crossing.

15. LANDSAT satellites orbit the earth at about 1000 km. They can examine electronically all parts of the earth at least once every 18 days.

16. Weather satellites have made weather forecasts quite accurate. They also can be used to reveal ocean currents.

17. The orbiter is one of three main parts of the space shuttle. The orbiter is carried into space by two solid fuel rockets and a liquid fuel rocket.

18. The space shuttle can release satellites in orbit, service satellites, and be used for scientific experiments.

19. It is first lifted off the earth by powerful rockets and placed in earth orbit. After several orbits it is aimed at the moon and the engines are started for a short time. The lunar module separates from the rocket engines. About midway the moon attracts the craft, pulling it in. After 60 hours, the lunar module's engines are pointed toward the moon and fired. This places it in orbit about the moon. A moon landing craft separates and with its own rockets guides itself to the moon's surface.

20. In synchronous orbit, the satellite stays in one spot over the equator, while the satellite in polar orbit eventually travels over every spot on earth. The two satellites have opposite characteristics.

21. Below 320 km there is enough atmosphere to cause a drag on a satellite in orbit. This drag slows the satellite, giving it a more elliptical orbit which increases drag. Eventually the satellite is going too slow to stay in any orbit and falls to earth.

22. These satellites use sensors that respond to infra-red and other wavelengths; therefore, they do not need daylight conditions.

references

Cooper, Henry S. F. *A House in Space.* New York: Holt, Rinehart and Winston, 1976.

Moore, P. *The Atlas of the Universe.* New York: Rand McNally Co., 1970.

Tamber, G. E. *Man's View of the Universe.* New York: Crown Publishers, 1979.

For additional information, write to NASA: Scientific and Technical Information Division, NASA, Washington, DC 20546)

audio-visual aids

16mm Sound Films

ERTS—Earth Resources Technology Satellite. NASA, 27 min., color, HQ223

Exploration of the Planets. NASA, 25 min., color

Pollution Below. NASA, 14 min., color, HQ247

(NASA will supply an up-to-date film catalog.)

6 models of the planet earth

objectives

When finished with this chapter, the student will be able to:
1. Compare the ancient beliefs about the earth's place in the universe with what is believed today.
2. List several reasons why the earth is believed to be a sphere.
3. Describe a method of measuring the size of the earth.
4. Describe how directions and locations can be determined on the earth.
5. Explain how map projections can be made.
6. List the uses of maps.
7. Describe what contour lines show on a topographic map.
8. Describe how topographic maps can be interpreted.

approach

If this chapter has been preceded by a study of Unit 1, which deals with the earth's position in the universe, most students will probably already be familiar with the concept of scientific models. Those that describe characteristics of the earth are the least abstract and therefore the most easily understood. Since physical models such as globes are familiar to students, students can be expected to easily identify other earth models they have seen or used. The existence of scientific models, like those representing the structure of atoms, should be mentioned in preparation for dealing with other models later in the course.

The importance of techniques used for locating positions on the earth's surface should be emphasized. The ability to locate places on the earth is no less important today than it was during the period of adventurous exploration when most lands and seas were unknown. Long distance travel has become routine, partly because modern methods of navigation are accurate. Yet the demand for precise techniques for locating places has steadily increased. Thus, the science of location, based on the principles described in the text, has become an important part of the technology of travel.

Generally, a study of the earth sciences will be more meaningful if students master basic map skills early in the course. The ability to use maps is an essential skill. Students with a strong interest in mathematics may want to explore in detail the methods used to make map projections. But all students should become reasonably competent in the fundamental skills of map reading, including determination of location, direction, distance, and relief.

activity hints

Studying a Topographic Map
(pages 131-134)

This activity concerns the study and interpretation of a topographic map.

Answers to activity questions :

1. Contour lines point upstream as they cross a body of running water. Therefore, the Neversink River flows in a general southerly direction.

2. The contour interval shown in the key is 20 feet. Starting at the 1600-foot contour line, just south of the bench mark, you must cross five closed contours in order to reach the triangle. Thus, the elevation of the bench mark to the closest contour line is 1700 feet.

3. The cross-hair marks located in the southeastern end of Wolf Reservoir designate a point situated five minutes of latitude and five minutes of longitude from the southeast corner. (Latitude—41°35′N; Longitude—74°35′W).

4. At different points along the railway you will find the letters BM followed by a numeral. This notation marks the exact elevation of a point above sea level. The elevation points indicate that a train would be ascending as it moved north toward Monticello.

5. The area would probably best be described as a valley. The limited number of contour lines indicates a generally flat area. The bench marks show that most of the area is not much more than 500 feet above sea level. The symbol in the northeast end of this strip represents a swamp or marshland. This can mean that the area is close to or just below the water table. Also quite apparent is the closeness of the contour lines on either side, and adjacent to the strip. This sharp rise in elevation on either side and the direction of the flow of streams in the general vicinity make the area in question appear to be a watershed or drainage basin with a valley-like structure.

6. 4½ to 5 miles.

7. Using the rule of contours as applied in question one, the contours appear to bend upstream in the direction of the reservoir. This indicates that the reservoir is the source for the water entering the stream and emptying into the lake. The stream would be an outlet from the reservoir.

8. The answer to this question is based on the same reasoning used in the previous question. The stream connecting Lake Anawana with the smaller lake to the south is an outlet stream. Water is flowing downhill from Lake Anawana toward the smaller lake.

other activities

Geology and Earth Sciences Sourcebook (2nd edition, 1970) Holt, Rinehart and Winston.
Chapter 16, Topographic Maps, pp. 387-405

Also see Chapter 6 in *Activities and Investigations for Modern Earth Science* (1983 edition) Holt, Rinehart and Winston.

answers—chapter 6

Vocabulary Review

1. e	5. k	9. d	13. o
2. a	6. i	10. m	14. q
3. f	7. g	11. h	15. j
4. b	8. c	12. p	

Questions: Group A

1. d	8. d	15. a	22. c
2. c	9. c	16. c	23. b
3. c	10. a	17. d	24. c
4. d	11. c	18. d	25. b
5. d	12. a	19. b	26. b
6. b	13. b	20. d	27. d
7. a	14. c	21. a	28. d
			29. c

Questions: Group B

1. An oblate spheroid is a spherical body flattened at the poles. Its cross section through the poles is an ellipse.

2. Latitude 0°, longitude 0°. The North Pole is at 90° north latitude. All meridians come together at the poles. For this reason, it does not make sense to speak in terms of longitude for the poles.

3. Relief is the difference in elevation between the highest and lowest points of the area being mapped.

4. Another name for a topographic sheet is a quadrangle. By either name it is a map. These sheets are produced by the Geological Survey and are an accurate, measured description of a land area.

5. California, North Dakota.

6. The answer to this question will depend on your city. It can probably be picked up from Figure 6-11 on page 124.

7. Latitude is measured by equally spaced parallel lines around the earth. The meridians come together at the poles and are most widely separated at the equator. There a degree of longitude equals a degree of latitude as a measure of distance, but nearer the poles the degree of longitude measures less distance.

8. Any arbitrary choice of reference points would allow directions to be established. A major city or prominent topographic feature would be satisfactory. Reference to the earth's magnetic field would be possible, assuming it was present and unchanging. Imagine the confusion if the reference point changed with each generation.

9. At the equator, the distance between a degree of longitude nearly equals the same distance for a degree of latitude. Near the poles, the distance between a degree of longitude decreases, since the parallels of longitude converge at the poles.

10. The answer is generally no. This would give a single place more than one elevation. There is an exception, however, and that is at an overhanging cliff. There are positions at the top level of the cliff, on the face of the cliff, and at the bottom of the cliff that all have the same location when viewed from above, as on a topographic map. Of course, they have different elevations.

11. 111 kilometers. 360 degrees represents 40,000 kilometers. This means that one degree represents $\frac{1}{360} \times$ 40,000 kilometers or 111 kilometers.

12. 1.85 kilometers. 60 minutes represents 111 kilometers. (See problem 11.) This means that one minute represents $\frac{1}{60} \times$ 111 kilometers or 1.85 kilometers.

13. 30.8 meters. 60 seconds represents 1.85 kilometers or 1850 meters. (See problem 12.) This means that one second represents $\frac{1}{60} \times$ 1850 meters or 30.8 meters.

14. This process would not affect the distortion. It would only make the map projection larger.

15. One nautical mile is 1.85 km or 1850 meters. Therefore a nautical mile is larger than a mile (1600 meters) by 250 meters.

16. In the Northern Hemisphere, the shorter arc of a great circle connecting two points at the same latitude is found north of the parallel of latitude. The longer arc is south of that latitude. In the Southern Hemisphere, the reverse is true.

17. $\frac{X}{2.4} = \frac{1}{24,000}$ X = .0001 km or .1 meter, which is 10 centimeters.

18. The angle measured by Eratosthenes bears the same relation to 360 degrees as the distance from the well to the earth's circumference.

19. $\dfrac{5 \text{ degrees}}{360 \text{ degrees}} = \dfrac{550 \text{ kilometers}}{\text{circumference}}$

$\text{circumference} = \dfrac{550 \times 360}{5} \text{ kilometers}$

$\text{circumference} = 39{,}600 \text{ kilometers}$

references

Brown, L. A. *Map Making: The Art That Became a Science.* Boston: Little, Brown and Co., 1960.

Greenhood, D. *Mapping.* Chicago: University of Chicago Press, 1963.

Kjellstrom, B. *Be Expert With Map and Compass.* Harrisburg, PA: Stackpole Books, 1967.

Thrower, Norman J. W. *Maps and Man.* Englewood Cliffs, NJ: Prentice-Hall, Inc. 1972.

U. S. Geological Survey. *Topographic Maps.* Washington, DC: Government Printing Office, 1967.

Weiss, M. P. *Topographic Maps and How To Use Them.* Englewood Cliffs, NJ: Prentice-Hall, Inc., 1968.

audio-visual aids

Filmstrips

Latitude and Longitude. EGH

Latitude, Longitude and Time. McGraw-Hill

8mm Film Loops

Map Projection Series. PHOENIX

7 earth chemistry

objectives

When finished with this chapter, the student will be able to:

1. List and describe the role of the three fundamental particles found in an atom.
2. Explain the role of atoms in the makeup of elements, and, when given atomic number and atomic mass of the atom of a given element, determine the number of fundamental particles it contains.
3. Describe two ways electrons are involved in chemical bonding.
4. Explain how forces that hold molecules together form a gas, liquid, or solid.

approach

The chemical principles developed in this chapter are those considered necessary for an understanding of the composition of minerals and other earth substances. If students already have some background in this topic, the chapter might be used as a brief review. For classes without any previous experience in basic chemistry, the chapter might be treated as a minimum foundation and further discussion of chemistry should be continued as the teacher wishes.

activity hints

Electron Structures (page 143)

Interpretation of the two figures gives the following answers to the questions asked in the activity:

1. yes
2. no
3. yes
4. yes; boron atoms have an electron that is not paired with another electron.
5. Increases the energy.
6.

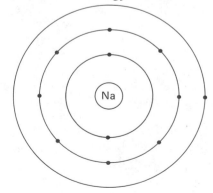

other activities

Geology and Earth Sciences Sourcebook
(2nd edition, 1970) Holt, Rinehart and
Winston.
Chapter 1, Earth Minerals, pp. 1-51
Presentation, pp. 2-27
Suggested problems, pp. 48-49

See also Chapter 7 in *Activities and Investigations for Modern Earth Science* (1983 edition) Holt, Rinehart and Winston.

answers—chapter 7

Vocabulary Review

1. h	**5.** b	**9.** i	**13.** k
2. e	**6.** f	**10.** n	**14.** q
3. a	**7.** j	**11.** p	**15.** m
4. g	**8.** d	**12.** l	

Questions: Group A

1. c	**9.** a	**17.** a	**25.** a
2. a	**10.** c	**18.** b	**26.** c
3. c	**11.** c	**19.** a	**27.** b
4. b	**12.** d	**20.** d	**28.** d
5. b	**13.** c	**21.** d	**29.** c
6. a	**14.** c	**22.** a	**30.** c
7. d	**15.** d	**23.** c	
8. b	**16.** a	**24.** b	

Questions: Group B

1. Oxygen, O; Silicon, Si; Aluminum, Al; Iron, Fe.
2. Protons and neutrons are found in the nucleus; electrons are in the outer region of an atom.
3. The mass of the electron in the lightest element, hydrogen, accounts for only .0054 (1/1840th) of the mass of the atom. In atoms of other elements, it accounts for even less.
4. An ion is an atom that has gained or lost an electron from its normal state, and therefore carries an electrical charge.
5. Atoms of the same element with differing numbers of neutrons in the nucleus are isotopes of that element. Although each one has a different atomic mass, they all behave the same chemically. The atomic number of an element is defined as the number of protons contained in the nucleus of the atoms of that element. The atomic mass of an atom is the sum of the number of protons and neutrons in its nucleus. A proton is arbitrarily assigned one atomic mass unit.
6. Atomic masses could be expressed in any unit desired. Grams are a possibility, but a simpler whole number system was devised by arbitrarily assigning one atomic mass unit for both protons and neutrons.
7. Atomic number 13. Atomic mass 27.
8. Atomic number 19. Number of neutrons 20.
9. The equality of protons and electrons gives matter its neutral charge. If the numbers were not usually equal, most objects would attract or repel other objects with tremendous force.
10. Electrical charges between atoms hold electrons in place. The electrical attraction between an electron and a proton is balanced by the tendency of the rapidly moving electron to fly off into space.
11. A diagram of a hydrogen atom should show 1 proton in the nucleus and 1 electron circling around it. A breakdown of oxygen should show 8 protons and 8 neutrons in the nucleus and 8 electrons outside the nucleus.
12. Naturally occurring elements are made up of several isotopes. Taking into account the abundance of each isotope, it is the average of the atomic masses that is usually given.
13. One hydrogen atom has a single electron. By sharing electrons, two atoms of hydrogen achieve the favorable number 2 for their electron structure.
14. Sodium atoms lose an electron while

chlorine atoms gain an electron. The resulting ions combine and are held together by electrical forces between the charges. This very strong link is called an ionic bond. Water molecules are formed by oxygen and hydrogen atoms sharing electrons. The term for this is covalent bond; it is weaker than the ionic bond of sodium chloride.

15. Hydrogen. They all have an atomic number of 1.

16. It is usually the outer electrons of an atom that determine how the atoms of one element will combine with atoms of other elements. The attempt of atoms to achieve a favorable electron number involves them in combinations of covalent as well as ionic bonds.

17. Oxygen (O), having an electron number 8, requires two electrons to achieve the closest favorable electron number, 10. A silicon atom has four electrons more than a favorable number (14). Thus, one silicon atom will provide the electrons needed for two oxygen atoms. The formula is SiO_2. This is the chemical formula for the mineral quartz.

18. As heat energy is added to solid water, its molecules move faster, causing it to melt to a liquid. If enough heat energy is added, the water molecules move fast enough to break away from the liquid as a gas.

references

Asimov, I. *How Did We Find Out About Atoms?* New York: Walker, 1976.

Bolton, Lamphere, Menesini, and Huang. *Action Chemistry*. New York: Holt, Rinehart and Winston, 1979.

Courneya and McDonald. *The Nature of Matter*. Toronto: D. C. Heath Canada Ltd., 1976.

Metcalfe, Williams, and Castka. *Modern Chemistry*. New York: Holt, Rinehart and Winston, 1982.

O'Connor, Davis, Haenisch, MacNab, and McClellan. *Chemistry: Experiments and Principles*. Lexington, MA: D. C. Heath & Co., 1977.

Page, Thomson, and Lichten. *Chemistry*. Reading, MA: Addison-Wesley Science Module, 1973.

Tracy, Tropp, and Friedl. *Modern Physical Science*. New York: Holt, Rinehart and Winston, 1974.

audio-visual aids

16mm Sound Films

Evidence for Molecules and Atoms. EBEC, 19 min., color

Explaining Matter, Atoms and Molecules. EBEC, 14 min., color

The Nature of Matter. McGraw-Hill

The Structure of Atoms. McGraw-Hill, 12½ min., color

What Are Things Made Of? Coronet, 11 min., color

Filmstrips

Atoms. Learning Arts, set of 2

Atoms, Molecules, Ions. SVE

Classification of Matter. EBEC

The Composition of Atoms. EBEC

Elements and Binary Compounds, P. H. Media

Elements, Compounds, Mixtures. SVE

Exploring Matter and Energy: Properties of Matter. SVE with cassette

Investigating Matter, EBEC, set of 6 with cassette

Molecules, Atoms and Simple Reactions. EBEC

Structure of Matter, Atoms and Molecules. SVE, set of 4

8mm Film Loops

Introducing Atoms. EBEC, color

Definite Proportions. Ealing, color

Mixtures, Solutions and Compounds. Rand McNally

The Building Blocks of Matter. Doubleday Multimedia

8 materials of the earth's crust

objectives

When finished with this chapter, the student will be able to:

1. Describe several ways silicon and oxygen combine to form silicates and list some representative silicate minerals.
2. List and describe the composition of several nonsilicate minerals.
3. Describe several properties of minerals and several tests that are used to identify a mineral.
4. Explain how rocks can be classified according to how they were formed in the crust and list several examples of each kind of rock.

approach

The study of minerals and rocks can be approached on many different levels. This text is designed to facilitate a relatively simple approach to this study. However, the material can be expanded to any degree that seems appropriate. For example, the topic of crystals could be explored in much greater depth, examining what causes the precise geometry of crystals and the relationship of these causes to mineral properties. Unless the students have a good academic background and are well motivated, it is usually wise to guide their interests in minerals and rocks toward collection and field activities. Most localities offer some opportunity to collect specimens that can be easily identi-

fied. The mineral key in Appendix E., pages 536-549, of the text can be useful for this kind of activity.

Since identification of specific minerals and rocks is usually difficult for the unskilled, students should be encouraged to collect specimens and make observations that will help them learn how rock masses were formed. A majority of surface exposures are likely to be sedimentary rocks. These often yield conspicuous clues to their origin. It is generally true that seeking evidence that reveals the way rocks and minerals were created can often prove to be more rewarding than attempts to actually identify them.

activity hints

Classification Key (page 163)

The purpose of this activity is to introduce the idea that organizing information about things is an excellent way to learn more about them. Letters of the alphabet are used in this activity, but you could develop the same idea using any common group of related items. Paper fasteners, wood fasteners (i.e., nails, screws, etc.) are items you can use if available. At the end of this activity, have the students classify this new letter: μ. After they do this, ask them which letter of the alphabet this new letter is most like. (U)

other activities

Geology and Earth Sciences Sourcebook (2nd edition, 1970) Holt, Rinehart and Winston
Chapter 1, Earth Minerals, pp. 1-57

Also see Chapter 8 in *Activities and Investigations for Modern Earth Science* (1983 edition) Holt, Rinehart and Winston.

Vocabulary Review

1. e	**5.** i	**9.** g	**13.** o
2. h	**6.** b	**10.** j	**14.** m
3. a	**7.** k	**11.** q	**15.** p
4. f	**8.** c	**12.** l	

Questions: Group A

1. a	**9.** c	**17.** d	**25.** c
2. b	**10.** d	**18.** b	**26.** d
3. d	**11.** a	**19.** c	**27.** b
4. d	**12.** b	**20.** a	**28.** a
5. c	**13.** d	**21.** d	**29.** b
6. d	**14.** b	**22.** c	**30.** c
7. c	**15.** d	**23.** c	
8. a	**16.** d	**24.** a	

Questions: Group B

1. A heated platinum loop is dipped into powdered borax and then held in a flame. This produces a colorless bead that is also dipped in the powdered mineral, tested, and again held in a flame. The resulting color of the bead helps to identify the metal in the mineral.

2. Igneous rocks are formed from the molten state of magma. Sedimentary rocks are a collection of sediments cemented into a series of layers to form new rocks. Metamorphic rocks are produced by heat and pressure below the earth's surface, or by chemical action on igneous or sedimentary rocks.

3. Metal ore minerals, gem minerals, and some rock-forming minerals other than silicates.

4. Impurities and surface tarnish frequently change the color of minerals.

5. When an object is viewed through a clear piece of calcite, it appears as a double image.

6. Color, streak, luster, crystal form, cleavage and fracture, hardness, specific gravity, magnetism, fluorescence, radioactivity, optical properties, chemical tests, and flame tests.

7. Streak is the color of a thin layer of the finely powdered mineral. A streak can be obtained by rubbing the mineral on a piece of unglazed porcelain.

8. Evaporites are produced when materials dissolved in water are deposited and the water evaporates. Halite and gypsum are examples.

9. The weight of an equal volume of water is 24-16 or 8 grams. The specific gravity is then $\frac{24 \text{ grams}}{8 \text{ grams}}$ or 3.

10. Metamorphic rocks are formed from igneous and sedimentary rocks. Pressure, heat, and chemical reactions produce changes in both structure and composition of the original rocks. New texture and compounds result from these actions.

11. See the text of this chapter.

12. Shale is changed to slate by metamorphism. Slate, in turn, changes to phyllite under extreme heat and pressure. With even more intense heat and pressure, phyllite becomes schist. Schist can also be formed from igneous rock. In some instances, the origin of schist is very difficult to determine.

13. Moh's scale is a list of ten selected minerals in order of increasing hardness. These standard minerals are assigned numbers from 1 to 10. If a mineral being tested scratches one of those numbers on Moh's scale, but will not scratch the next hardest mineral, the tested mineral has a hardness between the two standards.

14. One flame test uses a borax bead on a platinum wire. When the bead is dipped in the powdered form of a mineral and heated, the borax may

react with metals in the mineral, giving the bead a color that varies with different metal atoms. A second test involves heating a small amount of the mineral on a block of charcoal. The appearance of the material left on the block, after heating, can aid identification. In a third test, a bit of the powdered mineral is moistened with acid and held in a flame. The color of the flame is a clue to identifying the mineral's composition.

15. Cleavage refers to one or more well-defined directions or flat planes along which the mineral may split. Fracture means that no well-defined flat planes are present when the mineral is broken.

16. See page 154.

17. Intrusive volcanic activity produces rocks whose crystals can grow to large size. The resulting rocks thus have a coarse texture. Extrusive volcanic activity produces rocks that have smaller crystals or no visible crystals at all. A finer texture results from extrusive volcanic activity. Gases that may be trapped in extrusive volcanic rocks give it a sponge-like appearance known as pumice.

18. In the minerals quartz, mica, and hornblende, one or more of the oxygen atoms in the silicon-oxygen tetrahedron can be joined to silicons in other tetrahedrons. For olivine, the silicon-oxygen tetrahedrons remain separate but linked by magnesium and iron atoms.

19. Properties of silicate minerals are determined first by the way in which the silicon-oxygen tetrahedrons are joined, and also by other elements present in their structures.

20. The specific gravity is 3.5. Use any volume of the mineral. If you choose 3 cm^3, the mass will be 3×3.5 or 10.5 grams. The weight loss in water would be 3 grams. Therefore, the specific gravity would be $\dfrac{10.5 \text{ grams}}{3 \text{ grams}}$ or 3.5.

references

Arem, J. *Rocks and Minerals.* New York: Bantam Books, Inc., 1975.

Boyer, R. *Geology Fact Book.* Northbrook, IL: Hubbard, 1974.

Chesterman, C. W. *The Audubon Society Field Guide to North American Rocks and Minerals.* New York: Knopf, 1978.

Dana, E. S., and C. S. Hurlbut. *Minerals and How to Study Them.* New York: John Wiley & Sons, 1972.

Dietrich, R. V. *Stones: Their Collection, Identification, and Uses.* San Francisco: W.H. Freeman &.Co., 1980.

Gart, R. *Exploring Minerals and Crystals.* New York: McGraw-Hill, 1972.

Hamblin, W. K. *The Earth's Dynamic Systems.* New York: Burgess, 1978.

Leet, L. D., S. Judson, and M. E. Kauffman. *Physical Geology* (5th ed.) Englewood Cliffs, NJ: Prentice-Hall, 1978.

MacFall, R. *Gem Hunter's Guide: How to Find and Identify Gem Minerals.* New York: Thomas Y. Crowell Co., 1975

audio-visual aids

16mm Sound Films

Minerals. EBEC

Minerals and Rocks (2nd ed.). EBEC, 15 min.

Rocks and Minerals. PHOENIX

Rocks That Form on the Earth's Surface. EBEC, 16 min., color, No. 2198

Rocks That Originate Underground. EBEC, 23 min., color, No. 2403

Filmstrips

Evolution of Rocks. Edwin Shapiro Co.

Investigation of Rocks. EBEC, set of nine

Minerals and Rocks. Creative Visuals

8mm Film Loops

Formation of Sedimentary Strata. EBEC, color, S-80752

Igneous Processes. Listening Library

Metamorphism and Coal Formation. Listening Library, color

Sedimentation. EBEC, color

Sedimentation and Sedimentary Rocks. Listening Library, color

Volcanoes and Laccoliths. EBEC, color

35mm Slides

Ward's Color Slides for Mineralogy. (see catalogue for listing)

Rock, Thin Sections. World's, color, LW74

Rocks, Slide Program. Denoyer-Geppert, 3 carousels, 240 slides

Examining the Earth's Crust, A Study of Rocks. Inquiry A.V. (daylight projectible slide program) 3 carousels

9 the movement of continents

objectives

When finished with this chapter, the student will be able to:

1. Describe the theory of continental drift.
2. Discuss the theory of plate tectonics and what it suggests.
3. Draw and label a diagram showing the main layers of the earth and state the evidence for their existence.
4. Describe three kinds of crustal plate boundaries.
5. Compare the earth's magnetic field to a bar magnet and discuss its changing nature.
6. Discuss how volcanism is related to plate tectonics.
7. Describe the major causes, processes, and features of volcanic activity.

approach

The material in this chapter helps establish a foundation for a genuine understanding of the basic processes that give the earth its familiar shape and form. If students are aware of the continuous supply of energy that acts upon all parts of the earth, they should have more appreciation for the constant changes that take place in their environment. Such dramatic expressions of physical change as earthquakes and volcanoes are not difficult to observe, and will probably be familiar to almost all students through at least visual experience. The more subtle changes such as those caused by weathering and most types of erosion are likely to go unnoticed. It is important to convey the idea that all observed changes on the surface of the crust are the result of energy being transferred from one form to another. Understanding this concept is fundamental to a knowledge of the earth's processes. If this concept is developed in great detail, it will probably lead to deep student involvement with abstract physical principles that require a solid background in the basic sciences and mathematics.

activity hints

Volcanoes (page 186)

It will be necessary to provide students with maps of the world that show latitude and longitude, as well as tectonic plate boundaries. A master from which student copies can be made appears on page T76 in this Teacher's Edition. The map that identifies the position of the volcanoes appears on page T79.

Answers to questions follow:

1. Pacific Ocean.
2. Yes.
3. L. Mauna Loa, M. Kilimanjaro.

other activities

Geology and Earth Sciences Sourcebook (2nd edition, 1970) Holt, Rinehart and Winston.
Chapter 2, Volcanoes, pp. 59-77

Also see Chapter 9 in *Activities and Investigations for Modern Earth Science* (1983 edition) Holt, Rinehart and Winston.

answers—chapter 9

Vocabulary Review

1. l	4. a	7. i	10. h
2. b	5. g	8. c	
3. j	6. e	9. k	

Questions: Group A

1. c	9. b	17. d	25. c
2. b	10. b	18. a	26. b
3. b	11. a	19. b	27. a
4. d	12. c	20. c	28. d
5. d	13. a	21. b	29. d
6. c	14. a	22. a	30. c
7. d	15. c	23. c	
8. d	16. c	24. b	

Questions: Group B

1. The theory of continental drift suggests that the continents move slowly over the earth's surface.
2. The theory of plate tectonics involves the motion of the continents that probably fit together at a previous time.
3. The theory of plate tectonics has the earth divided into large connected plates that expand and push against each other. The theory of continental drift has the continents drifting like rafts on a lake.
4. See text, Figure 9-2, page 184.
5. The Moho is the dividing line between the crust and the mantle of the earth.
6. Some earthquake waves do not penetrate the core and others cause the formation of a shadow zone. This gives evidence of a core's existence.
7. The earth has three layers. The crust is the outermost layer. It is thin and solid. The mantle is next, and is partly melted and partly more rigid rock. At the center is the core, which is composed mostly of heavier materials such as iron with small amounts of lighter elements like silicon.
8. Mid-ocean ridges are the sites of new plate material that supports sea-floor spreading.
9. One kind of plate boundary is found at the mid-ocean ridge where new plate material is formed. A second kind occurs where plates collide and cause one plate to move beneath the other. The third boundary has plates that slide past each other in opposite directions.
10. The magnetic field of the earth has a shape similar to a bar magnet. A compass needle behaves the same with both of them. The poles are tilted with respect to the geographic poles.
11. The geographic poles are the points on the surface of the earth around which the earth rotates. The geomagnetic poles are points on the earth's surface, about 11.5° away from the geographic poles, that represent the magnetic north and south poles of the earth.
12. Magnetic declination is the angle between the direction to the geographic pole and the direction a compass needle points.
13. Scientists believe that the liquid parts of the earth's iron core may produce electric currents that cause the earth's magnetic field.
14. The earth's magnetic poles have

shifted, but always close to the geographic poles. The magnetism has reversed itself at least nine times, exchanging the magnetic north and south poles.

15. Stripes with the same magnetic directions recorded in their rocks are located symmetrically on both sides of mid-ocean ridges.

16. Near the surface the earth's temperature increases about 2-3°C for every 100 m of depth. Below about 100 km the temperature probably increases very slowly, reaching about 3000°C at the center.

17. The very fluid rock near the earth's surface is called *magma*. When magma reaches the surface, it is called *lava*. Activities caused by the movement of magma are called *volcanism*, while any opening in the earth's surface that gives off lava is a *volcano*.

18. (a) Shield cones: broad at the base with gentle slopes. (b) Cinder cones: narrow at the base with steep sides that do not pile high. Made up of loose rock fragments. (c) Composite cones: combination of lava and rock fragments structured in alternating layers. (d) Parasitic cones: small cones formed from the side of the larger cone. (e) Spatter cones: very small cones formed from lava some distance away from the main vent.

19. Most volcanoes are located near the edges of crustal plates.

20. Island arcs are a chain of volcanoes growing along the trench that marks the boundary of crustal plates.

references

Ballard, Robert D. "Window on Earth's Interior." *National Geographic Magazine.* (August 1976), pp. 228-249.

Boyer, R. *Geology Fact Book.* Northbrook, IL: Hubbard, 1974.

Decker, R., and B. Decker. *Volcanoes.* San Francisco: W. H. Freeman & Co., 1981.

Fodor, R. V. *Earth in Motion: The Concept of Plate Tectonics.* New York: William Morrow, 1978.

Heirtzler, James R. "Where the Earth Turns Inside Out." *National Geographic Magazine.* (May 1975), pp. 586-603.

Krafft, Maurice, and Katia. *Volcano.* New York: Harry N. Abrams, Inc., 1975.

Leet, L. D., S. Judson, and M. E. Kauffman. *Physical Geology* (5th ed.), Englewood Cliffs, NJ: Prentice-Hall, 1978.

Oakeshaft, G. B. *Volcanoes and Earthquakes: Geologic Violence.* New York: McGraw-Hill Book Co., 1976.

Sullivan, Walter. *Continents in Motion: The New Earth Debate.* New York: McGraw-Hill Book Co., 1974.

Tarling, T., and M. Tarling. *Continental Drift.* New York: Doubleday, 1975.

Wilson, J. T., comp. *Continents Adrift,* readings from *The Scientific American.* San Francisco: W. H. Freeman & Co., 1972.

audio-visual aids

16mm Sound Films

Continental Drift—The Theory of Plate Tectonics. EBEC, 22 min.

Fire Under the Sun, The Origin of Pillow Lava. Moonlight Productions, 20 min.

Heart of a Volcano. EBEC, 20 min.

Heartbeat of a Volcano. EBEC, 21 min., color

Not So Solid Earth. Time-Life, 25 min., color

Our Dynamic Earth. NGS

The Earth, Volcanoes. Coronet, 11 min.

The Restless Earth: Plate Tectonics Theory. Indiana University, 58 min., color

Volcanoes: Exploring the Restless Earth. EBEC, 18 min.

Filmstrips

Geology: Our Dynamic Earth—The Restless Earth. NGS, with cassette

The Earth and Its Wonders: Volcanoes. EBEC

This Restless Earth: A Study of Volcanoes and Continental Drift. Schloat, 4 sound filmstrips, color, T648

The Restless Earth: Understanding the Theory of Plate Tectonics. Science and Mankind, Inc., set of 4, with cassettes

Volcanism. Edwin Shapiro Co., color

8mm Film Loops

Anatomy of a Volcano. EBEC, S-80750

Changing Continents. PHOENIX

10 movement of the earth's crust

objectives

When finished with this chapter, the student will be able to:

1. Explain the theory of uniformitarianism.
2. Describe the effect of forces that bend rock. Present evidence that this occurs.
3. Discuss the origin of the forces that deform the crust.
4. Explain earthquakes and how they are studied.
5. Describe how most earthquakes are related to crustal plates.
6. Compare kinds of mountains and the processes that formed them.

approach

The largest land forms on earth were created by movements of the earth's crust. By far the most dramatic result of these movements are earthquakes. For this reason, the topic of earthquakes and their detection is emphasized in this chapter. An effort should be made to establish the idea that movements of the crust are responsible for a lengthier process, the lifting and lowering of great blocks of the crust. These processes are quite slow and tend to be overlooked. Earthquakes are actually only relatively minor results of the gradual adjustment of the earth's crust to unknown internal forces.

Careful study of the landscape of almost any region will reveal some evidence of movement, such as faulting, folding, uplift, or subsidence. If possible, students should be motivated to examine the countryside for evidences of such movements. Many times this evidence is present only on a small scale, as in the slight tilting of sedimentary layers or small faults.

activity hints

Earthquakes and Mountains (pages 222-223)

A master copy of the map students are to use in this activity appears on page T76 in this Teacher's Edition. A spirit master can be made from it to produce class sets. The map has crustal plates and volcanoes recorded on it. The map that identifies the position of the earthquakes and mountains appears on page T79.

The following are answers to activity questions:

1. The earthquakes are near the edges of crustal plates.
2. Most of the mountain ranges are near the edges of crustal plates.
3. Volcanoes, earthquakes, and many mountain ranges seem to be related to crustal plate boundaries.
4. No. Earthquake A in China is somewhat farther from the edge of a crustal plate than the others listed.

5. No. The Appalachian Mountains are not near a plate edge.

other activities
Geology and Earth Sciences Sourcebook (2nd edition, 1970) Holt, Rinehart and Winston.
Chapter 3, Earthquakes and Earth's Interior, pp. 79-97

Also see Chapter 10 in *Activities and Investigations for Modern Earth Science* (1983 edition) Holt, Rinehart and Winston.

answers—chapter 10

Vocabulary Review

1. e	5. i	9. f	13. r
2. g	6. d	10. h	14. q
3. a	7. b	11. n	15. o
4. c	8. l	12. j	

Questions: Group A

1. a	7. d	13. b	19. b
2. d	8. a	14. c	20. b
3. c	9. c	15. a	21. d
4. c	10. b	16. d	22. a
5. a	11. a	17. c	
6. d	12. d	18. a	

Questions: Group B

1. The principle of uniformitarianism states that the same forces that changed the rocks in the past are still operating today to cause the same kind of changes.
2. An anticline is the raised portion of a fold in rock. A syncline is the lowered portion of a fold in rock.
3. A rift valley forms when the blocks of a rock between a pair of parallel faults sink. The resulting broad valley with steep cliffs on both sides is also called a graben.
4. Evidences of motion along the sides of the fault can be seen.
5. Motion of the plates creates stress in the rock of the lithosphere as the plates rub together, collide, or spread apart.
6. Isostasy means that the heavier parts of the crust sink deeper toward the mantle. Mountains have deep roots, for example.
7. The elastic rebound theory is an explanation for earthquakes. Slow movements of rock on both sides of a fault build up stress in the rocks, forcing them to bend and stretch. When the stress becomes too great, the rocks spring back and release their stored energy as earthquake waves.
8. Earthquakes are usually one of the events connected with the motion along the boundaries of the crustal plates.
9. The speed with which primary and secondary waves travel through various materials is known. The time lag between the arrival of primary waves and the arrival of the secondary waves at a seismograph station can be used to compute the distance from the source of the waves.
10. The focus of an earthquake is the actual point along a fault where slippage occurs to cause the earthquake. The epicenter is the point on the earth's surface directly above the focus.
11. Scientists think earthquakes can be predicted. An early warning system for earthquakes could save many lives. Also, if the stress can be released slowly, the magnitude of earthquakes could be reduced.
12. Most earthquakes and volcanoes are related to the activity at the edges of crustal plates and therefore occur in the same general zones.
13. The mountain systems are all basically made up of very large and com-

plex folds. They also have much faulting and intrusive igneous activity.

14. Fault-block mountains are created when the crust breaks into large blocks at faults. These blocks are then tilted to create the mountains.

15. When a part of one crustal plate is pushed beneath the edge of another plate, the upper plate edge rises to form a mountain system. When plates collide head-on, the plates crumple to form mountains. The mountain system at the mid-ocean ridge results from volcanic activity at the crustal plate edges.

references

Boyer, R. *Geology Fact Book*. Northbrook, IL: Hubbard, 1974.

Calder, N. *The Restless Earth*. New York: Viking Press, 1972.

Lamber, P. *Earthquakes*. New York: Random House, 1973.

Leet, L. D., S. Judson, and M. E. Kauffman. *Physical Geology*, (5th ed.), Englewood Cliffs, NJ: Prentice-Hall, 1978.

Matthews, W. H. *Science Probes the Earth: New Frontiers of Geology*. New York: Sterling Publishing Co., 1969.

National Geographic Society. Powers of Nature. Washington, DC, 1978.

Oakeshaft, G. B. *Volcanoes and Earthquakes: Geological Violence*. New York: McGraw-Hill Book Co., 1976.

audio-visual aids

16mm Sound Films

An Approach to the Prediction of Earthquakes. AEF, 27 min., color
Birth of a Mountain. EBEC, 24 min.
Earth: The Restless Planet. NGS
Earthquake Below. NASA, 14 min., color
Earthquakes: Lesson of a Disaster. EBEC, 13 min., color
Fractured Look. NASA, 14½ min.
The San Andreas Fault. EBEC, 21 min., color

8mm Film Loops

Earthquake Mechanism. PHOENIX, color
Structure of the Earth. PHOENIX
Tectonic Movement. EBEC, color

forces that sculpture the earth

11 weathering and erosion

objectives
When finished with this chapter, the student will be able to:
1. Name, describe, and give an example of the two ways weathering occurs.
2. Explain how the nature of the rock, climate, and topographic conditions affect the rate of weathering.
3. List and describe the principal parts of mature soil and explain how soil is produced.
4. Describe the roles that gravity and wind play in erosion and give an example of each.
5. Describe several methods currently being used to conserve soil for agricultural use.
6. Discuss the life cycle of mountains, plains, and plateaus.

approach
Great emphasis should be placed on the slowness that is characteristic of the weathering processes. It should also be stressed that different kinds of rock weather at different rates. Students should be alerted to watch for evidence of unequal weathering among rocks in the same location. Examples of weathering are likely to be found in natural accumulations and in human-made structures of rock and masonry in the vicinity of your school.

It is usually possible to find some local spot that is obviously eroded and familiar enough to everyone so that it can be used as an example and basis for discussion of the symptoms and causes of erosion. If such a region is close by, a field trip to observe both weathering and erosion evidences would be helpful. Particular attention should be paid in the classroom and the field to the mass-wasting process because it is one of the most common but least conspicuous forms of erosion.

One of the most important understandings contained in the chapter is the relationship between weathering of rock and the formation of soil.

activity hints

Leaching (page 235)
The student is asked to describe what happens when a small amount of water is used to wash some table salt through a sand filter. The simplest test for checking where the salt has gone in question 4 is to taste the water that collects in the bottom cup. If you have a silver nitrate solution available, you may suggest its use as follows. A drop or two of silver nitrate solution would cause a white precipitate to form, indicating the presence of the chloride ion in salt ($NaCl$).

other activities

Geology and Earth Sciences Sourcebook (2nd edition, 1970) Holt, Rinehart and Winston.
Chapter 7, Erosion and Landforms, pp. 159-177

Also see Chapter 11 in *Activities and Investigations for Modern Earth Science* (1983 edition) Holt, Rinehart and Winston.

answers—chapter 11

Vocabulary Review

1. m	**5.** c	**9.** l	**13.** f
2. o	**6.** s	**10.** a	**14.** h
3. q	**7.** j	**11.** i	**15.** d
4. e	**8.** r	**12.** n	**16.** b

Questions: Group A

1. b	**9.** c	**17.** a	**25.** b
2. d	**10.** c	**18.** b	**26.** d
3. a	**11.** a	**19.** d	**27.** a
4. c	**12.** d	**20.** c	**28.** d
5. a	**13.** b	**21.** c	**29.** d
6. b	**14.** c	**22.** b	**30.** a
7. c	**15.** b	**23.** a	
8. d	**16.** a	**24.** d	

Questions: Group B

1. For erosion to take place, the earth's crust must already be broken into small particles. Weathering, the process that created these small particles, therefore must occur before erosion.
2. It may contain loose scales or chips that are easily stripped off; its surface may crumble under pressure; its surface color may differ from the interior; and, if broken open, stain marks may appear around cracks.
3. Mechanical weathering processes produce a change only in the shape or size of the rock. Chemical weathering alters the rock's composition.
4. The three examples of mechanical weathering in rock are: (a) Freezing water in existing cracks expands, causing particles to chip and break off. (b) In drier climates, water may evaporate from the rock, leaving behind salt crystals that build up pressure in the cracks strong enough to chip the rock. (c) Clay particles in rock expand as they absorb water, thus also removing particles.
5. For carbon dioxide to dissolve limestone, feldspar, or calcite, it must be dissolved in water.
6. The mineral would have reddish colored streaks or perhaps even be reddish colored throughout since this is the color of iron oxide, a product of weathering.
7. Roots of plants work their way into cracks of rocks and wedge them apart; the activities of burrowing animals, gophers for example, constantly expose new rock surfaces; earthworms uncover fine rock particles and expose them to the air. Burrows are passageways for water and air to enter the upper layers, causing more weathering.
8. (a) The nature of the rock (b) climate (c) topographic conditions.
9. The first step involves the weathering of bedrock, the stage at which the mineral particles that go into the makeup of soil are formed; the second step involves growing plants and animals whose remains form the humus of mature soils.
10. A typical soil profile consists of three distinct layers called horizons. Horizon A, the topsoil made up of rather loose material, contains many living organisms. Horizon B is the compact

subsoil that contains the minerals leached from above and below. Horizon C, the deepest layer, consists of partly weathered bedrock.

11. All three are obvious examples of erosion by gravity. Soil creep and slumping are rather slow, downhill movements that under certain conditions may become very rapid. At this speed it becomes a landslide.

12. Falling rock, running water, glacial movement, blowing wind, and all other erosion agents are caused by gravity.

13. Particles that are large enough to cause such features are windborne only very close to the ground.

14. In a wind that persists for some length of time, sand grains are constantly being blown up over the windward side and deposited on the leeward side. The continual removal and deposition of sand a little farther along pushes the sand dune along in the direction of the wind.

15. Sand dunes are formed from the coarser material of the desert; loess develops out of the very fine dust that is blown higher and is generally deposited at a great distance in large amounts.

16. Natural erosion takes place very slowly and is kept in balance by the formation of new soil by the process of weathering. Humans, with their unwise clearing of land for growing crops, building structures, etc., hinder soil formation and by these activities actually speed up erosion.

17. Igneous and metamorphic rock are very resistant to mechanical weathering. Therefore, any appreciable weathering of these rocks is chiefly by chemical means. Sedimentary rocks are generally held together loosely and undergo weathering quite easily by mechanical means.

18. (a) Contour plowing: plowing in such a way that the furrows that follow the contour of the land slow down the flow of water on hillsides. (b) Strip cropping: arrangement of two crops in alternate bands, always using one crop as a brace to hold the soil. (c) Terracing: step-like ridges that follow the land contours on steep hills allow water to seep in on the steps and slowly run off. (d) Crop rotation: alternating rows of crops that expose the soil one year and provide a cover crop that protects the soil the next year.

19. Monadnocks are rocks that are more resistant to erosion than the surrounding material. They appear as knobs on peneplains.

20. Both mesas and buttes are the remnants of a plateau and both have steep walls and flat tops. Mesas have broad tops, while buttes have narrow tops.

21. Peneplains look very much like true plains and might be mistaken for them. Peneplains have low rolling hills. Under the peneplain surface are the folded, twisted, or tilted rocks that are unlike the horizontal layers of true plains.

references

Boyer, R. E. *Field Guide to Rock Weathering*. Boston: Houghton-Mifflin Co., 1971.

Foster, Robert J. *Physical Geology* (2nd ed.). Columbus, OH: Charles Merill Publishing Co., 1975.

Foth, Henry, and H. S. Jacobs. *Field Guide to Soils*. Boston: Houghton-Mifflin Co., 1971.

Hunt, C. B. *Geology of Soils*. San Francisco: W. H. Freeman & Co., 1972.

Leet, L. D., S. Judson, and M. E. Kauffman, *Physical Geology* (5th ed.). Englewood Cliffs, NJ: Prentice-Hall, 1978.
Strahler, A. N. *The Earth Sciences* (2nd
ed.). New York: Harper & Row, 1971.

audio-visual aids

16mm Sound Films
Erosion. PHOENIX
Erosion—Each Moving Grain. MEVA
Erosion: Leveling the Land. EBEC, 14 min.,
color
From the Ground Up. USSCS, 13 min.,
color
Why Do We Still Have Mountains? EBEC,
20 min., color
Conserving Our Soil Today. Coronet, 11
min., color
The Earth: Changes in its Surface. Coro-
net, 11 min., color
The Story of Soil. Coronet, 11 min., color

Filmstrips
Changing the Face of the Earth. SVE
Chemical Changes in Rocks and Minerals.
Denoyer-Geppert
Conservation. Edwin Shapiro Co.
How the Earth's Surface is Worn Down.
McGraw-Hill, with cassette
The Earth's Surface Wears Down. Eye Gate,
with cassette
Weathering and Erosion. Edwin Shapiro
Co.
What is Soil? EBEC
Work of Running Water. SVE
Work of Wind. SVE

8mm Film Loops
Alluvial Fan. Ealing, color
Cavern Formations. PHOENIX, super 8,
color
Investigations in Science. PHOENIX,
Earth Science Series, set of 7 on ero-
sion, weathering, soil, super 8, color

Stream Erosion Cycle. Ealing, color
Erosion. PHOENIX, color
Weathering, PHOENIX, color
Soil, Parts 1 and 2, PHOENIX, color
Humus, PHOENIX, color
Mountains and Plateaus. EBEC, color
Water-Soil. Hubbard, set of 8, super 8

12 water and rock

objectives
When finished with this chapter, the
student will be able to:
1. Explain the operation and importance
of the hydrologic cycle.
2. Describe the importance of water conservation.
3. Trace the steps in the development of
a river system.
4. Describe erosion caused by moving
water.
5. Explain how the water table is related
to the water in the ground.
6. Describe the formation of springs and
geysers.
7. Describe the formation of sinks and
caverns.

approach
Local conditions should greatly influence the general treatment and emphasis
of the material in this chapter. The variety of local situations should be closely
related to the text material and the topics of greatest local interest and importance expanded as much as possible in
supplemental work.

Students might be interested in con-
tacting local well-drilling companies to
collect information about depth and rate
of flow for wells dug in the region. If sufficient information can be obtained, the

wells may be located on a map. This will give some indication of local ground water supplies. The subject of "water-witching" or "divining" wells will usually provoke a spirited discussion and an opportunity to illustrate the characteristics of a scientific attitude.

activity hints

Capillarity (page 267)

Tubes with a diameter larger than 1 cm can be used. Of course, more sand will be needed. One-half inch, or larger, plastic tubing may be available in the plumbing department of a home repair store. Sand could be sorted into coarse and fine grain using a sieve. Students could graph the results of their measurements.

Answers to activity questions follow:
1. Capillarity. **2.** Fine; fine. **3.** Yes.
4. Capillarity would bring water to the surface where it could evaporate easily.

other activities

Geology and Earth Sciences Sourcebook (2nd edition, 1970) Holt, Rinehart and Winston.
Chapter 6, Waters of the Continents, pp. 137-157

Also see Chapter 12 in *Activities and Investigations for Modern Earth Science* (1983 edition) Holt, Rinehart and Winston.

answers—chapter 12

Vocabulary Review

1. c	**5.** a	**9.** h	**13.** k
2. g	**6.** i	**10.** l	**14.** n
3. j	**7.** d	**11.** q	**15.** p
4. e	**8.** b	**12.** o	

Questions: Group A

1. b	**8.** c	**15.** c	**22.** b
2. c	**9.** d	**16.** b	**23.** d
3. d	**10.** a	**17.** a	**24.** b
4. a	**11.** c	**18.** c	**25.** a
5. c	**12.** c	**19.** c	**26.** c
6. b	**13.** b	**20.** d	**27.** b
7. d	**14.** d	**21.** a	

Questions: Group B

1. The hydrologic cycle is a series of events including evaporation, cloud formation, precipitation, run-off or sinkage, and storage in the ground until evaporation takes place again.
2. Rate of flow depends upon: amount of water present in the ground, pressure, the permeability of the underlying material, and the gradient.
3. If the gradient of a stream is made more steep by some process such as bending or faulting, the stream is said to be rejuvenated.
4. Stalactites—a calcite deposit on the cavern's ceiling formed by water dripping into the spaces inside a cavern. Stalagmites—a calcite formation left on the cavern floor by dripping water.
5. Well-established tributaries; largest volume of water; less tendency to deepen its channel; wider valley with a relatively flat floor, river channel normally occupying only a small part of the valley floor.
6. The flood plain is a relatively flat area having rich soil and usually an abundance of water (good farm land).
7. During the driest part of the year, probably late summer.
8. A water budget consists of income in the form of precipitation. Outgo is the water that evaporates. Over the entire earth, it is probably in balance, but run-off, dry and rainy seasons, and

ground water all keep the budget from being balanced at any particular place.

9. A curve in a river results in water traveling faster on the outside edge of the curve. This speeds up erosion of the outer bank; sediments build up along the slower moving inside edge. Over a period of time, looping curves called meanders are formed.

10. A hot spring is a place where water, heated by volcanic activity, finds its way to the surface. Because hot water dissolves minerals more easily than cool water, hot water is richer in mineral content. A geyser is a hot spring that throws water and steam from its underground structure high into the air.

11. Gradient becomes less steep; V-shaped valley is broader and the floor relatively flat; waterfalls and rapids usually disappear; a few tributaries develop into many and become well-established; small volume of water increases the tendency to deepen its channel and has a widening effect as a river matures.

12. By removing cover plants, logging and fires cause greater run-off than normal and increase flood possibilities. Artificial levees, forest and soil conservation, and dams are methods of preventing floods.

13. Examples: steepness of the gradient, amount of precipitation, size of watershed, and composition of the land surface over which the water flows.

14. A tributary could form a delta if the water in the stream into which it feeds is not moving too rapidly.

15. The zone of aeration is the uppermost region where spaces between the rock contain both water and air. Below this is the capillary fringe where water is drawn up from below and spreads out on the surface of the rock particles. The zone of saturation lies below the capillary fringe. Here all spaces are completely filled with water. The top of the zone of saturation is called the water table.

16. When a meander becomes so curved that the river cuts across the narrow neck of land, the meander is isolated from the river. If the water remains in the isolated meander, it will become an oxbow lake.

17. The amount of run-off varies, and so far no predictable pattern of change has been discovered. Climatic changes are also unpredictable at present.

18. Alluvial fans would contain coarser materials because they are formed by a sudden slow-down in high velocity streams. The water with greater velocity would carry coarser materials than rivers that form deltas.

19. The answers will be varied. A simple thing like turning off the shower while you lather could save several gallons of water. If each individual saved 10 gallons of water each day, this would amount to about 11,000,000,000,000 gallons per year. It might be interesting to find what part of the total water available this accounts for. (7.5 gallons = 1 cu. ft.)

references

Birkeland, P. W., and E. E. Larson, *Putnam's Geology* (3rd ed.). Oxford Univ. Press, 1978.

Leet, L. D., S. Judson, and M. E. Kauffman, *Physical Geology* (5th ed.). Englewood Cliffs, NJ: Prentice-Hall, 1978.

Leopold, L. *Water: A Primer.* San Francisco: W. H. Freeman & Co., 1974.

16mm Sound Films

A Drop of Water. ABP, 14 min., color

Erosion, Leveling and Land. APB, 14 min., color

Finding Out About the Water Cycle. UWF, 13 min., color

The Making of a River. Coronet, 10 min., color

Water, The Common Necessity. MIS, 11 min., color

The Water Cycle. EBEC, 14 min.

Filmstrips

Enough Water for Everyone. EBEC, color, sound

Running Water. Edwin Shapiro Co., color, sound

Streams and Rivers. Wards, color, sound

8mm Film Loops

Rivers, Meanders. Gordon Flesch, color, 35155S2

Stream Erosion Cycle. Gordon Flesch, color, P85-0024/1S2

13 ice and rock

objectives

When finished with this chapter, the student will be able to:

1. Explain how glaciers form and describe the two kinds of glaciers presently found on earth.
2. Describe how glaciers erode the land. List the two general kinds of glacial deposits and give a land feature of each kind of deposit.
3. List the conditions necessary for a lake to form and describe the life history of a lake.
4. Describe the conditions that existed during the ice ages and give several possible explanations for their cause.

approach

Point out either on a globe or a map the extent of the earth's surface that is still covered by ice. Show the class that the region around the North Pole has no land surface and is covered with a floating ice pack. Also point out the high mountain ranges spread over the earth that are still heavily glaciated. The idea that the sculpting effects of ice are indeed continuing and are a powerful force in changing parts of the earth's landscape is an important one. In this context, it might be useful and interesting to consider the effects on the earth's surface if the present glaciers should suddenly melt.

In discussing various glacial deposits, emphasis should be placed on their origins and how their present form reflects these origins.

Since it is unlikely that students will have personally seen a glacier, it is important to use films, slides, and other visual aids to insure a clear visual understanding of the processes described.

activity hints

Glacial Flow (page 279)

Silicone putty can be obtained from toy stores in an egg-shaped container under the name "Silly Putty." It is a very viscous fluid whose flow is markedly similar to glaciers. A transparent variety is available from science equipment distributors. One "egg" could be used for two groups to cut down on cost. After using the material, make certain it is placed back in the egg for storage.

other activities

Geology and Earth Sciences Sourcebook (2nd edition, 1970) Holt, Rinehart and Winston.
Chapter 9, Glaciation, pp. 197-221

Also see Chapter 13 in *Activities and Investigations for Modern Earth Science* (1983 edition) Holt, Rinehart and Winston.

answers—chapter 13

Vocabulary Review

1. d	4. m	7. n	10. h
2. k	5. f	8. l	11. e
3. i	6. c	9. a	12. b

Questions: Group A

1. c	9. c	17. c	25. d
2. a	10. b	18. d	26. c
3. c	11. d	19. b	27. c
4. b	12. a	20. c	28. d
5. d	13. a	21. d	29. d
6. d	14. c	22. a	30. c
7. a	15. c	23. c	
8. b	16. a	24. b	

Questions: Group B

1. Glaciers are formed from yearly accumulations of layers of snow. Piled up layers of snow press down on the lower layers, turning them to firm or grainy ice. As more layers of snow are added, the lower layers become a solid mass of ice. Eventually, when the weight of snow and ice accumulated over the years becomes too great, parts of the body of ice begin to slowly move. It can then be called a glacier.
2. At present, glaciers are found in high mountains in middle or high latitudes and in the polar regions where more snow falls each year than melts.
3. Alpine and continental.
4. A sliding motion of the upper layers and a flowing motion of deep layers.

Near the edges and bottom of a glacier the movement is slower than near the middle. The most rapid movement occurs at about two-thirds the depth of the glacier.

5. The uneven movement of different parts of the glacier.
6. Rocks are broken off the walls of a valley by frost, then dropped onto the glacier. The glacier itself pulls loose rocks from the sides and floor of the valley.
7. The valley walls and floor are ground into a U-shape. The sharp bends are somewhat straightened.
8. Continental glaciers smooth the landscape by grinding down projections, whereas the alpine glaciers help make valley walls steeper and in the lowlands deposit huge amounts of debris, giving them a rough terrain.
9. Stratified drift is made up of sorted debris and is always found in layers, whereas till is unsorted debris and appears in unstratified piles.
10. (a) Lateral moraine—formed by rock debris along the edges of the valley.
(b) Medial moraine—formed by the joining of two or more lateral moraines from several merging valley glaciers.
(c) Terminal moraine—formed by debris pushed ahead of the glacier and debris dumped at the end of the glacier.
(d) Ground moraine—debris deposited on the valley floor below the glacial ice sheets.
11. (a) A wide, flat deposit at the end of a glacier, formed out of drift from an outwash plain.
(b) Kettles are depressions in the outwash plain. They are frequently filled with water.
(c) Eskers are ridges of stratified drift extending some distances over flat,

glaciated country. They look much like a crooked, raised roadway.

12. (a) Basins eroded in the earth's surface. An example is the Finger Lakes of central New York.

(b) From ground moraine that produces the effect of an uneven surface. Lakes in northern North America and Northern Europe were formed this way.

(c) The damming of existing streams by terminal and lateral moraines, as in Minnesota, the Dakotas, Wisconsin, Indiana, Ohio, and northern Illinois. The Great Lakes were formed by such action.

13. The uplift of the land surface due to decrease in pressure from the melting of the continental glacier makes the Great Lakes drain northward through the St. Lawrence River.

14. Water may drain away or evaporate.

15. Soil washing down from neighboring slopes. Wave action against lake banks. Windblown dust and landslides. Growth of plants around the edge of the lakes.

16. All of Canada and the mountainous regions of Alaska. Western mountains of the United States west to the coast and east to the foothills. A continental glacier centered in Hudson Bay spread as far as the Missouri and Ohio Rivers.

17. (a) Alternating cold and warm periods in response to changes in the amount of energy received from the sun.

(b) Changes such as mountain building altered the topography of the earth and could have lowered the temperature slightly. An increase in glaciation due to the added height of mountains was the logical outcome.

(c) Melting ice in the Arctic regions brought heavier snowfall to the existing glacial area. The ocean levels dropped and the ice caps in the polar regions appeared again. Thus the amount of snowfall was decreased and once again the earth underwent the gradual disappearance of glaciers.

18. This was because the last ice age took place prior to recorded history and we have no direct evidence of what caused it. There is no overall increase or decrease in the amount of glaciation going on at present. Therefore, we cannot answer all the questions involved in developing a sound theory.

references

Birkeland, P. W., and E. E. Larson. *Putnam's Geology* (3rd ed.). Oxford Univ. Press, 1978.

Leet, L. D., S. Judson, and M. E. Kauffman. *Physical Geology* (5th ed.). Englewood Cliffs: Prentice-Hall, 1978.

Tallcott, Emogene. *Glacier Tracks*. New York: Lothrop, Lee and Shepard Co., 1970.

audio-visual aids

16mm Sound Films

Continental Glaciers. OSU, 13 min., color
Evidence for the Ice Age. EBEC, 19 min., color
Glaciation. McGraw-Hill, 12 min., color
Glacier on the Move. EBEC, 11 min., color
Glacier National Park. MTPS, 22 min., color
The Rise and Fall of the Great Lakes. National Film Board of Canada, 16 min., color

Filmstrips

The Antarctic. Time-Life
Evolution of Glaciers. Edwin Shapiro Co.
Glaciers. Wards
Glaciers and the Ice Age. EBEC, 4 sound filmstrips

How a Glacier Shapes Its Valleys. EBEC
Investigating a Glacier. EBEC
Origin of Lakes. Edwin Shapiro Co.
Reconstructing the Ice Age. EBEC
Some Side Effects of the Ice Age. EBEC
The Glacial Valley. SVE

8mm Film Loops
Arctic Thaw—Part 1. Ealing, color
Arctic Thaw—Part 2. Ealing, color
Evolution of a Lake. EBEC, color
Glacial Movement. EBEC, color, No. S-80749
Origin of Moraines. EBEC, color

14 shorelines

objectives
When finished with this chapter, the student will be able to:
1. Describe four ways that rock is weathered by sea waves and give some examples of formations made by wave erosion.
2. Describe the makeup of a typical beach and the movement of sand grains along the beach.
3. Given a set of shoreline features, explain the likelihood of the features being the result of emergence or submergence.
4. List the kinds of coral reefs.
5. Explain how human use affects shorelines.

approach
For schools located inland where students do not have an easy opportunity to observe shoreline features, much of the material in this chapter might be related to features they have seen along lake shores. Wide use of visual aids during class discussion will help to overcome

the possible handicap of distance from the seashore.

A point that should be emphasized in the study of shoreline development is that the cycle of advance and retreat seldom proceeds without interruption. Thus a given coastline will seldom exhibit the classic cycle of development described in the text. This point should be emphasized and related to the difficulty the class will experience if they examine a particular shoreline with the intention of fitting it neatly into one particular stage of development.

activity hints

Shorelines (page 304)
This activity asks the student to use the location of a photograph and information of how the area is changing to predict the area's appearance in several thousand years. Students will choose a rate of rise or fall based on where they estimate the photograph was taken. In turn, their description or diagram of the area in several thousand years will depend on their estimate. Students should be the ones to decide.

other activities
Geology and Earth Sciences Sourcebook (2nd edition, 1970) Holt, Rinehart and Winston.
Chapter 8, Oceans, pp. 179-195

See also Chapter 14 in *Activities and Investigations for Modern Earth Science* (1983 edition) Holt, Rinehart and Winston.

answers—chapter 14

Vocabulary Review

1. e	**4.** i	**7.** a	**10.** c
2. h	**5.** d	**8.** g	
3. l	**6.** f	**9.** b	

Questions: Group A

1. c	**9.** d	**17.** b	**25.** b
2. d	**10.** b	**18.** d	**26.** d
3. a	**11.** a	**19.** a	
4. b	**12.** a	**20.** c	
5. d	**13.** c	**21.** c	
6. b	**14.** c	**22.** d	
7. a	**15.** a	**23.** c	
8. c	**16.** d	**24.** c	

Questions: Group B

1. (a) Water and air forced into the cracks under pressure cause the rock to break. (b) Pieces of rock flung against the shore chip and wear away the surface. (c) Erosion at sea level weakens overhanging rock and it breaks loose. While falling, the rocks grind against each other. (d) Small rocks rub against one another in the tumbling water.

2. The attack of salt water upon many rocks is sufficient to decompose them. A term for this is chemical weathering. One can generally view this kind of weathering above the zone of wave action.

3. A wave-cut terrace is formed by wave action cutting back the cliff of a shoreline. The rock below the wave-line does not erode, leaving behind a terrace. Rock and debris pile up as a cliff erodes, then are carried seaward and deposited at the edge of the wave-cut terrace. Thus, the sediment eroded from the wave-cut terrace becomes a wave-built terrace.

4. Terraces tend to slow down the action of waves on the shoreline since most of the energy of the waves is spent in the shallow water. If subsidence results, thus increasing the depth of the water, erosion may speed up again. Erosion of the terrace may also result, which would further increase erosion of the shoreline.

5. Uneven erosion of sea cliffs often results in the formation of sea arches, stacks, and caves.

6. Beach sand is produced by wave action on small rock particles. The sand is then washed up onto the beach, also by wave action, and deposited on top of the existing shoreline material. Frequently sand is produced in rivers and finds its way into the ocean. This sand may also be washed up on the beach by wave action.

7. The general pattern is one in which the rock fragments are lifted by the wave and given the same circular motion as the water. The motion is first toward the beach and then away from the beach. The return motion, however, is along the bottom where the particles cannot move as readily. Thus each particle is moved very slightly forward by each succeeding wave until it reaches the beach.

8. The steeper the slope of the berm, the larger the waves have been.

9. Parallel movement results when waves consistently approach the shore at an angle, causing longshore currents that carry the sand.

10. The melting or forming of glaciers; Scandinavia is rising due to the melting of glacial ice. Movement of crustal plates; the eastern U.S. coast is sinking due to being on the trailing edge of a crustal plate.

11. Cliffs are typical of an emergent shoreline, whereas bays and inlets are prominent features of a submergent shoreline.

12. (a) submergence
    (b) emergence
    (c) submergence

13. (a) Miami Beach was formed by the emergence of an offshore sand bar.
    (b) The barrier islands off the south coast of Long Island were formed as a result of submergence when a gla-

cial outwash plain was drowned.

(c) Atolls are formed by deposits of coral followed by erosion of the central island or by submergence of the original land.

14. (a) This sand came about by the erosion and weathering of granite. Granite is found along the coastline of North America. As a result, the sand of beaches here tends to be predominantly light in color. The fact that the sand is coarse would indicate that it is not old geologically and comes from some nearby granite outcrop.

(b) This sand would indicate the presence of basalt and/or lava flows that upon erosion and weathering produce dark sand. Again, this sand is young, geologically speaking.

(c) Fine sand is either old or has been transported over great distances, thus being ground down until quite small. Since the material of which it is composed is not related to that of the shoreline, the sand was probably washed down a river and deposited in the ocean, then washed up on nearby beaches.

15. When glaciers form, the land upon which they rest sinks due to the increased weight of the water (ice), while the ocean floor rises. Just the reverse occurs when the glaciers melt. The land rises while the ocean floor sinks. This could happen only if the base upon which the land and sea floor rests is elastic.

16. Some evidences are wave-cut terraces containing similar type fossils and discovery of submerged beaches on continental shelves that contain similar fossils.

17. When the sea level rises, it tends to fill in all the lower levels, causing islands to disappear and great bays to form. But a drop in sea level would form islands, cliffs, swamp lands, etc.

18. The coastline can be harmed by each specific example given as follows: *ports,* use by ships could result in oil and waste pollution; *shipbuilding,* wastes such as paint and oil residues, etc., could pollute water; *industrial and residential development,* landfills could change drainage and stop the life cycle in estuaries; *recreation,* the change of the shoreline to form marinas, etc., could destroy nearby beaches by changing the sand transport patterns; *commercial fishing,* could destroy link in sea life food chain; *dumping ground for wastes,* could cause extensive damage to sea life as well as causing basins to fill with toxic sludge and silt.

19. Knowledge of the changes in shorelines that might be destructive, and a single authority for an organized development of coastal resources.

references

Barton, R. *Oceanography Today—Man Exploits the Sea.* New York: Doubleday, 1972.

Bascom, W. *Waves and Beaches* (Science Study Series). New York: Doubleday, 1964.

Birkeland, P. W., and E. E. Larson. *Putnam's Geology* (3rd ed.). Oxford Univ. Press, 1978.

Boyer, R. E. *Oceanography Fact Book.* Northbrook, IL: Hubbard, 1974.

Carlisle, N. *The New American Continent—Our Continental Shelf.* Philadelphia: J. B. Lippincott, 1973.

Hawkins, M. E. *Vital Views of the Environment.* Washington, DC: National Science Teacher's Assoc., 1970.

Idyll, C. P. *Exploring the Ocean World, A History of Oceanography.* New York: Thomas Y. Crowell Co., 1972.

Mars, W. *The Protected Ocean, How to Keep*

the Seas Alive. New York: Coward, McCann & Geoghegan, 1972.

Pennington, H. *The New Ocean Explorers.* Boston: Little, Brown and Co., 1972.

Taber, R. W. *1001 Questions Answered About the Oceans and Oceanography.* New York: Dodd, Mead & Co., 1972.

Thurber, Walter A., et al. *Oceanography.* Boston: Allyn & Bacon, Inc., 1976.

audio-visual aids

16mm Sound Films

The Beach—A River of Sand. EBEC, 17 min., color

The Earth—Coastlines. Coronet, 11 min., color

The Earth—Its Oceans. Coronet, 13½ min., color

Marine Erosion Processes. MFI, 11 min.

Oceanography: A Voyage to Discovery. VEVA, 20 min., color

Filmstrips

Lakes and Oceans. Ward's

Evolution of Shorelines. Edwin Shapiro Co.

Work of the Sea. SVE

Miracle of the Sea. Time-Life

Landscapes of the Sea. Time-Life

8mm Film Loops

Shorelines. Ealing, color

Beach Formation and Elevated Beaches. Ealing, color

Shore Drift. Ealing, color

Ocean Shores. EBEC, color

The Beach I: Source of Sand. EBEC, color

The Beach II: Profile Study. EBEC, color

The Beach III: Longshore Transport. EBEC, color

The Beach IV: Formation of a Sand Spit. EBEC, color

15 the unknown sea

objectives

When finished with this chapter, the student will be able to:

1. Define the term "oceanography."
2. Name the seven seas and describe several methods used to measure their characteristics.
3. Describe, using the correct terms, the shape of the earth's crust beneath the sea.
4. Describe the different kinds of sea-floor sediments.

approach

The study of the oceans is an excellent opportunity to stress the interdependence of the basic sciences. The importance of the sea to the future welfare of all nations cannot be overemphasized. There are bountiful resources in the oceans that must be eventually utilized if the world's exploding population is to be adequately fed. Whether these resources become available depends upon how science succeeds in understanding the sea.

activity hints

Settling Rates (page 328)

25mm × 200mm test tubes work well for this activity; however, almost any clear-sided container will do.

Answers to activity questions follow:

1. The bottom ones are larger.
2. The largest must settle first.
3. Smallest particles settle last.
4. (Answers depend on soil used.)
5. Very small particles.
6. Near shore since they settle out first.
7. Very small since they settle out last and would have traveled farthest.

other activities

Geology and Earth Sciences Sourcebook (2nd edition, 1970) Holt, Rinehart and Winston.
Chapter 8, Oceans, pp. 179-195

See also Chapter 15 in *Activities and Investigations for Modern Earth Science* (1983 edition) Holt, Rinehart and Winston.

answers—chapter 15

Vocabulary Review

1. f	**5.** j	**9.** g	**13.** e
2. l	**6.** d	**10.** a	**14.** m
3. h	**7.** p	**11.** q	**15.** k
4. n	**8.** c	**12.** i	

Questions: Group A

1. d	**9.** a	**17.** d	**25.** c
2. c	**10.** c	**18.** b	**26.** c
3. a	**11.** d	**19.** c	**27.** a
4. c	**12.** a	**20.** a	**28.** c
5. d	**13.** c	**21.** a	**29.** b
6. a	**14.** b	**22.** c	**30.** c
7. c	**15.** a	**23.** d	
8. d	**16.** a	**24.** b	

Questions: Group B

1. Seventy percent of the earth's surface is covered by water. This is the greatest coverage known for a planet.

2. North Atlantic, South Atlantic, North Pacific, South Pacific, Indian, Arctic, and Antarctic.

3. *H.M.S. Challenger* made the first extensive voyage for the purpose of oceanographic research. The voyage spanned the years from 1873 to 1876.

4. Depth, temperature, currents, and chemical properties.

5. Oceanography makes use of the combined knowledge of physics, chemistry, geology, and biology.

6. Bathythermography—used to measure water temperature. Nansen bottle—used to take water samples at various depths. Electronic wave sensors—used to record wave activity. Scuba—underwater breathing apparatus. There are also a variety of submersible research vessels.

7. The Arctic Ocean is surrounded by land. The Antarctic Ocean surrounds a continent.

8. Gasoline is used in the flotation tanks in the bathyscaph because, being a liquid, it has low compressibility. Of course, if it were not less dense than water it would not help float the vessel.

9. Mountains and valleys in the ocean are as high and as deep and as numerous as those on land.

10. The continental shelf is part of the continent and receives most sediment from the land. The continental slope is the edge of the continent. Sediments slide down the slope and form a pile called the continental rise.

11. The surface features of the sea floor become covered with sediments, thus smoothing them into abyssal plains.

12. A seamount is one of a number of separate volcanic mountains under the ocean surface.

13. A seamount whose top has been worn flat by wave action and then sunk into the sea is called a guyot.

14. The volcanoes that form seamounts move with the crustal plate, but the hot spot remains fixed and may form another volcano.

15. Different parts of the mid-ocean ridge produce different amounts of material, thus different rates of spreading. Due to unequal spreading, the ridge breaks into sections.

16. Particles from land, organic sediments, and chemical sediments.

17. Water that runs off the continents as natural drainage and in rivers carries large and small rock fragments with it. Large fragments are deposited near shore, while smaller fragments are carried farther out on the continental shelf and even beyond the continental slope.

18. Particles from land. Large and small rock fragments carried by water runoff. Fine particles are sometimes carried great distances by ocean currents. Organic sediments. The fragments of hard skeletons of plants and animals that live in the sea. Chemical deposits. These are mainly oxides of manganese, nickel, and iron. Chemical reactions in the sea produce nodules or lumps of these materials.

19. The total volume of sediments on the sea floor seems to be about one-tenth that calculated to have been carried by rivers into the ocean. One possible answer to this problem is that older sediments may be packed together and appear to be part of the crustal foundation of the sea floor.

20. Red clay is composed of clay particles, silt, and organic materials. It is brown in color and is found on the sea floor.

references

Boyer, R. E. *Oceanography Fact Book.* Northbrook, IL: Hubbard, 1974.

Carlisle, N. *The New American Continent—Our Continental Shelf.* Philadelphia: J. B. Lippincott, 1973.

Cousteau, Jacques-Yves. *Three Adventures: Gallapagos, Titicana, The Blue Holes.* New York: Doubleday, 1973.

Davies, E. *Ocean Frontiers.* New York: Viking, 1980.

Gross, M. Grant. *Oceanography: A View of the Earth.* Englewood Cliffs, NJ: Prentice-Hall, 1977.

Groves, D. G., and L. M. Hunt. *The Ocean World Encyclopedia.* New York: McGraw-Hill, 1980.

Heintal, C. *The Bottom of the Sea and Beyond,* Nashville, TN: Thomas Nelson, Inc., 1975.

Idyll, C. P. *Exploring the Ocean World: A History of Oceanography.* New York: Thomas Y. Crowell Co., 1972.

Parker, S. *Encyclopedia of Ocean and Atmospheric Sciences.* New York: McGraw-Hill, 1980.

Pirie, R. Gordon. *Oceanography: Contemporary Readings in Ocean Sciences.* Englewood Cliffs, NJ: Prentice-Hall, 1977.

Ross, David A. *Introduction to Oceanography.* Englewood Cliffs, NJ: Prentice-Hall, 1977.

Waters, J. *The Continental Shelves.* New York: Abelard-Schuman, 1975.

————. *The Rand McNally Atlas of the Oceans,* Chicago: Rand McNally, 1977.

audio-visual aids

16mm Sound Films

The Beach—A River of Sand. EBEC, 20 min., color

Oceanography: A Voyage to Discovery. UEVA, 20 min., color

Waves on Water. EBEC, 16 min.

Water. EBEC

Filmstrips

Exploring the Oceans. P.H. Media, set of 5 with cassettes.

Oceanography: A Developing Science. Denoyer-Geppert

Oceans and Their Histories. Creative Visuals

Understanding Oceanography. Hubbard

8mm Film Loops

Formation of Sedimentary Strata. EBEC, color

Ocean Basin Topography. Listening Library, color

Ocean Bottoms. PHOENIX, color

Sediment Deposition. EBEC, color

Sedimentation. Gordon Flesch, color

---

16 sea water

objectives

When finished with this chapter, the student will be able to:

1. Describe the physical properties of sea water and several conditions that change these properties.
2. Describe the chemical properties of sea water.
3. Explain how the salt content and the dissolved gases in sea water affect life in the sea.
4. Explain how the sea can be a valuable resource.

approach

All significant characteristics of the sea are representative of the properties of sea water and the materials dissolved in it. Water is a substance of many distinctive properties, all of which contribute to the unique features of the sea. In creating a foundation for understanding the many characteristics of the sea, the most fundamental concept is that the oceans would be static and lifeless without the continuous addition of solar energy. Almost all physical and chemical properties of sea water are founded upon the interaction of water with energy from the sun. This fact should be established firmly in the minds of students and constantly referred to in the study of the physical and chemical features of sea water.

activity hints

Density Currents (page 340)
Part I

Care must be taken not to get water in the straw or subsequent readings will be in error. You may want to prepare the substitute sea water in advance.

Part II

Plastic shoe boxes work well for the second part of the activity. Deep baking dishes can also be used.

Answers to the activity questions follow:

Part I
1. It floats highest in the sea water and lowest in the distilled water.
2. Sea water.

Part II
1. No.
2. Near the bottom.

other activities
Geology and Earth Sciences Sourcebook (2nd edition, 1970) Holt, Rinehart and Winston.
Chapter 8, Oceans, pp. 179-195

Also see Chapter 16 in *Activities and Investigations for Modern Earth Science* (1983 edition) Holt, Rinehart and Winston.

Vocabulary Review

1. f	4. h	7. l	10. n
2. j	5. d	8. e	11. p
3. b	6. a	9. i	12. g
			13. k

Questions: Group A

1. c	6. a	11. c	16. d
2. a	7. b	12. d	17. d
3. d	8. c	13. a	18. a
4. c	9. c	14. d	19. b
5. b	10. d	15. b	20. c
			21. c

Questions: Group B

1. Temperature conditions on the earth are such that liquid water can exist at most locations at all times. On any planet with an average temperature much less than 0°C, the oceans would be frozen. If the average temperatures were much greater than 100°C, the boiling of the oceans would cause rapid evaporation. These are the conditions that exist on the other planets in the solar system.
2. Chemical properties of sea water are those that are associated with dissolved solids. Physical properties are those that are associated with effects such as temperature and pressure.
3. There is plenty of fresh water on the earth, but it is not where it is needed. Desalting is only one solution to this problem, but it is probably the simplest. It would be possible to transport fresh water to areas that need but do not have it.
4. Water molecules are in constant motion. If, as a result of this motion, molecules escape into the atmosphere, they become water vapor. This process of escape is called evaporation.
5. Phytoplankton require sunlight for

growth. Zooplankton feed on this phytoplankton and therefore are found primarily in areas where sunlight can penetrate. Small fish feed on the zooplankton and stay primarily where there is food. Since sunlight only penetrates the sea to about 200 meters, most fish are found in this region.

6. The wavelengths of blue light from the sun are the last color to be absorbed by sea water. Therefore, the ocean appears to be this color.

7. One kilogram is 1000 grams. 34 grams in 1000 grams of water yields a salinity of 34‰.

8. $D = \dfrac{mass}{volume} = \dfrac{350 \text{ grams}}{70 \text{ cm}^3} = \dfrac{5 \text{ grams}}{\text{cm}^3}$

9. Living things in the sea must receive all of the materials required for growth from the water. Their life processes throw a great variety of by-products into the water. Finally, the substances in their bodies are returned to the water.

10. Light rays from the sun fall more nearly perpendicularly on the ocean near the equator. Therefore, there are more direct rays striking a given area and the amount of heat received is greater.

11. Heating by the sun is the most obvious reason for the upper part of the sea being warmer than the lower part. Another reason is that if the temperature of the lower part becomes higher than the water above, the water becomes less dense also and therefore rises.

12. The ocean has apparently reached a state of equilibrium. Dissolved materials are being deposited as sediments as rapidly as other materials are carried to the ocean by rivers.

13. Salinity is the number of parts of dissolved materials per thousand parts of the sample. 72 grams in 2000 grams of the sample would be 36 parts per thousand. The salinity is 36‰.

14. Water reaches its greatest density at 4°C. It is less dense at 0°C. When a pond freezes, the water at the surface is 0°C, and at the bottom 4°C.

15. The average density of the steel ship must be less than that of the ocean in order for the ship to float. The ship is filled mostly with air and its average density is less than 1.

16. $D = \dfrac{M}{V}$; $V = \dfrac{4\pi r^3}{3} = \dfrac{4\pi 2^3}{3} = \dfrac{32\pi}{3} = 33.5$

$D = \dfrac{200 \text{ grams}}{33.5 \text{ cm}^3} = \dfrac{5.97 \text{ grams}}{\text{cm}^3}$

17. The cost of obtaining gold from sea water is too great. But it is quite economical to obtain bromine and magnesium from this source.

18. It is the only source of obtaining enough protein food for supplying the entire population of the world.

19. See Table, page 339. Rain dissolves many minerals that are carried to the oceans by the continental streams. Magnesium: used in airplane manufacturing. Bromine: used in the manufacture of high test gasoline and photographic film.

20. Ice floes are drifting sections of arctic ice; icebergs are broken-off sections of tidal glaciers. Floes are smooth and flat; icebergs are jagged. Floes are composed of frozen salt water; icebergs are composed of frozen fresh water.

21. The thermocline is produced because the water of the upper layers becomes lighter as a result of being heated. This layer tends to float on the cooler, denser lower layers. This difference in density prevents the mixing of the two layers. In most places 45 to 360 meters below the surface, there is a sharp drop in temperature. One effect is to separate the

currents in the warm water above the thermocline and the deep currents in the cold water beneath.

22. Plants remove substances containing carbon, hydrogen, oxygen, sulfur, nitrogen, phosphorous, and silicon. Animals eat the plants and each other, returning the elements to the sea when they die and decay.

23. Plankton are free-floating microscopic living forms. Phytoplankton are plant plankton, while zooplankton are animal plankton.

24. Plankton are eaten by larger life forms such as small fish and squid. These in turn are eaten by larger fish and marine animals.

references

Barton, R. *Oceanography Today—Man Exploits the Sea.* New York: Doubleday, 1971.

Bhatt, J. J. *Oceanography: Exploring the Planet Ocean.* New York: D. Van Nostrand Co., 1978.

Boyer, R. E. *Oceanography Fact Book.* Northbrook, IL: Hubbard, 1974.

Fairbridge, R. W. (ed.). *The Encyclopedia of Oceanography.* New York: Van Nostrand-Reinhold Books, 1966.

Groves, D. G., and L. M. Hunt. *The Ocean World Encyclopedia.* New York: McGraw-Hill, 1980.

Idyll, C. P. *Exploring the Ocean World: A History of Oceanography.* New York: Thomas Y. Crowell Co., 1972.

————. *The Rand McNally Atlas of the Oceans.* New York: Rand McNally, 1977.

audio-visual aids

16mm Sound Films

New Water for a Thirsty World. Sterling, 22 min., color

Oceanography: A Voyage to Discovery. UEVA, 20 min., color

Filmstrips

Oceanography, A Developing Science. Denoyer-Geppert

Oceans and Their Histories. Creative Visuals

Understanding Oceanography. Hubbard

8mm Film Loops

Formation of Sedimentary Strata. EBEC, color

Ocean Basin Topography. Listening Library, color

Ocean Bottoms. PHOENIX, color

Sediment Deposition. EBEC, color

Sedimentation. Gordon Flesch, color

Water Cycle on the Ocean. PHOENIX, color

---

17 motions of the sea

objectives

When finished with this chapter, the student will be able to:

1. Identify the causes of sea water movements.
2. Describe the patterns of circulation near the sea surface.
3. Describe the characteristics and effects of the Gulf Stream.
4. Compare deep currents with surface currents.
5. Describe, using the correct terms, the characteristics of ocean waves.
6. Explain how ocean waves change near shore.
7. Identify the causes and effects of a tsunami and tides.

approach

The theme of this chapter is an extension of the study of the effects of energy imparted to the sea by the sun and the moon through their gravitational influence. The sun's heat energy is absorbed by the atmosphere and generates winds that in turn produce waves and currents. This is the most important cause of the continuous movement of water near the sea surface. It is difficult to account for the pattern of surface currents without a knowledge of the wind belts over the earth's surface. (See Chapter 20.) A general discussion of these wind belts might be introduced at this time, without going into details of their causes or actually naming the various belts. It is usually sufficient to sketch in the direction of the prevailing winds on an outline map or slated globe, and then to relate these markings to the surface currents shown in the text illustrations.

activity hints

Penetration of Radiation (page 354)

Thermometers should read about the same before starting since the temperature differences may be small.

If a high wattage lamp is used, caution students that it will get very hot and be very bright.

A data table with two columns is needed. One column would be labeled "time" and would show times starting at zero and every two minutes for 10 to 30 minutes, depending on the light used. The second column would be labeled "temperature" and would contain the appropriate measurements.

The results of this activity will show that the greatest heating occurs near the surface of the water. Care must be exercised in this interpretation since convection currents will cause the hottest water to be at or near the surface.

Temperature Effects on Density (page 358)

This density current activity is very simple and should not cause any problems.

Answers to activity questions follow:
1. No, they remain separated.
2. The colder water sinks and moves to the lower part of the container.
3. Near the bottom. (This is true above 4°C only. Below 4°C, the surface water is colder.)

other activities

Geology and Earth Sciences Sourcebook (2nd edition, 1970) Holt, Rinehart and Winston.
Chapter 8, Oceans, pp. 179-195

Also see Chapter 17 in *Activities and Investigations for Modern Earth Science* (1983 edition) Holt, Rinehart and Winston.

answers—chapter 17

Vocabulary Review

1. f	**5.** k	**9.** h	**13.** o
2. i	**6.** b	**10.** d	**14.** n
3. l	**7.** c	**11.** p	**15.** m
4. a	**8.** e	**12.** q	

Questions: Group A

1. a	**7.** d	**13.** c	**19.** a
2. c	**8.** c	**14.** d	**20.** c
3. b	**9.** b	**15.** b	**21.** d
4. b	**10.** b	**16.** a	**22.** d
5. a	**11.** a	**17.** d	**23.** b
6. d	**12.** b	**18.** b	**24.** a
			25. c

Questions: Group B

1. (a) The main source of energy is the sun.
   (b) The water receives energy mostly from the atmosphere via winds, and to a lesser degree by direct absorption of solar radiation.
2. Movements of the ocean floor, such as earthquakes, volcanic activity, and underwater landslides, cause temporary ocean currents.
3. (a) Ocean currents are deflected from

their straight-line path to the right in the Northern Hemisphere.

(b) Ocean currents are deflected to the left in the Southern Hemisphere.

4. Water can be made to spiral differently, depending on the shape of the sink, the original motion of the water, and the speed at which the water drains. In the Northern Hemisphere, water is most likely to spiral clockwise as it goes down a drain. To prove whether this is due to the Coriolis effect is somewhat difficult. If it could be established that water spirals downward in a sink usually in a counterclockwise direction in the Southern Hemisphere (as indeed it does), this would help.

5. See Figure 17-3, page 355 of the text.

6. The general current flow in all the oceans is similar in that its direction of movement depends upon global wind patterns, the position of land masses, the earth's rotation, and the Coriolis effect. In addition, each of the major oceans possesses two equatorial currents and an equatorial counter current. The warm and cold currents are located at approximately the same latitude in each ocean.

   Among the obvious differences is the fact there is no counterpart to the North Atlantic Drift in the North Pacific. Also the Atlantic's Gulf Stream breaks up into several currents as it proceeds northward. The corresponding Japan Current in the Pacific does not break up into well-defined currents but instead tends to proceed as the North Pacific Current. The size of the Pacific Ocean requires that the corresponding currents cover much greater distances.

7. Deep currents are the result of the sinking of the more dense cold water masses near the polar regions. In general they flow toward the equator. In some instances they flow opposite the direction of the surface current.

8. As the North Equatorial Current moves westward, it tends to pile up water in the Gulf of Mexico. The stored up water finds an outlet at the northern edge of the Gulf, where it turns northward, and flows up the east coast of the North American continent.

9. The results are: (a) a very swift movement of the water out of the northern part of the Gulf of Mexico (b) the eastern edge of the Gulf Stream being approximately 48 centimeters higher than the westward side (c) the deflection of the current eastward toward the European continent.

10. The main cause of surface waves is wind. The larger waves expose more surface for the wind to push against. Therefore they are quite efficient in trapping energy from the wind and tend to grow larger as a result. Small waves have a smaller surface; therefore, the energy of the wind tends to discriminate against small waves.

11. The height, wavelength, and period.

12. Using: $\text{speed} = \dfrac{\text{wavelength}}{\text{period}}$ gives

$$\text{speed} = \frac{200 \text{ meters per wave}}{10 \text{ seconds per wave}}$$
$$= 20 \text{ meters per second}$$

13. There are two possibilities here, depending on whether the motion is more up and down or more back and forth. If the motion is elongated up and down (vertical), the waves become quite tall and may come close to forming whitecaps in the open sea. If the motion is elongated back and forth (horizontal), the wave is also tall but ready to form a breaker due to the dragging effect of the shallow ocean near a beach.

14. The portion of the wave reaching the beach first also tends to slow up first, thus causing the wave to bend (refract) toward the beach. This brings it into line with the beach and it thereby approaches the beach head-on.

15. The undertow becomes a rip current when channels that develop in the sand bars at the beach tend to funnel the water back to sea in a swift localized current.

16. It was found that the length of the vessels was very near one half of the wavelength of many of the larger waves encountered on the seas. This meant that the vessel was constantly rocking on crests or in troughs or was perched with the stern high on a crest and the bow pointed downward in the trough. This put a tremendous strain on these ships and many unfortunately broke up and sank rapidly as a result.

17. Tidal wave is an erroneous term, indicating incorrectly that the tide is responsible for the type of wave produced. Tsunami does not imply this connection. A tsunami is caused by a shift in a portion of the ocean floor and not by gravitational effects of the moon or sun.

18. The shape and size of the Bay of Fundy is such that great tidal oscillations are set up that nearly empty the bay in places. Ships must enter at high tide, and at low tide rest on large concrete slabs with practically no water underneath.

19. The purpose the Russians have in mind is to melt much of the polar ice cap in order to open up the Arctic Ocean for shipping. It would also make the climate of Siberia a good deal warmer. It would be interesting to explore the effect this project would have on the climate of the eastern North American continent, which brings up a reason why we are reluctant to allow such a project.

references

Carters, S. *The Gulf Stream Story*. Garden City, NY: Doubleday, 1970.

Groves, D. G., and L. M. Hunt. *The Ocean World Encyclopedia*. New York: McGraw-Hill, 1980.

Idyll, C. P. *Exploring the Ocean World: A History of Oceanography*. New York: Thomas Y. Crowell Co., 1972.

Pennington, H. *The New Ocean Explorers*. Boston: Little, Brown and Co., 1972.

Taber, R. W. *1001 Questions Answered About the Oceans and Oceanography*. New York: Dodd, Mead & Co., 1972.

audio-visual aids

16mm Sound Films

How Level Is Sea Level. EBEC, 13 min., color

Ocean Tides. EBEC, 14 min., color

The Beach, A River of Sand. EBEC, 20 min., color

Waves on Water. EBEC, 16 min., color

Filmstrips

Evolution of Shorelines. Edwin Shapiro Co., captions

Exploring the Oceans, P.H. Media, set of 5 with cassettes

Ocean Currents, Prentice-Hall, set of 2

Ocean Currents, S.P., set of 2

Filmloops

Growth of Sea Waves. PHOENIX

Orbital Motion Beneath Waves. PHOENIX

Origin of Waterwaves. PHOENIX

Surface Currents on the Ocean. PHOENIX

The Moon and Tides. PHOENIX

the record of earth history

18 the rock record

objectives

When finished with this chapter, the student will be able to:

1. Explain the principle of superposition and describe how it relates to conformity, unconformity, and disconformity.
2. Explain how the half-life of a radioactive element may be used to obtain the actual age of rocks, and name several radioactive elements used for this purpose.
3. Describe three means other than radioactivity that have been used to measure geologic time.
4. Describe several ways fossils are formed and explain how they can be used to relate rock layers to one another.

approach

The study of this chapter might begin with a discussion of the concept: "The present is the key to the past." This statement makes clear the foundation on which all of our knowledge of earth history rests. Changes in the earth are the result of forces that operated in the past and are continuing today. All evidence uncovered from the study of rocks must be interpreted in light of present conditions that we know alter rock structures. The ability of scientists to analyze and interpret the rock record is largely determined by the extent of their knowledge of these timeless laws and processes.

Students should clearly understand that of the billions of animals and plants that have lived and died on the earth, only a tiny fraction have left any evidence of their existence. This means that our knowledge of past life forms is incomplete. However, the fossil evidence available does clearly show a series of evolutionary changes throughout the vast span of geologic time, with the simplest forms appearing in the oldest (lower) rock layers, and with gradual changes in the direction of complex, more recent types.

activity hints

Simulation of Radioactive Decay (page 381)

Poker chips spray painted on one side are good discs for this activity. Washers, coins, etc., may be used but are more expensive to replace. At least 100 items are needed for the laws of chance to work well and to give enough half-lives for a good graph.

other activities

Geology and Earth Sciences Sourcebook (2nd edition, 1970) Holt, Rinehart and Winston.

Chapter 13, Paleontology, pp. 297-321

Also see Chapter 18 in *Activities and Investigations for Modern Earth Science* (1983 edition) Holt, Rinehart and Winston.

answers—chapter 18

Vocabulary Review

1. e	**5.** n	**9.** a	**13.** g
2. h	**6.** k	**10.** d	**14.** c
3. f	**7.** m	**11.** p	
4. j	**8.** l	**12.** o	

Questions: Group A

1. b	**9.** b	**17.** a	**25.** b
2. a	**10.** a	**18.** b	**26.** a
3. c	**11.** b	**19.** d	**27.** c
4. d	**12.** c	**20.** b	**28.** d
5. a	**13.** c	**21.** a	**29.** c
6. c	**14.** b	**22.** b	**30.** a
7. b	**15.** d	**23.** a	
8. a	**16.** c	**24.** b	

Questions: Group B

1. Sedimentary rock layers are formed from the weathering and erosion of surface rocks. These processes have been in existence during most of the earth's history. Therefore any such rock layer may contain evidences of geological activity during that time.
2. It is believed that when layers are horizontal, or nearly so, each bed is younger than the one beneath it.
3. The main source of error lies in the fact that periods of violent crustal movements, called revolutions, may have caused older rock layers to be pushed up over younger ones.
4. An unconformity is produced when an erosion surface is covered by younger deposits that appear in an uninterrupted sequence. Examination of the younger rock layers would show that the top layers are considerably younger than those beneath the erosion surface, rather than merely following them in time. The problem, therefore, lies in recognizing the unconformity and determining the length of delay before the younger layers were deposited.
5. (a) Dikes and sills in sedimentary rocks are younger than the layers they invade. (b) Lava flows are younger than the rocks they cover.
6. Since the earliest rocks were either deeply buried or have been lost through erosion and metamorphism, they are not readily available for study.
7. An atomic nucleus becomes more stable by throwing off electromagnetic energy and beta or alpha particles. This takes place one step at a time until a stable nucleus results.
8. Half-life is a term meaning the length of time necessary for any given sample of a radioactive substance to decay to a point at which only half the original substance remains.
9. Knowing the half-life, the amount of the original element, and the amount of decay product present in a radioactive substance, one can calculate how many years it took for the original material to decay into the final product. The rock then is at least this age if it contained the original element at the start of this process.
10. The carbon-14 dating method makes use of the fact that carbon-14 is taken into an animal or plant's cells at a given rate while it is alive and then not at all after it dies. Rocks rarely contain any previously living organisms, so this dating method can not be used on most rocks. In addition, the carbon-14 dating process is useful only to about 6000 years, and most rocks are much older than this.

11. (a) Rate of erosion: the rate at which a stream cuts its channel or erodes other features can be determined, and a calculation made to determine the stream's age. (b) Rate of deposition of sediments: the average rate at which sedimentary rocks such as limestone, shale, and sandstone are formed is used to calculate the length of time necessary for a given layer to form. (c) Amount of salt in the sea: the amount of salt found in sea water is calculated, and the average amount of salt carried into the sea by all the rivers and streams is calculated. From this information, the age of the sea can be determined.

12. The paleontologist is a specialist in the study of fossils and how they are related to the earth's history; the geologist is mainly interested in the structure of the earth.

13. (a) It must be found over a wide area of the earth's surface. (b) It must have features which clearly distinguish it from other forms. (c) The organism which formed the fossil must have lived during a relatively short span of geologic time. (d) It must occur in fairly large numbers.

14. Gas, oil, and coal are actually the remains of ancient plants and animals. Although they are considerably altered, they are still considered organic fossils.

15. Fossils give important clues to environments that existed in the past. The location of ancient seas, the types of climates that existed, and the shapes of the ancient continents are examples of information that can come from such clues.

16. For the entire animal or plant to be preserved, it must be protected from attack by other animals and from bacterial decay. Very low tempera-tures and other special properties of oil-soaked soils, etc., are required for this protection.

17. Minerals such as silica, calcite, and pyrite slowly replace, molecule by molecule, every part of the organism, literally turning it to stone.

18. (a) Imprints such as leaf outlines, dinosaur footprints, etc. (b) A mold of a seashell. (c) Preserved trails and burrows. (d) Waste products such as castings and coprolites. (e) Gizzard stones.

19. Any gap in the normal sequence of deposition of sedimentary layers is called an unconformity. If, as shown in the diagram, the sedimentary formations labeled D and E have been tilted and otherwise disturbed before being eroded to a nearly horizontal surface, the overlying layers produced by a second period of deposition (layers F, G, H) are separated by an unconformity. The unconformity exists between layers E and F.

20. The correct order of deposition from oldest to youngest would be: D, E, A, B, F, G, H, C. The oldest formations are those that were deposited first and are, therefore, those that are found at the lowest levels. Since the igneous intrusions (formations A, B, C) are always younger than the layers they intrude, the rock layer that appears at the lowest level is D. Layer E would have been deposited next, then tilted, faulted, and partly eroded at the same time as D. The igneous intrusions A and B preceded the deposition of layer F. Note the eroded surface of formation B. Formation B is shown intruding A; thus A is older than B. The layers above the unconformity (F, G, H) were next to be deposited as successive layers.

Igneous formation C must be the

youngest in the diagram because it came from below and penetrated even the youngest (layer H) of the sedimentary layers.

references

Beerbower, J. *Field Guide to Fossils*. Boston: Houghton Mifflin Co., 1971.

Blackbeer, Lawrence. *The Student Earth Scientist Explores the Geologic Past*. New York: Richards Rosen Press, 1976.

Dott, R. H., and R. L. Batten. *Evolution of the Earth*. New York: McGraw-Hill, 1971.

Harbough, J. W. *Stratigraphy and the Geologic Time Scale*. Dubuque, IA: William C. Brown Co., 1974.

Major, Alan. *Collecting Fossils*. New York: St. Martin's Press, 1974.

U.S. Geological Survey. *Geologic Time: The Age of the Earth*. Washington, DC: U.S. Government Printing Office, 1970.

Wyckoff, J. *The Story of Geology*. New York: Golden/Western, 1976.

audio-visual aids

16mm Sound Films

Fossils, Clues to Prehistoric Times. Coronet, 11 min., color

Fossils Are Interesting. FA, 9 min., color

Prehistoric Animals of the Past. FA, 12 min., color

Radioactivity. McGraw-Hill, 12 min., color

Rocks and the Record. McGraw-Hill, 28 min., color

Story in the Rocks. Shell, 17½ min., color

The Fossil Story. Shell, 19 min.

The Story of Two Creeks. University of Wisconsin, 30 min., color

This Land (a geologic history of the North American continent). Shell, 41 min., color

Filmstrips

Dating the Past, EBEC, set of 3

Discovering Fossils. EBEC, with cassette

Earth's Biography. Learning Arts, set of 4 with sound

Geology from Space, NASA, sound, color

Life Long Ago. SVE, set of 6 with cassette

Putting Fossils to Work, EBEC, set of 5

8mm Film Loops

Grand Canyon: Record of the Past. Gordon Flesch, super 8, color, PB5-0263/182

35mm Color Slides

Invertebrate Fossils. Ward's, set of 201 35mm slides, color, 174W5003

Teaching Aids

Stratigraphic Fossiliferous Rock Collection. Ward's

Student Stratigraphic Collection. Ward's

Earth History Model. Hubbard

Rock Fossil Collection. Hubbard

19 a view of the earth's past

objectives

When finished with this chapter, the student will be able to:

1. List the kinds of information that may be obtained from exposed rock layers.
2. Explain how the Geologic Column is used.
3. Name the geologic eras in order of their age beginning with the oldest.
4. Explain why maps of the earth's surface would not look the same throughout its history.
5. Give a brief description of the conditions and life that existed on North America during each geologic era.
6. Describe the tectonic activity associ-

ated with the Mesozoic era and its effect on earth conditions of that time.
7. List the two periods of the Cenozoic era and, using Appendix C, describe the general conditions and prominent life features that existed during each period.

approach

A strong effort should be made to give students some real appreciation for the tremendous period of time involved in geologic history. It is easy to speak of millions and hundreds of millions of years without much real feeling for the length of time described. Any technique that will help to convey the vastness of such time intervals will make the study of the material in this chapter more meaningful.

Students can become thoroughly familiar with the geologic time scale only by memorizing its main divisions along with the time period represented. It is desirable to have them do this early in the study of this chapter to make certain that the class work and text reading will be understood.

activity hints

Geologic History (page 409)

On page T77 in this teacher's edition, you will find a master copy (for class sets) of a diagram of a hypothetical cross-section of rock containing fossils. On page T80 is a correctly labeled copy for your use.

The activity uses this and many other figures in Chapter 19 to describe a core of fossils that might be found somewhere in the southwestern part of the United States. The clues given to the student should lead to the conclusion that the general region of the state of New Mexico is the location for the events described by the fossils.

other activities

Geology and Earth Sciences Sourcebook (2nd edition, 1970) Holt, Rinehart and Winston.
Chapter 14, Sedimentation and Stratigraphy, pp. 323-360

Also see Chapter 19 in *Activities and Investigations for Modern Earth Science* (1983 edition) Holt, Rinehart and Winston.

answers—chapter 19

Vocabulary Review

1. f	4. j	7. g	10. e
2. d	5. a	8. c	
3. h	6. b	9. l	

Questions: Group A

1. b	9. c	17. d	25. d
2. c	10. b	18. c	26. d
3. a	11. a	19. d	27. a
4. c	12. c	20. a	28. d
5. d	13. a	21. b	29. b
6. d	14. c	22. c	
7. b	15. d	23. a	
8. a	16. c	24. b	

Questions: Group B

1. The clues left in the rock layers are difficult to interpret; pieces of history are missing in the layers and most events happened before anyone was around to observe them.
2. Sedimentary rocks. They make up about 75% of the kinds of rocks found close to earth's surface, and only they contain fossils.
3. By natural weathering and erosion of an existing hillside or mountain, by cuts made by railroads or highway construction, and by the natural forces which force up huge blocks of rock.
4. A small valley was originally formed by the Colorado River. Slowly, the land was lifted to form a high pla-

teau. The Colorado River was able to cut through the plateau as it formed, thus keeping the water level near its original position.

5. First, the position of the layers allows the principle of superposition to give the relative ages of all the exposed layers. Second, the composition of the layers gives information suggesting their sources and the kinds of climates that existed at the time. Third, the fossils found in many of the layers give evidence of the plant and animal life that existed.

6. During the past 150 years scientists have found that layers of rock of the same age are arranged in the same pattern no matter where on earth they are found. Although nowhere on earth are all the layers found together, the scientists have put together all the layers into one column. This allows them to compare any rock layers they find with this Geologic Column and get the correct ages for the rock layers.

7. The Geologic Calendar was constructed directly from the Geologic Column once the age of the rock layers could be determined.

8. Eras, periods, and epochs.

9. Early Precambrian and Late Precambrian. It enables scientists to indicate that the simplest life forms originated in a particular era, the Late Precambrian era.

10. Precambrian—3.9 billion
Paleozoic—375 million
Mesozoic—155 million
Cenozoic—70 million

11. There were six continents, a very large one and five smaller ones. These continents slowly moved in different directions, colliding with each other until eventually they all joined into the supercontinent Pangaea.

12. Cratons are the very ancient rock around which the continents formed.

13. There were very few. Most were small and without skeletons.

14. The Paleozoic era ended with the formation of giant mountain ranges and severe climate changes. The Mesozoic era ended with the creation of the Rocky Mountains and the re-elevation of the ancient Appalachians.

15. A trilobite was a very common invertebrate during the Cambrian period. It had a segmented body covered by an exoskeleton. Today the trilobite is a useful index fossil.

16. They have the highest degree of intelligence of all living organisms.

17. Since the temperature was fairly constant, simple body systems were all that was needed.

18. Pangaea began breaking up, forming Laurasia and Gondwanaland. These in turn broke up to form the general outlines of the modern continents.

19. The Jurassic period. The climate was warm with many shallow seas and abundant plant life of the type eaten by the reptiles.

20. It is during these two periods that many great coal beds were laid down.

references

Beerbower, J. Field Guide to Fossils. Boston: Houghton Mifflin Co., 1971.

Dott, R. H., and R. C. Batten. Evolution of the Earth. New York: McGraw-Hill, 1971.

Harbough, J. W. Stratigraphy and the Geologic Time Scale. Dubuque, IA: William C. Brown Co., 1974.

Silverberg, R. Clocks for the Ages: How the Scientists Date the Past. New York: Macmillan, 1971.

Stearn, C. W., R. L. Carroll, and T. H. Clark. Geological Evolution of North

America (3rd ed.). Somerset, NJ: John Wiley & Sons, 1979.

U.S. Geological Survey. *Geologic Time, The Age of the Earth.* Washington, DC: U.S. Government Printing Office, 1970.

audio-visual aids

16mm Sound Films

The Dinosaur Age. FA, 13 min., color

From Water to Land. McGraw-Hill, 28 min., color

In the Beginning. Hennessy, 28 min., color

Story in the Rocks. Shell, 17½ min., color

The Story of Two Creeks. University of Wisconsin, 30 min., color

This Land (a geologic history of the North American continent). Shell, 41 min., color

8mm Film Loops

Grand Canyon, Record of the Past. Gordon Flesch, super 8, color, PB5-0263/1S2

Filmstrips

Earth's Biography. Learning Arts, set of 4, color

Fossils. EBEC, set of 5, color

Geology from Space. NASA, sound, color

Prehistoric Life. EBEC

Prehistoric Animals. McGraw-Hill

The Earth's Diary. SVE, set of 4

unit 6
the earth's atmosphere

20 air and its movements

objectives

When finished with this chapter, the student will be able to:

1. List the materials that make up the earth's atmosphere.
2. Describe air pressure and the instruments used to measure it.
3. Describe the chemical balance of the earth's atmosphere.
4. Describe some of the types of air pollution and their effects on earth.
5. Describe the effect of the atmosphere on the solar energy received by the earth.
6. List the layers of the atmosphere and identify the method by which they are determined.
7. Explain how movement of the air is caused by heating the atmosphere and describe how the rotation of the earth affects the motion.

approach

Comparing the atmospheres of the different planets is an interesting way to introduce a study of the earth's atmosphere. This will show that the composition of the earth's atmosphere is unique. From there, you could proceed to a consideration of the reasons the earth now has its particular atmosphere. Students may already have a good background of study on the atmosphere, particularly weather, from earlier work in other classes. A pretest might be in order with particular classes to establish the extent of previous knowledge.

The study of air pollution should be as closely related to the local situation as possible. Schools that are located in urban areas may wish to encourage a detailed study of the existing air conservation problems and the potential for control.

activity hints

Conduction, Convection, Radiation (page 433)

Any six-inch metal rod could be used, but nails are cheapest. Birthday candles will work, but are more expensive.

Make sure you illustrate how much 40 cm is. Students are to hold their hands 40 cm above the lighted candle. If they mistakenly hold their hands only 40 mm above, they may get burned.

Answers to activity questions follow:

1. Answers will vary, but probably less than 30 cm.
2. Convection.
3. Answers will vary, but about 10 cm.
4. Radiation.
5. Answers vary, but usually more than 15 sec.
6. Conduction.

other activities

Geology and Earth Sciences Sourcebook (2nd edition, 1970) Holt, Rinehart and Winston.
Chapter 5, The Atmosphere, pp. 115-134

Also see Chapter 20 in *Activities and Investigations for Modern Earth Science* (1983 edition) Holt, Rinehart and Winston

answers—chapter 20

Vocabulary Review

1. f	5. b	9. k	13. p
2. d	6. g	10. o	14. q
3. a	7. c	11. j	15. l
4. h	8. e	12. m	

Questions: Group A

1. b	6. d	11. d	16. a
2. c	7. d	12. c	17. c
3. c	8. c	13. b	18. a
4. a	9. a	14. d	19. c
5. c	10. a	15. d	20. d
			21. a

Questions: Group B

1. (a) troposphere
   (b) tropopause
   (c) stratosphere
   (d) stratopause
   (e) mesosphere
   (f) mesopause
   (g) thermosphere
   (h) exosphere
   The region of the thermosphere and upper mesosphere is often called the ionosphere.

2. Since blue light is scattered by the atmosphere more than other colors, a clear sky has a blue color due to the visible scattered rays. Sunsets are red because you are looking almost directly at the sun. The blue is mostly scattered, while the red colors come through strongly.

3. Shorter wavelength infrared rays penetrate the glass in a greenhouse and are absorbed by the plants and earth inside. In turn, the absorbed rays are re-emitted as longer wavelength infrared rays. The glass blocks these rays and energy is trapped inside. A similar process causes the atmosphere to trap the sun's heat.

4. 30 inches $\times \dfrac{2.54 \text{ cm}}{1 \text{ inch}} = 76.2$ cm

5. Certain microscopic plants take nitrogen from the air and chemically change it into nitrogen compounds. These compounds are removed from the soil by plants, which are in turn eaten by animals. The compounds are returned to the soil by excretions from these animals and the decay of their bodies after death. In the soil, decay processes release nitrogen and return it to the atmosphere.

6. Combustion, breathing, and decay of vegetable and animal matter consume oxygen and release carbon dioxide to the atmosphere. Land and water plants use carbon dioxide in photosynthesis and release oxygen to the atmosphere.

7. Air pressure presses down on the surface of the mercury which is exposed to the atmosphere. The mercury rises in the tube until the pressure due to the mercury column just balances the atmospheric pressure. It is a mistake to say the vacuum above the column draws the mercury up the tube. The vacuum merely allows the outside pressure to push mercury into the empty space.

8. A temperature inversion occurs when warm air exists above a layer of cool air next to the ground. Since cool air has a tendency to sink, the polluted air close to the ground is trapped.

9. The makeup of dry air is nitrogen, 78%; oxygen, 21%; argon and carbon dioxide, 1%. Water vapor may be present up to about 3%, allowing nitrogen and oxygen to take up about 76 and 20% respectively. There are also traces of other gases.

10. The atmosphere is heated by radiation due to direct absorption of the sun's radiant energy. Conduction causes heating by direct contact with warmer material. Convection involves the movement of fluids that are unevenly heated.

11. The barograph makes a permanent, continuous record of the pressure at all times, while the barometer must be read by someone.

12. The *trade winds* are between the equator and about 30 degrees north and south latitude. These winds blow from the east. The *westerlies* are between about 40 and 60 degrees latitude in both the Northern and Southern hemispheres. The *polar easterlies* are between about 70 degrees latitude and the poles.

13. Most ultraviolet light is absorbed in the upper atmosphere and by the ozone layer. Much of the visible light is scattered in passing through the atmosphere. Infrared rays may be partly absorbed by water vapor in the atmosphere. Those rays that reach the earth are absorbed and reradiated as longer wavelengths, which in turn are absorbed by water vapor and carbon dioxide.

14. Around the equator temperatures are relatively high. This produces a low pressure area into which winds blow. The polar regions, on the other hand, have relatively low temperatures that produce a high pressure area, causing air to move away from these regions.

15. The simplest way is to imagine an object moving above the earth's surface from near the north pole toward the south. As it moves to the south, the earth beneath it would be turning to the east. Since the earth's rotation carries points nearer the equator around faster than points near the poles, the earth would move faster to the east as the object moved south. The effect is for the object to appear to have been pushed to the right (west) during its journey. If the object is moving north it is carried farther east as it moves because its speed toward the east is greater than the ground over which it must travel. It too appears as if it were pushed to the right (east).

16. The tilt of the earth's axis causes the sun's vertical rays to shift north and south of the equator during the year. The positions of the pressure belts and wind belts also shift, but much less than the 23½° movement of the sun's vertical rays.

references

Elliott, S. M. *Our Dirty Air.* New York: Julian Messner, 1973.

Gedzelman, S. D. *The Science and Wonders of the Atmosphere.* New York: John Wiley & Sons, 1980.

Kaufman, J. *Winds and Weather.* New York: William Morrow & Co., 1971.

Neiburger, M., et al. *Understanding Our Atmospheric Environment.* San Francisco: W. H. Freeman & Co., 1973.

O'Donnel, P. *Air Pollution.* Menlo Park, CA: Addison-Wesley, 1971.

Parker, S. *Encyclopedia of Ocean and Atmospheric Sciences.* New York: McGraw-Hill, 1980.

Reiter, Elmar R. "Rapid Rivers of Air." *Natural History* (October, 1973), pp. 46-51.

Riehl, H. *Introduction to the Atmosphere.* New York: McGraw-Hill, 1978.

Woodurn, J. H. *The Whole Earth Energy Crisis.* New York: G. P. Putnam's Sons, 1973.

audio-visual aids

16mm Sound Films

Air All Around Us. McGraw-Hill, 11 min., color

Atmosphere and Its Circulation. EBEC, 11 min.

The Noise Boom. NBC, 26 min., color

Solar Radiation 1: Sun and Earth. American Meteorological Society, 20 min., color

Something in the Air. Caterpillar, 28 min., color

What Makes the Wind Blow. EBEC, 16 min., color

Filmstrips

Air Pressure and Winds. Edwin Shapiro Co.

Meteorology. Eye Gate, set of 4 with cassettes

Water Vapor in the Atmosphere. Edwin Shapiro Co.

8mm Film Loops

Air Expansion by Heat. EBEC, color
Global Circulation, Model I. Thorne, color
Global Circulation, Model II. Thorne, color
Why Air Circulates. EBEC, color

---

21 water in the atmosphere

objectives

When finished with this chapter, the student will be able to:

1. Explain how water vapor enters the atmosphere and how it is measured.
2. Describe how water leaves the atmosphere.
3. Explain the conditions necessary for cloud formation.
4. Describe the various types of clouds based on altitude.
5. Describe several ways precipitation may occur.
6. Describe the conditions needed for the various types of precipitation to occur and how the amount of precipitation may be measured.

approach

The foundation upon which all understanding of the behavior of water molecules in the atmosphere rests is the concept of latent energy changes. An increase in the energy content of air tends to increase its humidity, while a decrease in energy tends to lower the water content. This is a fundamental concept and it should be emphasized in connection with the study of each of the processes described in the text. Any demonstrations or experiments that will bring out this concept will strengthen understanding of the basic causes of the continuous change in moisture content of the air.

activity hints

Measuring the Dew Point (page 442)

Shiny, clear, thin plastic glasses make an excellent substitute for the shiny tin can traditionally used. You must caution the students to remove the ice as soon as any condensation occurs and to take the temperature immediately. If a heavy layer of condensation occurs before removal of the ice cube, it will take a very long time for it to disappear. It is better in this case to empty the container and start over again.

other activities

Geology and Earth Sciences Sourcebook (2nd edition, 1970) Holt, Rinehart and Winston.
Chapter 5, The Atmosphere, pp. 115-134

Also see Chapter 21 in *Activities and Investigations for Modern Earth Science* (1983 edition) Holt, Rinehart and Winston.

answers—chapter 21

Vocabulary Review

1. n	**5.** b	**9.** c	**13.** h
2. f	**6.** l	**10.** g	**14.** e
3. a	**7.** q	**11.** d	**15.** k
4. p	**8.** o	**12.** i	

Questions: Group A

1. c	**9.** d	**17.** d	**25.** d
2. b	**10.** b	**18.** d	**26.** b
3. d	**11.** c	**19.** b	**27.** a
4. a	**12.** d	**20.** c	**28.** d
5. a	**13.** b	**21.** d	**29.** d
6. d	**14.** b	**22.** c	**30.** b
7. b	**15.** a	**23.** b	
8. c	**16.** a	**24.** a	

Questions: Group B

1. Life-supporting weather would be missing. The planet's surface would probably be a barren desert swept by clouds of dust. Temperatures would range from very hot in the day to bitter cold at night.
2. One gram of water would require 600 calories for evaporation; therefore, 454 grams would require:
   $454 \times 600 = 272,400$ calories.
3. By far the largest amount of moisture in the air comes from the evaporation of sea water. Substantial amounts also come from lakes, rivers, and soil. A lesser amount is derived from plants carrying out their life processes. An even smaller amount comes from burning fuels and from volcanoes.
4. The absolute humidity is 0.01 gram/liter.
5. Relative humidity
   $$= \frac{\text{absolute humidity}}{\text{saturation value}} \times 100$$
   $$= \frac{8.0}{10} \times 100 = 80\%$$
6. (a) There is no effect on absolute humidity. (b) Since warmer air can hold more moisture, the saturation value is decreased. (c) The relative humidity is decreased because the saturation value is increased while the absolute humidity remains the same.
7. The three ways are: (1) Using the wet and dry bulb psychrometer; the difference in the two thermometer readings for any temperature varies according to the humidity of the air. (2) The hair hygrometer measures the stretch of human hair as the humidity of the air changes. (3) By observing the change in electrical conductivity of certain chemicals as the humidity of the air changes.
8. Air may be cooled when radiation takes heat away from the earth at night, by mixing with colder air, and by the upward motion of air.
9. Each 1°C change is equal to 100 meters altitude; therefore, 5 × 100 is 500 meters or $\frac{9°}{5\frac{1}{4}°F} \times 100$ ft. This is about 1600 feet.
10. The temperature must be at or below the dew point; condensation nuclei must also be present.
11. (a) cumulonimbus capped with cirrus
    (b) nimbostratus
    (c) stratus
    (d) cirrostratus
    (e) cirrus
    (f) cirrocumulus
12. The seacoast fog is probably advec-

tion fog, produced when the moist air from the sea moved in over the cold land and buildings of the city. The valley fog is most likely radiation fog, produced when the land cooled down and the air above it also cooled.

13. Liquid droplets may grow by the process of coalescence, in which small droplets become large enough to fall by colliding with others. The second process involves the growth of small ice crystals by the freezing of water vapor from the cloud at the expense of the water droplets present.

14. Powdered dry ice may be scattered into the cloud, causing condensation into large droplets. A smoke generator producing silver iodide crystals will cause similar condensation when the silver iodide crystals reach the clouds. The benefits would be the production of rain over selected areas when it is most needed, the clearing of clouds from airfields, etc. Some problems that may arise are: production of more rainfall than anticipated; limitation of rainfall area; depletion of normal rainfall from other areas.

15. The tipping bucket relies on mechanical tipping to release small quantities of water. When snow falls, the temperature is usually low enough to freeze water, rendering the mechanism inactive.

16. In the tropical oceanic regions, enormous quantities of water are evaporated, carrying with the vapor tremendous amounts of solar energy. This energy is then used to generate the violent storms called hurricanes and typhoons.

17. The temperature is about 95°F when the winds leave the desert area. After passing over the mountains, they descend from about 3000 feet to sea level. This sinking of the air causes an ad-

ditional increase in temperature. If the sinking is adiabatic, it would be 3 × 5½°F or 16½°F. This is the reason temperatures of 110-115° can be attained.

18. (a) Since air will only be lifted when it is warmer than its surroundings, the smoke-laden air warmed by the surface will not rise above the warmer layer above it. Thus the smoke and noxious gases generated in the city continue to pile up until winds or a cool air mass cause the warm layer to break up.

(b) Inversion means "upside down." Normally the temperature pattern in the atmosphere indicates that as the altitude increases, the temperature decreases. In the case of an inversion layer, the temperature close to the ground is cooler than in the overlying warmer layer. This is just opposite the normal pattern. This condition is thus called a temperature inversion.

references

Barry, R. G. *Atmosphere, Weather and Climate*. New York: Holt, Rinehart and Winston, 1970.

Committee on Atmospheric Sciences. *Atmospheric Sciences and Man's Needs: Priorities for the Future*. Washington, DC: National Academy of Science, 1971.

Gedzelman, S. D. *The Science and Wonders of the Atmosphere*. New York: John Wiley & Sons, 1980.

Good, R. M., and J. Walker. *Atmosphere*. Englewood Cliffs, NJ: Prentice-Hall, 1972.

Kaufman, J. *Winds and Weather*. New York: William Morrow & Co., 1971.

Neiburger, M., et al. *Understanding Our Atmospheric Environment*. San Francisco: W. H. Freeman & Co., 1973.

Riehl, H. *Introduction to the Atmosphere*. New York: McGraw-Hill, 1975.

audio-visual aids

16mm Sound Film

Rainshower. Churchill, 17 min., color

Storms, The Restless Atmosphere. EBEC, 22 min., color

The Formation of Raindrops. American Meteorological Society, 15 min., color

The Origins of Weather. EBEC, 13 min., color

Tornado Below. NASA, 14½ min., color, HQ246

Weather: Understanding Precipitation. Coronet, 11 min., color

What Makes Clouds. EBEC, 19 min., color

Filmstrips

Condensation and Precipitation. Edwin Shapiro Co., captions

Meteorology from Space. NASA, color, sound

Moisture in the Atmosphere. Filmstrip House

The Mystery of Rain. Time-Life

Understanding Weather and Climate. SVE, set of 6

Water Vapor in the Atmosphere. Edwin Shapiro Co.

Weather and Climate. Creative Visuals

8mm Film Loops

Condensation of Water Vapor. EBEC, color

Formation of a Cloud. EBEC, color

Flash Flood. Ealing, color

Hydrology Series. PHOENIX, set of 12, color.

---

22 weather

---

objectives

When finished with this chapter, the student will be able to:

1. Describe the origin of air masses and name the air masses that influence North American weather.

2. Explain how weather fronts and the storms associated with them form.

3. Describe the methods and instruments used to observe weather.

4. Describe how to make a weather map and the principles of weather forecasting.

5. Explain how local weather phenomena develop.

approach

Topics studied in this chapter should be closely related to existing weather conditions. Newspapers and U.S. Weather Bureau maps are useful in locating and identifying the types of air masses involved in daily weather. In addition, the existing fronts and the weather patterns associated with them will be shown on these maps. This information can be displayed in the classroom, possibly as large outline maps on which the air masses and fronts may be drawn in from day to day. A good deal of classroom discussion can be built around this type of display and will make the study of weather origins more meaningful.

activity hints

Isotherms (page 471)

On page T78 in this teacher's edition, you will find a master copy of the map needed for this activity. You can use it to make a spirit master for class sets of the map. On page T80, you will find a completed map for your reference.

By making an overhead transparency of the map for this activity, you can show the first couple of isotherms. This will help facilitate the map-drawing process.

The following are answers to activity questions:

1. 9°C; 25°C (23°C if the tip of Florida is ignored).

2. Low temperature.

3. Cold air mass.

other activities

Geology and Earth Sciences Sourcebook (2nd edition, 1970) Holt, Rinehart and Winston.
Chapter 5, The Atmosphere, pp. 115-134

Also see Chapter 22 in *Activities and Investigations for Modern Earth Science* (1983 edition) Holt, Rinehart and Winston.

answers—chapter 22

Vocabulary Review

1. e	**5.** d	**9.** m	**13.** p
2. c	**6.** b	**10.** k	**14.** n
3. a	**7.** i	**11.** o	**15.** l
4. g	**8.** f	**12.** q	

Questions: Group A

1. a	**7.** b	**13.** d	**19.** c
2. b	**8.** b	**14.** a	**20.** b
3. c	**9.** a	**15.** c	**21.** d
4. d	**10.** c	**16.** b	**22.** a
5. a	**11.** d	**17.** c	**23.** c
6. a	**12.** b	**18.** a	

Questions: Group B

1. Satellites provide information regarding cloud formations, storm areas, carbon dioxide, ozone, and ions in the atmosphere. This information is of great importance in weather forecasting.
2. Photographs taken from satellites in polar orbits give pictures of the entire earth. Other orbits cover only parts of the earth.
3. Polar Canadian (cP); Polar Pacific (mP); Polar Atlantic (mP); Tropical Continental (cT); Tropical Gulf (mT); Tropical Atlantic (mT); Tropical Pacific (mT).
4. The rotation of the earth seemingly causes winds to be turned to the right as they move in the Northern Hemisphere and to the left in the Southern Hemisphere.

5. Observations are made and sent to a central office; station models are plotted; isobars are drawn and fronts are located and labeled.
6. A warm front is formed when a warm air mass advances over a mass of colder air. A gentle slope generally produces stratus clouds and heavy precipitation over a large area.
7. A cold front is formed when a cold air mass overtakes a warm air mass. A steep slope will cause heavy cloud formation if the warm air is moist. Storms created along a cold front are usually short and violent.
8. $16 \text{ knots} \times \dfrac{1.15 \text{ mi/hr}}{1 \text{ knot}} = 18.4 \text{ mi/hr}$
9. $29.75 \text{ inches} \times \dfrac{33.86 \text{ millibars}}{1 \text{ inch}} = 1007.3 \text{ millibars}$
10. The station model should use symbols shown in diagrams on pages 469 and 470.
11. $C = 5/9 (F - 32)$
    $C = 5/9 (98.6 - 32)$
    $C = 5/9 (66.6)$
    $C = 37.0°$
12. $-40°$; this is the only place on both the Celsius and Fahrenheit temperature scales at which the numerical values are the same.
13. $1015.8 \text{ millibars} \times \dfrac{1 \text{ inch}}{33.86 \text{ millibars}} = 30.00 \text{ inches}$
14. One primary advantage of the Celsius temperature scale is that it is widely used throughout the world and therefore probably better understood.

references

Battan, L. J. *Weather*. Englewood Cliffs, NJ: Prentice-Hall, 1974.

Boesen, V. *Doing Something About the Weather*. New York: G. P. Putnam's Sons, 1975.

Breuer, G. *Weather Modification: Prospects and Problems.* New York: Cambridge University Press, 1980.

Buehr, W. *Storm Warning: The Story of Hurricanes and Tornadoes.* New York: William Morrow & Co., 1972.

Constant, Constantine. *The Student Earth Scientist Explores Weather.* New York: Richards Rosen Press, 1975.

Frazier, K. *The Violent Face of Nature: Severe Phenomena and Natural Disasters.* New York: William Morrow & Co., 1979.

Gedzelman, S. D. *The Science and Wonders of the Atmosphere.* New York: John Wiley & Sons, 1980.

Goody, R. M., and J. Walker. *Atmospheres.* Englewood Cliffs, NJ: Prentice-Hall, 1972.

Lehr, P. E., R. W. Barnett, and H. S. Zim. *Weather: Air Masses, Clouds, Rainfall, Storms, Weather Maps, Climate.* New York: Simon and Schuster, 1971.

McAdie, A. G. *Man and Weather.* Detroit, MI: Gale Research, 1975.

Middleton, W. E. K. *Invention of Meteorological Instruments.* Baltimore, MD: Johns Hopkins Press, 1969.

Neiburger, M., et al. *Understanding Our Atmospheric Energy.* San Francisco: W. H. Freeman & Co., 1973.

Sahler, H. K. *Nature's Weather Forecasters.* New York, 1979.

audio-visual aids

16mm Sound Films

Hurricane Below. NASA, 14 min., color
The Origins of Weather. EBEC, 13 min., color
Storms, The Restless Atmosphere. EBEC, 22 min., color
Story of a Storm. EBEC
The Weather Watchers. NASA
Tornado Below. NASA, 14 min., color

Weather Forecasting. EBEC, 16 min., color
What Makes Clouds. EBEC, 19 min., color
What Makes Weather. EBEC, 14 min.

Filmstrips

Condensation and Precipitation. Edwin Shapiro Co.
Meteorology. Eye Gate, set of 4 with cassettes
Meteorology from Space. NASA
Understanding Weather and Climate. Hubbard
Weather and Climate. Creative Visuals

8mm Film Loops

Condensation of Water Vapor. EBEC, color
Formation of a Cloud (laboratory experiment). EBEC, color
Gathering Information for Weather Forecasting. Doubleday Multimedia, color
Meteorology Station. EBEC, color
Story of a Storm. EBEC, color
The Cold Front. PHOENIX, color
The Warm Front. PHOENIX, color
Tornado. Thorne, color

---

23   elements of climate

objectives

When finished with this chapter, the student will be able to:

1. Define climate and explain how temperature, moisture, and air movements control climate.
2. Name and briefly describe earth's three main climatic zones.
3. Describe the characteristics of several different climates found in each of the three main climatic zones.
4. List the three major climate controls

and describe how they influence North American climates.

5. Describe several controls that influence major climate areas that tend to produce a great number of local climates.

approach

Most students will already have a basic knowledge of climates from work in geography in lower grade levels. However, their previous work will probably have been based on description of climates rather than on the causes of basic climate types. These students will be accustomed to describing climates in terms of crops and other human activities that predominate in a given climatic zone. It should be emphasized that climate is merely the result of a set of climatic conditions and that the study of this chapter is aimed toward an understanding of the causes for a given climate.

activity hints

Local Temperatures (page 480)

The student prepares a graph of the local reported high and low temperatures during days of 14 or more daylight hours. Another graph of local reported high and low temperatures during days of 10 or less daylight hours is prepared. If an almanac is not available, suggest the student research the local newspapers.

other activities

Geology and Earth Sciences Sourcebook (2nd edition, 1970) Holt, Rinehart and Winston.
Chapter 5, The Atmosphere, pp. 115-134

Also see Chapter 23 in *Activities and Investigations for Modern Earth Science* (1983 edition) Holt, Rinehart and Winston.

answers—chapter 23

Vocabulary Review

1. e	**4.** j	**7.** b	**10.** i
2. h	**5.** a	**8.** d	
3. f	**6.** g	**9.** k	

Questions: Group A

1. c	**9.** d	**17.** c	**25.** a
2. a	**10.** b	**18.** d	**26.** a
3. a	**11.** d	**19.** d	**27.** b
4. b	**12.** a	**20.** c	**28.** c
5. d	**13.** c	**21.** c	**29.** b
6. c	**14.** d	**22.** b	**30.** a
7. d	**15.** c	**23.** a	
8. b	**16.** c	**24.** b	

Questions: Group B

1. Local conditions such as mountains or large bodies of water may change the temperature and moisture conditions of the air, thus effecting changes in climate.

2. The chief influence on average temperature is the total amount of solar energy received at any one location.

3. Warm surface water tends to mix with cooler water below. The heat capacity of water is higher than that of land. Evaporation from the surface of a body of water tends to keep its temperature down. This does not occur to any great extent with land surfaces.

4. The principal cause is the greenhouse effect; the fact that solar radiation is converted into heat in the lower air layers is due mainly to moisture content.

5. This is due to the prevailing westerly winds blowing off the warm North Atlantic Drift.

6. Monsoon weather is of two typical kinds: (a) A summer monsoon blows moist warm air in from the ocean and and is characterized by long periods of heavy precipitation. (b) A winter monsoon is characterized by very dry

weather, since the winds blow consistently from the land toward the ocean.

7. As the seasons change, the belts of precipitation move in a north-south direction. They move northward in summer and southward in winter.

8. (1) Warm, humid climate, rising air masses, annual rainfall greater than 250 cm.

(2) Dry, tropical, sinking air masses, annual rainfall less than 25 cm.

(3) Wet and dry climate cycle (wet in summer, dry in winter) results from poleward shifts of the precipitation belts with the seasons, called a savanna climate.

9. Tropical deserts are associated with subtropical highs where cT air is sinking and heating adiabatically. This warm, dry air picks up moisture from the surface of the land, drying it out, and thus causing the desert-type environment.

10. The ground is frozen most of the year. Therefore, plants such as mosses and certain shrubs that are active for only a very short time can exist.

11. The prevailing westerlies of the middle latitudes, the large land masses of this region, and the cyclonic storms which regularly cross the continents combine to give day to day and seasonal changes in weather.

12. The moist air blowing in over the western coast is relatively cool. Therefore it must be cooled even more for precipitation to occur. Thus, winter months are best for this condition. The winter air continuing on across the continent becomes drier as it sweeps nearly to the Gulf Coast and swings off over the East Coast. However, during the summer months moist tropical air reaches up quite close to Canada. As cyclonic winds carrying cooler air collide with moist tropical air, precipitation occurs.

13. Many local conditions may modify greatly any general classification of a given climate. Some of these are: altitude; exposure to air masses from the north, south, east, or west; presence of lakes, forests, or even large cities.

14. There may be many ideas brought out here such as extreme coldness, etc.

15.

Climate	General Location	Type Air	Annual Precipitation	Plant Growth	Areas
Marine West Coast	Pacific coast	Moist westerlies	50-75 cm	Heavy cone forests	Northwestern North America
Mediterranean	Pacific coast	Moist westerlies and dryer subtropical	25 cm	Small trees, shrub	Southern California coast
Middle Latitude Deserts	Interior continent	Dry westerlies	Less than 25 cm	Little or none	Far western United States
Steppes	Interior continent	Dry westerlies and moist tropical	25-50 cm	Grasses	Midwestern United States
Humid Continental	Eastern continent	Cold, dry polar and moist tropical	75 cm	Heavy hardwood and softwood forests	Northwestern North America
Humid Subtropical	Southeast continent	Moist tropical	75-165 cm	Heavy forests	Southeastern United States

Probably the most difficult problem would involve the melting of the top portion of the permanently frozen ground. This melting forms a very plastic mud, generally from .5 to 3 meters deep. Either pylons anchored deep into the frozen earth or construction of a large flat foundation insulated from the rest of the building would be necessary to keep the building from sinking or floating away.

16. The rising moist air at and near the equator gives greater than 250 cm annual rainfall. Farther north or south of the equator, the sinking air masses of the subtropical highs give less than 25 cm annual rainfall. Between these two regions are found an annual wet-and-dry cycle: dry in winter and wet in summer. Middle latitudes are under the influence of cyclonic storms producing variable climates with annual rainfall from less than 25 cm to 165 cm. Farther north the subarctic climate (Alaska) has 25-50 cm annual rainfall due mostly to moist polar masses. Near the Arctic Circle little rain falls, since polar air masses are descending.

17. Northwest United States has a marine west coast climate, which has abundant rainfall of 50-75 cm annually, caused by moist air from the Pacific Ocean being blocked by mountains. Southern California has a much drier, warm climate with about 25 cm rainfall annually due to the drier air of the subtropical high.

18. After passing over the coastal mountains of the west, the air is dry, thus causing deserts. As it moves eastward, it comes in contact with moist tropical air, which can then give precipitation to the midwestern states.

19. Temperature decreases with altitude; thin air retains less heat at night.

20. They reduce loss of ground heat.

references

Barry, R. G. *Atmosphere, Weather and Climate*. New York: Holt, Rinehart and Winston, 1971.

Gedzelman, S. D. *The Science and Wonders of the Atmosphere*. New York: John Wiley & Sons, 1980.

Lehr, P. E., R. W. Burnett, and H. S. Zim. *Weather: Air Masses, Clouds, Rainfall, Storms, Weather Maps, Climate*. New York: Simon and Schuster, 1971.

Neiburger, Morris, et al. *Understanding Our Atmospheric Environment*. San Francisco: W. H. Freeman & Co., 1973.

audio-visual aids

16mm Sound Films

Above the Timberline. National Film Board of Canada, 16 min., b/w-color

Climates of North America. EBEC, 12 min., b/w-color

The Desert. Barr, 11 min., color

High Arctic: Life on the Land. National Film Board of Canada, 22 min., b/w-color

The Spruce Bog. National Film Board of Canada, 23 min., color

Life in a Tropical Forest. Time-Life, 30 min., color

Climate and the World We Live In. Coronet, 13 min., color

Filmstrips

Arctic Tundra. Time-Life

The Desert. Time-Life

Rain Forest. Time-Life

Understanding Weather and Climate. Hubbard, WFS-763, set of 6

Weather and Climate. Creative Visuals, MH642327-3

World Climate. P.H. Media, set of 2 with cassettes

8mm Film Loops

Geographic Causes of Deserts. Ealing, color

Global Circulation Model I & II. AGS

Northern Hemisphere Cloud Patterns I & II. AGS

North America Cloud Patterns I & II. AGS

the earth community

24 people and the planet

objectives

When finished with this chapter, the student will be able to:

1. Distinguish between renewable and nonrenewable resources.
2. Describe how the fossil fuels were formed and the methods used to extract them from the earth's crust.
3. Describe several ways geothermal and tidal energy can be utilized and their present and future roles as energy sources.
4. Identify the most important future energy problems.
5. Define the term "ore" and describe several ways an ore may be formed.
6. List some nonmetallic mineral resources and explain how they are used.

approach

Gold is used as an example of how a resource is located, its source rapidly consumed and depleted, and the area abandoned. You may wish to point out that this same series of events was repeated some years later when in 1870 thousands of diggers descended on South Africa in a diamond rush. There is a similarity between these events and the discovery of new petroleum deposits, such as those on the north slopes of Alaska.

A resource is defined as a natural material found in the earth that can be used in some way. The idea that some resources are renewable while others are not becomes a convenient and presently important means of classifying all resources. The resources of the earth's crust are broken down into metallic minerals and nonmetallic minerals. If your local area is a source of any of the resources discussed, it would be interesting to bring in samples and discuss the local operations. Students having relatives working in the mining industry might want to bring in information and samples.

activity hints

Pyrolysis of Wood (page 501)

The students should be warned to set the apparatus up correctly, as shown. There is a danger of plugging up the glass tubing and blowing the stopper out of the test tube. Be aware of this as you oversee the setups. The gases being evolved will most likely turn litmus red. The gases consist mainly of CO, CO_2, vaporized hydrocarbons, methanol, and pyroligneous acid. Most of the odor is due to the presence of pyroligneous acid, which is a mixture containing about 6% acetic acid. It is sometimes called wood vinegar. If coal is available, the same process could be used for its destructive distillation, and the products compared.

other activities

Geology and Earth Sciences Sourcebook (2nd edition, 1970) Holt, Rinehart and Winston.
Chapter 10, Mineral Resources, pp. 223-241

Also see Chapter 24 in *Activities and Investigations for Modern Earth Science* (1983 edition) Holt, Rinehart and Winston.

answers—chapter 24

Vocabulary Review

1. n	**5.** h	**9.** o	**13.** c
2. l	**6.** i	**10.** k	**14.** f
3. j	**7.** b	**11.** e	**15.** g
4. m	**8.** a	**12.** d	

Questions: Group A

1. c	**9.** a	**17.** a	**25.** c
2. a	**10.** b	**18.** c	**26.** d
3. b	**11.** c	**29.** c	**27.** b
4. d	**12.** d	**20.** d	**28.** c
5. c	**13.** b	**21.** c	**29.** c
6. a	**14.** c	**22.** d	
7. b	**15.** a	**23.** b	
8. d	**16.** d	**24.** a	

Questions: Group B

1. As magma cools, the more dense crystals form and settle to the bottom. Magma may come into contact with surrounding rock, depositing bands of ore minerals around the magma as it cools. In later stages of cooling, magma may produce mineral-bearing fluids that enter surrounding rock, forming veins.

2. The concentration of native metals or ore in layers of gravel by streams or ocean waves. Gold and platinum.

3. Depositing of ores by action of streams or waves along the shore. Water moving down through the ground.

4. (a) sand, gravel, crushed rock, blocks of stone, limestone, clay, gypsum (b) sulfur (c) sulfur, phosphate rock (d) halite

5. The recycling of metals and the use of less rich ores will help extend the amount of metals available.

6. Plant remains accumulated in ancient swamps forming a soft brown or black material called peat. Layers of sediment were deposited in the peat causing a slow change to lignite, a brown coal. As more sediment caused higher pressure, the lignite became bituminous. Much higher pressures produced the very hard coal, anthracite.

7. The organic remains of plants and animals accumulated on the sea floor, producing tiny drops of oil in the sediment. As sediments changed to rock, oil pools and natural gas were trapped under a solid, nonporous layer of rock.

8. Oil shale is crushed and heated.

9. A nucleus of a large atom splits into two or more smaller ones, giving off energy.

10. The U-235 is extracted from natural uranium; the purified U-235 is made into rods; these fuel rods are put together in a bundle that starts the fissioning.

11. U-235 fissions in fuel rods, causing a chain reaction. The fuel rods heat up. A liquid is circulated around the rods, carrying the energy away. The hot liquid or steam is used as an energy source for running electrical generators.

12. It would utilize the hydrogen from the vast supply of water in the oceans.

13. Volcanoes, hot springs, earthquakes, etc.

14. Deep wells may be drilled into a body of hot rock. Water pumped down these wells would be changed to steam and returned to the surface for use.

15. A dam could be built across the mouth of a bay. After high tide the dam would be closed, trapping the water behind it. At low tide, the water could then be allowed to run out through the dam. The moving water would then be used to turn turbines that generate electrical power.

16. Windows that let the sun's rays heat a house; solar collectors for heating water; solar cells for producing electricity; use of mirrors to generate steam for producing electricity.
17. Windmills, wind generators, warm sea water used to vaporize a liquid to run a generator, converting biomass to energy.
18. By using the agricultural wastes produced from its crops for energy.

references

Branley, F. *Energy for the 21st Century.* New York: Thomas Y. Crowell Co., 1975.

Cardwell, Harvey. *Earth Science at Crisis.* New York: Vantage Press, Inc., 1976.

Carter, Anne (ed.). *Energy and the Environment: A Structural Analysis.* New Hampshire: The University Press of New England, 1976.

Dubos, R. *The Wooing of Earth.* New York: Charles Scribners' Sons, 1980.

Gadler, S. J., and W. W. Adamson. *Sun Power: Facts About Solar Energy.* Minneapolis: Lerner Publications, 1978.

Kaplan, W., and M. Lebowitz. *The Student Scientist Explores Energy and Fuels.* New York: Richards Rosen Press, 1976.

Marshall, James. *Going, Going, Gone.* New York: Coward, McCann and Geoghegan, Inc., 1976.

Skinner, B. J. *Earth Resources.* Englewood Cliffs, NJ: Prentice-Hall, 1976.

Note: Creating Energy Choices for the Future. Energy Research and Development Administration, Office of Public Affairs, Washington, DC 20545, 1975 (free to schools).

audio-visual aids

16mm Sound Films

Air Pollution: Take a Deep Deadly Breath, Parts I & II. McGraw-Hill

Energy: The Fuels of Man. NGS, 23 min.

Energy: The Problems and the Future. NGS, 23 min.

Energy: The American Experience. U.S. Dept. of Energy

Energy Crisis. JFI, 13 min., color

Energy, New Sources. CF, 20 min., color

ERTS-Earth Resources Technology Satellite. NASA, 27 min., color, HQ 223

Geothermal, Nature's Boiler. U.S. Dept. of Energy

Pollution Below. NASA, 14 min.

Saving Energy at Home. Ramsgate, 13 min., color

The Power Game. EMC, 28 min.

The Trouble with Trash. Caterpillar, 28 min., color (free loan)

Water: A Precious Resource. NGS, 23 min.

Filmstrips

Earth's Resources. Learning Arts, set of 6, color

Ecological Crisis. Learning Arts, set of 6, color

Energy. Learning Arts, set of 6, color, sound

Energy and the Earth. EAV, color, 2 sound filmstrips

Energy for Tomorrow. Educational Materials and Equipment Co., color

Geology, Rocks and Minerals. Hubbard, set of 4, color

Land, Uses and Values. P.H. Media, set of 3 with cassettes

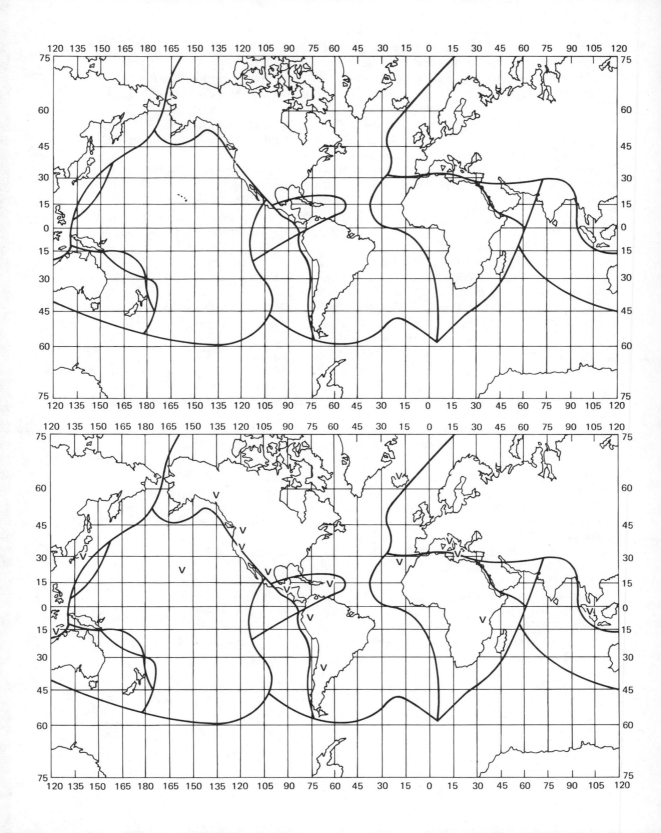

Geologic Period	Fossils		Land/Water
_____		human	_____
		cat	
_____		camel	_____
		early horse	
		fish	
_____		sea snail	_____
		shark	
		dinosaur	
_____		giant fern	_____
		land plants	
_____		dinosaur	_____
		salt layer	
_____		early reptile	_____
		amphibian	
_____		fish	_____
		shark	
		land plants	
_____		early amphibian	_____
		millipede	
_____		scorpion	_____
		ancient fish	
_____		branchiopod	_____
		cephalopod	
_____		trilobite	_____
		sea snail	

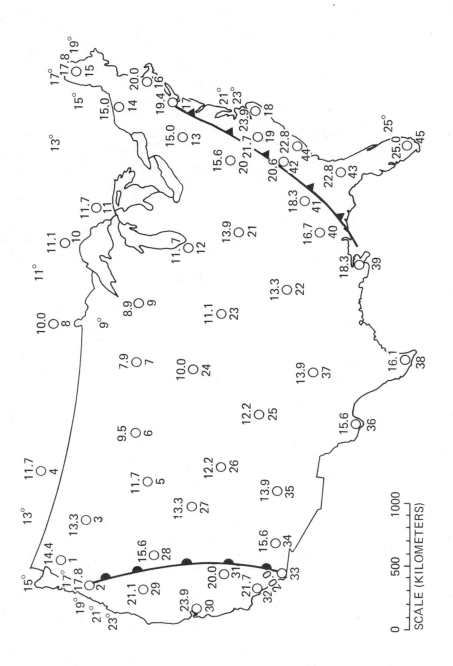

SCALE (KILOMETERS)

0   500   1000

T78

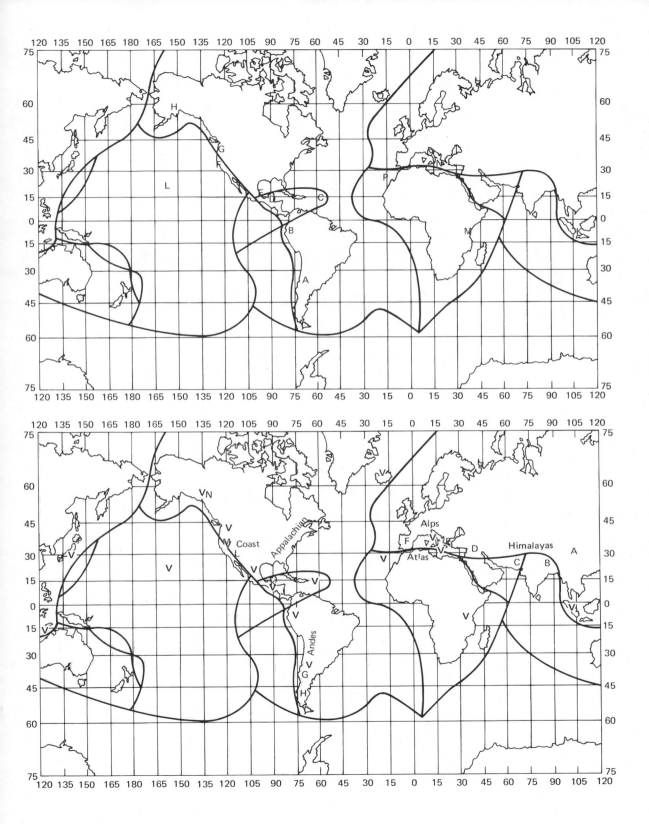

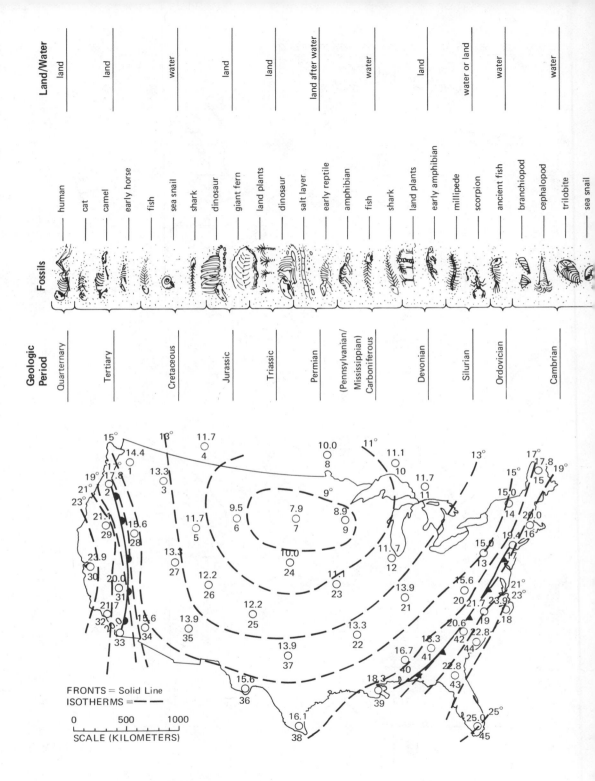

Land/Water

land | land | water | land | land | land after water | water | land | water or land | water | water

Fossils

human | cat | camel | early horse | fish | sea snail | shark | dinosaur | giant fern | land plants | dinosaur | salt layer | early reptile | amphibian | fish | shark | land plants | early amphibian | millipede | scorpion | ancient fish | branchiopod | cephalopod | trilobite | sea snail

Geologic Period

Quarternary | Tertiary | Cretaceous | Jurassic | Triassic | Permian | (Pennsylvanian/Mississippian) Carboniferous | Devonian | Silurian | Ordovician | Cambrian

FRONTS = Solid Line
ISOTHERMS = ----

0   500   1000
SCALE (KILOMETERS)

T80

Modern Earth Science

Ramsey
Phillips
Watenpaugh

ABOUT THE COVER
The cover of your MODERN EARTH SCIENCE text captures the awesome beauty and excitement of one of nature's marvels. This volcano, named Surtsey after the Norse god of fire, rose from the Atlantic Ocean just south of Iceland on November 14, 1963. Surtsey is part of an underwater belt of volcanoes located along the mid-Atlantic ocean ridge, whose eruptions usually are hidden beneath the sea. However, some may produce volcanic cones that rise above the sea surface.

This photo was taken by Ernst Haas in May 1965, as the quiet lava flow continued. By the end of 1967, Surtsey was 173 meters (567 ft) high, covered over 2.6 square kilometers (1 sq mi), and exhibited plant life on the island. It is interesting to note that before the initial eruption, fishermen observed unusually warm water temperatures, leading scientists to believe a chain of erupting volcanoes was located here.

Modern
Earth
Science

William L. Ramsey
Clifford R. Phillips
Frank M. Watenpaugh

Holt, Rinehart and Winston, Publishers
New York • Toronto • Mexico City • London • Sydney • Tokyo

THE AUTHORS

WILLIAM L. RAMSEY
Head of the Science Department,
Helix High School, La Mesa, California

CLIFFORD R. PHILLIPS
Science Department Chairman,
Mount Miguel High School,
Spring Valley, California

FRANK M. WATENPAUGH
Science Department, Helix High School,
La Mesa, California

Editorial Development	William N. Moore, Roland Cormier
Editorial Processing	Margaret M. Byrne, Regina Chilcoat
Art, Production, and Photo Resources	Vivian Fenster, Russell Dian, Paula Darmofal, Ira A. Goldner, Susan Gombocz, Mary Ciuffitelli, Richard Haynes, Joan McNeil
Product Manager	John Cooke
Researchers	James R. George, Erica Felman
Advisory Board	Rhenida Bennett, Thomas Glunt, David J. Miller, George Salinger

Cover design by Caliber Design Planning Inc.

Cover photo by Ernst Haas

Photo credits appear on page 580.

There is a new earth. An explosion of scientific knowledge about the earth and its neighbors in space has occurred during the past few decades. This knowledge has resulted in a new scientific view of our home planet. For the first time, scientists have a general understanding of how the earth's surface has been shaped into the familiar continents and oceans. Earthquakes, volcanoes, and the creation of huge mountain ranges are now seen as different parts of the same great process that continuously alters the face of the earth. Better understanding of how the earth may have come into existence and developed into a unique planet has come from exploration of other members of the sun's family of planets. The fruits of this new scientific era have been incorporated into this edition of MODERN EARTH SCIENCE.

The position of the earth in the universe is first established in Unit 1. Unit 2 explores the general characteristics of the earth as a planet. Unit 3 deals with the ways by which the earth's surface is sculptured into its many landforms. The characteristics of the oceans, one of the most vital of all the earth's surface features, are described in Unit 4. Unit 5 is devoted to the earth's history, with a more detailed description of the development of North America. Unit 6 is concerned with the atmosphere, emphasizing the nature of weather and climate. The final unit, Unit 7, includes one chapter in which the relationship between energy supplies and the resources needed to support the world's population is explored. Following most of the seven units is a special section that introduces interesting discoveries, phenomena, and careers in earth science.

At the beginning of each chapter, objectives have been provided to guide the student in the basic concepts of that chapter.

To assist students in the mastery of technical terms associated with the text discussion, the terms are printed in *italics* where they occur and are clearly defined. In order to help evaluate behavioral skills, each chapter includes a number of marginal notes to the student. These notes are intended to extend the learning process beyond the realm of the textbook. At the end of each chapter the vocabulary is reviewed in a section that students may use as a self-quiz to test their understanding. A complete *Glossary* also appears at the back of the book. In addition to the *Vocabulary Review* at the end of each chapter, there are two groups of *Questions.* Those in *Group A* are based directly on the text and may be used as a self-quiz by the student. The questions in *Group B* are more difficult and often require interpretation of the text material.

The *Appendix* includes a section listing the national parks and correlating their important geologic features to the text. Also included in the *Appendix* are a geologic calendar, a short guide to common stones, a mineral key, and tables of measurement and temperature units.

In addition to the text, there is a workbook, *Activities and Investigations* for MODERN EARTH SCIENCE. The activities give further insight into the basic concepts of the chapter. The laboratory experiments give direct experience with the spirit of scientific inquiry.

In regard to the sections in MODERN EARTH SCIENCE dealing with geologic evolution and the age of the earth, the authors have used scientific data to present this material as theory rather than fact. The information as presented allows for the widest possible interpretation. Every attempt has been made to present this material in a non-dogmatic way.

ACKNOWLEDGMENTS

The authors wish to thank the following consultants for their valuable suggestions and criticisms regarding MODERN EARTH SCIENCE.

Dr. Robert W. Ridky, Office of the President, University of Maryland, Adelphi, Maryland.

Dr. J. Craig Wheeler, Department of Astronomy, University of Texas, Austin, Texas. (Unit 1)

Mr. James Helwig, Mt. Healthy High School, Cincinnati, Ohio.

Mr. Daniel B. Kujawinski, North Collins, New York.

contents

Unit 1   THE EARTH'S PLACE IN SPACE

1 □ a small planet in space   2
2 □ the sun   21
3 □ design of the solar system   44
4 □ the moon   71
5 □ probing the secrets of space   93
Comets   50 – 51
Career Paths in Earth Science   112 – 113

Unit 2   THE PLANET EARTH

6 □ models of the planet earth   116
7 □ earth chemistry   138
8 □ materials of the earth's crust   153
9 □ the movement of continents   182
10 □ movement of the earth's crust   207
Our Dynamic Planet   228 – 229

Unit 3   FORCES THAT SCULPTURE THE EARTH

11 □ weathering and erosion   232
12 □ water and rock   253
13 □ ice and rock   275
14 □ shorelines   296

Unit 4   THE EARTH'S ENVELOPE OF WATER

15 □ the unknown sea   316
16 □ sea water   335
17 □ motions of the sea   352
Earth, the Water Planet   372 – 373

Unit 5   THE RECORD OF EARTH HISTORY

18 □ the rock record 376
19 □ a view of the earth's past 396

Unit 6   THE EARTH'S ATMOSPHERE

20 □ air and its movements 420
21 □ water in the atmosphere 439
22 □ weather 457
23 □ elements of climate 479
Ice and Fire 494–495

Unit 7   THE EARTH COMMUNITY

24 □ people and the planet 498
Searching for Energy 520–521

Appendix: A. International System of Measurement 522
B. Geology in the National Parks 523
C. Geologic Calendar 530
D. Stones 532
E. Identification of Minerals 536
F. Temperature Scales 550

Glossary 551
Index 566

how to use this book

objectives

☐ Name and describe two ways that weathering occurs.

☐ Explain how the nature of rock, climate, and topographic conditions affect the rate of weathering.

☐ List and describe the layers of a mature soil.

☐ Explain how soil is produced.

Rocks appear to us to are almost always und exposed to air and wa different conditions fro formed. Gases of the at ture changes bring abc composition of the roc into small pieces as the All changes that destrc called *weathering*. Wh ing processes, the form sult. *Soil* is a mixture als produced by living t

Weathering of large of resulting smaller fi of rocks by the proces:

Objectives
A list of objectives appears on the first page of each chapter. The objectives state the important concepts to be learned in that chapter.

Italic Words or Phrases
Words or phrases associated with the development of new subject material in a chapter are printed in *italics*.

THE BODY OF THE EARTH

Inside the earth. According to the theory of plate tectonics, the earth's surface is made up of a number of separate rigid plates. The plates are slowly moving relative to each other. This movement has been responsible for creating the familiar land and ocean features on the earth's surface. "Tectonics" refers to the forces that move and shape

Major Topic Headings
Major topic headings introduce the general theme of that section of the text.

Boldface
Boldface phrases at the beginning of each new section point out the main idea of that section.

Ocean. Older island olcanic islands that

g plate boundaries. riors of plates where rd from the asthenom within plates are e occurrence of hot

Interpret
From the graph in Figure 9–14, determine the temperature and pressure at the boundary between the mantle and the core.

Student Marginal Notes
These notes are intended to extend the learning process beyond the realm of the text. They are designed to offer the student the opportunity to examine, discover, construct, experiment, discuss, observe, and interpret many interesting ideas in earth science.

Activity

A marginal activity appears in every chapter. The purpose of the activity is to further develop a concept or several concepts covered in the chapter.

activity

Most volcanic activity occurs in certain regions of the earth's surface. In this activity you will locate several volcanoes on a map of the world.

Obtain a map of the world that has the crustal plates outlined. Your teacher may provide you with one.

Volcano	Name	Longitude	Latitude
A	Aconcagua	70W	35S
B	Tungurahua	80W	0N
C	Pelée	61W	15N

Tables

Tables present information in a concise way for easy reference.

Table 14 − 1 Beach Materials

Description	Size (millimeters)
Boulders	More than 200 (over 8 in)
Cobbles	76 to 200 (3 to 8 in)
Gravel	
Coarse	19 to 76
Fine	5 to 19
Sand	
Coarse	2 to 5
Medium	0.4 to 2
Fine	0.07 to 0.4
Silt	less than 0.07

Vocabulary Review

The vocabulary review at the end of each chapter covers key words used in the chapter.

VOCABULARY REVIEW

Match the word or words in the column on the right with the correct phrase in the column on the left. *Do not write in this book.*

1. Very large body of solidified magma extending to unknown depths.
2. Sent out when an earthquake occurs on or near the earth's surface.
3. Activities caused by movement of magma.
4. The modern theory stating that the earth's crust is made up of several rigid parts.
5. The location of the spreading boundaries of the crustal plates.

   a. plate tectonics
   b. seismic waves
   c. crust
   d. mantle
   e. Moho
   f. lithosphere
   g. mid-ocean ridges
   h. magnetic declination
   i. magma

Questions—Group A
Group A questions are based directly on the text and may be used as a self-quiz.

QUESTIONS

Group A

Select the best term to complete the following statements. *Do not write in this book.*

1. The theory of continental drift is supported by the (a) size of continents (b) equal spacing of mountains (c) shapes of the continents (d) thickness of the mantle.

2. In his theory on continental drift, Wegener claimed there once was a superconti-

Questions—Group B
Group B questions are more difficult and often require interpretation of the text material.

Group B

1. Briefly describe the theory of continental drift.

2. How is the theory of plate tectonics similar to the theory of continental drift?

appendix

A. International System of Measurement

B. Geology in the National Parks

C. Geologic Calendar

D. Stones

Appendix
The Appendix contains important information that may be useful while you are using the text. It contains a key that can be used to identify minerals. It also contains a section on the National Parks, metric measurement, stones, and the Geologic Calendar.

glossary

aa. Black lava.
absolute humidity. The weight of water vapor actually contained in a given quantity of air.

Glossary
The glossary is an alphabetical list of important words used in the textbook. Next to each word is its definition.

index

aa lava, 197
absolute humidity, **441**

aluminum, recycling of, *498* – **499**
amber, 390
Andes Mountains, 223, 290
Andromeda, **16,** 17

Index
Page numbers in boldface type indicate illustrations and those in italic type indicate definitions.

unit

1

*the earth's
place in space*

1

a small planet in space

objectives

☐ Explain the importance of constellations.

☐ Compare the ancient Greek and Copernican models.

☐ List Kepler's Laws of Planetary Motion.

☐ Use Newton's Law of Gravitation to describe the size of a gravity force.

☐ Describe the electromagnetic spectrum.

☐ Describe the Milky Way Galaxy.

☐ Name three types of galaxies.

Late afternoon on a summer day in the desert of southern Utah, a small group of scientists gather in a special chamber in the ruins of an ancient Indian castle. As the sun sets on a distant ridge of mountains, a single shaft of light from a narrow opening in the castle's western wall suddenly lights the chamber. The scientists congratulate each other.

Measurements and computer analysis had suggested to scientists that the old ruin held evidence that its Indian builders were skilled observers of the sky. The opening in the castle wall was carefully lined up to allow sunlight to enter the chamber only on two days during the year. Those were the longest day during summer and the shortest day of midwinter. Hundreds of years ago the castle's builders knew that the sun's position on those special days marked the turning points in the yearly cycle of the seasons. This knowledge allowed them to establish a calendar to guide the planting of crops and determine the time for ceremonies.

Other Indian ruins in the American Southwest also contain evidence that astronomy, the science of the sky, is very old. Ancient civilizations all over the world have left a record of the knowledge of astronomy in their structures, art, and written history. In Europe there are many prehistoric stone monuments that seem to have been built as astronomical observatories. One of the best known of these monuments is Stonehenge, in England, shown in Figure 1–1. In Mexico the ancient Mayans developed an accurate calendar able to predict eclipses many centuries into the future. The positions of the stars, sun, and moon were marked with precision by the Chinese thousands of years ago. Ancient Greeks were able to make a surprisingly accurate measurement of the earth's circumference by observing stars from different locations.

Like the people of the past, we also look into the sky and wonder about the meaning of what we see. As we observe the moving sun, moon, and stars, it may seem as if we are on a planet suspended at the center of the universe. All of our senses seem to tell us that the earth is fixed in space with the heavenly bodies moving around it. It has taken many centuries of careful scientific investigation to reveal the truth: The earth is only one of a family of planets moving around an ordinary star in a universe without known limits. One way to begin the study of this small planet is to first become familiar with this vast universe.

THE SKY GLOBE

Observing the sky. The brightest and largest object seen in the sky is the sun. It appears to rise in the east, move to its highest point in the sky, then set in the west. However, careful observation shows that the path of the sun as seen from the Northern Hemisphere never crosses the sky directly from east to west. After rising in the east, the path of the sun makes an arc across the sky in a southward direction until it reaches its highest point in the sky. Then it moves down to finally set in the west. See Figure 1–2. In winter, the sun's path shifts south to cross the sky in a low arc that makes days short and nights long. In summer, the path shifts north again to give longer days and shorter nights. See Figure 1–3.

FIG. 1–1. Stonehenge is an ancient stone monument in England, believed to have been used as an astronomy observatory. (Arthur Krasinsky/KPI, Inc.)

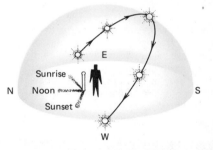

FIG. 1–2. The shadow cast by an upright stick shows the motion of the sun as it crosses the sky from east to west.

FIG. 1–3. In the summer the sun's path carries it higher in the sky than in winter. During which season of the year would an upright stick cast its longest shadow?

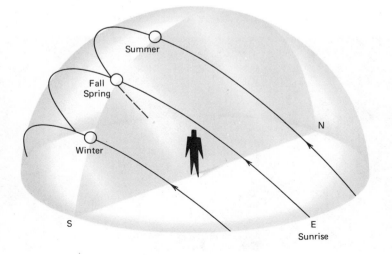

Like the sun, the moon also appears to move across the sky in a regular pattern. It rises in the east and sets in the west, but not at the same time the sun does. Sometimes the moon is seen during the night and other times it is seen during the day. Its appearance regularly changes as the earth's shadow slowly covers, then again reveals its face. If the moon's position is observed against the background of stars, it can be seen to return to the same spot in the sky every 27⅓ days.

The night sky. If you aimed a camera at the northern sky on a clear evening, and left the shutter open for an hour or so, the result would be a photograph like the one shown in Figure 1–4. The circular streaks or trails left by most stars appear as if they were carried on a rotating ceiling over the entire earth. One star in the northern sky, however, remains nearly stationary while the others circle around it. This star is called *Polaris,* or the North Star. It is located almost directly above the earth's North Pole. This point in the sky marked by Polaris is the *Celestial North Pole.* "Celestial" refers to any object or position in the sky. To an observer in the northern half of the earth, all stars except Polaris seem to move in circles around the Celestial North Pole.

As the stars move across the sky, their positions relative to each other do not appear to change. Their unchanging patterns in the sky have been observed and recorded from the earliest times. Although seeming to be closely arranged in a pattern, the stars are not all the same distance from the earth, and are very far from each other. Some of these star patterns, called *constellations,* were given names suggested by their arrangements.

FIG. 1–4. Star tracks made by aiming a camera toward the Celestial North Pole and leaving the shutter open. Can you find Polaris? (David Healy)

FIG. 1–5. As the earth moves in its orbit around the sun, the part of the sky visible from the dark side of the earth changes from one part of the year to another. Thus the pattern of stars seen in summer is different from that seen in winter. (Roland J. Cormier)

For example, certain constellations have names of real or imaginary animals, such as *Ursa Major*, the great bear; *Leo*, the lion; *Scorpius*, the scorpion; and *Draco*, the dragon. Other constellations are named for gods or legendary heroes, such as *Hercules* and *Orion*. Most constellations do not look much like the figures for which they are named.

Modern astronomers use the constellations to describe certain regions of the sky. In the same way as we might locate a volcano as being in the state of Washington, an astronomer might locate a star as being in the constellation Scorpius. That star as seen from the earth appears as part of the pattern forming the constellation Scorpius. The entire sky can be covered by describing 88 constellations. Figure 1–6A, B, C shows how one constellation is described.

Stars differ greatly in brightness. Within a particular constellation, the stars are labeled with the letters of the Greek alphabet — alpha (α), beta (β), gamma (γ), and so on, according to their brightness. See Figure 1–6C. Thus the brightest star in Scorpius is called Alpha (α) Scorpii. Many of the brightest stars also have names that are not connected with the constellation in which they are found. For example, Alpha (α) Scorpii is also called *Antares*. The five brightest stars in the northern sky, in order of decreasing brightness, are called *Sirius*, *Vega*, *Capella*, *Arcturus*, and *Rigel*.

The objects that appear to be the brightest "stars" seen in the night sky are not stars at all. They are the planets: each following its own path among the fixed background of stars. The word "planet" means "wanderer." Some of the planets seem to follow odd, looping paths that carry them forward and backward among the stars. Other planets never seem to move high in the sky, but rise and set close to the horizon. The planets, sun, and moon all follow paths within a narrow belt across the sky. There are twelve constellations that appear within this same narrow band in the sky. They are often called the "Signs of the Zodiac." Some ancient people believed that the positions of the sun, moon, and planets among the constellations of the Zodiac could be used to help predict the future. The exact positions of the heavenly bodies at the time a person was born were supposed to control that person's life. This belief is called astrology and is still accepted by some people today. Although astrology is not scientific, the observations on which it was based did help to advance the science of astronomy.

FIG. 1–6A. The brightest stars in this picture make up the constellation called Scorpius, the scorpion.

FIG. 1–6B. A drawing of Scorpius from an old star chart.

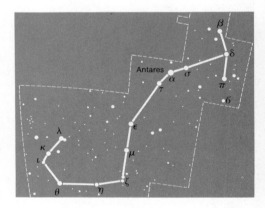

FIG. 1–6C. Modern astronomers show Scorpius on star charts with definite boundaries.

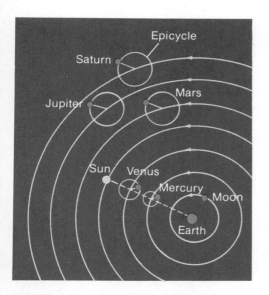

FIG. 1–7. Ptolemy's model showed the various celestial objects orbiting the earth in what he called "epicycles." Why wouldn't the sun have an epicycle in this model?

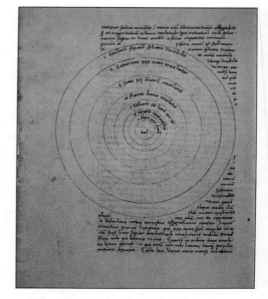

FIG. 1–8. This book published by Copernicus in 1543 began a scientific revolution. (The Granger Collection)

FIG. 1–9. Johannes Kepler (1571–1630). He used the measurements of Tycho Brahe, shown in the background painting, to determine the laws of planetary motion. (Jean Leon Huens)

THE SUN IS MADE TO STAND STILL

The ancient universe. Ancient civilizations had wisemen and priests who made or invented models of the universe. When faced with a problem needing some explanation, people often invent models to help their thinking. The word "model" used this way means a representation of an idea. A physical or mental model is used to describe some observable part of nature. Models used by ancient people to explain the universe took many forms. Most of them had one thing in common: The earth was thought to be at the center of the universe. The ancient Greeks, for example, thought of the heavenly bodies as being fixed on a series of transparent globes one inside the other. On their separate spheres, the sun, moon, planets, and stars were thought to be carried around the earth. The most complete model of an earth-centered universe was developed by the astronomer Ptolemy, who lived around 120 A.D. See Figure 1–7. Using the system constructed by Ptolemy, it was possible to predict accurately the positions of the various heavenly bodies. This earth-centered model became firmly established in the western world as the correct explanation of the universe. It was generally accepted for the next two thousand years.

A new model. Ptolemy's earth-centered model of the universe was not seriously challenged until the 1500s. A Polish mathematician, Nicolaus Copernicus (1473–1543 A.D.), began a revolution in which the idea that the earth was the center of the universe was rejected.

Copernicus did not accept the complicated system of moving transparent spheres used by Ptolemy. He believed that the true explanation was much simpler. After years of study, Copernicus wrote a book stating his belief that the earth was one of a family of planets all revolving around the sun. This was such a revolutionary idea that Copernicus hesitated to publish his work until 1543, the year of his death. Copernicus was correct when he predicted that his theory would create bitter argument. The old idea of an earth-centered universe did not die easily, but the evidence that came from observations of the sky was on the side of Copernicus. Within a hundred years his view of the earth as a planet circling the sun became established in scientific thought.

Copernicus laid the foundation for the model we now call the *solar system.* The sun and a large number of objects of all sizes orbiting in regular paths around the sun make up the solar system. Copernicus did not describe the exact way in which these parts move around the sun. This was done by Johannes Kepler (1571–1630), a German mathematician and astronomer. Kepler used observations of the motions of planets to set forth three main rules that describe the way objects in the solar system orbit around the sun. These rules are called Kepler's Laws of Planetary Motion. *Scientific laws* are rules that have been shown to correctly describe some part of the natural world. Kepler's laws can be stated as follows:

1. The orbits of the planets are not perfect circles, but have the shape of an *ellipse.* An ellipse is the shape formed when a cone is cut through at a certain angle, as shown in Figure 1–10. You can draw an ellipse with a loop of string around two pins, as shown in Figure 1–11. Each pin will be one of the focus points of the ellipse drawn. If the ellipse represents a planetary orbit, the sun would be located at one of the focal points. This means that a planet's distance from the sun constantly changes as it moves toward, then away from, the sun at one focus of its elliptical orbit. When a planet is closest to the sun it is said to be at *perihelion.* At its farthest distance it is at *aphelion.* Usually a planet's distance from the sun is given as the average of the perihelion and aphelion distances. For example, the earth's aphelion distance is

FIG. 1–10. An ellipse is the shape produced when a cone is sliced through at an angle.

Measure
How far north or south does the sunset-point move on the horizon in one week? You might want to set up a Polaroid camera to photograph the sun each day from the same position, to obtain a permanent record.

FIG. 1–11. An ellipse can be drawn in this way. Each pin is located at a focus of the ellipse drawn.

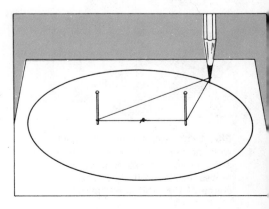

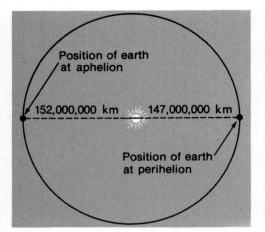

FIG. 1–12. This is a drawing of the earth's elliptical orbit made to the correct scale.

Explain
Look up information on Bode's Law and explain how it was used to predict the presence of a planet between Mars and Jupiter.

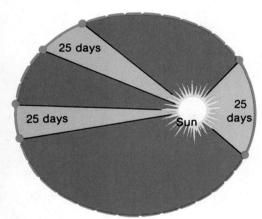

FIG. 1–13. A line joining a planet with the sun will sweep out equal areas in equal times as the planet moves in its orbit. The elliptical shape of the orbit is exaggerated.

152,000,000 km (94,500,000 mi). Its perihelion distance is 147,000,000 km (91,300,000 mi). The average distance between the earth and sun is 150,000,000 km (92,900,000 mi). These figures indicate that the earth's orbit is only slightly elliptical. If the orbit is drawn to the correct scale, as in Figure 1–12, it is seen to be very close to a perfect circle.

2. Kepler's second law describes the fact that the planets do not move in perfect circles around the sun. In a perfectly circular orbit, a planet would always travel at the same speed. When moving in an elliptical orbit, its speed will be greater at perihelion than at aphelion. Kepler expressed this relationship between a planet's position in its orbit and its speed in this way: When a planet is closest to the sun, a line joining it with the sun sweeps out a short and wide triangle in a certain period of time. At aphelion, the planet moves more slowly, making the triangle swept out at the same time longer and thinner. But the areas included in both triangles are the same. Thus Kepler's second law says that a line joining a planet with the sun will sweep out equal areas in equal times. See Figure 1–13.

3. Kepler's third law relates the distance between a planet and the sun to the time it takes the planet to make one revolution in its orbit. If the time of a planet's revolution is given in years and the distance between the sun and earth is given the value of 1, then Kepler's third law can be written as

$$(\text{distance from sun})^3 = (\text{time of revolution})^2$$

For example, suppose that a planet makes one revolution every eight years. Then

$$(\text{distance})^3 = (8)^2 = (8 \times 8)$$
$$(\text{distance})^3 = 64$$
$$\text{distance} = \sqrt[3]{64} = 4$$

Using the earth-sun distance as 1, this planet would be four times farther from the sun than the earth is.

The discovery of Kepler's laws was a great step forward in understanding the solar system. However, these rules did not explain why the regular and predictable motions of the planets took place. The model that pictured the earth and other planets moving around the sun was not complete. Kepler's laws represented clues to the mystery of planetary motion. A deeper rule of nature had as yet to be discovered.

FIG. 1–14. Isaac Newton is shown surrounded by some of his instruments. The symbols in the background represent his discoveries that laid the foundation for the modern view of the universe. (Jean Leon Huens)

Gravity. Kepler's laws showed that planetary motion could be explained by some as yet undiscovered principle. Kepler believed that the sun must have some kind of control over the way the planets move. But he could not discover the secret. It remained for a man born twelve years after Kepler's death to discover the mysterious force by which the sun controls the solar system. That man was Isaac Newton (1642–1727).

Newton provided the answer to the question that had puzzled all scientists since Copernicus. What provides the push or pull to keep the planets moving around the sun in endless cycles? Newton's answer was that the planets do not need a push or pull to keep moving. He demonstrated that any moving body will change its motion only if some outside force acts upon it. Our experience indicates that moving objects eventually stop due to friction. When a moving object is in contact with another surface, friction acts to take away its energy of motion. However, a moving planet meets almost no friction in space. Thus a planet moving through space does not need a push or pull to keep it going at almost constant speed.

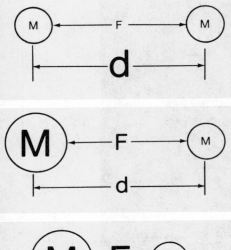

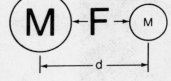

FIG. 1–15. The gravitational force between two objects increases in direct proportion to their masses. Changing the distance between the objects has an even greater effect.

Discover
Consult physics reference materials to find out how Lord Cavendish used a laboratory balance to determine the mass of the earth.

Newton also showed that the planets would move in a straight path unless an outside force existed to cause them to curve. Newton identified and described this force as gravity. Without the force of gravity acting upon them, once in motion the planets would follow straight paths off into space.

Newton's Law of Gravitation states that a force exists that pulls every particle of matter toward every other particle. The size of this attracting force is the result of two factors. The gravitational force becomes greater if the amount of matter, or mass, in the two bodies is greater. The other factor is the distance between the two objects. Newton's Law of Gravitation may be expressed in mathematical form as:

Gravitational force (F) is proportional to

$$\frac{\text{Mass of one object } (M_1) \times \text{Mass of other object } (M_2)}{\text{Distance separating objects}^2 \ (d^2)}$$

This means that two automobiles a certain distance apart attract each other with greater force than two baseballs the same distance apart. Changing the distance separating the objects has a large effect on the gravity force between them. Doubling the distance reduces the gravity force by four times. Halving the distance increases the gravity force by four times. The masses of automobiles and baseballs are, of course, very small when compared with the mass of the earth. Thus we are much more aware of the large gravity force pulling these objects toward the earth than the small gravity force acting between them.

In the solar system, the huge mass of the sun provides the gravity force that controls the movements of the planets. The planets also attract the sun, but their masses are so comparatively small that their gravity has little influence on the massive sun. However, the attraction of the planets for each other does have a small effect on how they move in their orbits.

Gravity and the earth. The gravity of a large round object such as the earth is best understood if the planet is imagined to be made of layers like an onion. The particles in each layer attract each other; however, the outer layers are largest and contain the greatest number of particles. Thus the outer shells are pulled toward the center with greater force than the inner layers. See Figure 1–15. On the earth's surface, every object acts as if it is a particle in the outer layer. Thus a force pulls all objects on the earth's surface toward the center of the earth. The effect of this force is called the *weight* of an object. According to Newton's Law of Gravitation, the earth's gravity force on an object increases as the object's mass increases. Therefore, an object with a large mass, such as an automobile, weighs more than an object with a small mass, such as a baseball.

The weight of an object with a certain mass is not the same everywhere on the earth's surface. For example, your weight is likely to be slightly greater at ground level than in a flying aircraft. Your weight is a little less in the aircraft because you are at a greater distance from the earth's center. Your weight at a point on the earth's equator would be about 0.3 percent less than your weight at the North Pole. This is the result of the rotation of the earth on its axis. At the equator you would be carried around at a speed of about 1609 km/sec (1000 mi/hr). The motion that causes a tendency for objects to fly off into space cancels some of the pull of gravity and makes your weight slightly less than at a pole. This effect also causes the earth to bulge out slightly near the equator.

Gravity force also differs slightly all over the earth's surface because of the composition of the materials beneath the surface. For example, a large mass of iron causes a slightly higher gravity force on the surface above it. Scientists can use these slight gravity differences as clues to the kind of rocks beneath the earth's surface.

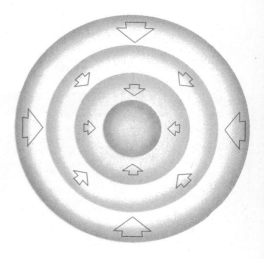

FIG. 1–16. If the earth is imagined to be made of layers, each layer would be attracted inward toward the lower layer.

MESSAGES FROM THE UNIVERSE

Energy from space. The only direct way to learn about the universe beyond the earth is to study the energy received from space. A form of energy able to travel through space is commonly called "light." Scientists use the term light to mean more than the form of energy that can be seen with the eyes. Light also includes all those forms of energy such as radio signals that are able to move through space in the form of waves. All such forms of energy are called *electromagnetic radiation*. The energy waves that make up electromagnetic radiation are alike in the very high speed with which they move through space. They travel at the speed of 300,000 km/sec (186,000 mi/sec). The various kinds of electromagnetic radiation do, however, differ in some important ways, such as the length of their waves. The distance from one wave crest to the next,

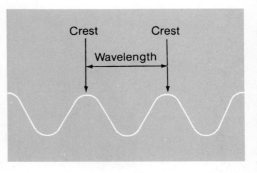

FIG. 1–17. Wavelength measures the distance between wave crests.

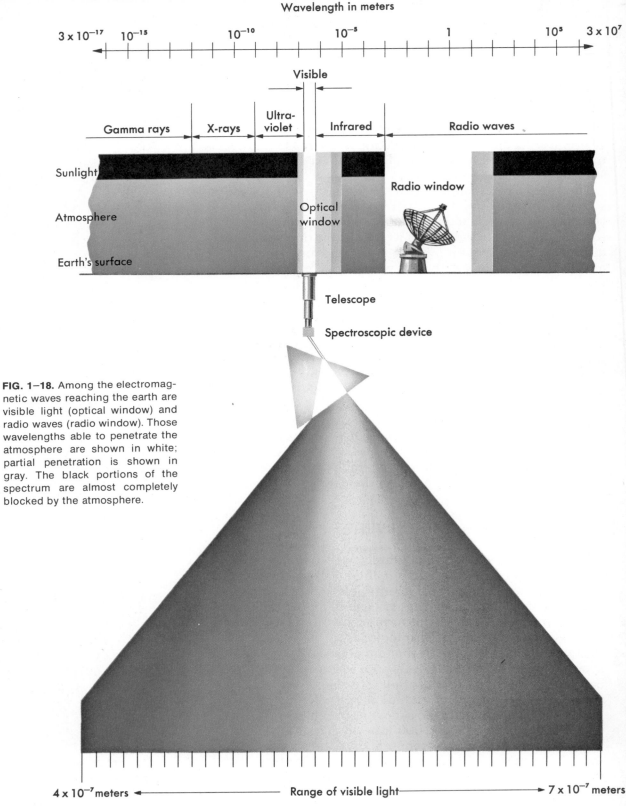

Wavelength in meters

3×10^{-17} 10^{-15} 10^{-10} 10^{-5} 1 10^{5} 3×10^{7}

Visible

Gamma rays | X-rays | Ultra-violet | Infrared | Radio waves

Sunlight

Atmosphere

Earth's surface

Optical window

Radio window

Telescope

Spectroscopic device

FIG. 1–18. Among the electromagnetic waves reaching the earth are visible light (optical window) and radio waves (radio window). Those wavelengths able to penetrate the atmosphere are shown in white; partial penetration is shown in gray. The black portions of the spectrum are almost completely blocked by the atmosphere.

4×10^{-7} meters ◄—— Range of visible light ——► 7×10^{-7} meters

FIG. 1–19. A refracting telescope such as the one shown at the left uses lenses to gather and focus light rays. It is difficult to make large lenses that do not produce a fuzzy image. (Lick Observatory)

Explore
Look in the library for information on advances made in radio telescopes. How do they work and what can be learned from using them?

FIG. 1–20. The reflecting telescope uses a curved mirror to focus light. The largest telescopes are reflectors. (Kitt Peak National Observatory)

as shown in Figure 1–17, is called the *wavelength* of a wave. Visible light, for example, is made up of radiations with wavelengths from about 0.0000004 meter, or 4×10^{-7} m, (about 1/70,000 in) to about 0.0000007 meter, or 7×10^{-7} m. Each separate wavelength within this band of radiations is seen by the eye as a particular color. Wavelengths shorter than those of visible light, down to about 3×10^{-17} m, are called ultraviolet, X-rays, and gamma rays. Longer wavelengths, up to 3×10^{7} m, are known as infrared and radio waves. The complete range of wavelengths makes up the *electromagnetic spectrum*. See Figure 1–18.

Windows in the sky. When you look into the night sky, your eyes use one of the "windows" available for observing the outside universe. You see the stars and other celestial bodies because there is a window in the earth's atmosphere that admits visible light. Stars actually give off radiations covering all parts of the electromagnetic spectrum, but many of the wavelengths, such as X-rays and gamma rays, are blocked by our atmosphere. Most of the wavelengths making up visible light are able to pass through to create what is called the *optical window*. Telescopes that concentrate visible light are referred to as *optical telescopes*.

FIG. 1–21. These antennas on the New Mexico desert are linked to a central computer that allows them to operate as a single huge radio telescope. (Steve Northup/Black Star)

Observe
On a clear, moonless night, locate the Milky Way in the sky. Describe its appearance.

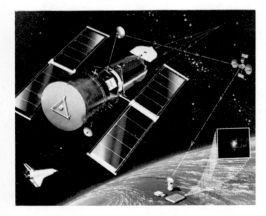

FIG. 1–22. A space telescope can be put in place and serviced by a space shuttle. (NASA)

There are two types of optical telescopes. In one kind, light waves pass through lenses and are bent, or refracted, to produce an image. Telescopes using lenses are called *refractors*. Perhaps you have looked through a simple refractor made of lenses fitted into a tube. The refracting telescopes used by astronomers are much larger. See Figure 1–19. The other kind of optical telescope uses a curved mirror to reflect light rays and produce an image. This type of telescope is called a *reflector*. See Figure 1–20. Very large telescopes are reflectors, since it is simpler to build large mirrors than very large lenses. Astronomers do not use large optical telescopes to magnify images of stars. A telescope with a large lens or mirror is able to gather more light to produce brighter images. Thus large telescopes are able to "see" faint objects at very great distances. Electronic light detectors may also be used with the telescopes to help detect faint objects.

Scientists are also able to use a second window in the atmosphere, the *radio window*. Some kinds of radio waves are able to reach the earth's surface. Radio telescopes can be used to investigate the objects producing the radio signals. Most radio telescopes have a dish-shaped antenna that gathers the faint radio waves. See Figure 1–21. The signal is then amplified and fed into a computer for analysis. A single antenna for a radio telescope is not able to accurately locate the direction from which radio signals come. Often radio telescopes are built with a number of antennas spread out over a large area to help solve this problem.

Space telescopes are carried in satellites that orbit the earth high above the atmosphere. They are able to detect all of the wavelengths that cannot reach the earth's surface. The orbiting telescopes are controlled from the earth by scientists using radio commands. Radiations detected by a telescope in space are changed into radio signals and sent down to the ground observatory. Most space telescopes are carried by small satellites. They are usually designed to detect only a certain kind of radiation, such as X-rays. A large telescope carried into orbit by a space shuttle is able to observe most kinds of radiation. It can detect objects located 50 times farther away than objects seen with the largest earth-based telescopes. See Figure 1–22.

THE SCIENTIFIC UNIVERSE

The Milky Way. Look into the night sky on any clear, dark night. You will see a cloudlike mass of lights crossing the middle of the sky like part of a great circle. Because of its milky appearance, this part of the sky is called the *Milky Way*. If you look at the Milky Way with the help of a magnifier such as binoculars, you can see that it is made up of millions of stars, along with dark areas of dust clouds and bright regions of glowing gas clouds. Scientists have determined that the sun is one member of this enormous community of stars. Gravitational attraction between the stars causes them to form a huge system called a *galaxy*. The galaxy to which the sun belongs is called the *Milky Way Galaxy*.

The Milky Way Galaxy, also called Our Galaxy, is so huge that its size is usually expressed in light-years. A *light-year* is the distance light can travel in one year, moving at its speed of 300,000 km/sec. At this speed light can travel around the earth seven times during one second. In one light-year, light can cover 6 trillion miles. The average diameter of the Milky Way Galaxy is 100,000 light-years. At its center it is about 2000 light-years in thickness. The sun is about 30,000 light-years from the center of the galaxy.

The Milky Way Galaxy slowly rotates like a wheel. The sun is carried around in one complete rotation in 200

activity

To draw an ellipse the same shape as the orbit of the earth around the sun, do the following.

Get a piece of string about 30 cm long. Make a loop 121 mm around. (Use a slip knot if possible.)
Place a piece of paper on cardboard and stick two pins near its center so that they are 2 mm apart (or as close as you can get them.)
Place the string loop around the pins and draw the ellipse by moving the pencil so that it always keeps the string tight, as shown in Fig. 1–11. This ellipse represents the path of the earth around the sun. Label it "Earth."

1. Without using the word "ellipse," how would you describe the path of the earth around the sun?

To draw an orbit of Mars to the same scale, move one of the pins until it is 17 mm from the other. The untouched pin represents the sun. Label its location "Sun." Make a loop 199 mm around. Place the loop around the pins and draw an ellipse as before.

2. How would you describe the orbit of Mars?

To draw the shape of the moon's orbit to another scale, get another piece of paper.
Draw an ellipse using two pins 7 mm apart and a loop 135 mm around. (One of the pins represents the earth.)

3. How would you describe the moon's orbit?

4. Astronomers had difficulty finding that planets move in elliptical paths. Explain.

FIG. 1–23. The probable appearance of the Milky Way Galaxy is shown from top (right) and side (above). The arrow indicates the location of the solar system.

FIG. 1–24. The Andromeda Galaxy has the same shape as our own galaxy. (David Healy)

million years. Within the disk of the galaxy, each star seems to have its own motion. Some stars close to the sun seem to be moving toward us. Others seem to be moving away. Altogether, the galaxy to which the earth belongs appears to be an enormous and complex community of stars with clouds of gas and dust all controlled by their mutual gravity.

Many years of study of the Milky Way Galaxy have shown that it is made up of about 100 million stars. Most of these stars are spread out fairly evenly in a pattern that resembles a dinner plate. Great clouds of glowing gases form curving or spiral arms. If you could see the Milky Way Galaxy from a point far away in space, these spiral gas clouds would make the galaxy look like a giant pinwheel. The sun seems to be located between two of the spiral arms, about two thirds of the way from the center. See Figure 1–23. From the earth, as you view the stars in either direction away from the Milky Way, the number you can see in a given area decreases. The dim light in the region of the Milky Way is not spread evenly throughout the sky. This suggests that most of the visible stars lie along a flat circular disk. You see many stars of the Milky Way when you look out through the universe in a direction along the plane of the flat disk. Looking at right angles to the plane of the Milky Way, you see fewer stars and a darker sky. Surrounding the entire disk of the Milky Way Galaxy is a globe-shaped cloud of very thin gas. Within this gas cloud are numerous clusters of stars along with individual stars all moving independently of the rest of the galaxy.

Other galaxies. Almost all of the objects in the night sky that you can see with the eye alone belong to our galaxy. Telescopes, however, reveal that the sky is filled with other galaxies. Two of these at a distance of about 150,000 light-years are neighbors of our own galaxy. A total of about twenty other galaxies of various sizes are found within 3 million light-years of our own. These are called the Local Group. Only one of the Local Group resembles Our Galaxy in size and shape. It is the spiral Andromeda Galaxy, located about 2 million light-years away. See Figure 1–24. Telescopes show that all galaxies occur in clusters such as the Local Group. Some clusters that contain as many as a thousand galaxies have been observed. Galaxies seem to form clusters as a result of their gravitational attraction for each other.

All of the vast number of observed galaxies have one of four general shapes. About 20 percent are spirals like the Milky Way and Andromeda galaxies. Another type also has a spiral shape, but with the spiral arms coming from a bar running through the center of the galaxy. See Figure 1–25. These are called barred spirals and make up about 10 percent of the total observed galaxies. A third type has no spiral shape and is without the clouds of dust and gas found in spirals. These galaxies are called elliptical because they usually appear pumpkin-shaped. See Figure 1–26. Sixty percent of all observed galaxies are ellipticals. They include both the largest and smallest galaxies known to exist. A fourth type, making up 10 percent of the total, has no particular shape. See Figure 1–27. They are called irregular galaxies. The distorted shape of irregular galaxies is believed to be the result of gigantic explosions at their centers that occurred when they collided with neighboring galaxies or as an accident of birth.

Earth and the universe. Through scientific discoveries, it is now known that the earth is only a very small part of one of the billions of galaxies in a universe whose size is unknown. Natural forces such as gravity that control the earth also appear to operate throughout the entire universe. The earth, its companions in the solar system, and the sun itself are the laboratories in which we search for understanding of a puzzling universe.

FIG. 1–25. A barred spiral galaxy. (Yerkes Observatory)

FIG. 1–26. An elliptical galaxy. (Hansen Planetarium)

FIG. 1–27. An irregular galaxy. (Lick Observatory)

VOCABULARY REVIEW

Match the word or words in the column on the right with the correct phrase in the column on the left. *Do not write in this book.*

1. A pattern formed by a group of stars.
2. A point about which the stars in the northern half of earth's sky rotate.
3. The sun and a large number of objects orbiting it.
4. Rules of the natural world that have been shown to be correct.
5. Energy that travels through space as waves.
6. The shape of the path of a planet orbiting the sun.
7. Closest approach of a planet to the sun.
8. The complete range of wavelengths of electromagnetic radiation.
9. A measure of the earth's gravitational pull.
10. The point in a planet's path where it is farthest from the sun.

a. weight
b. scientific laws
c. Celestial North Pole
d. ellipse
e. solar system
f. aphelion
g. constellation
h. electromagnetic radiation
i. perihelion
j. electromagnetic spectrum
k. Milky Way Galaxy
l. Andromeda Galaxy

QUESTIONS

Group A

Select the best term to complete the following statements. *Do not write in this book.*

1. Evidence left by ancient cultures suggests that astronomy was first used to (a) accurately tell the change of seasons (b) prove that the sun was the center of the universe (c) tell how long a day lasts (d) predict when wars would occur.
2. If the moon's position is observed against the star background, it can be seen to return to the same spot in the sky every (a) 30 days (b) 31 days (c) $27\frac{1}{3}$ days (d) $28\frac{1}{2}$ days.
3. The star in the night sky that does not form circular streaks when photographed over a period of time is (a) Antares (b) Polaris (c) Sirius (d) Vega.
4. Which one of the following is not a constellation? (a) Hercules (b) Draco (c) Orion (d) Sirius.
5. The brightest star in the constellation Scorpius is (a) Alpha-Scorpii (b) Beta-Scorpii (c) Gamma-Scorpii (d) Vega.
6. The twelve constellations called the Signs of the Zodiac are located (a) in the Northern Sky (b) in the Southern Sky (c) in a narrow band through which the sun, moon, and planets pass (d) near the Celestial North Pole.
7. The Ptolemaic model of the universe (a) placed the sun, moon, planets, and stars on separate spheres (b) placed the earth at the center of the universe (c) was accepted by the western world for two thousand years (d) all these are true.

8. Copernicus made a change in the early Greek model of the universe by (a) describing exactly how the planets move about the sun (b) replacing the earth with the sun as the center of the universe (c) placing all the planets on one sphere (d) placing all the stars on one sphere.

9. The idea that is not a part of Kepler's Laws is that (a) the planets move in elliptical orbits (b) a planet's speed varies as it orbits the sun (c) the planets rotate on their axes at various rates (d) the squares of the periods of revolution are proportional to the cubes of their average distances from the sun.

10. Kepler's Laws accurately described the motions of planets, but were not complete in that they (a) did not explain why the planets move (b) were contrary to the accepted model of the solar system (c) required the earth to move in a manner different from the other planets (d) did not describe the motion of all members of the solar system about the sun.

11. According to Sir Isaac Newton, any moving body will (a) slow down and come to rest (b) be acted upon by friction (c) change its motion only if some outside influence acts on it (d) eventually lose energy.

12. According to Newton, a moving body will move in a curved path only if (a) it is a planet (b) a force acts continuously on it (c) it is slowing down (d) it is speeding up.

13. Gravity, a force studied by Newton, explains (a) why planets move through space without noticeable friction (b) why the sun is the center of the solar system (c) why planets move in orbits about the sun (d) why planets rotate on their axes.

14. The force of gravity between any two objects depends upon their masses and (a) their size (b) their speed (c) the material of which they are made (d) the distance between them.

15. The gravitational force that pulls all objects on the earth's surface toward the earth is called the object's (a) mass (b) weight (c) gravity (d) volume.

16. The electromagnetic radiation that has a wavelength longer than visible light is (a) X-rays (b) ultraviolet (c) infrared (d) gamma rays.

17. The electromagnetic radiation given off by our sun contains (a) visible light only (b) mostly X-rays (c) mostly ultraviolet (d) a mixture of radiation from X-rays to radio waves.

18. The two windows to the universe available to us from earth are (a) X-rays and radio waves (b) ultraviolet and visible light (c) visible light and radio waves (d) infrared and visible light.

19. The largest optical telescopes collect light with (a) mirrors (b) lenses (c) antennas (d) amplifiers.

20. A light-year is (a) 300,000 km (b) seven times around the earth (c) 365 days (d) the distance light travels in one year.

21. Which one of the following is not a characteristic of Our Galaxy? It (a) is composed of about 100 million stars (b) contains a large amount of matter between the stars (c) is about 100,000 light-years across and about one-fifth as thick (d) has our sun located near the center of the galaxy.

22. The galaxy of which our sun is a member is thought to be similar in shape to the galaxy named (a) Andromeda (b) Polaris (c) Cygnus (d) Perseus.

23. Our Galaxy belongs to a cluster of galaxies called the Local Group, which con-sists of Our Galaxy and (a) three others (b) eleven others (c) seventeen others (d) twenty others.

24. The type of galaxy to which our own galaxy belongs is called a(n) (a) spiral galaxy (b) barred galaxy (c) elliptical galaxy (d) irregular galaxy.

25. The most common type of galaxy is the (a) spiral (b) barred (c) elliptical (d) irreg-ular.

26. Almost all the objects in the night sky that you can see with the eye alone belong to our (a) solar system (b) galaxy (c) Local Group (d) universe.

Group B

1. Describe the path of the sun as viewed from earth in the Northern Hemisphere on a typical winter day. How does the sun's path differ from this on a typical summer day?

2. Name four constellations and tell what they are supposed to look like.

3. Star patterns in the night sky called constellations were recognized and named by people living in ancient times. What value do they have today?

4. Name the five brightest stars in the northern sky in order of decreasing bright-ness.

5. How does the motion of the planets differ from that of stars?

6. What is meant by the term "astrology?"

7. Why was the Ptolemaic model of the universe so widely accepted for two thou-sand years?

8. In what way did Copernicus lay the foundation for the "solar system" model as we know it today?

9. State briefly, in your own words, Kepler's Laws of Planetary Motion.

10. Why are Kepler's laws considered scientific laws while Copernicus' description is termed a scientific model?

11. Write out Newton's Law of Gravity. Use it to explain what happens to the gravi-tational force on an object when it is moved twice as far from the center of the earth.

12. Explain what is meant by the "optical window."

13. Use Figure 1–18 to list the six kinds of electromagnetic radiations that make up the electromagnetic spectrum of the sun.

14. What advantage does a large telescope carried into orbit by a space shuttle have over an earth-based telescope?

15. Convert the average diameter and thickness of the Milky Way Galaxy given in your text into miles. Write the distances by name and number.

16. Describe the make-up of the Milky Way Galaxy.

17. Describe the four general shapes of the galaxies.

2
the sun

At some time you have probably enjoyed the warmth and glow of a cheerful fireplace or campfire. Making use of the heat and light energy from a fire is an experience that you share with your ancestors, going back to the beginning of civilization. Fire has always been a source of energy needed for human survival. When something burns, like a piece of wood, it releases the energy stored within. That stored energy originally came from the sun.

Throughout most of human history, the sun itself was believed to obtain its energy from fire. It was a natural conclusion to people whose only source of energy came from burning some kind of fuel. Only during the past century has the true source of the sun's energy been revealed. Scientists now know that the sun is a star that produces energy in the same way as billions of other stars. Because the sun is so close, details can be observed on the sun that cannot be seen in remote stars. This chapter discusses the sun as a representative of all the stars that populate the universe.

THE SUN'S FIRE

Nuclear reactions. As discussed in Chapter 1, gravity draws all the layers of a round body like the earth toward the center. This action creates pressures at the earth's core that are millions of times greater than the pressure at the surface. The sun has a total mass greater than 300,000 times that of the earth. Pressure near the center of the sun is enormous. When atoms that make up matter are pressed together at such very high pressures, they are transformed. Under the pressure and temperature conditions normally found on earth, atoms consist of a small

objectives

☐ Identify and give an example of the kind of reaction that produces the sun's energy.

☐ Describe two ways solar energy may be used.

☐ List each part of the sun and describe its role.

☐ Compare and contrast sunspots, prominences, and solar flares.

☐ Explain a method for measuring the distance to a star.

☐ Show how spectroscopy is important in the study of stars.

☐ Classify stars according to temperature and brightness.

☐ Describe the life history of a typical star.

☐ Give two observations that support the Big Bang theory.

Investigate
Use magazine articles to find
out about the current status of
fusion reactions in solving our
energy problems

central nucleus surrounded by one or more electrons. The nuclei of these atoms remain intact even if other parts of the atom are changed. However, within the deep interior of the sun, electrons are stripped away from the atomic nuclei. The exposed nuclei can then be changed by *nuclear reactions*. One kind of nuclear reaction commonly taking place inside the sun causes atomic nuclei to join together. Any nuclear reaction in which two atomic nuclei are combined is called *nuclear fusion*.

Source of the sun's energy. The sun contains more atoms of hydrogen than any other kind of atom. Hydrogen is the simplest of all kinds of atoms. Most hydrogen atoms consist of a nucleus that is only a single *proton*. A proton is a positively charged atomic particle. Near the center of the sun almost all hydrogen atoms have lost their electrons, leaving only the proton nucleus. It becomes possible for a kind of nuclear fusion known as the proton-proton reaction to take place in the sun's deep interior.

The proton-proton reaction usually involves three steps. First, two protons collide and join together. One of these protons in the pair then changes into a *neutron*. A neutron is an atomic particle without an electrical charge. The second step occurs when another proton combines with the proton-neutron combination already formed. This change leaves a group made up of two protons and one neutron. In the third step, two of the groups each made up of two protons and one neutron undergo fusion. The resulting cluster throws off two protons, leaving a total of two protons and two neutrons. The three steps of the proton-proton reaction are shown in Figure 2–1.

The final product of the proton-proton reaction is a nucleus of another kind of atom called helium. The proton-proton reaction may also take place in different steps from those described, but the final product is always helium. Where there were four hydrogen nuclei in the sun's interior before, there is left one helium nucleus after the proton-proton reaction. The helium nucleus, however, has slightly less mass than the four original hydrogen nuclei. It weighs 0.7 percent less than the combined weight of four protons. According to a scientific rule first developed by Albert Einstein, when mass disappears during a nuclear reaction energy must appear in its place. This rule is often written as the equation $E = mc^2$, where E is energy, m is mass, and c is the speed of light. This equation shows that a small mass can be

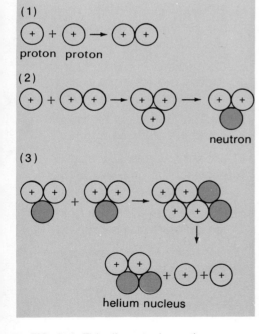

FIG. 2–1. This diagram shows the steps in the proton-proton nuclear reaction.

changed into a large amount of energy. Thus the proton-proton reaction in the sun is able to produce much of the huge energy output because mass is converted into energy as hydrogen is changed into helium. Each second the sun changes more than 600 million tons of hydrogen nuclei into helium nuclei. The mass lost during this change appears as most of the energy released by the sun each second.

Other kinds of nuclear fusion reactions also help supply the sun's energy. For example, the nuclei of carbon, nitrogen, and oxygen can be involved in a complicated reaction that also produces helium. All fusion reactions believed to take place in the sun produce helium as a final product as well as a large amount of energy.

On the earth it is very difficult to produce nuclear fusion reactions because these reactions need very high pressures and temperatures. The kind of nuclear weapons called "hydrogen bombs" obtain their energy from nuclear fusion. No way now exists to control a reaction so that a continuous energy supply is produced by combining atomic nuclei. At some time in the future methods may be developed that will provide a continuous flow of energy from nuclear fusion.

Solar energy and the earth. Every 15 minutes the earth receives enough energy from the sun to meet the energy needs of the entire world for one year. Two weeks of sunshine falling on only one square meter of the earth's surface is equal to the energy in 4 liters (1 gal) of gasoline or 6.5 kg (about 14 lb) of coal. Our energy problems would be solved if we could capture and use only a part of the solar energy received by the earth.

Some difficult problems must be solved before solar energy can become an important part of our energy supply. For example, the energy received from the sun is spread over the entire lighted half of the earth's surface. To be used, it must be collected and converted into some concentrated form of energy such as heat or electricity.

There are two general ways of converting solar energy into usable forms. One method is called *passive* because it requires no mechanical systems or special equipment. An example of a passive solar energy design would be a house with windows placed to catch the warming rays of the winter sun. See Figure 2–2. The other general method of capturing solar energy is called *active*. Active systems require some kind of device to concentrate solar energy

Investigate
Study your school building and describe any ways solar energy is used as part of the energy supply. Suggest how the building could be changed to make better use of solar energy.

FIG. 2–2. This building uses a passive solar design, which allows direct rays of the sun to warm the interior. (Mark Antman)

FIG. 2–3. This specially built aircraft is powered by electricity produced by solar cells. (Randa Bishops/Contact Press Images) •

FIG. 2–4. A mirror at the top of this solar telescope reflects a beam of sunlight down the shaft to another mirror that reflects the image back up to the observation room. (Hansen Planetarium)

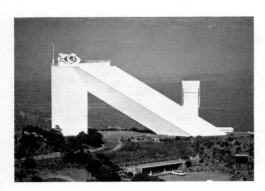

and convert it into another form. For example, when sunlight falls on the surface of an electronic device known as a solar cell, electricity is produced. See Figure 2–3. At the present time the high cost of solar cells makes this a very expensive way to generate electricity. Most other kinds of active solar energy systems also are very costly. It will take many years of research and development for scientists and engineers to learn how to make active solar energy systems that can produce inexpensive energy. The first steps in the long-range goal of using solar energy to meet a large share of our energy needs should involve the use of passive solar designs in homes and other buildings. The earth's total energy resources, including solar energy, are discussed in Chapter 24.

ANATOMY OF THE SUN

The body of the sun. The sun appears in the sky as a dazzling, brilliant ball without any visible features. NEVER LOOK DIRECTLY AT THE SUN. Even a brief glance directly at the sun may damage your eyes. Looking directly at the sun with optical aids such as binoculars will instantly and permanently damage your sight. Scientific investigation of the sun requires the use of special instruments combined with knowledge of the way atoms are changed by the pressure and heat of the sun's interior. The model of the sun that comes from the work of scientists shows it to be an organized system in which matter is changed into energy according to Einstein's equation, $E = mc^2$. The nuclear reactions that power the sun take place in its very hot and dense *core*. Within the sun's core the temperature is around 15,000,000°C. Matter in the core is pressed together so tightly that it is about ten times denser than iron.

Energy flowing out from the core is absorbed by a thick surrounding layer called the *zone of radiation*. The atoms in this region absorb energy and then pass it on to neighboring atoms. Thus the energy produced in the core is slowly transferred toward the outside.

The solar atmosphere. Gases in the upper parts of the zone of radiation boil upward and give off energy in the form of electromagnetic waves. Near the surface the gases become cooler. The cooled gases become denser and sink back toward the interior to repeat the process.

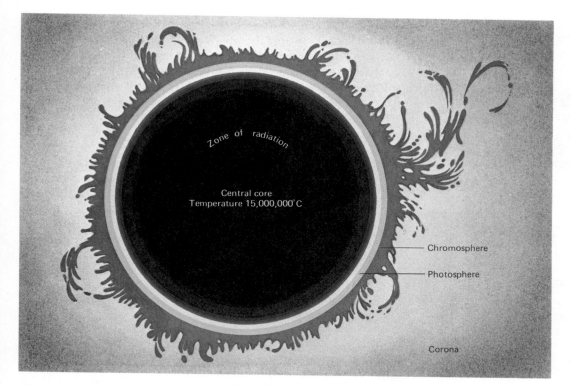

Zone of radiation

Central core
Temperature 15,000,000°C

Chromosphere

Photosphere

Corona

The thin outer layer of gases bubbling up from below make up the sun's *photosphere* (light sphere). This layer appears brilliant to us because much of the energy given off from the photosphere is in the form of visible light. The photosphere might be thought of as the surface of the sun made up of gases at a temperature of about 6000°C. It is a surface in constant motion.

Above the photosphere lies a thin layer of gases whose radiations include a weak red visible light. This is the *chromosphere* (color sphere). The gases of the chromosphere are also continuously moving up and down as they pick up energy from below and carry it outward. Narrow jets of hot gas lasting only a few minutes constantly shoot up in the chromosphere and then die away.

Finally the gases of the sun blend into space in the form of the *corona* (crown). The sun's corona is a huge, very thin, and hot cloud of gas extending far out into space until it finally disappears. Most of the atomic particles boiled off the sun's surface are blocked from escaping into space by the corona. But holes appear in the corona allowing electrically charged particles to stream out into space to make the *solar wind*. This invisible wind of

FIG. 2–5. Parts of the sun are shown in this diagram. The interior of the sun is dark because visible light is not produced until radiation from the interior reaches the photosphere.

FIG. 2–6. At the top is a telescopic photograph of the photosphere showing many bubbles of hot gases rising from the interior. Below is a drawing of how the sun's surface might look near some sunspots. (Observatoires du Pic du Midi et de Toulouse)

FIG. 2–7. A telescopic photograph of a group of sunspots. (NASA)

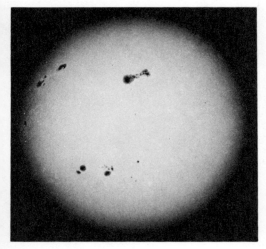

atomic particles shed by the sun blows throughout the solar system, reaching to the most distant planets. Both the chromosphere and corona are normally not seen on earth because the weak light they produce is lost in the photosphere's total brilliance. Only during the brief minutes of total eclipse are the chromosphere and corona revealed without use of special instruments.

The active sun. The atoms and electrically charged atomic particles that make up the gases of the sun are never still. Energy pouring out from the sun's interior sets every part into violent motion. See Figure 2–6. In addition, the sun rotates. Observations have shown that the sun turns on its axis about once every twenty-seven days. Because the sun is a ball of hot gas, it does not have the same speed of rotation at all points. Different regions of the sun rotate at different speeds. Places closest to the sun's equator take only 25⅓ days to make one rotation. A point halfway between the equator and the poles takes 28 days, while points near the poles take about 33 days.

Scientists believe that any body of very hot gas experiencing both the up-and-down boiling motions and the turning motion of rotation will produce magnetic fields. Such magnetic fields are detected in the sun. Much of the activity observed in the sun seems to be controlled by these magnetic fields. For example, dark areas called *sunspots* are often seen within the photosphere. See Figure 2–7. Usually sunspots first appear in groups about midway between the equator and poles, and then slowly disappear. During a cycle of years, new spots appear in large numbers closer to the equator; then for several years the sun's face is almost free of spots. Each cycle of the growth and disappearance of sunspots averages about eleven years.

Every sunspot marks the location of a powerful magnetic field. This prevents the usual upward flow of hot gas from below. As a result, a cooler, darker region is seen in the photosphere. Large sunspots are several times the size of the earth.

The same magnetic fields responsible for sunspots also create other disturbances in the solar atmosphere. Great clouds of glowing gases, called *prominences*, form huge arches reaching high above the surface. See Figure 2–8. Each prominence follows curved lines of magnetic force from one sunspot area to another. Some prominences are seen for several months. Others die out quickly.

The most violent of all solar storms occur near sunspots. These sudden eruptions are called *solar flares*. A flare sprays a fountain of electrically charged atomic particles upward at very high speed. Many of these electrified particles are flung out so violently that they escape into space. This creates sudden increases in the strength of the solar wind. When violent bursts of solar wind reach the earth, a magnetic storm is generated in the earth's atmosphere.

The most spectacular effect of a magnetic storm is an increase in the occurrence of *auroras* (the northern and southern lights). When electrically charged particles approach the earth, they are guided toward the poles by the earth's magnetic field. The particles strike the gas molecules in the upper atmosphere, and green, red, and occasionally blue or violet light is produced. This light is seen in the night sky as glowing curtains or sheets that shift and change. See Figure 2–9. Auroras are most commonly seen near the poles. However, large magnetic storms cause displays that can be seen over greater areas. Figure 2–10 shows the average frequencies per year for auroras seen in the Northern Hemisphere. Very severe magnetic storms can interrupt some radio communications, disrupt telephone service, or even cause electrical power failures.

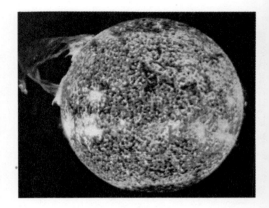

FIG. 2–8. A solar prominence is usually seen as a huge arch or part of an arch above the sun's surface. (NASA)

FIG. 2–9. An aurora is usually seen as a glowing curtain of colored light. Sometimes auroras fill the entire sky with light. (Dr. E. R. Degginger)

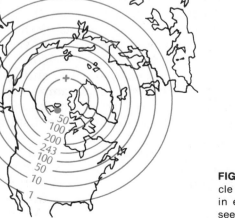

+ Geographic pole
● Magnetic pole

50
100
200
243
100
50
10
7

FIG. 2–10. The number of each circle is the average number of days in each year that auroras can be seen if you are north of the circle.

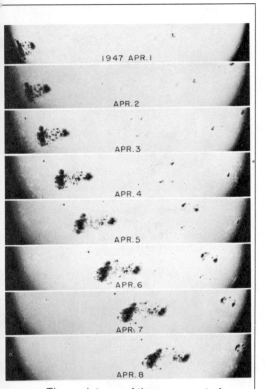

1947 APR.1

APR.2

APR.3

APR.4

APR.5

APR.6

APR.7

APR.8

These pictures of the movement of sunspots were taken at the same hour on successive days during the sun's rotation. (Mt. Wilson and Palomar Observatory)

activity

Use the sequence of photos above to answer the following questions.

1. What is your best estimate of the fraction of the distance around the sun the largest sunspot moves in 7 days?

2. At this rate of movement, how many days would it take the largest sunspot to move completely around the sun?

3. On the basis of the sunspots' motion, does this place the sunspots near the equator, near the pole, or halfway between? Explain.

OTHER SUNS

How far away are stars? On a clear dark night you may be able to see about 2000 stars with your eyes alone. Some stars appear bright, while others appear dim. Suppose that you needed to compare the brightness of all these stars visible to your naked eye. You would need some way to compare the brightness of each star with all the others. Astronomers use instruments attached to telescopes to measure a star's brightness as seen from the earth. The measurement is then given a number called the star's *apparent magnitude.* The brighter stars are assigned small numbers, while dim stars are assigned larger numbers. For example, the very faintest star you can see with your eyes alone is given an apparent magnitude of 6. A star that is about 2.5 times brighter will have an apparent magnitude of 5. This is called a fifth magnitude star. A fourth magnitude star will then be about 2.5 times brighter, and so on. A first magnitude star will appear among the brightest seen in the sky. To describe the very brightest stars and other celestial objects such as planets, moon, and sun, it is necessary to use negative apparent magnitudes. For example, the brightest star, Sirius, has an apparent magnitude of -1.4. That of the sun is -26.5. The biggest telescopes can detect stars with an apparent magnitude of about 23. Such a star is about 100 million times fainter than the average star seen with the naked eye.

The apparent magnitudes of stars can be deceiving. A star that appears bright from the earth can actually be a faint star close to us. On the other hand, a dim star might really be very bright but very far away. Before the true brightness of a star can be known, its distance from the earth must be found. One way to measure the distance to a far object like a star is to use parallax. You see examples of parallax every day. When you look at objects from a moving car, those closeby seem to shift positions against the distant background. In the same way, when viewed from the orbiting earth, some nearby stars seem to change positions when seen against the distant stars. See Figure 2–11.

By measuring the parallax angle observed for a star and applying some principles of geometry, the star's distance can be calculated. However, only the stars within about 300 light-years of the earth have parallax angles large enough to be accurately measured. More distant

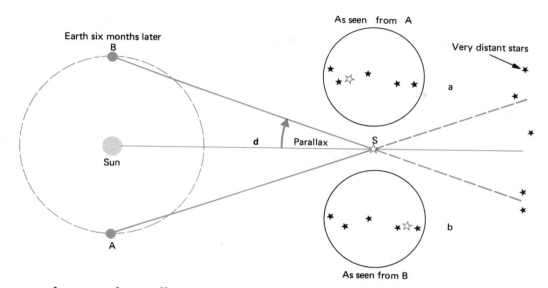

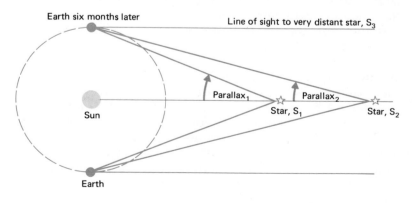

stars have such small apparent movements that their parallax angles cannot be measured. See Figure 2–12.

If the distance to a star can be found, then its true brightness can be easily determined. For example, Polaris is seen as a dim star in the sky. But the distance to Polaris has been found to be 680 light-years. Thus the star must actually be very bright in order to be seen at all. Using its apparent magnitude and known distance, astronomers have calculated that Polaris actually shines with a brightness 10,000 times greater than that of the sun.

When astronomers want to express the true brightness of a star, they refer to its *absolute magnitude*. Finding the absolute magnitude of a star involves calculating how bright it would appear if seen from a certain distance.

FIG. 2–11. The parallax angle of a star measures its apparent change of position when seen from different points in the earth's orbit. Can you see a right triangle formed by the star (S), the sun, and the earth? What is the length of the base of this triangle along a line joining the sun with either A or B? Find out how to use this information to calculate the distance to the star.

Measure
Use a photographic light meter to measure how the apparent brightness of a light bulb varies with distance. Determine an algebraic relationship.

FIG. 2–12. This diagram shows how the parallax angle of a star grows smaller as the star's distance from the earth increases.

Table 2−1 Scale of Magnitudes

Object	Apparent Magnitude
Sun	−26.5
Full Moon	−12.5
Venus (at brightest)	−4
Jupiter; Mars (at brightest)	−2
Sirius	−1.4
Aldebaran; Altair	1.0
Naked-eye limit	6.5
Binocular limit	10
15 cm (6 in) telescope limit	13
5 m (200 in) visual limit	20
5 m (200 in) photographic limit	23.5

Increasing Brightness

Observe

Hold a finger at arm's length. Close one eye, then open that eye and close the other eye. As you switch eyes, notice the apparent motion of your finger against the other things you can see. This apparent motion is called parallax and can be used to measure distances.

That distance has been arbitrarily chosen to be about 32.6 light-years. If the sun were about 32.6 light-years away, it would be about a fifth magnitude star. Thus the absolute magnitude of the sun is about 5. Most stars have absolute magnitudes falling between −5 and +15. This means that the sun falls in about the middle of the range of true star brightness. The sun's absolute magnitude is a clue that it is an ordinary star.

If a star is too far away for its parallax to be measured, knowing its absolute magnitude provides another way to determine its distance. The star's apparent magnitude is measured with an instrument attached to a telescope. Then how far away the star must be to have the brightness observed is determined. The same principle is used when you judge the distance of an approaching car at night by the brightness of its headlights. In order to use this method to measure the distance to a star, the absolute magnitude of the star must first be determined. With one type of star, called a *cepheid variable* (*sef*-i-id), the true brightness is easily found. Cepheid variables are examples of a type of star that suddenly becomes brighter and then gradually dimmer in a regular pattern. This is caused by a rhythmic swelling and shrinking of the star. Some cepheids complete a cycle of bright to dim in one day, while others have a cycle of more than a month. But for all cepheids there is a direct relationship between the time taken to complete one cycle and their absolute magnitude. Astronomers have been able to find that relationship. Measurement of the time taken for a cepheid to complete its cycle of bright to dim allows its true brightness to be determined. Thus astronomers are able to use cepheids as celestial "meter sticks." For example, if cepheids can be seen in a distant galaxy, the distance to that galaxy can be measured. This is the reason that astronomers consider cepheids, along with other kinds of variable stars used in the same way, to be among the most important in the sky.

Reading starlight. The star nearest our solar system, other than the sun, is over four light-years away. The only contact we have with these distant bodies is through the small trickle of electromagnetic waves traveling through space. All that we can learn about stars' composition or motions must come from analysis of that electromagnetic radiation. One of the most effective ways of performing such an analysis is through a method called *spectroscopy.*

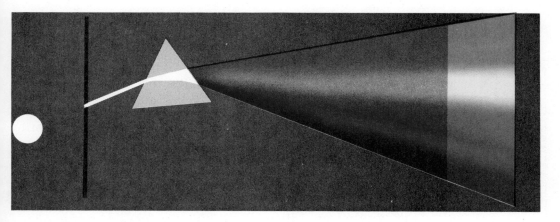

FIG. 2–13A. A continuous spectrum is produced by glowing solids, liquids, and gases under high pressure.

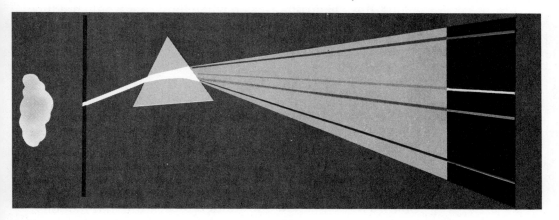

FIG. 2–13B. A bright-line spectrum is produced by glowing gases under low pressure. Each kind of atom in the gas produces a particular pattern of line.

FIG. 2–13C. A dark-line spectrum results when light passes through a cool gas.

A   B   C   D

activity

The bright-line spectrum shown in A is that of sodium chloride. The chlorine part of the compound does not have any lines in the visible part of its spectrum. Use this information to do the following scientific deduction.

1. Which of the spectra is that of sodium nitrate?

2. Which is of cupic nitrate?

3. Which is of cupric chloride?

The science of spectroscopy began when Isaac Newton found that visible light can be separated into a spectrum, or band of colors. In 1666, he passed a beam of sunlight through a triangular-shaped glass prism to produce an unbroken band of colors called a continuous spectrum. See Figure 2–13A. After Newton's pioneering work, other scientists discovered that three different types of spectra can be produced. A continuous spectrum like the one discovered by Newton is produced by glowing solids, liquids, or gases under high pressure. The white-hot piece of wire in an ordinary light bulb produces a continuous spectrum. A second type of spectrum, the bright-line spectrum, is produced by glowing gases under low pressure. See Figure 2–13B. The third kind of spectrum, the dark-line spectrum, is seen when light that would produce a continuous spectrum passes through a relatively cool gas. A pattern is produced in which dark lines separate the colors. See Figure 2–13C.

Along with the discoveries of the types of spectra, scientists also developed an understanding of their meaning. They were able to show that the patterns of spectral lines were the signatures of the atoms making up the chemical elements. Common elements found on the earth, such as hydrogen, helium, iron, and calcium, can be detected in the sun and other stars by their characteristic spectral lines.

The spectra can also show more than a star's composition. They can show the star's temperature and motions.

Astronomers separate starlight into different wavelengths by passing light from a telescope through a *spectrograph*. Each wavelength is recorded on photographic film or on magnetic tape for playback into a computer. Spectrographs are used with telescopes at ground level or in earth orbit to detect all forms of electromagnetic radiation, including radio, infrared, ultraviolet, X-ray, and gamma ray wavelengths. The various wavelengths recorded are compared with the spectral lines produced by chemicals in laboratory experiments. Astronomers also consult references to determine what substances produce the observed spectral lines. The relative brightness of the various colors in the spectra provide information about the temperature of the star's surface. A cool star will have brighter reds, while a hot star will show stronger blues.

When astronomers compare lines in the spectra from stars with those in references, they usually find a shift in the wavelengths from stars. This is caused by the *Doppler effect*. The Doppler effect causes waves from a moving source to crowd together or move apart as the source moves toward or away from an observer. A common example of the Doppler effect occurs with sound waves. The sound of a locomotive horn rises as the train approaches, and then falls as the train passes. The sound of the horn changes because sound waves are crowded closer together as the train approaches. As the train

FIG. 2–14. A Doppler shift is seen in the dark lines of a star's spectrum as the star moves toward or away from the earth. Below is the spectrum of the star Arcturus, comparing the position of its dark lines with reference spectra (above and below) as the star approaches (a) and recedes (b). (Mount Wilson and Palomar Observatories)

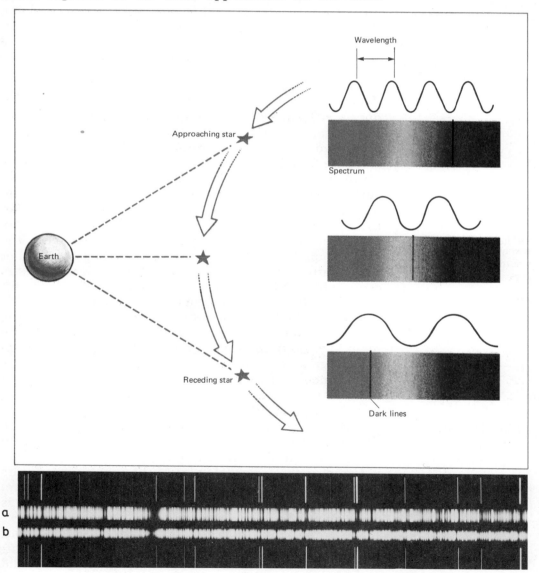

Experiment
Use a tape recorder to record
sounds that illustrate the Doppler
effect. A car passing while
blowing its horn and a fire
engine passing with siren going
are examples. How does the
pitch of a given sound change
with speed?

FIG. 2–15. The Doppler effect causes the sound waves from a moving locomotive horn to change in pitch as the train passes.

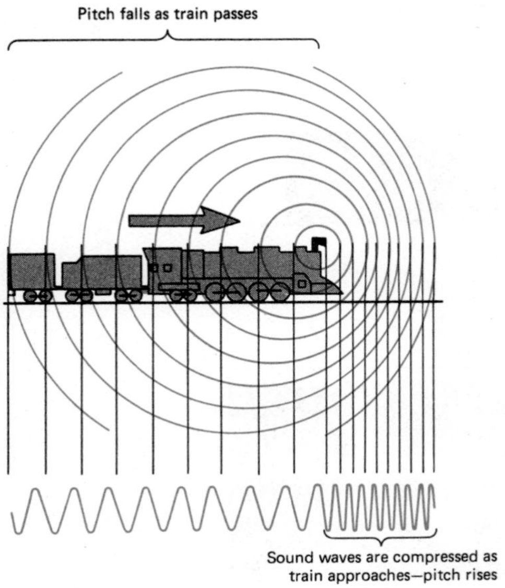

Pitch falls as train passes

Sound waves are compressed as train approaches—pitch rises

moves away, the waves are farther apart. See Figure 2–15. The Doppler effect is observed when either the observer or the source of a wave is moving, or both.

Light waves also show the Doppler effect. When a spectrograph is used to analyze starlight, the Doppler effect can be seen as a shift in the lines of the spectrum from their expected positions. As a star or other light source moves toward the earth, spectral lines are shifted toward the blue end of the spectrum. For a source moving away from the earth, light is shifted toward the red end of the spectrum. See Figure 2–14. Astronomers can determine the speed and direction of motion of a source relative to the earth by measuring the amount of the shift.

Kinds of stars. Stars can be observed to differ in color. Some shine with a brilliant blue-white light. Others produce a reddish light. Spectrographic analysis of starlight shows that the color of a star's light is a result of its surface temperature. Temperatures of the outer layers of stars are found to range from about 3000°C to as high as 50,000°C. Astronomers have discovered that plotting the surface temperature of stars against their absolute magnitude reveals an important pattern among stars. An example of this kind of graph is shown in Figure 2–16. The most important feature of this kind of graph is the way the circles representing stars fall into groups rather than scattering in a random way. The majority of stars occur within a line from cool, dim red stars at lower left to hot, bright blue stars at upper right. Stars within this band are called *main sequence* stars. The sun is in the main sequence along with almost every star visible in the night sky. Near the upper right corner of the graph appears a group of cool but bright stars. They must be very large, since a cool star can be very bright only if it has a large surface area to give off light. These huge stars are called *red giants*. Some are so large that they are known as *supergiants*. If a supergiant star were located at the center of the solar system, the edge of the star would extend beyond the earth's orbit. The third group of stars is found in the lower left part of the graph. These stars are hot but dim because of their small size. They are known as *white dwarfs*. A typical white dwarf star is about the same size as the earth.

The graph shows that almost all stars observed in our own galaxy as well as other galaxies are one of three different kinds. Stars in the main sequence, red giants, and

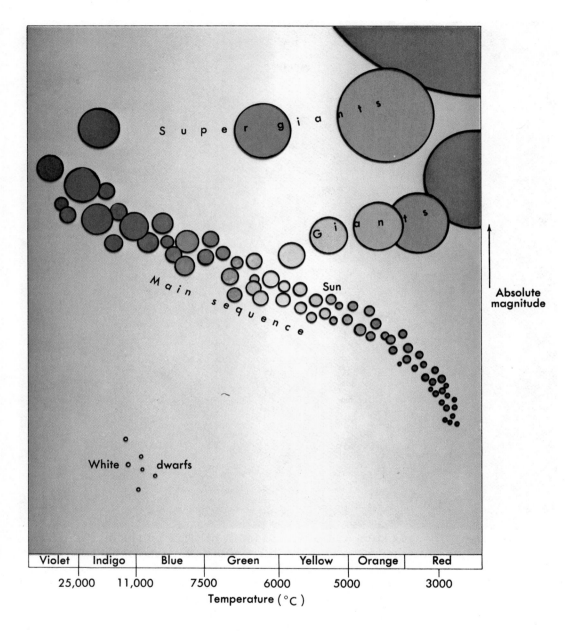

Super giants

Giants

Main sequence

Sun

White o dwarfs

↑ Absolute magnitude

| Violet | Indigo | Blue | Green | Yellow | Orange | Red |

25,000   11,000   7500   6000   5000   3000

Temperature (°C)

white dwarfs are very different from each other. The reason for the differences is a question that scientists try to answer.

Life history of a star. There is no way to observe the entire lifespan of a typical star such as the sun. However, the

FIG. 2–16. A graph showing how stars appear in certain groups when plotted according to their surface temperatures (horizontal) and absolute magnitudes (vertical).

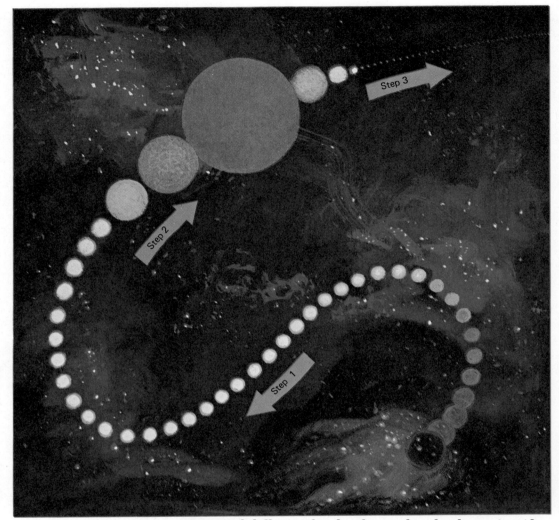

FIG. 2–17. Three stages in the life cycle of a star.

existence of different kinds of stars has lead to scientific theories about the life history of stars that help explain those differences. The test of any theory that attempts to explain how stars are born, develop, and finally die is how well that theory fits with observations. Scientists now believe that the following life story best explains their observations and accounts for the formation of typical stars like the sun. The steps that make up the life history of many stars are shown in Figure 2–17.

Step 1 begins with the birth of a star from a cloud of gas and dust. The space between existing stars is not really empty. It is filled with a very thin scattering of gas and dust. About three quarters of this material is hydrogen gas. The remaining one quarter is mostly helium with small amounts of the heavier chemical elements. Astronomers often observe some of these clouds as dark areas called *nebulas*. See Figure 2–18. Many nebulas appear dark because they cut off the light from more distant stars. Each particle within a nebula has a very weak gravitational attraction for other particles. If some of the particles happen to crowd together, their attraction for each other increases according to Newton's Law—gravity force increases as distance decreases. Slowly a globe of matter grows within the cloud. As the mass of the ball of matter increases, its gravitation steadily increases the pressure and temperature in the core region. The infant star begins to glow with a dull red light. When the temperature in the core finally reaches millions of degrees, the nuclear reactions that will power the star for the rest of its life begin. The star now ends the first stage of its life by moving into the main sequence.

Many of the clouds from which stars are formed produce more than a single star. Stars are very commonly found in pairs or groups of three or four. More than half the stars that you can see in the night sky are *double stars*. Mutual gravity makes these stars move around each other.

Step 2 in the life history of a star begins when the star uses up most of its hydrogen fuel. Up until this time, the

FIG. 2–18. A cloud of gas and dust in the shape of a horse's head forms this dark nebula called the Horsehead Nebula. (California Institute of Technology and Carnegie Institute of Washington)

Eclipse by dim star

Eclipse by bright star

FIG. 2–19. The star called Algol is a double star with the pair orbiting each other. From the earth, the dim star of the pair is seen crossing in front of its brighter partner every 69 hours. Thus Algol is seen to change rapidly from dim to bright, making it a kind of variable star.

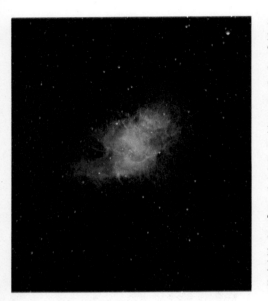

FIG. 2–20. The Crab Nebula is a great expanding cloud of gas flung off from a supernova explosion that was recorded by observers in 1054 A.D. A neutron star lies in the center. (NASA)

star has been in a state of balance between gravity, causing it to shrink, and its internal heat, causing it to expand. This stable condition ends when a star has used up most of the original hydrogen within its core. Loss of its source of nuclear fuel causes the core to collapse under its own gravity. Two things then happen inside the star. Hydrogen around the core begins to undergo the energy-releasing nuclear reactions formerly occurring only within the core. Also, the higher pressures created by the collapsing core start nuclear reactions in the still plentiful helium. Both of these actions release a new flood of energy from the star's interior. The outer layers of the star expand greatly, but grow cooler as a result of their expansion. The star is now a red giant with a very hot interior, but a cool, red outer shell. Scientists think that the sun, now a main sequence star, will become a red giant in a few billion years. At that time, the earth, along with most of the solar system, will be destroyed.

Step 3 marks the slow death of a typical star. The supply of helium fuel is gradually used up. The star is no longer stable, and it begins to shrink and then expand in a series of pulses. These pulses grow stronger until the outer layers of the star are completely flung off into space. The hot dense core is then exposed. The star is now too small to support nuclear reactions. Finally the nuclear furnace at the core shuts off. The only energy source for the dying star is the heat produced by gravity pulling its matter inward. The star turns into a white dwarf, shining with a faint white light. After billions of years it cools off completely and becomes a black dwarf, which is invisible to the astronomer's telescopes.

Stars that are much larger than the sun go through their life histories more rapidly. Very massive stars can also end their existence by becoming supergiants that collapse suddenly in a gigantic explosion called a *supernova*. A supernova may leave behind a very small but incredibly dense ball of neutrons. A spoonful of matter from one of these objects, known as *neutron stars*, would weigh millions of tons on the earth. Neutron stars are very small and usually produce no visible light. Their diameters are about 32 km (about 20 mi). They rotate very rapidly, at the same time giving off two thin beams of radiation that sweep across the sky like the light beams from a lighthouse. Astronomers detect this radiation as a pulse of radio waves called a *pulsar*. Thus we know that a pulsar marks the usually hidden location of a neutron star.

Before the discovery of pulsars, many scientists found it hard to believe that neutron stars could really exist. As they studied the neutron stars, many scientists began to wonder if the strangest of all predicted objects, the *black hole*, could also exist. Scientific theory predicts that the collapse of stars even larger than those that create neutron stars can leave behind a black hole. A black hole is created when gravity becomes so great at a point that space is bent in on itself. If black holes exist, nothing, including light, can escape from them. Thus a black hole could be located only by radiation given off as nearby matter plunges into the hole to disappear forever. Some astronomers believe that such radiation has been detected, and the existence of a black hole has been proven. See Figure 2–21.

Black holes are not the only strange objects found in the universe. Objects known as *quasars* are still unexplained. A quasar (for "quasi-stellar radio source") usually appears as a small, faint star in an optical telescope. But measurements show the total energy output of a quasar to be 100 times greater than that of a large galaxy. The spectra of quasars show very large red shifts, indicating that they are very far away and moving at very high speed. At the present time, no one knows what quasars really are.

FIG. 2–21. An object called Cygnus X-1 may be a black hole pulling gases off a companion star. The gases swirling into the black hole would produce the X-rays known to be coming from this object.

DOES THE UNIVERSE HAVE A LIFE HISTORY?

A scientific view of the universe. One observation has had a great deal of influence on the way most scientists now think of the universe. The light reaching us from all other galaxies shows a red shift. This apparent Doppler effect seems to indicate that all other galaxies are moving away from us at very high speed. The most distant galaxies show the greatest red shift and seem to be moving away at the greatest speed. The galaxies appear to be moving as if thrown out from some central point. Many scientists now favor the *Big Bang theory* to explain how the galaxies might have been given their motion.

According to the Big Bang theory, about 20 billion years ago the universe began as an explosion of a very hot and dense fireball that contained all of the energy and

FIG. 2–22. According to the Big Bang theory, about 12 to 20 billion years ago matter sent out by a huge explosion eventually formed galaxies.

space now in the entire universe. This primitive universe exploded outward, beginning the expansion of space that is still going on. At the moment of explosion, the temperature was thought to be several hundred million degrees. A few minutes after the explosion, the temperature began to drop. Atomic particles then came into existence and began to form atoms. Only hydrogen and helium atoms were first created as the cooling continued. After a few billion years, the gases began to condense into clouds that formed the galaxies. In time, stars formed within the galactic clouds. It is thought that most of these early stars were very large and that most of them became supernovas. Their explosions scattered the heavier atoms that had been created by their internal nuclear reactions. Stars forming later, such as the sun, were created from clouds containing a variety of atoms in addition to the abundant hydrogen and helium. The universe we now observe is populated by a huge number of galaxies that were formed from the matter generated by the Big Bang. The space between the galaxies is still expanding, thus carrying the galaxies away from each other like raisins in an expanding loaf of raisin bread. There seems to be nothing to prevent the expansion from continuing forever.

How can a theory such as the Big Bang theory be tested? One experiment is based on the prediction that at the time of the Big Bang, the universe would have been filled with intense electromagnetic radiation. If the Big Bang theory is correct, remains of that radiation should still exist today as certain wavelengths found everywhere in space. Experiments have shown that radiation exactly matching the predicted wavelengths is actually present throughout the universe. Scientists think that these electromagnetic waves are the fingerprints left behind by the Big Bang.

Many scientists accept the Big Bang theory because it best explains the facts now known about the universe. But science is more than a collection of facts and theories. It is, first of all, a way of investigating. Results of scientific investigations constantly reveal new facts that change existing theories or cause them to be discarded. Science offers no answer to the question of what existed before the Big Bang or how the universe will end. The answers to those questions lie beyond the evidence on which scientists now base their view of the universe.

VOCABULARY REVIEW

Match the word or words in the column on the right with the correct phrase in the column on the left. *Do not write in this book.*

1. The northern or southern lights.
2. Analyzing light by passing it through a glass prism.
3. A system that concentrates solar energy and converts it to another form.
4. A pattern found in plotting magnitude versus temperature of stars.
5. A reaction in which two atomic nuclei combine.
6. Dark areas within the photosphere.
7. The source of the sun's visible light.
8. Explains why galaxies appear to have been thrown out from a central point.
9. The true brightness of a star.
10. A shift in wavelength due to a moving source.

a. nuclear fusion
b. active
c. photosphere
d. sunspots
e. auroras
f. apparent magnitude
g. absolute magnitude
h. spectroscopy
i. Doppler effect
j. main sequence
k. nebulas
l. Big Bang theory

QUESTIONS

Group A

Select the best term to complete the following statements. *Do not write in this book.*

1. The reaction taking place in the sun's core involves the continuous fusing of atomic nuclei due to (a) extremely high pressures (b) extremely high pressures and temperatures (c) presence of radioactive materials (d) presence of heavy nuclei.
2. The basic fuel for the thermonuclear reactions going on within the sun is (a) hydrogen (b) helium (c) carbon (d) oxygen.
3. In Einstein's equation, $E = mc^2$, the letters E, m and c, in that order, stand for (a) energy, speed of light, and mass (b) speed of light, mass, and energy (c) energy, mass, and speed of light (d) mass, speed of light, and energy.
4. Solar energy can be used directly in the home by using solar collectors that produce heat or solar cells that produce (a) microwaves (b) radio waves (c) heat (d) electricity.
5. The part of the sun that is made of relatively cool gases which give off a weak red light is the (a) corona (b) chromosphere (c) zone of radiation (d) photosphere.
6. Reactions that produce the sun's energy take place within its (a) core (b) zone of radiation (c) photosphere (d) chromosphere.
7. The earth receives enough energy from the sun to supply the energy needs of the entire world for one year every (a) 15 seconds (b) 15 minutes (c) 15 hours (d) 15 days.
8. The layer of the sun that appears most brilliant to us is the (a) core (b) zone of radiation (c) photosphere (d) corona.

9. The photosphere of the sun appears to us as it does because much of the energy given off is in the form of (a) X-rays (b) visible light (c) hot gases (d) weak red light.

10. The sun turns on its axis about once every (a) 27 minutes (b) 27 hours (c) 27 days (d) 27 months.

11. Observations of sunspots indicate that the sun rotates on its axis (a) at a rate slower at its equator than at its poles (b) at the same rate of 28 days everywhere (c) at a rate that is faster at its equator than at its poles (d) at the same rate of 25.33 days everywhere.

12. Dark regions on the photosphere that mark the locations of powerful magnetic fields are called (a) sunspots (b) prominences (c) solar flares (d) auroras.

13. Electrically charged particles, guided by the earth's magnetic field, react with the atmosphere to cause (a) sunspots (b) prominences (c) flares (d) auroras.

14. The brightness of a star as it appears at a distance of 32.6 light-years is called its (a) magnitude (b) apparent magnitude (c) absolute magnitude (d) maximum magnitude.

15. Which of the following is *not* used to find the distance to stars? (a) apparent brightness (b) regular variation in brightness (c) parallax (d) elements on the star.

16. When sunlight passes through a triangular glass prism, an unknown band of colors is produced which is called (a) an unbroken spectrum (b) a continuous spectrum (c) a color band (d) a colorful spectrum.

17. A spectrograph can tell us (a) the elements on a star (b) the size of a star (c) the mass of a star (d) the origin of a star.

18. Which of the following is *not* a type of spectrum? (a) continuous spectrum (b) bright-line spectrum (c) nuclear spectrum (d) dark-line spectrum.

19. The lowering of pitch when a locomotive horn passes is due to the (a) spectral effect (b) Doppler effect (c) pitching effect (d) shift law.

20. The group of stars that would fit into the main sequence are the (a) red giants (b) white dwarfs (c) supergiants (d) stars like our sun.

21. The three different kinds of stars are (a) red dwarfs, main sequence, and white giants (b) white dwarfs, main sequence, and red giants (c) blue giants, main sequence, and red dwarfs (d) red giants, main sequence, and blue dwarfs.

22. Our sun is thought to be an average star because it is (a) in the main sequence (b) a red giant (c) a yellow dwarf (d) a white dwarf.

23. When clouds of cosmic dust and gases begin to shrink, becoming more dense, (a) a light is given off (b) heat is generated (c) gravity is decreased (d) the gases become cool.

24. Stars whose light appears to pulsate in a cyclic manner are called (a) nebulae (b) variable stars (c) double stars (d) star clusters.

25. A new star is born when cosmic dust and gases shrink (a) to a solid ball (b) sufficiently to burn (c) sufficiently to cause nuclear reactions (c) to a very large but thick gas ball.

26. A supernova explosion is connected with the last stages in the formation of a (a) neutron star (b) red giant (c) blue giant (d) white dwarf.

27. Neutron stars are produced when a normal star collapses to the size of (a) the earth (b) 20 to 40 kilometers in diameter (c) a red dwarf (d) a white dwarf.

28. Two observations that support the Big Bang theory are that certain wavelength radiation is found everywhere in space and (a) we have found the origin (b) there are many galaxies (c) the Milky Way is so close (d) most galaxies are moving away from us.

Group B

1. What conditions are found in the core of the sun that can cause nuclear reactions to occur?

2. Describe the difference between passive and active methods of converting solar energy into usable forms.

3. Describe the sequence of events by which the sun appears to produce energy. Use the proper names for the parts of the sun involved.

4. How did Einstein's discovery of the relationship of matter and energy enable scientists to better understand how the sun could continue to give off such huge quantities of energy?

5. How are sunspots related to the magnetic disturbances found on the sun?

6. Describe the relationships and differences among sunspots, prominences, and solar flares.

7. What is parallax and for what can it be used?

8. How can the apparent magnitude and absolute magnitude of a star be used to determine its distance?

9. How can it be determined that the sun contains most of the same chemical elements as the earth?

10. What information can be obtained by the use of a spectrograph in astronomy?

11. Scientists constantly seek patterns in the data they collect. What pattern is found when the absolute magnitude is compared to the surface temperature for a large number of stars?

12. What relationship is felt to exist between the main sequence grouping of stars and the life history of stars?

13. How can we explain the existence of multiple star systems?

14. Describe the life history of a typical star such as our own sun.

15. Describe the nature of a pulsar.

16. Describe two observations that support the Big Bang theory to explain galaxy motion.

3

design of the solar system

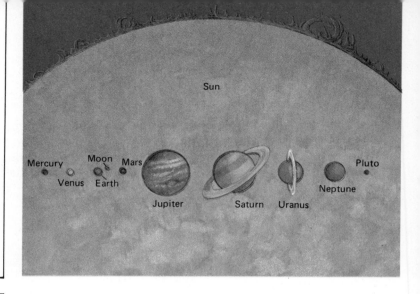

objectives

- ☐ Describe how the solar system may have had its origin in a cloud of gas and dust.

- ☐ Compare asteroids, meteorites, and comets.

- ☐ List the main differences between the inner and outer planets.

- ☐ Explain the summer and winter solstices and the vernal and autumnal equinoxes.

- ☐ Compare apparent solar time and mean solar time.

- ☐ Explain why we use standard time and have the International Date Line.

Maps used by Columbus had most of the world labeled "Terra Incognito," meaning "Unknown Land." Exploration in those days meant crossing oceans to add more of the earth's surface to the known world. Today there is almost no part of the earth's land and oceans left to explore. The new frontiers of exploration are beyond the earth. The solar system is the target of the modern Age of Discovery. During the past few decades, astronauts have landed on the moon. Unmanned space vehicles have landed on some of the planets closest to earth. Other space probes have flown near the distant planets.

Scientific exploration of the solar system has revealed an astonishing variety of objects and environments. Surfaces that are as hot as an oven or covered by a solid sheet of ice or an unusual film of liquid, as well as giants made of gas without any apparent solid surface at all, are found among the bodies that make up the solar system. What theory can scientists propose to account for the origin of a system containing such different parts? At the present time, most scientists believe that the beginnings of the solar system can be traced back to the birth of the sun. Many details of this story are missing, but most of what has been discovered seems to support the view that the sun and all of the objects in orbit around it were born together.

BIRTH OF THE SOLAR SYSTEM

A cloud of gas and dust. If modern scientific theory is correct, the exact place where you are now sitting was once very cold, nearly empty space. About five billion years ago,

44

FIG. 3–1. This photograph of a part of the Milky Way shows dark areas caused by clouds of dust and gas. Stars in these dark regions are between the earth and the cloud. (Hale Observatory)

the small part of the Milky Way Galaxy now occupied by the solar system was a frigid, almost perfect vacuum between old, widely scattered stars. For every cubic centimeter of that space there were only a few atoms. By comparison, each cubic centimeter of the air you are now breathing contains about thirty million trillion atoms. Hydrogen atoms made up most of this very thin, ancient gas cloud. About one out of every dozen atoms was helium. A very few atoms of heavier elements also could be found. Between the atoms of the gases, the cloud also contained tiny dust particles. Each cubic kilometer of the cloud probably contained only a hundred of these dust particles. Most of the dust particles were made of the kinds of atoms now found in a common stone you might pick up anywhere on the earth. However, some of the chemicals present in the cloud were those now found in living things. Thus the chemical building blocks necessary for life seem to have been present even before the solar system began.

If something had not happened, the ancient gas and dust cloud would have continued as one of the dark nebulas that can be seen in the Milky Way. But something made the cloud that was to become the solar system begin to shrink. Some astronomers think that a supernova explosion of a nearby large star may have sent a shock wave through the cloud. See Figure 3–2. As the atoms and particles within the cloud were crowded together, gravity began to take control. The cloud was pulled in on itself by

FIG. 3–2. The solar system may have begun to grow from a nebula when a nearby supernova exploded.

FIG. 3–3. A young star, the sun, begins to grow at the center of the solar nebula.

FIG. 3–4. The protoplanets were formed from a flat disk of matter around the young sun.

the mutual gravitational attraction between atoms and dust particles. One part of the shrinking cloud happened to have a little more matter and became a separate clump. Contraction speeded up within the clump of matter, which also began to slowly rotate. Its rotation eventually turned this region of the cloud into a disk. This was the *solar nebula*. See Figure 3–3. It was about the same diameter as the present solar system, but contained twice as much matter. Matter continued to fall toward the center of the solar nebula, making the central parts hotter. The sun was beginning to grow at the center of the disk.

At the same time, smaller clumps were forming in the thin disk rotating around the young star. Gravity slowly pulled these small clumps together to form larger bodies called *protoplanets*. See Figure 3–4. Four protoplanets grew close to the central sun. These were to become the inner planets: Mercury, Venus, Earth, and Mars. Four other protoplanets were formed farther from the growing sun. These are now the outer planets: Jupiter, Saturn, Uranus, and Neptune. As you will see later, Pluto, the smallest planet now in the solar system, was probably formed as a satellite of Neptune.

Their positions close to or far from the developing sun had a great influence on the protoplanets. The matter of inner protoplanets was rich in the heavier elements, such as iron, common to the center of the solar nebula. Radioactive elements in the inner protoplanets heated them until their matter was molten. The heavier substances, including large amounts of iron, sank to the centers of the protoplanets. The lighter materials were pushed to the outer layers. Thus the inner planets developed with dense iron cores covered by lighter sheets of rock.

It was a different story with the outer protoplanets. The cold regions of the solar nebula far from the sun contained large amounts of ices made of methane and ammonia, as well as water. Outer protoplanets thus grew as small dense cores surrounded by thick layers of ices, all covered by a deep atmosphere. Far from the heat of the sun, the outer planets were able to grow into giants that swept up huge amounts of hydrogen and helium left in the outer edges of the solar nebula. Jupiter and Saturn grew to enormous size. They developed like miniature solar systems, each accumulating a number of smaller satellites.

While the protoplanets formed in the flat sheet making

up the disk of the solar nebula, the sun continued its growth at the core. Temperatures rose at the very center as matter continued to condense. Finally, the temperature became high enough to start the fire of nuclear fusion. The sun then became a newborn star. Astronomers have found that very young stars finally adjust to their birth by throwing off some of their matter. The infant sun probably sent off such a blast during the very early history of the solar system. It must have stripped away the original atmospheres of the inner planets, leaving only the solid bodies. Outer planets were struck with less force and held their thick atmospheres. Eventually the sun settled down to its present quiet existence as a main sequence star. The solar system had reached the general form that we find today. Planets would continue to develop in their own ways, but their basic paths were set by events of their birth.

Minor members of the solar system. More than 99 percent of all the matter in the original solar nebula ended up in the sun. Less than 1 percent was left to make the planets. The protoplanets grew by sweeping up smaller chunks of matter as they moved in their orbits. All of the solid, orbiting bodies of the solar system show surface craters as evidence of collisions with smaller objects in their early histories. The moon, for example, is covered with craters. Materials swept up by the young planets became part of their bodies. During the long history of the solar system, the planets have acted like great vacuum cleaners, sweeping their neighborhoods clean of the small bits of matter still left from the earliest times.

There is one large region of the solar system, however, that has no planets to sweep up the leftover matter. Orbiting in this zone between Mars and Jupiter are many chunks of material that range in size from small lumps to fragments 100 kilometers or more in diameter and thousands of tons in weight. All of these bodies are called *asteroids.* The total number of asteroids is not known. Only the largest can be seen in telescopic photographs. See Figure 3–5. About 230 asteroids with diameters greater than 100 kilometers have been discovered. Astronomers do not consider an asteroid to be discovered until its orbit has been determined. Altogether, about 2,000 asteroids have been offficially discovered by having their orbits calculated. Thousands more with diameters of a few kilo-

FIG. 3–5. Asteroids are seen as streaks in telescopic photographs such as this one exposed for 2½ hours. (Lowell Observatory)

FIG. 3–6. Barringer Crater in Arizona was made about 20,000 years ago when a meteorite about the size of a railroad boxcar collided with the earth. (Meteor Enterprise)

FIG. 3–7. This stony meteorite has been sliced through to show the round inclusions found in most meteorites made of stone. (Institute of Meteoritics, University of New Mexico, Albuquerque)

meters or less must exist. The total mass of all asteroids combined would not be equal to even a very small planet.

All but a very few asteroids follow orbits that fall in a wide belt beginning about 100 million kilometers from the orbit of Mars and stretching on about 160 million kilometers toward Jupiter's orbit. Sometimes an asteroid passes close enough to Mars or Jupiter to allow the gravity of one of these planets to change the asteroid's orbit. The asteroid is then tugged into a very elliptical orbit that causes it to cross the orbits of Mars and earth during each revolution. Such orbits permit collisions to occur between asteroids and the earth. Hundreds of asteroids actually do collide with the earth every year. They enter the earth's atmosphere traveling at between 10 and 30 kilometers/second. By comparison, a rifle bullet has a speed less than one kilometer/second. Friction between the asteroid and the atmosphere creates high temperatures that melt and vaporize outer parts of the asteroid, producing a brilliant flash of light and a loud noise. Any part of the asteroid that survives its plunge into the atmosphere and finally falls to the ground is called a *meteorite*. Fortunately, most meteorites are small, with a mass of less than one kilogram. Very rarely a large meteorite punches through the earth's atmospheric shield and strikes the surface with the force of an exploding bomb. About thirty large asteroids, each the size of a small mountain, are known to have orbits carrying them across the path of the earth. The chance of a collision with one of these flying mountains is very small, but there is no way to predict when it might happen. If such a collision should occur, it would cause a gigantic explosion that would leave a crater hundreds of meters deep and as wide as a large city. See Figure 3–6.

Examination of meteorites shows that they can be classified into three basic types. First, there are stony meteorites made of material much like ordinary rocks found on the earth. More than 90 percent of all meteorites are stony. See Figure 3–7. A few stony meteorites also contain carbon-bearing substances similar to the materials found in living things. A second type of meteorite is made of iron. See Figure 3–8. Iron meteorites are commonly found because they are less likely to break up as they pass through the atmosphere. The third and rare type of meteorite contains both iron and stone. Scientists believe that the different kinds of meteorites are samples of materials that originally condensed out of the solar nebula. The oldest meteorites seem to be about 100 million years older than the earth or moon. Thus meteorites are clues to the very early chapters of earth history.

Comets make up another group of objects thought to be remains of the beginning of the solar system. They are believed to be icy lumps thrown to the outer edges of the solar system by the powerful gravity forces of the outer planets. Most were lost forever in deep space. However, some remain in huge orbits beyond the orbit of Pluto. Occasionally, the gravity of a distant star passing our region of space will cause a comet to fall into a long orbit that carries it around the sun. This kind of comet is discussed on page 50. Almost any night you can see evidence of comets having crossed the earth's orbit. Meteors, small streaks of light seen briefly in the dark sky, are caused by tiny particles entering the atmosphere at high speed and burning up at high altitudes. Most meteors are believed to be small bits of matter left behind by passing comets. Large numbers of meteors, called meteor showers, are seen when the earth passes a point in its orbit known to have been crossed by a comet in the past.

FIG. 3–8. These scientists are collecting meteorites on the ice in Antarctica. Meteorites that have fallen on the Antarctica ice sheet over long periods of time accumulate and are exposed in some places where the ice is slowly removed by wind erosion and evaporation. (Dr. W. A. Cassidy)

A CATALOG OF THE PLANETS

Mercury. This barren little planet, about the same diameter as the moon, has an orbit that keeps it at an average distance of only 58 million kilometers (36 million miles) from the sun. It moves swiftly, completing one trip around its orbit in 88 days. During two revolutions around the sun, it turns on its axis three times. This means that it rotates on its axis once every 58.65 days. If you were on Mercury, the sun would rise and remain in the sky for nearly three earth months before setting. Such a long period of sunlight and the planet's closeness to the sun cause the temperatures on Mercury to rise above 400°C. Only a tiny amount of gases covers Mercury's surface. The absence of a heavy protective atmosphere allows the full force of the sun to heat the lighted side. The opposite dark side escapes the sun's force. As a result, the temperature on Mercury's dark side plunges to about −200°C.

Mercury's position between the earth and sun makes it very difficult to observe the planet from the earth. To view this planet, telescopes must always look in the direc-

Where do comets come from?

Comets come from a vast spherical cloud of material that surrounds the entire solar system. This cloud lies beyond the outermost planet and extends out trillions of kilometers from the sun. It may have been formed from matter remaining after the solar system was formed. Or its origin may be in matter captured later by the sun's gravity. Once in a while the gravity of a passing star causes one of the chunks of matter in the cloud to fall into an orbit that carries it toward the sun. It may fall into a long, cigar-shaped orbit with the sun at one end and the other end in the cloud. Such comets are called long-period comets because they take several thousand to several million years to make one revolution in their extremely elliptical orbits.

Sometimes a comet is captured by the gravity of a giant planet such as Jupiter. It falls into an orbit with the sun at one end and the planet at the other. This type of comet becomes a short-period comet. Each of the giant outer planets has a family of comets orbiting around it. These comets complete their orbits in a few years. Many short-period comets are seen at regular intervals when their orbits bring them close to the sun. Halley's Comet, which returns every 76 years, is an example of a short-period comet. It should next be seen in 1986.

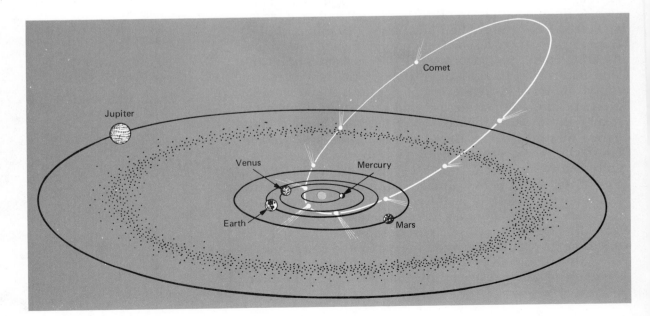

comets

Comet West as it made its two-week passage across predawn skies in the early part of 1976.

What is a comet made of?

The solid part, or *nucleus*, of a comet is a chunk of rock, dust, and ice. It is no more than 80 km (50 mi) in diameter. Comets have been called "dirty snowballs." When a comet passes inside the orbit of Mars, the sun's heat and solar wind are strong enough to cause a bright *coma*, or tail, to develop. The coma may be as much as one million kilometers in length. It is made of gas and dust streaming away from the nucleus. Its light may come from sunlight reflected by the dust, or from gas made to glow by solar radiations. Some comets develop two separate comas, one of dust and the other of gases. Because it is blown by the solar wind, the coma always points away from the sun. Sometimes the nucleus of a comet breaks up to become a collection of smaller comets that all follow the same orbit.

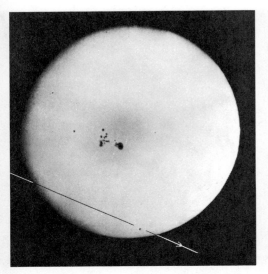

FIG. 3–9. The arrow in the lower left shows the path of Mercury as it transits the sun. The planet itself is the small black dot near the bottom edge of the sun. (Yerkes Observatory)

tion of the sun's glare. At certain times of the year, it is seen in the night sky for about 1½ hours after sunset when it is farthest east of the sun. At other times of the year, Mercury rises about 1½ hours before sunrise when it is farthest west of the sun. At other times, Mercury is not visible because it is behind or in front of the sun. However, Mercury does not usually pass directly across the face of the sun because its orbit is tilted with respect to the earth's. Thus it usually crosses a little above or below the sun. About 14 times during each century, Mercury can be seen crossing the face of the sun. Such a crossing is called a *transit*. See Figure 3–9. A transit of Mercury can be seen in 1986 and 1993.

The best telescope photographs of Mercury show no details of its surface. However, spacecraft carrying cameras and instruments have flown within a few hundred kilometers of Mercury. The close-up photographs show a landscape covered with craters, jumbled hills, plains, and cliffs. See Figure 3–10. Without an atmosphere or running water to alter its surface, Mercury's surface has probably changed very little during the past three billion years. Its craters are a record of the time during the solar system's early history when the young planets were bombarded with debris still in orbit around the sun. Like the other inner planets, Mercury shows evidence of the early separation of its light and heavy materials. It is very dense for its size, with a total mass equal to $\frac{1}{18}$ of the earth's mass, although it is not much larger than the moon, which is only $\frac{1}{81}$ of the mass of the earth. This means that Mercury must have a large, dense core, probably of iron, with rocky outer layers. It too is surrounded by a magnetic field characteristic of planets with a metal core.

FIG. 3–10. The cratered surface of Mercury is photographed from a spacecraft at a distance of 200,000 km (124,000 mi). (NASA)

Less than half of the surface of Mercury has been photographed from spacecraft. Much more is still to be learned about this small, sun-scorched planet.

Venus. Although it is the earth's closest neighbor planet, until recent years Venus was a mystery. Telescopes show it only as a bright disk that passes through phases like the moon. See Figure 3–11. None of the surface features of Venus can be seen because the planet is always completely covered by clouds. As the second closest planet to the sun, at an average distance of 108 million kilometers (67 million miles), Venus receives about twice as much sunlight as the earth. Much of the sunlight is reflected by its clouds, making Venus one of the brightest objects seen in the sky. Its position between the earth and sun (like Mercury) causes it to be seen during certain times of the year in the west after sunset or in the east before sunrise. Sometimes Venus is 42 million kilometers from earth, which is only about one hundred times farther away than the moon. At that time Venus, the always bright evening and morning "star," is especially brilliant.

In the past, Venus was often called the "earth's twin." With a diameter of 12,100 kilometers (7,250 miles) and a mass about four-fifths of the earth, Venus does resemble the earth. However, information gathered by spacecraft sent to Venus indicates that the planet is very different from the earth. Its surface is blistering hot, with temperatures around 475°C. These oven-like conditions are caused mainly by the thick atmosphere made almost entirely of carbon dioxide that surrounds the planet. This gas lets short wavelength radiations from the sun pass through the cloudy atmosphere. When these radiations strike the surface of the planet, they are changed into heat. The carbon dioxide will not allow the heat to escape. Thus heat is continuously trapped, turning the surface of Venus into a furnace. Since it turns very slowly on its axis, once every 243 days, days and nights are very long on Venus. Its rotation is opposite to the direction it moves in its orbit. On Venus, the sun would rise in the west and set in the east. Even during the long night, the trapped heat does not escape. On the dark side of Venus, the rocks on the surface would probably be seen to have a dull red glow like the coils of an electric heater.

Spacecraft have gathered information that shows the upper layers of the clouds of Venus to be made mainly of droplets of sulfuric acid. Swift winds carry these upper

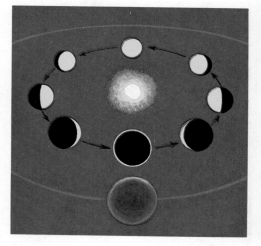

FIG. 3–11. Because Venus follows an orbit between the earth and sun, we see this planet with both its lighted and dark sides facing us.

Interpret
Only Mercury and Venus can be observed passing through phases like the moon. If you were on Mars would there be other planets that would exhibit phases? Explain your answer.

FIG. 3–12. The clouds of Venus show a swirling pattern when photographed in ultraviolet wavelengths by cameras on spacecraft. In ordinary light the clouds appear uniformly pale yellow. (NASA)

FIG. 3–13. This huge volcano on the surface of Mars is almost as large as the state of Texas. It is 27 km (about 17 mi) in height. (NASA)

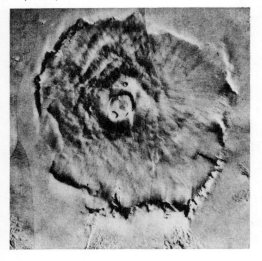

clouds completely around the planet every four days. See Figure 3–12. Below the clouds, the winds are gentle and the atmosphere is clear. At the surface, the pressure created by the dense atmosphere is 90 times greater than that on the earth's surface.

Spacecraft in orbit around Venus have been able to use radar to penetrate the clouds and scan most of the surface. The resulting radar maps show that two-thirds of the planet is covered by a relatively smooth plain. There are some features similar to the continents, mountains, and valleys found on the earth's surface. However, the radar maps are not able to show small details of the planet's face. The variety of features suggests that with the exception of running water, all of the forces that have changed the earth's surface have also been at work on Venus. The exact history of Venus, however, can only be told from future explorations of this hidden planet.

Earth. The third planet from the sun is the one object in the universe we know best. The characteristics of our home planet are discussed in detail beginning with Unit 2. Two general features of earth set it apart from all other known planets. It is the only planet whose surface is covered mostly by liquid water. And it is the only location in the universe where life is known to exist.

Mars. The fourth planet from the sun has been studied more than any other planet except earth. A series of space missions to Mars has made this planet almost as well known as the moon. It is a smaller planet than the earth with a diameter of 6,787 km (4,208 mi), which is about half that of the earth. Its mass of only one-tenth that of the earth indicates that there must be only a small, dense core within the planet. Mars rotates on its axis once every 24.6 hours. The length of a Martian day is nearly the same as earth's. Mars also has seasons like the earth because the tilt of its axis is nearly the same. However, it takes 687 days to make one revolution around the sun, making each season last nearly twice as long. Mars is a cold planet, since the solar energy it receives is not effectively held by its thin, carbon dioxide atmosphere. The air pressure on the surface of Mars is less than $1/100$ of the normal pressure at the earth's surface. During the Martian summer, the temperature near the equator may approach 20°C. Winter temperatures near the poles drop as low as −130°C.

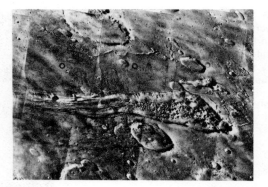

The surface of Mars is covered with features that show it is not a dead planet. Many craters that were apparently created during the early bombardment of the planets are still visible. However, many of the craters have been changed by later activity on the surface of Mars. There is evidence of widespread volcanic activity on the Martian surface. Large areas are covered with plains created when lava flowed out onto the surface. In addition, there are many volcanoes. Some of these volcanic mountains are huge. See Figure 3–13. One part of the Martian surface is covered by a system of deep canyons. See Figure 3–14. Scientists believe that these canyons were created by great cracks in the planet's crust caused by Marsquakes.

One of the most surprising discoveries on the Martian surface was the presence of many features that seem to have been made by running water. See Figure 3–15. Since the low pressure and temperatures now found on Mars do not allow for the existence of liquid water, the evidence gathered indicates that flowing water was part of Mars' ancient past. Long ago, Mars apparently had a warmer and wetter climate. Then, for unknown reasons, the planet became cold and dry. The water that remains is now trapped in polar ice caps or frozen beneath the surface. Much of the dry surface is covered with loose reddish soil. See Figure 3–16. Strong winds that occur during the Martian summer often create great dust storms. Some of these violent dust storms cover most of the planet for several months at a time. Loose soil moved by the winds collects in large regions of sand dunes near the polar regions.

Like earth, Mars is an active planet whose surface is affected by volcanoes, water, and wind. However, there is no evidence that Mars shelters living things.

FIG. 3–14. (Left) A system of deep canyons stretches along the Martian equator, reaching a depth of more than 4 km (2½ mi) in places. (NASA)

FIG. 3–15. (Right) Channels that were apparently made by running water are seen in places on the Martian surface. This channel, about 20 km (12½ mi) wide, may have come from water reaching the surface from below. (NASA)

FIG. 3–16. The surface of Mars as photographed from a spacecraft resting on the Martian surface. The white patches are frost, probably made of a mixture of frozen carbon dioxide and water. (NASA)

FIG. 3–17. Jupiter as seen from a spacecraft at a distance of 28.4 million km (17.6 million mi). Three of Jupiter's moons are visible. The moons revolve in the same plane but appear in these positions because the spacecraft was above their plane of revolution around the planet. (NASA)

Compare
Determine the speed of each plane in kilometers per day

$$\left(\text{speed} = \frac{\text{distance}}{\text{time}} = 2\,\pi\,R/\text{time},\right.$$

where R is average distance between planet and the sun). How does a planet's distance from the sun affect its speed?

FIG. 3–18. The Great Red Spot on Jupiter seen from a distance of 6 million km (about 4 million mi). (NASA)

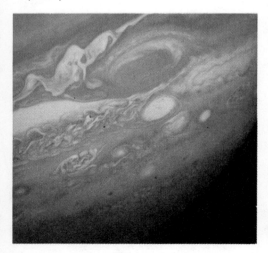

Jupiter. The first of the outer planets is many things. It is a giant among the planets. With a diameter of 143,100 km (88,930 mi) at the equator, it is more than eleven times larger than the earth. If Jupiter were hollow, 1,400 earths could fit inside. Its mass is 318 times greater than the mass of the earth. This single, huge planet has a mass 2½ times greater than that of all the rest of the solar system, except the sun, put together. Jupiter is made of 82 percent hydrogen, 17 percent helium, and 1 percent other substances. It seems to contain basically the same materials present in the solar nebula that gave birth to the solar system.

All we can see of Jupiter is a thin outer layer of colored clouds. See Figure 3–17. The colors result from the presence of substances such as ammonia, methane, and water vapor. Lightning flashes through these clouds. Some scientists think that simple forms of life may have developed as the lightning created the necessary chemical building blocks in Jupiter's atmosphere. The colorful clouds form only a thin layer covering a body made mostly of hydrogen. There may be a small rocky core at the planet's center, but the bulk of Jupiter is made of hot, liquid hydrogen. The interior is still very hot from heat created at its birth.

Heat and pressure change the hydrogen deep within the planet into a substance that acts like a metal. This causes a huge magnetic field to be created as Jupiter rotates rapidly on its axis. It completes one rotation in only 9 hours and 55 minutes. However, at an average distance of 778 million km (483 million mi) from the sun, Jupiter takes 12 years to move once around the sun. Its rapid rotation is responsible for the colored bands seen in the atmosphere. Warm gases rising from below create dark-colored belts. These gases spill over into neighboring light-colored zones as they cool. Rapid rotation of the planet creates strong winds between the belts and zones. These winds mix the gases in swirling patterns. See Figure 3–18. Although most of Jupiter's atmosphere appears as churning bands of clouds, the Giant Red Spot, also shown in Figure 3–18, is always seen at the same position. It has been seen in telescopic views of Jupiter for at least 300 years. Scientists think that it must be caused by a permanent storm held in place for some unknown reason.

Among its many other unique features, Jupiter is also the center of a miniature solar system. So far fifteen moons have been discovered following various orbits around the giant planet. There is also a faint ring made of very small solid particles. Four of Jupiter's moons are the size of small planets. They are often called the Galilean moons because they were first seen by Galileo in 1610. Spacecraft have shown the Galilean moons, called Io, Europa, Ganymede, and Callisto, to be unusual objects. Each of these planet-sized bodies has different characteristics, which are shown in Figure 3–19. Future study of these dramatic worlds in orbit around Jupiter promises a better understanding of all solid planets, including earth.

Saturn. Early astronomers called Saturn the wonder of the heavens. Even low-power telescopes show Saturn as a bright globe surrounded by a spectacular system of rings. See Figure 3–20. But closer study by spacecraft shows that, except for its dramatic rings, Saturn is in many ways a smaller version of Jupiter. Its equatorial diameter is 120,660 km (74,930 mi). Like Jupiter, Saturn rotates rapidly, turning on its axis once every 10 hours and 30 minutes. Both Saturn and Jupiter bulge out at their equators and are flattened at the poles as a result of their rapid rotation. As the sixth planet from the sun, Saturn orbits the sun at a distance 9½ times greater than the

Jupiter

Europa

FIG. 3–19.
Europa: diameter of 3,130 km (1,940 mi); average distance of 670,000 km (416,400 mi) from Jupiter; icy crust.

Io

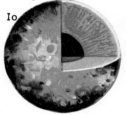

Io: 3,640 km (2,260 mi) in diameter; average distance of 421,000 km (261,000 mi) from Jupiter; outer covering of sulfur and frozen sulfur dioxide; active surface volcanoes.

Callisto

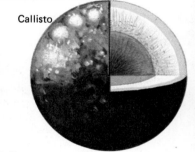

Callisto: 4,850 km (3,012 mi) in diameter; average distance of 1,880,000 km (1,170,000 mi) from Jupiter; thick ice crust with water or soft ice layer beneath.

Ganymede

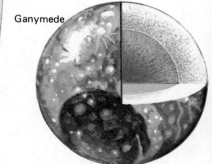

Ganymede: largest moon with a diameter of 5,270 km (3,270 mi); average distance of 1,070,000 km (664,500 mi) from Jupiter; icy crust over thick layer of circulating water or soft ice, solid core.

FIG. 3–20. Saturn and some of its larger moons are shown in this photograph assembled from separate pictures made from spacecraft. Clockwise from upper left, the moons are Rhea, Titan, Mimas, Thethys, Dione, and Enceladus. (NASA)

FIG. 3–21. Top: The view of Saturn's rings changes as Saturn moves around the sun. Bottom: When Saturn is in the positions A, B, and C, its rings will be seen at those angles in the years indicated.

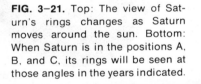

Earth's orbit

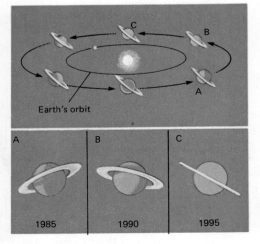

earth, and receives about 1 percent of the heat and light that falls on the earth. Thus the outer layers of Saturn are very cold. Its clouds are hidden beneath a thick layer of haze that gives the planet a bright face showing little detail. Beneath the layer of haze, Saturn's clouds show nearly the same pattern of belts and zones found on Jupiter. Very powerful winds driven by heat flowing from Saturn's interior circle the planet, as in the case of Jupiter. The body of Saturn is composed mostly of liquid hydrogen with a small solid core. Its total mass is 95 times that of the earth.

Saturn is not the only planet with a system of rings circling above its equator. Most scientists think that planetary rings form in a planet's early history either by break-up of an existing moon or failure of material to gather together to form a moon. The rings of Saturn are the largest, brightest, and most complicated known to exist in the solar system. They are made of billions of chunks of ice. The pieces of ice range from dust-like particles to chunks the size of a house. Each follows its own individual orbit around Saturn. All together, the ring system fills a band about 5 km (about 3 mi) thick and about 67,000 km (about 42,000 mi) wide. The system is divided into several large bands, which in turn are made of hundreds of narrow ringlets. Saturn's rings are not always clearly seen in telescopic views from earth. This is because only their edges are visible at times during the nearly 30 years it takes Saturn to complete one revolution around the sun. See Figure 3–21.

In addition to the billions of bodies in the rings, Saturn has at least fifteen moons that have been discovered. Most are small with diameters less than 100 km (621 mi). However, five of Saturn's moons are fairly large. One, called Titan, has a diameter of more than 5,000 km (3,100 mi). See Figure 3–22. Thus Titan is between the size of Mercury and Mars. Most of Saturn's moons are very cold, icy bodies showing many craters. Titan is different. It has a thick atmosphere composed mainly of nitrogen that conceals the surface. Titan may be a frozen example of a very early stage in the development of earth-like planets. Future exploration of this moon of Saturn may provide important clues about the way earth came to be the kind of planet we now find.

Uranus. A journey to this seventh planet from the sun would take you about half way out of the solar system. The sun, nearly 3 billion kilometers (1¾ billion miles) away, would appear as a very bright yellow star in a black sky. Because it lies in the most distant parts of the solar system, there is still much to be learned about Uranus. It is known to be about four times larger than the earth, with a diameter of about 51,800 km (about 32,000 mi). Even the largest telescopes show no details on its greenish face. Without visible markings, its rotation rate is difficult to establish. It seems to be between 13 and 25 hours. Uranus is very unusual in the way its axis is tilted. Other planets spin like tops with their axes tilted only slightly. Uranus is tilted so far that its rotation gives it the appearance of rolling as it completes its orbit once every 84 years. See Figure 3–23. The greenish color of Uranus seems to result from the presence of methane, along with hydrogen and helium, in its atmosphere. Judging from its size and mass, which is about 14 times greater than the earth's, scientists think that Uranus is made of mainly liquid hydrogen with a small solid core.

Five moons are known to orbit Uranus with others probably yet undiscovered. A system of nine very faint rings has also been discovered around the planet.

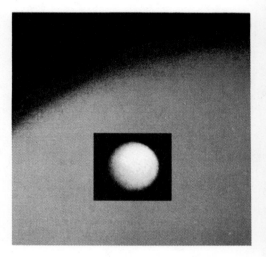

FIG. 3–22. Saturn's largest moon, Titan, is covered with a thick atmosphere that conceals its surface (inset). Layers of haze appear as blue in a close-up photograph. (NASA)

FIG. 3–23. Uranus' axis is tilted 98° causing first one pole and then the other to point toward the sun as the planet moves around its orbit.

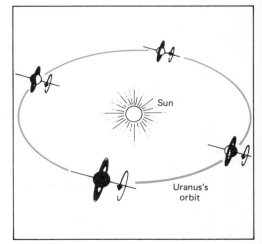

Neptune. Except for its greater distance from the sun, 4½ billion km (2¾ billion miles), Neptune appears to be a twin of Uranus. Its size, mass, and period of rotation are like those of Uranus. Neptune is orbited by two moons. One, called Triton, is very large, with a diameter of 6,000 km (nearly 4,000 mi). Unlike all other moons in the solar system, Triton moves backward around Neptune as the planet completes its orbit around the sun once every 165 years.

Pluto. Several features of this small body make scientists wonder if it is actually a planet. Pluto follows a very elliptical orbit that brings it within 4.4 and 7.4 billion km (2.7 and 4.6 billion mi) of the sun. Thus it is sometimes closer to the sun than Neptune. The orbit also tilts much more than the orbits of all other planets. Pluto is also very small, with a diameter about the same as that of the earth's moon. However, it has only about one-sixth of our moon's mass. Apparently it is a ball of ice. A single moon, called Charon, moves in an orbit very close to Pluto — so close, in fact, that both Pluto and Charon circle each other as they move around the sun once every 248 years.

Some astronomers think that Pluto was once a moon of Neptune. Pulled away by the gravity of a neighboring planet, perhaps also breaking off a piece that became Charon, Pluto may have then fallen into its own strange orbit. This small, distant body and its companion may also be relics from the formation of the solar system.

THE MOVING EARTH

Orbiting the sun. According to modern scientific theory, the sun and planets were formed from a disk-shaped cloud of gas and dust that was turning like a giant wheel. The protoplanets forming from separate clumps within the cloud also began to rotate as the cloud grew. Thus the basic plan of the solar system was established. Each planet follows its own orbit as it moves around the sun. At the same time, the planets rotate around an axis. Even as you read this sentence, you are traveling with the earth in an elliptical orbit around the sun at a speed of 106,000 km/hr (66,000 mi/hr). But the motion of the earth in its orbit would have little effect on your life if it did not produce the change of seasons.

If the orbit of each planet is drawn on a level or plane

like a flat sheet of paper, then the axis of rotation of the planet will have a certain tilt when compared to the plane surface. For example, the earth's axis is tilted 23½° from a perpendicular to the plane of its orbit. As the earth revolves around the sun, the tilt of its axis is always the same. This means that the North Pole is tilted first toward the sun, then away from it. See Figure 3–24. When the North Pole is tilted toward the sun, the northern hemisphere receives more of the sun's direct rays. The sun's direct rays deliver more energy than slanting rays. This produces the warm summer season in the northern hemisphere. See Figure 3–25.

When the North Pole is tilted most toward the sun, the direct rays fall on a line north of the equator called the Tropic of Cancer. See Figure 3–25A. The time of the year when the sun's direct rays fall on the Tropic of Cancer is called the *summer solstice*. "Solstice" means "sun stand

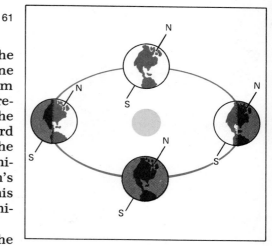

FIG. 3–24. The tilt of the earth's axis remains the same throughout its orbit.

FIG. 3–25. This diagram shows how the sun's direct rays fall on the Tropics at the solstices and on the equator at the equinoxes.

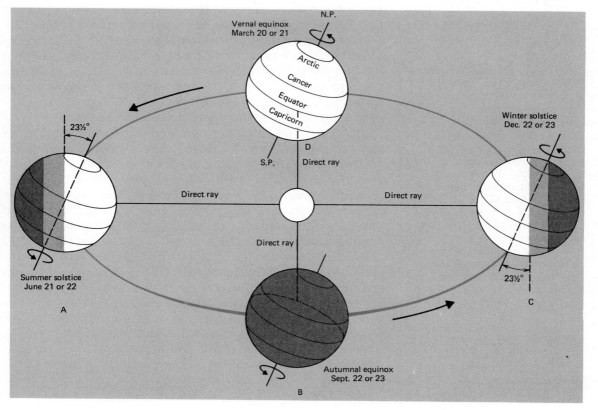

activity

Mercury and the earth move around the sun in their own orbits. See the diagram below. Only part of the orbits is shown.

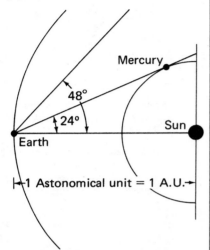

The line to the sun from the earth makes an angle with the line to Mercury from the earth. This angle is never more than 24° and can be found by careful observation and measurement. Be very careful not to look at the sun.

A. Make a drawing like the one above, so that the earth-to-sun distance is 10 cm. The circle representing the orbit of Mercury must just touch the side of the 24° angle. See the diagram. At one time, it was impossible to measure the distance between the earth and the sun. So the distance was simply called "one astronomical unit" (1 A.U.). In your drawing, 1 A.U. would be represented by 10 cm.

B. Measure the distance between Mercury and the sun in centimeters.

1. What is the length of this line? (Continued on page 63.)

still." It refers to the fact that on the day of the summer solstice, everyone north of the equator sees the sun in its most northerly position in the sky. Following the summer solstice, which occurs on June 21 or 22, the sun is seen a little farther south each day. Regions of the earth's surface within the Arctic Circle have 24 hours of sunlight at the time of the summer solstice, as shown in Figure 3–25A.

By September, the earth is one-fourth farther around its orbit than it was in June. On September 22 or 23, the sun's direct rays fall on the equator. See Figure 3–25B. This time is called the *autumnal equinox* and marks the beginning of the fall season. "Equinox" means "equal night" and refers to the fact that the hours of daylight and darkness are equal everywhere on the earth at this time of the year. At the beginning of fall, the Northern Hemisphere is no longer receiving most of the sun's direct rays. This area begins to cool off.

By December, the earth has moved halfway around its orbit. The North Pole is tilted most away from the sun. The sun's direct rays fall on a line south of the equator called the Tropic of Capricorn. See Figure 3–25C. This is the *winter solstice* and occurs on December 21 or 22. It begins the winter season north of the equator. As seen from the Northern Hemisphere, the sun is at its most southerly position in the sky at the winter solstice. Regions in the Antarctic Circle have 24 hours of daylight, and those within the Arctic Circle have 24 hours of darkness.

By March, the earth has completed three-quarters of its trip around the sun. The sun's direct rays again fall directly on the equator on March 21 or 22. This time of the year is called the *vernal equinox* and marks the beginning of spring. The Northern Hemisphere again begins to warm up as it receives more of the sun's direct rays. In June, the summer solstice is again reached, and the cycle of the seasons is complete.

A *year* is the length of time it takes the earth to complete one revolution in its orbit. The most common way to measure a year is to determine the exact length of time between two consecutive vernal equinoxes. That length of time is found to be 365 days, 5 hours, 48 minutes, and 46 seconds. Actually this is a little short of the time it takes the earth to make one revolution in its orbit. The error is a result of a slow wobbling of the earth as it turns on its axis. This motion, called *precession*, is caused mainly by the moon's gravitational effects on the earth. Precession causes the earth's axis to move in a slow circle, completing one turn every 26,000 years. See Figure 3–26. A spinning top also shows precession when its upper end moves in a small circle much more slowly than the top spins. Precession of the earth's axis causes two successive vernal equinoxes to occur about 20 minutes sooner than the time it takes the earth to make one revolution. The true period of revolution of the earth can be measured by reference points among the stars. Stars are so far away that earth's precession makes little difference in their observed positions. A year measured by the stars is called a *sidereal year* (sy-*deer*-ee-al) and is sometimes used by astronomers.

Rotation. For people on the earth, the most important effect of the earth's rotation on its axis is the rising and setting of the sun. Since the earth rotates from west to east, the sun appears to rise in the east, travel across the sky, and set in the west. The sun's path across the sky shifts northward in summer and southward in winter. This is a result of the changing angle at which we see the sun as the poles tilt toward and away from the sun with the changing seasons.

The daily apparent movement of the sun across the sky is the basis for all of our common ways of measuring the passage of time. For example, a sundial uses the changing position of a shadow to follow the sun across the sky. The position of the shadow on the face of the sundial shows the time. See Figure 3–27. Time measured with a sundial is called *apparent solar time* because it uses a shad-

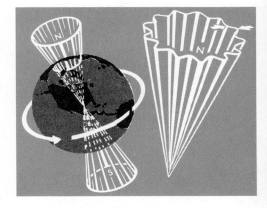

FIG. 3–26. Precession causes the earth's poles to trace small circles as the earth turns on its axis. The circles are not smooth, but have a slight in-out motion as shown above.

FIG. 3–27. A sundial shows apparent solar time by the position of the sun's shadow on the face of the dial. (John V.A. Neal/Photo Researchers)

2. If one A.U. is 10 centimeters, how many A.U. is the Mercury-to-sun distance?

3. If the earth-to-sun distance is 150 million kilometers, what is the Mercury-to-sun distance in kilometers?

Venus has the same kind of limit, but the maximum angle is 48° between the earth-sun line and the earth-Venus line.

C. On your drawing of the orbit of Mercury, add the 48° angle and draw the circle to represent the orbit of Venus.

4. What length is the Venus-to-sun distance in your drawing?

5. How many A.U. is the Venus-to-sun distance?

6. Using 150 million kilometers as the earth-to-sun distance, what is the Venus-to-sun distance?

Investigate
Determine how a sundial works.
Why is it tilted like it is? Will any
tilt work? Set up a sundial and
compare its timekeeping
accuracy with other clocks.

ow that moves as a result of the apparent change of the sun's position in the sky. It would seem that a sundial is a good way to measure the length of "one day," or the time needed for the earth to make one rotation on its axis. Each time the earth turns once, the sun should return to the same position in the sky. If this were true, then the time taken for the sundial shadow to return to the same place on the face of the sundial should measure the length of one day of 24 hours. However, if you try to use a sundial in this way for a year, you will find that the days you measure are not all the same length. Your sundial will run too fast during some months of the year and too slow during others. This means that either the earth's rotation is faster in winter and slower in summer, or the sun does not cross the sky at the same speed all year. If the speed of the earth's rotation is measured against the stars, it is found to change by only about 0.001 second per day. Thus the much larger error in a sundial must result from the apparent change in the speed of the sun.

There are two reasons why the sun does not cross the sky at the same rate during the year. First, the earth moves faster at perihelion when its elliptical orbit brings it closest to the sun. This happens during the winter season in the Northern Hemisphere. When the earth is moving fastest in its orbit, it moves farthest ahead during one rotation. Thus it must rotate farther to bring the sun back to a given position in the sky. This makes the sun appear to move more slowly across the sky. Six months later, when the earth is at aphelion, the sun appears to cross the sky more rapidly.

Second, the angle at which we view the sun makes it appear to change its speed. Near the solstices in summer and winter, the sun appears to move almost directly across the sky from east to west. Thus it will appear to cover a given distance in a short time. On the other hand, near an equinox in spring or fall, the sun appears to move farthest to the north or south at the same time it crosses the sky. To cover the extra distance, the sun needs more time to complete a given part of its daily trip and appears to move more slowly. When the effects of the speed of the earth in its orbit and the angle at which we see the sun move are combined, shadows cast by the sun move across the face of a sundial faster in summer and winter than in spring and fall. Sundials measure off days that differ in length from one part of the year to another.

Clocks are set to measure a day that is the average, or mean, of the length of the days measured by sundials throughout a year. The time measured by clocks is called *mean solar time*. The use of mean solar time means that all days in the year have the same length.

Suppose that clocks where you live were set to read 12 noon when the sun was nearly overhead. If another person in a city or town to the west of you did the same thing, that clock would read 12 noon a little later than yours. The difference would be the time it takes the sun to move west and appear overhead. Even places close together would have clocks set a few minutes apart. To prevent this problem, the earth's surface is marked off into 24 *standard time zones.* Standard time in each zone is defined as the mean solar time for the center of that zone. When you cross the boundary of a zone going east, you set your clock one hour ahead. Going west, clocks are set back one hour. In the United States, with the exception of Alaska and Hawaii, there are four standard time zones, as shown in Figure 3–28.

During the summer, when there are more hours of daylight, many states adopt *Daylight Saving Time*. Clocks are set one hour ahead of standard time in April, and back again one hour in October. This gives an extra hour of daylight during the summer evenings. A simple rule for knowing how to change clocks is: "spring" ahead and "fall" back.

Since there are 24 standard time zones, travelers going

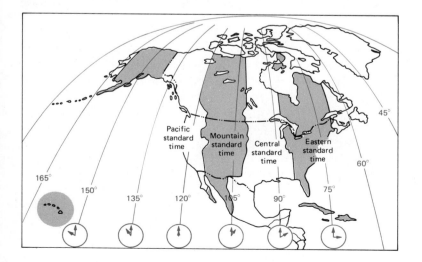

FIG. 3–28. Standard time zones for North America and Hawaii are shown on this map. Time zone boundaries are arranged so that all of each state has the same standard time. Most parts of Alaska and Hawaii are two hours earlier than Pacific Standard Time.

all the way around the world would have to set their clocks ahead or back by a total of 24 hours. They would lose or gain a day and end up a day ahead or behind of the place where the trip began. To prevent such confusion, the *International Date Line* has been established. See Figure 3–29. To anyone crossing this imaginary line, a day is immediately gained or lost. Going east to west, you switch to the next day; crossing west to east, a day is repeated.

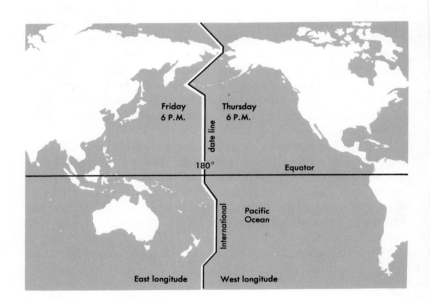

FIG. 3–29. The International Date Line runs through the Pacific Ocean, bending to avoid crossing land areas and inhabited islands.

VOCABULARY REVIEW

Match the word or words in the column on the right with the correct phrase in the column on the left. *Do not write in this book.*

1. A disk-shaped cloud from which it is thought solar systems are formed.
2. The earliest stage in the formation of a planet.
3. Chunks of matter found orbiting mostly between Mars and Jupiter.
4. Any part of an asteroid that survives to reach the earth's surface.
5. The day the sun is in its most northerly position in the sky.
6. The beginning of fall in the Northern Hemisphere.
7. The wobbly motion of the earth as it rotates on its axis.
8. Time measured with a sundial.
9. Time measured by clocks.
10. A time used in summer to give an extra hour of daylight in the evening.
11. The length of time it takes for the earth to complete one orbit around the sun.
12. When a planet crosses the sun's disk as viewed from earth.
13. The clock time of any one of the Standard Time Zones.
14. Marks the beginning of spring.
15. Traveling east to west, you switch to the next day.

a. meteorite
b. summer solstice
c. winter solstice
d. vernal equinox
e. mean solar time
f. solar nebula
g. asteroids
h. protoplanets
i. autumnal equinox
j. apparent solar time
k. Daylight Saving Time
l. precession
m. standard time zones
n. comets
o. International Date Line
p. standard time
q. year
r. transit

QUESTIONS

Group A

Select the best term to complete the following statements. *Do not write in this book.*

1. The theory accepted by most scientists of how the solar system was formed starts with (a) a cloud of dust and gas (b) the collision of two stars (c) two stars becoming companions (d) a small dark star.

2. One cubic centimeter of the air we breathe contains how many atoms? (a) 5 to 10 (b) 100 to 1,000 (c) 1,000,000 to 1,000,000,000 (d) over 1,000,000,000,000,000,000.

3. A shock wave sent out by a nearby exploding star is thought by some astronomers to have (a) made the dust and gas (b) caused the collision of two close stars (c) pushed the cloud of dust and gas closer together (d) pushed two stars close together, forming companions.

4. The inner protoplanets (a) grew to a large size (b) formed iron cores covered with rock (c) formed before the outer ones (d) formed from the light elements.

5. The four inner protoplanets became (a) Mercury, Venus, Earth, and Jupiter (b) Mercury, Venus, Earth, and Mars (c) Mercury, Venus, Mars, and Jupiter (d) Jupiter, Saturn, Uranus, and Neptune.

6. Which one of the following properties was not associated with the outer proto-planets? (a) small dense core (b) thick layers of ices (c) thick atmosphere (d) large light core.

7. Only the sun in our solar system had a core that was (a) made of iron (b) hot enough to cause nuclear fusion (c) made of helium (d) dense enough to collect gases.

8. It is thought that the original atmospheres of the inner planets were stripped away by (a) explosions on the sun (b) the sun's gravity (c) near collisions of other stars (d) near collisions of comets.

9. The sun ended up with (a) about 50% of the matter in the solar nebula (b) more than 50% of the matter in the solar nebula (c) about 90% of the matter in the solar nebula (d) more than 99% of the matter in the solar nebula.

10. It is thought that the asteroid belt exists because (a) no protoplanet formed between Mars and Jupiter to sweep the area clean (b) twin planets collided, forming many pieces of matter (c) a comet collided with a moon of Jupiter (d) a comet collided with a moon of Mars.

11. The largest asteroids have a size of more than (a) 100 km (b) 1000 km (c) 10,000 km (d) 100,000 km.

12. Some asteroid orbits are so elliptical that (a) they pass through Saturn's orbit (b) they are considered comets (c) they pass through the earth's orbit (d) they became lost from the solar system.

13. The three basic types of meteorites are (a) stony, iron, and carbon (b) stony, iron, and stone-iron mixed (c) stony, carbon, and stone-iron mixed (d) stony, iron, and iron-carbon mixed.

14. More than 90% of the meteorites are (a) carbon (b) iron (c) stone-iron mixed (d) stony.

15. Comets are thought to be (a) chunks of the sun (b) icy lumps found in the outer regions of the solar system (c) asteroids that escaped the asteroid belt (d) chunks from other stars that passed close to the solar system.

16. Most meteors are thought to be (a) tiny bits of comets (b) asteroids (c) dust from other solar systems (d) tiny bits of asteroids.

17. The smallest of the inner planets is (a) Mercury (b) Venus (c) Earth (d) Mars.

18. The number of days required for Mercury to make one revolution around the sun is (a) 59 (b) 70 (c) 88 (d) 370.

19. When we use a telescope, most characteristics of Venus are concealed from us because it is (a) too near the sun (b) too far away from the earth (c) too far away from the sun (d) covered by heavy clouds.

20. The planets that have orbits smaller than the earth's and therefore make transits of the sun are (a) Mercury and Venus (b) Mercury and Mars (c) Mars and Venus (d) Jupiter and Mars.

21. The diameter of Venus is (a) much greater than the earth's diameter (b) about the same as the earth's diameter (c) much less than the earth's diameter (d) not known.

22. The temperature on the surface of Venus is about (a) 200°C (b) 500°C (c) 1000°C (d) 2000°C.

23. Venus (a) has three moons (b) rotates in the same direction as the earth (c) rotates in the opposite direction from the earth (d) doesn't rotate.

24. The diameter of Mars is about half that of the earth, while its mass is about (a) one-half (b) one-fourth (c) one-fifth (d) one-tenth.

25. The period of rotation of Mars on its axis (a) is about the same as that of the earth (b) is much less than that of the earth (c) is much greater than that of the earth (d) cannot be determined.

26. Mars has seasons like the earth because (a) its days are nearly the same as the earth's (b) its axis of rotation is tilted similar to the earth's (c) its years are the same length as the earth's (d) it receives the same amount of energy.

27. Temperatures on Mars range from 20°C in summer to winter temperatures of (a) 100°C (b) 0°C (c) −40°C (d) −130°C.

28. The giant planet Jupiter (a) is 318 times more massive than earth (b) would hold 1400 earths (c) has a mass 2½ times greater than all the rest of the solar system, except the sun (d) has all these characteristics.

29. The colored bands of gas seen in Jupiter's atmosphere are caused mostly by (a) the rapid rotation of the planet (b) the rapid revolution of the planet about the sun (c) the Giant Red Spot (d) the rings that surround the planet.

30. The largest, brightest, and most complicated rings in the solar system are found around (a) Jupiter (b) Saturn (c) Neptune (d) Uranus.

31. The two main reasons the sun appears not to cross the sky at the same rate during the year are (a) the rotation of the earth is not constant; the earth varies its speed around the sun (b) the earth varies its speed around the sun; the angle we view the sun varies from season to season (c) the sun moves faster in summer than winter; the angle we view the sun varies from season to season (d) the rotation of the earth varies; the angle we view the sun varies from season to season.

32. Seasons occur on earth because its axis is tilted. This inclination of the axis is (a) 3 degrees (b) 12.5 degrees (c) 23.5 degrees (d) 31.5 degrees.

33. Apparent solar time is not used for clocks because it (a) never changes (b) changes constantly (c) changes occasionally (d) depends on the moon phase.

34. The number of standard time zones in the entire world is (a) 12 (b) 24 (c) 36 (d) 360.

35. One reason for establishment of the International Date Line was to keep one location from having (a) two dates at one time (b) two times a day (c) too long a day (d) too short a day.

Group B

1. Describe how it is thought the solar nebula formed.

2. What were protoplanets? How and when did they form?

3. Describe how the cores of the inner planets are thought to have been formed.

4. How did the sun affect the atmosphere of the protoplanets?

5. Why is Mercury visible only shortly before sunrise and shortly after sunset?

6. Discuss the length of time it takes for Mars to go around the sun and its effect on the planet's temperature.

7. State the evidence that shows that the surface of Mars is changing.

8. List the planets in order of size, starting with Jupiter. Also give the number of natural satellites of each.

9. Explain why some scientists think that an additional planet may have failed to develop when the solar system formed. If their theory is true, where would it be located?

10. Why do the rings of Saturn occasionally fail to show up in photographs?

11. What is a transit of the sun? Which planets viewed from the earth exhibit this phenomenon?

12. Describe the relationship between asteroids, comets, meteors, and meteorites.

13. Name several ways the inner planets are alike. Describe some of the differences.

14. Compare in a general way the outer planets with the inner planets with respect to (a) distance from sun (b) sunlight received (c) makeup of atmosphere (d) makeup of the planet sphere (e) time to complete an orbit of the sun (f) rotation on axis.

15. Discuss the following, concerning asteroids: (a) where found (b) their makeup (c) size (d) importance for scientific study (e) effect on the earth when collisions occur.

16. Why does apparent solar time change continuously?

17. At what time of year are apparent solar time and mean solar time nearly the same? Why?

18. Explain standard time zones and why we have them.

19. Describe the apparent north-south motion of the sun during a single year. Give dates that help locate its position.

20. In your own words describe each of the following kinds of time. (a) Apparent Solar Time (b) Mean Solar Time (c) Standard Time.

21. Describe Daylight Saving Time. What are its benefits? Does it have any bad features?

22. Discuss the effect on seasons of the year if the earth's axis were not tilted to the plane of its orbit.

"I'm at the foot of the ladder. That is one small step for man, one giant leap for mankind. Yes, the surface is fine and powdery. I can kick it up loosely with my toe. I can see the footprints. There seems to be no difficulty moving around. No trouble." With these words, the American astronaut Neil Armstrong stepped from a spacecraft onto the surface of the moon on July 20, 1969. A new era of history began with the first visit by an earth dweller to another member of the solar system.

In the years following the first landing, a total of twelve astronauts walked on the lunar surface. They brought back to earth about 383 kg (843 lbs) of moon rocks for analysis. They left on the moon equipment for radioing back to earth reports of moon-quakes and meteorite impacts. Manned explorations of the moon ended in 1972, but scientists have continued to study lunar rocks and data from experiments left on the moon. As a result, the moon has become, next to the earth, the best understood body in the solar system. Many questions still remain unanswered, but the moon is now a more familiar place. It is seen as a barren world very different from the earth.

EXPLORING THE MOON

The visible moon. Unlike the other heavenly bodies, the moon's surface features can be seen from the earth without a telescope. This is possible because the moon is on the average only 384,400 km (238,700 mi) away. Also, it has no atmosphere to act as a cover. With the help of a little previous knowledge, you can make out much of the lunar

objectives

☐ Describe the surface of the moon.

☐ Compare the conditions on the surface of the moon with those on the earth.

☐ List four stages in the moon's history.

☐ Draw a diagram showing positions of the sun, moon, and earth during each of the moon's phases.

☐ Explain how solar and lunar eclipses occur.

☐ Show how the moon and sun produce ocean tides.

☐ Compare the Julian, Gregorian, and World calendars.

FIG. 4–1. The full face of the moon as it appears to an earth-based telescope. The bright crater at the bottom is Tycho. (NASA)

Observe
Look at the moon through a telescope or binoculars. How does the appearance differ from naked-eye observations? What can you see that you could not see before?

surface with your naked eye. To begin, go out on a night when the moon is nearly full. Then the moon's disk is seen as in Figure 4–1.

You will immediately notice that there are light and dark regions on the moon. The light areas are rough highlands with many sloping surfaces that catch and reflect the sunlight. Dark areas, on the other hand, are smooth and reflect little light. These dark regions are called *maria*, from the Latin "mare" meaning "sea." The maria are actually dry, nearly level surfaces. There is no liquid water on the moon. The eyes of the "Man in the Moon" are formed by maria with round shapes. The nose and mouth are formed by other maria.

It is easy to see that the moon's surface is covered by craters. If you imagine the moon's disk as a clock face, at about 7 o'clock there is a large crater with bright streaks called *rays* extending out from it. It is called Tycho (*Ti-ko*). The moon's largest craters, like this one, are given names. At about the 9 o'clock position on the moon's face is the large crater called Copernicus. This crater also has rays. Large craters such as Tycho and Copernicus have diameters of 100 km (60 mi) or more. Close-up examination of the lunar surface shows millions of smaller craters, many overlapping one another. See Figure 4–2. Even moon rocks are often covered with tiny craters that become visible when seen under a microscope. See Figure 4–3.

Scientists believe that most of the moon's craters were made about 4 billion years ago. Like the other members of the solar system, the moon was bombarded by chunks of matter still remaining after the birth of the planets. The effect of such collisions is shown in Figure 4–4. Rays now seen around many lunar craters were made by material thrown out when the crater was created. Craters similar to moon craters must have once existed on the earth, but they have almost completely disappeared. The earth's surface has been continuously changed during the past 3 billion years. There has been little change on the moon because it lacks the atmosphere necessary to wear away its surface. The lunar landscape is like a museum display of a solid planet existing 3 billion years ago.

The surface of the moon. How would your life be different if you lived in some future colony on the moon? It would be a very strange kind of life, since conditions on the lunar surface are very different from those on earth. The lack of an atmosphere, for example, would require you to always be in the artificial environment of a space suit or some kind of shelter. The temperature can reach 134°C (273° F) during periods of daylight, and plunge to −170° C

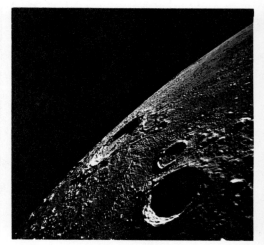

FIG. 4–2. The surface of the moon as seen from a spacecraft. (NASA)

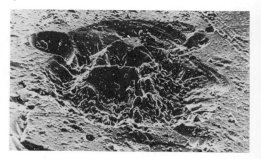

FIG. 4–3. Tiny craters made by meteorites can be seen when lunar rocks are examined under a microscope. (NASA)

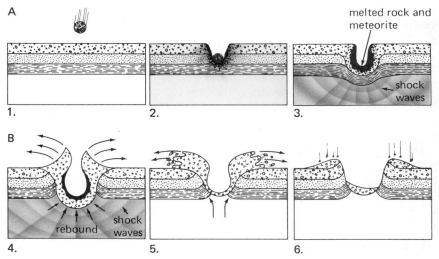

FIG. 4–4. (Left) These diagrams show how a crater is made when a meteorite strikes a solid surface. (A) The energy of the meteorite is changed into heat and shock waves. (B) The rebound of the shock waves causes debris to be thrown out as solid molten rock. Material splashed out of the craters can form smaller secondary craters and rays.

FIG. 4-5. The moon turns on its axis once in the same time it takes to make one revolution around the earth. This means that a spaceman on the moon would always face the earth and spend about half of the time taken to make one revolution in daylight and half in darkness.

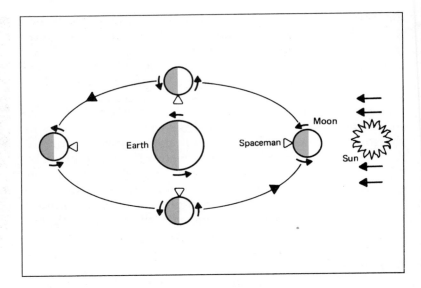

FIG. 4-6. The far side of the moon photographed from a spacecraft. (NASA)

(−338° F) at night. These temperatures are partly the result of the lack of a lunar atmosphere. Also, lunar days and nights last for about two weeks because the moon turns on its axis once every 27.3 earth days. The lighted side of the moon receives the full blast of solar radiation during the long period of daylight. At the same time, heat is rapidly lost from the dark side.

The moon's slow rotation on its axis is a result of the early effect of the earth's gravity. When the moon was young, the gravitational pull of the earth caused a bulge in the part of the moon facing the earth. Since the bulge was strongly attracted by the earth's gravity, it had to always face toward the earth. Thus the moon started rotating in a way that kept the same side facing the earth. The moon finally became locked into a period of rotation equal to the 27.3 days it takes to complete its orbit around the earth. See Figure 4-5. Today we always see the same side of the moon from the earth. Photographs taken by spacecraft of the far side of the moon show the surface to be heavily cratered highlands with none of the large maria seen on the earth-facing side. See Figure 4-6.

Although the moon's extreme temperatures might make living there difficult, another characteristic could make it rather pleasant. The moon's mass is 1/81 that of the earth, and its diameter is about one-quarter the size of the earth's. These two factors combine to give the moon a sur-

FIG. 4–7. This footprint of an astronaut on the moon shows that much of the surface is covered by a fine dust. (NASA)

face gravity of only ⅙ that of the earth. This means a person weighing 72 kg (158 lb) on the earth would weigh only 12 kg (26 lb) on the moon. A hop on the moon would carry you four times farther up than on the earth. A fall would be gentle and cause no harm. The moon's low surface gravity is probably responsible for its lack of an atmosphere. Gases that may have formed an atmosphere could not be held by the weak gravity and were allowed to escape into space.

Almost all of the lunar surface is covered by a layer of loose, gray dust. See Figure 4–7. In areas of the moon that have been explored, the depth of the layer of gray dust has been found to be about 3 m (about 10 ft). This dust has been formed as the rocks exposed on the moon's surface have been broken up through millions of years of bombardment by small meteorites. The many tiny "shooting stars" that we see burning up high in the earth's atmosphere reach the moon's unprotected surface like a very slow but steady rain.

Examination of lunar rocks brought back by astronauts shows that the moon is made of the same materials as the earth, but not in the same proportions. Like rocks in the earth's crust, lunar rocks near the surface are composed mainly of oxygen and silicon. Rocks taken from the lunar highlands are also rich in calcium and aluminum. They are light colored, as shown in Figure 4–8. Rocks from the

FIG. 4–8. A sample of the light colored rock found in the lunar highlands. (NASA)

FIG. 4–9. A sample of the dark colored rock found in the maria of the moon. (NASA)

maria are dark colored and contain larger amounts of titanium, magnesium, and iron. See Figure 4–9. All lunar surface rocks are missing the lighter materials, such as hydrogen, that can easily escape the weak gravity. For reasons not yet discovered by scientists, the moon seems to have much less iron than the earth. This might mean that the moon has only a small, heavy iron core. A small core could account for the moon's relatively low mass compared with its size, as well as its lack of a magnetic field.

History of the moon. Study of moon rocks along with the data gathered by astronauts has provided scientists with the knowledge needed to develop a general understanding of the moon's history. Research has led most scientists to the conclusion that the moon has gone through several separate stages in its development. The *first stage,* the period about which the least is known, is its origin. Before astronauts visited the moon, there were three conflicting theories about the moon's origin; that the moon split off from a very fast-whirling young earth; that it was captured by the earth after being formed elsewhere in the solar system; that it was formed at about the same time and from the same materials as the earth.

Analysis of lunar rocks does not rule out any of these possibilities. However, the evidence now available seems to point toward the idea that the moon and earth were formed at about the same time. Among the evidence supporting this view of the moon's birth is the fact that both moon and earth rocks are similar in composition. Also, the age of the oldest known lunar rock is about 4½ billion years. This is approximately the age of the oldest rocks on the earth. Scientists believe that they can determine the age of rocks by measurement of radioactive substances found in the rocks. Both the earth and moon could have been formed at nearly the same time in the same region of space within the cloud from which the solar system grew. Or the moon may have formed when a swarm of lumps of material orbiting the young earth was pulled together by mutual gravity. The final answer to the question of the moon's origin cannot be provided until further research supplies the missing information.

However the moon came into existence, it seems certain that the *second stage* of its history began with its surface

covered by an ocean of white-hot molten rock. The heat
could have come from the energy of the smaller bodies
that came together to form the moon. If the moon grew
rapidly, the colliding bodies would have created more heat
than could be lost by cooling. The accumulated heat may
have melted the entire moon. However, there is evidence
to show that a molten layer at least several hundred kilometers deep was created.

Melting of the lunar material created the *third stage* of
the moon's development. During this stage, the light and
heavy materials in the molten rock were separated. Lighter materials floated up, while the heavier materials sank.
The result was the formation of an outer layer made of
lighter material. Below this was a thick layer of more
dense material. If the entire moon became molten, the
heaviest materials, such as iron, probably sank to the center to form a small, heavy core.

The *fourth stage* began when the molten surface of the
moon cooled to form a thick, solid crust. Then the intense
bombardment by meteorites began to mark the lunar surface with craters. Some very large impacts tore huge circular basins in the lunar crust. Rock that was still molten
flooded out from beneath the cooling crust to fill these
huge basins. As the rock cooled, it formed the dark-colored
maria now seen on the moon. Apparently the far side of
the moon has a thicker crust, since the collisions on that
side were not able to break through the crust to form maria. The impacts of the large meteorites also seem to have
caused mountains to be thrust up in many parts of the
lunar surface. Finally, bombardment of the moon grew
less and almost ceased as the solar system was almost
swept clean of loose fragments. Then, about 3 billion
years ago, the moon became quiet. With the exception of a
few craters, a picture of the moon taken 3 billion years ago
would look the same as one taken today.

MOTIONS OF THE MOON

Orbit of the moon. If you were on Mars or Venus studying
the earth with a telescope, your attention might be drawn
as much to the moon as to its parent planet, earth. You
would see what appears to be a large and a small planet.
As both traveled around the sun, the smaller body would

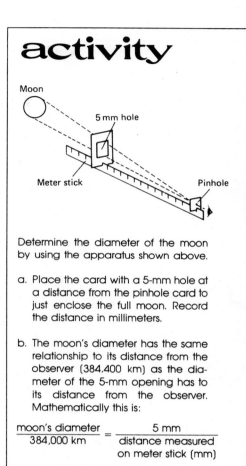

activity

Determine the diameter of the moon
by using the apparatus shown above.

a. Place the card with a 5-mm hole at
a distance from the pinhole card to
just enclose the full moon. Record
the distance in millimeters.

b. The moon's diameter has the same
relationship to its distance from the
observer (384,400 km) as the diameter of the 5-mm opening has to
its distance from the observer.
Mathematically this is:

$$\frac{\text{moon's diameter}}{384,000 \text{ km}} = \frac{5 \text{ mm}}{\text{distance measured on meter stick (mm)}}$$

Interpret
**What are rills on the moon? Was
Galileo right when he stated
that rills on the moon are the
remains of ancient stream
channels? Use your library.**

move first ahead, then behind the larger. From a distance, you might logically decide that the earth and moon are a double-planet system. No other planet in the solar system has a companion so near its own size. The moon's diameter is 3476 km (2159 mi), compared to the earth's diameter of 12,756 km (7921 mi). The earth and moon revolve around the sun as a double-planet system tied together by their mutual gravitational attraction.

To understand how the earth and moon affect each other's motions, imagine that the two bodies are connected by a weightless rod similar to the one shown in Figure 4–10. The balance point for this uneven, dumbbell-shaped system is closer to the earth's center than it is to the moon's. The actual location of the balance point is below the earth's surface on the side facing the moon. See Figure 4–10. The effects from the gravity of the earth and the moon cause both bodies to circle around this point as they revolve around the sun. At the same time the moon circles the earth, the earth moves in a small orbit around the balance point. Only the center of the earth-moon system follows a smooth orbit around the sun. See Figure 4–11.

As we observe the moon from the earth, it appears to follow an orbit around the earth moving at a speed of about 3665 km/hr (about 2278 mi/hr). At the same time, it turns on its axis in the same direction as the earth rotates (from west to east). Its orbit around the earth has an elliptical shape, which brings it a little closer to the earth at one point in its orbit than at any other. When the moon is closest to the earth, it is said to be at *perigee* (*pehr*-i-gee). At perigee, the distance from the center of the earth to the

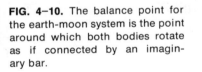

FIG. 4–10. The balance point for the earth-moon system is the point around which both bodies rotate as if connected by an imaginary bar.

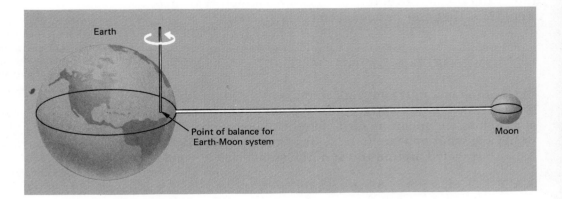

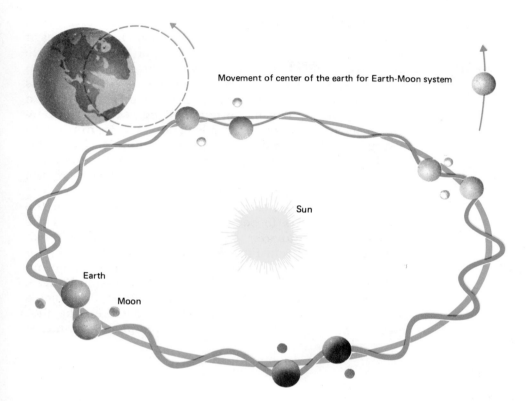

Movement of center of the earth for Earth-Moon system

Sun

Earth

Moon

center of the moon is 360,000 km (225,000 mi). At *apogee* (*ap*-o-gee), the farthest point, it is 404,800 km (253,000 mi) from center to center. The moon's average distance from the earth is 384,400 km (238,700 mi). By bouncing laser light off reflectors left on the moon by astronauts, scientists can measure the distance to the moon from any place on the earth with an accuracy of about 10 cm (about 4 in). See Figure 4–12.

FIG. 4–11. Only the balance point of the earth-moon system actually follows a smooth orbit around the sun. The path of the earth is slightly wavy as it turns around the balance point as shown at top.

FIG. 4–12. A laser beam sent to the moon and reflected back by equipment left on the moon by astronauts is used to measure the exact distance between earth and moon. (University of Texas McDonald Observatory at Mount Locke)

Explain
The phrase "dark side of the moon" refers to the portion of the moon's surface that faces away from the earth. What is wrong with this term?

Changing appearance of the moon. As the earth-moon system moves around the sun, the moon's path takes it first between the sun and earth and then to the dark side of the earth. See Figure 4–13. When the moon is between the sun and earth (position 1 in Figure 4–13), we see its unlighted side. The appearance of the moon is said to be in the *new phase*. The moon can still be seen in the new phase because its unlighted side is not dark. It shines with a dim gray light that comes from sunlight reflected off earth's clouds and oceans onto the moon, and then off the moon's surface back to earth. The light reflected off the earth is called *earthshine*.

As the moon moves in its orbit around the earth, the lighted half starts to become visible. This produces the *crescent, first quarter, gibbous,* and *full phases* (positions 2, 3, 4, and 5 in Figure 4–13). At the full phase, the moon appears to be very bright; however, it is reflecting only 7% of the sunlight falling on it. Black asphalt pavement reflects about the same amount of light. The moon's brightness does increase very rapidly between the first quarter and full phases. At its quarter phases, sunlight

FIG. 4–13. As the moon revolves around the earth, one half is always lighted by the sun. Phases of the moon are the result of our changing view of the lighted side.

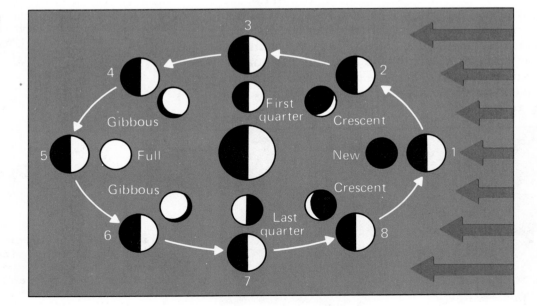

falls on the rough surface of the moon at right angles. As a result, the sunlight falls straight on the most visible part, making few shadows. After the full phase, the moon then passes through gibbous, last quarter, and crescent phases (positions 6, 7, and 8).

While the moon takes 27⅓ days to make one revolution around the earth, it takes slightly longer to go through a complete series of its phases. The period from one new moon to the next is 29½ days. The reason for this time difference can be seen in Figure 4–14. The moon at position A is in its new phase. After 27⅓ days it has moved into position B, completing one revolution and one rotation on its axis. However, the earth has also moved farther around the sun. Thus the moon must go on to position C before it can return to its new phase again. This accounts for the difference in time between 27⅓ days for one revolution and 29½ days for one cycle of phases.

If you watch the moon rise on successive nights, you will find that it rises later each night. The rising and setting of the moon is the result of the rotation of the earth on its axis. As the moon revolves around the earth from west to east, the earth rotates in the same direction. Because the moon is always moving ahead in its orbit, the earth needs to rotate a little longer each day to "catch up." This extra turning takes the earth an additional 50 minutes. Thus the moon rises about 50 minutes later each night. But it also sets 50 minutes later. The times of moon rise and setting and the phases seen at those times are shown in Figure 4–15.

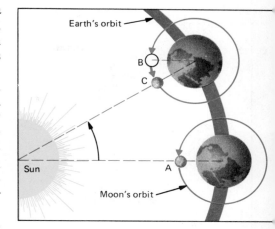

FIG. 4–14. For the moon to travel from new moon position (A) to the next new moon position (C), it must complete one revolution plus the distance from B to C.

FIG. 4–15. This diagram shows the time of day when the moon will be seen to rise and set with the phase given at the bottom. For example, the new moon rises near dawn and crosses the sky during daylight. The full moon rises near sunset and is seen during the night.

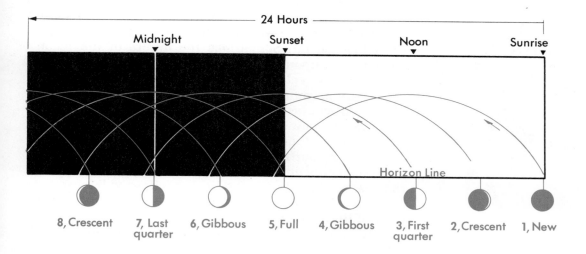

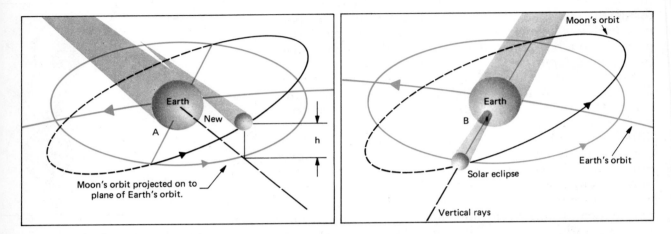

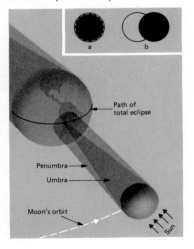

FIG. 4–16. Top left diagram shows that the tilt of the moon's orbit (h) causes the moon's shadow to fall above or below the earth. An eclipse occurs when the moon happens to pass directly between the sun and earth, as shown in the right diagram.

FIG. 4–17. An observer in position (a) on the earth will see a total eclipse, while those in position (b) will see a partial eclipse.

Eclipses of the sun and moon. All of the bodies moving around the sun, such as the earth and moon, cast long shadows into space. An eclipse occurs when either the earth or the moon passes through the other's shadow. Most of the time, however, they move together around the sun without an eclipse taking place. This is because the path of the moon around the earth is tilted about 5° to the plane of the earth's orbit around the sun. See Figure 4–16. This means that the moon usually crosses between the earth and the sun either too high or too low for its shadow to fall upon the earth. Likewise, when the moon passes on the side of the earth opposite the sun, it usually misses the earth's shadow.

Shadows cast by the earth and moon are in two parts. One is a cone-shaped inner part completely cut off from the sun. This is called the *umbra*. The other is an outer part called the *penumbra,* in which sunlight is only partially blocked off. Sometimes the umbra of the moon's shadow falls upon the earth. All people within the umbra will see a *total eclipse of the sun,* or a *solar eclipse.* That is, the sun is completely covered by the moon. An eclipse of the sun occurs only during the new moon phase. This is when the shadow cast by the moon crosses the earth. See Figure 4–17. Total eclipses of the sun do not occur very often. At any particular place a total solar eclipse is likely to be seen only once every several hundred years. Before the year 2000, only one total solar eclipse will be seen in the entire United States. This will be visible from many parts of the country on July 11, 1991.

Only a small part of the world is able to see any particu-

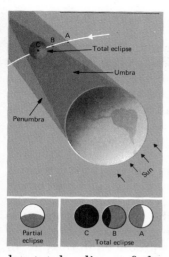

FIG. 4–18. An eclipse of the moon occurs when the moon passes into the earth's shadow. A total eclipse is seen if the moon passes through the umbra. A partial eclipse is produced if the moon passes only through the penumbra.

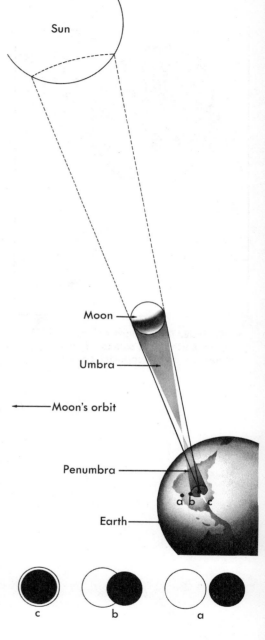

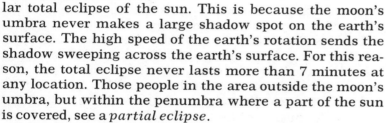

lar total eclipse of the sun. This is because the moon's umbra never makes a large shadow spot on the earth's surface. The high speed of the earth's rotation sends the shadow sweeping across the earth's surface. For this reason, the total eclipse never lasts more than 7 minutes at any location. Those people in the area outside the moon's umbra, but within the penumbra where a part of the sun is covered, see a *partial eclipse.*

An *eclipse of the moon,* or *lunar eclipse,* occurs during the full moon phase, when the earth's shadow may cross the lighted half of the moon. To produce a total eclipse, the moon must pass completely into the earth's umbra, as shown in Figure 4–18. A partial eclipse of the moon occurs when the earth's shadow passes across one edge of the moon. Though eclipses of the moon occur only about as often as eclipses of the sun, the lunar eclipses are seen by more people. An eclipse of the moon can be seen everywhere on the dark side of the earth. But a total solar eclipse is seen only by those in the small shadow path of the moon's umbra as it glides across the earth's surface.

Sometimes the moon passes directly between the earth and the sun in a position that could produce an eclipse of the sun. However, the moon's umbra may be too short to reach the earth. This will take place only if the moon is at or near apogee when it comes between the earth and the sun. If the moon's umbra fails to reach the earth, an *annular* (ring-shaped) *eclipse* will occur. In an annular eclipse, the sun is not completely blotted out. Instead, it shows a thin ring of light around the outer edge, as shown in Figure 4–19 (Stage c).

FIG. 4–19. During an annular eclipse the moon is at its greatest distance from the earth so that the umbra of its shadow falls short of the earth's surface. An observer at position c would see the edges of the sun.

FIG. 4–20. Tidal bulges are raised on the earth on its side facing the moon and on the opposite side.

TIDES

The moon's influence on the oceans. The earth and the moon have a gravitational influence on each other. All parts of the moon are attracted toward the earth. In the same way, all parts of the earth are attracted toward the moon. It is this mutual attraction between the two bodies that causes the earth and moon to move as a single system. But the moon's gravity has one very obvious effect on the earth.

On the side of the moon facing the earth, the moon's gravity pulls out a bulge in the earth. The solid part of the earth bulges out only very slightly. But the water of the oceans moves more easily than the solid earth. Thus the bulge produced in the parts of the earth covered by water is very noticeable. A *tidal bulge* is produced in the part of the sea facing the moon. This is a clear example of the effect of the moon's gravity on the earth.

There is also a tidal bulge on the side of the earth opposite the moon. See Figure 4–20. It is probably less clear why the moon's gravity can also be responsible for a tidal bulge on the opposite side of the earth. This opposite tidal bulge can be explained as a result of smaller gravitational pull by the moon on the far side of the earth. The water there is farther from the moon and is affected less by its gravity. This allows the forces caused by the rotation of the earth and moon around a common center to push water away from the earth. Thus a tidal bulge is created there also. The bulge on the side facing the moon is called the *direct tide*. The bulge on the other side is called the *opposite tide*.

The two tidal bulges follow the moon as it moves around the earth. If the earth did not turn on its axis, the tidal bulges would produce a high tide about every two weeks as they passed each location on the ocean shores. But the earth does rotate, once every day. Thus we should expect two high tides every 24 hours as the earth rotates and the tidal bulges pass by. However, the bulges also follow the moon's progress around the earth, so the earth's rotation has to "catch up" with the moving tides. It takes about 24 hours and 50 minutes for both high tides to pass a given spot. The high tides actually do not occur when the moon is directly overhead, but several hours later. This is because the earth's rotation carries the tidal bulges ahead of their expected position in a direct line with the moon.

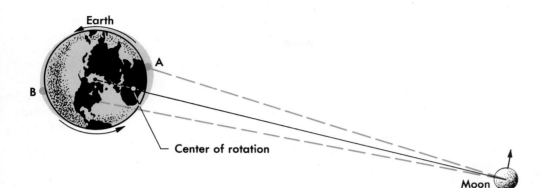

The tides in the earth's oceans also affect the speed of the earth's rotation. The tidal bulges are pulled around by the moon more slowly than the earth rotates. Friction against the sea floor from the moving masses of water in the tidal bulges tends to slow down the earth's rotation. The effect of this friction makes each day slightly longer than the previous one. In 1000 years, the length of each day will be 0.2 seconds longer than it is now.

At the same time the tides are slowing down the rotation of the earth, they are causing the moon to revolve more rapidly. This is because the earth's rotation carries the tidal bulges forward from their expected position directly under the moon. Figure 4–21 shows this effect. As you can see in the diagrams, bulge A, which is closer to the moon, has a greater attraction for the moon than does bulge B. The attraction of bulge A for the moon tends to pull the moon ahead, speeding it up slightly. As the moon increases in speed, it also moves farther from the earth. If the moon did exist during the earth's early history, it was probably much closer to the earth than it is now.

The sun also influences the tides. The sun's gravitational force also raises tidal bulges on the earth. However, these tides are very small because of the sun's greater distance from the earth. Twice each month the sun and moon are in line with each other (at full moon and new moon). At such times, the tidal effects of both the sun and moon are combined. The high tides produced as a result of this combined force are higher than usual and are known as *spring tides*. (The term is not related to the spring season.) See Figure 4–22.

FIG. 4–21. Due to the earth's rotation the tidal bulges (A and B) do not occur directly between the earth and moon. This effect causes the moon to speed up very slightly.

FIG. 4–22. Spring tides occur when the sun, moon, and earth are lined up.

FIG. 4–23. Neap tides occur during the new moon phase when sun, moon, and earth are lined up.

FIG. 4–24. The level of the water along a shoreline at various tides.

In the first and last quarter phases of the moon, the sun and moon are at right angles to each other. This produces lower than normal tides, called *neap tides*. See Figure 4–23.

The influence of the moon on the tidal bulges also changes as the moon's distance from the earth changes. At perigee the moon is closer than at apogee, and its gravitational pull is greater. If the moon is at perigee at the time of its new or full phase (spring tides), the tides will be much higher than usual. See Figure 4–24. This event does not occur very often and can be predicted in advance. Coastal regions may have flooding problems during perigee spring tides, particularly if the weather is stormy.

The actual tides that are experienced at any location along the ocean shore are the result of many influences. Usually the rise and fall of the tides at any particular place does not follow the expected pattern of two high and two low tides in 24 hours and 50 minutes. Some other factors that determine tidal action at different places are discussed in Chapter 17.

THE CALENDAR

The moon and the calendar. The calendar month was once based on the 29½ day cycle of the phases of the moon.

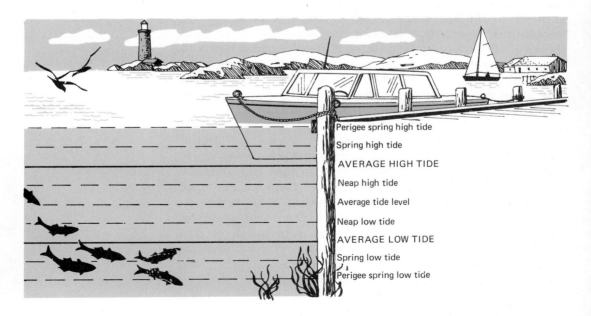

Perigee spring high tide
Spring high tide
AVERAGE HIGH TIDE
Neap high tide
Average tide level
Neap low tide
AVERAGE LOW TIDE
Spring low tide
Perigee spring low tide

This was a natural way of marking the passage of the year, since the phases of the moon are very obvious and regular. But a year is 365¼ days, a number that cannot be evenly divided by 29½. This made it impossible to develop a calendar in which the length of the months is based on the phases of the moon.

The calendar of the Romans, which laid a foundation for our modern calendar, once had six months of 30 days and six of 29 days. However, this still left each calendar year about eleven days short of the actual year. The Romans finally solved the problem by adding an extra month every few years to catch up again. During the reign of Julius Caesar, around 45 B.C., the Roman calendar had become so confused that Caesar ordered a complete change. The length of the months was increased to 30 or 31 days, with one month having only 28 days. This system ignored the phases of the moon and resulted in 12 calendar months out of a yearly total of 365 days. Since a year is actually about 365¼ days long, it was necessary for the new calendar to put in an extra day every four years. Leap year, which has one extra day, occurs whenever the last two digits of the calendar year are evenly divisible by four. This calendar became known as the Julian (after Julius Caesar) calendar.

Although it was a great improvement, the Julian calendar still was not quite correct. A year is actually 11 minutes and 14 seconds short of exactly 365¼ days. After the Julian calendar had been in use for hundreds of years, the months began to fall behind the seasons of the year. The vernal equinox had moved from March 21 back to March 11. In 1582, Pope Gregory ordered a change to bring the dates for church celebrations such as Easter back to their proper time of the year. The change was accomplished by taking out of the year 1582 the ten days that were gained. By official decree of the Pope, October 5 became October 15. See Figure 4–25.

At the same time, the century years (1800, 1900, etc.) were made leap years only if they could be evenly divided by 400. This omits three leap years every 400 years and satisfactorily keeps the months in agreement with the seasons. This form of the calendar is called the Gregorian calendar and is used by almost all countries.

A problem with the Gregorian calendar is that a normal year has 52 weeks and one extra day. A leap year has 52 weeks and two extra days. Thus a calendar date always falls one day later in the week each successive year. On

1582			October		1582	
Sun.	Mon.	Tue.	Wed.	Thu.	Fri.	Sat.
	1	2	3	4	15	16
17	18	19	20	21	22	23
24	25	26	27	28	29	30
31						

FIG. 4–25. When most of Europe changed to the Gregorian calendar in 1582, ten days were dropped from October. On this calendar page, colored numbers represent Julian dates and black numbers Gregorian dates.

FIG. 4–26. Some of the advantages of a World Calendar as shown below are:

1. A year of twelve months is easily divisible since all dates always fall on the same days.
2. Each year begins on the first day of the week, Sunday, January 1.
3. Seasons are equalized into quarters, each having three months or 13 weeks.
4. Each quarter-year is arranged in the same pattern of three months of 31, 30, and 30 days.
5. Each season begins on a Sunday and ends on a Saturday.
6. Holidays always fall on the same day and date.

leap year, the date "leaps" over a day. A more orderly calendar with respect to the dates and days of the week would have many advantages.

Many new kinds of calendars have been invented. For example, a proposed Universal Calendar has 13 months of 28 days each. This amounts to 364 days. An extra day without a date is therefore added at the end of each year. For leap year, two extra days are added. The chief objection to this calendar is that a year of 13 months cannot be divided evenly in halves or quarters.

A World Calendar with 12 months has also been proposed. The first month of each quarter (January, April, July, and October) has 31 days. All the rest have 30 days. This system requires the addition of one extra day to the end of the year. The extra day would be a holiday known as "World Day." It would arrive between December 30 and January 1. In a leap year, a second World Day would be added between June 30 and July 1. See Figure 4–26.

Changes in the calendar are not easily made. For that reason, it seems likely that the calendar will remain as it is for some time.

JANUARY	FEBRUARY	MARCH	APRIL	MAY	JUNE
S M T W T F S	S M T W T F S	S M T W T F S	S M T W T F S	S M T W T F S	S M T W T F S
1 2 3 4 5 6 7	1 2 3 4	1 2	1 2 3 4 5 6 7	1 2 3 4	1 2
8 9 10 11 12 13 14	5 6 7 8 9 10 11	3 4 5 6 7 8 9	8 9 10 11 12 13 14	5 6 7 8 9 10 11	3 4 5 6 7 8 9
15 16 17 18 19 20 21	12 13 14 15 16 17 18	10 11 12 13 14 15 16	15 16 17 18 19 20 21	12 13 14 15 16 17 18	10 11 12 13 14 15 16
22 23 24 25 26 27 28	19 20 21 22 23 24 25	17 18 19 20 21 22 23	22 23 24 25 26 27 28	19 20 21 22 23 24 25	17 18 19 20 21 22 23
29 30 31	26 27 28 29 30	24 25 26 27 28 29 30	29 30 31	26 27 28 29 30	24 25 26 27 28 29 30 W

JULY	AUGUST	SEPTEMBER	OCTOBER	NOVEMBER	DECEMBER
S M T W T F S	S M T W T F S	S M T W T F S	S M T W T F S	S M T W T F S	S M T W T F S
1 2 3 4 5 6 7	1 2 3 4	1 2	1 2 3 4 5 6 7	1 2 3 4	1 2
8 9 10 11 12 13 14	5 6 7 8 9 10 11	3 4 5 6 7 8 9	8 9 10 11 12 13 14	5 6 7 8 9 10 11	3 4 5 6 7 8 9
15 16 17 18 19 20 21	12 13 14 15 16 17 18	10 11 12 13 14 15 16	15 16 17 18 19 20 21	12 13 14 15 16 17 18	10 11 12 13 14 15 16
22 23 24 25 26 27 28	19 20 21 22 23 24 25	17 18 19 20 21 22 23	22 23 24 25 26 27 28	19 20 21 22 23 24 25	17 18 19 20 21 22 23
29 30 31	26 27 28 29 30	24 25 26 27 28 29 30	29 30 31	26 27 28 29 30	24 25 26 27 28 29 30 W

W—World Day, December W (365th day), a world holiday, follows December 30th every year. Leap-year Day, June W, another world holiday, follows Easter Sunday, April 8th.

VOCABULARY REVIEW

Match the word or words in the column on the right with the correct phrase in the column on the left. *Do not write in this book.*

1. Relatively smooth areas on the moon's surface.
2. Light streaks on the moon as if material had been thrown out.
3. Moon is closest to earth.
4. Darkest part of shadow cone.
5. Moon's umbra fails to reach the earth.
6. Moon passes through the earth's shadow.
7. Tidal bulge on the side of the earth facing the moon.
8. Occurs when the sun, earth, and moon are in line.
9. Most countries of the world use this calendar.
10. The earth passes through the moon's umbra.

a. apogee
b. lunar eclipse
c. rays
d. perigee
e. annular eclipse
f. maria
g. umbra
h. Gregorian
i. penumbra
j. total eclipse
k. spring tides
l. direct tides

QUESTIONS

Group A

Select the best term to complete the following statements. *Do not write in this book.*

1. When looking at the full moon, the dark regions that form the face of the "Man in the Moon" are called (a) maria (b) craters (c) rays (d) rills.

2. It is possible to see features on the moon's surface with the naked eye because it (a) is so bright (b) has an atmosphere that reflects light (c) is so close to the earth and has no atmosphere (d) gives off its own light.

3. Possible evidence for collisions of the moon with other bodies is (a) craters (b) rills (c) volcanoes (d) mountains.

4. The lighted side of the moon has temperatures as high as (a) −173°C (b) −100°C (c) 0°C (d) over 100°C.

5. On the moon, a boy weighing 72 kg would weigh (a) 18 kg (b) 66 kg (c) 12 kg (d) 33 kg.

6. Other than the earth, no planet in the solar system has (a) an atmosphere (b) a companion so nearly its own size (c) storms (d) gravity.

7. In one of its early stages, the heat retained from its origin and from colliding bodies caused the moon's surface to be covered with (a) mountains (b) molten rock (c) dust (d) rills.

8. Scientific evidence gathered on the moon suggests that its history consisted of (a) one (b) two (c) three (d) four or more stages of development.

9. The moon's diameter is (a) 2160 km (b) 12, 640 km (c) 3476 km (d) 7900 km.

10. The moon orbits around a point that is (a) at the earth's center (b) nearer the earth's surface than its center (c) half way between the earth and moon (d) near the moon's surface.

11. At perigee, the moon is how many kilometers from the earth? (a) 360,000 (b) 404,800 (c) 225,000 (d) 253,000.

12. The moon rises at a different time on consecutive days because (a) it is slowing down (b) the earth is moving around the sun (c) the earth rotates from east to west (d) the moon moves from west to east around the earth.

13. Each night the moon rises (a) a little earlier (b) about the same time (c) a little later (d) in the west.

14. We always see the same side of the moon because (a) the moon rotates on its axis at the same rate it moves around the earth (b) the earth rotates on its axis (c) the earth revolves around the sun (d) none of these.

15. The length of time it takes the moon to make one complete revolution around the earth is (a) 31 days (b) 29½ days (c) 27⅓ days (d) 30 days.

16. The time needed for the moon to go through a complete cycle of phases is (a) 31 days (b) 29½ days (c) 27⅓ days (d) 30 days.

17. The reason for the difference in the answers to questions 15 and 16 is (a) the earth's motion around the sun (b) the rotation of the earth on its axis (c) the rotation of the moon on its axis (d) the tilt of the earth's axis.

18. In each of its revolutions around the earth, the moon does not come directly between the sun and the earth because the (a) earth's axis tilts (b) moon's orbit tilts (c) moon does not rotate (d) earth's orbit tilts.

19. An annular eclipse could occur when the moon is (a) at apogee (b) full (c) at first quarter (d) at perigee.

20. The phase of the moon during an eclipse of the sun is (a) full (b) half (c) new (d) in any phase.

21. To produce a total eclipse of the moon (a) the earth must pass through the moon's umbra (b) the moon must pass through the earth's umbra (c) the moon must be at apogee (d) the moon must be at perigee.

22. The tidal bulge on the earth is (a) caused mostly by the sun (b) the result of the earth's rotation (c) a clear and noticeable effect of the moon's gravity (d) found only on the side of the earth facing the sun.

23. The tidal bulge on the side of the earth opposite the moon is caused by (a) the sun's gravity (b) the moon's rotation (c) the motion of the earth in its orbit (d) a smaller gravitational pull by the moon on the far side of the earth.

24. The moon is most likely to be overhead (a) at high tide (b) after high tide (c) before high tide (d) at low tide.

25. Tides always (a) increase in height each day (b) decrease in height each day (c) speed up the earth's rotation (d) slow down the earth's rotation.

26. The tidal bulge tends to (a) speed up the moon (b) slow down the moon (c) bring the moon closer (d) have no effect on the moon.

27. Spring tides occur at (a) new moon and first quarter (b) new moon and last quarter (c) first quarter and last quarter (d) full and new moon.

28. Neap tides occur at (a) new moon and first quarter (b) new moon and last quarter (c) first and last quarter (d) full and new moon.

29. The problem of developing a calendar based on phases of the moon is that (a) the moon's motion is not regular (b) 29½ is not evenly contained in 365¼ days (c) the months are too long (d) the moon's motion is too regular.

30. The Roman calendar fell short of a year by how many days? (a) 7 (b) 9 (c) 11 (d) 13.

31. The Julian calendar was not quite correct because it was (a) 11 minutes 14 seconds short each year (b) 11 minutes 14 seconds long each year (c) short by one day each 4 years (d) long by one day every 4 years.

32. The Gregorian calendar has a year that is (a) exactly 52 weeks long (b) 52 weeks and one day in length (c) 52 weeks and 2 days in length (d) short of 52 weeks by one day.

33. The proposed Universal calendar has how many months? (a) 13 (b) 28 (c) 12 (d) 11.

34. The calendar used in almost all countries today is the (a) World calendar (b) Universal calendar (c) Julian calendar (d) Gregorian calendar.

35. The World calendar is probably preferable to the Universal calendar because it has (a) equal months (b) 365 days per year (c) 12 months (d) more quarters.

Group B

1. Describe how the dust on the moon's surface was formed.

2. Why does the moon have such a large temperature range?

3. If the moon turned on its axis at twice its present rate, what would be the probable effect on its temperature?

4. Give a brief description of the different ways the moon may have come into existence.

5. Describe the four stages in the moon's development.

6. Suppose the moon rotated on its axis every 14 days instead of every 27. What effect would this have on the way we see the moon?

7. To a person on the moon, how would the earth appear at full moon, new moon, and during the first quarter?

8. How does the rotation rate of the moon on its axis affect our knowledge of the moon's surface?

9. List the conditions that must exist for a total eclipse of the sun.

10. Why can everyone on the dark side of the earth see an eclipse of the moon?

11. Why can't there be an eclipse of the moon at first quarter?

12. Why can't everyone see a total eclipse of the sun?

13. Why do the spring tides occur at full moon?

14. Give a simple explanation of opposite tides.

15. How do tides slow down the earth?

16. If the moon were farther from the earth than it is now, would high tides occur more often? Why?

17. Why do we have leap years in the century years only when they are evenly divisible by 400?

18. Why was the Julian calendar an improvement over the Roman calendar?

19. Describe the moon's orbit around the earth. Include speed, perigee distance, apogee distance, and general shape of the orbit.

20. The average distance between the earth and moon is 384,400 km. Calculate:
(a) the difference between the average distance and the apogee distance;
(b) the difference between the average distance and the perigee distance;
(c) a and b as percentages of the average distance according to the formula % = difference/average distance x 100.

21. What do the results of the calculations in question 20 tell you about the shape of the moon's orbit?

22. Explain why the Gregorian calendar was developed to replace the Julian calendar.

Suppose that you were born beneath the sea. Your entire life would be spent trapped in a watery world. As an intelligent being, you would be aware that a different realm existed above the surface of the water. But knowledge of that other world could only be in the form of theories based primarily on blurred images seen overhead. Then a vehicle is invented that allows escape from the sea. For the first time it is possible to visit dry land. All of the things that could not be studied beneath the sea, like fire and air, are suddenly discovered. The world as it really exists becomes known for the first time.

For most of our history, the human inhabitants of the earth have been similar to dwellers in the sea. We live beneath a blanket of air that forms a barrier between us and the universe beyond the earth. To begin to study and understand the universe, and the true nature of our home planet, instruments and people must go about 150 km (93 mi) above the earth's solid surface. Here the atmosphere blends into the almost complete vacuum of space. During the past few decades, vehicles that are able to travel into space have been developed. In this chapter you will learn about some of the problems associated with space travel. You will also learn how the earth can be better understood by probing the secrets of space.

GETTING INTO SPACE

The gravity barrier. Any object on the earth's surface is pulled toward the center by a gravity force. To move away from the earth, energy must be spent to produce an opposite force greater than the gravitational force. For exam-

objectives

- ☐ Explain how gravity affects a rocket leaving the earth.

- ☐ Compare flights to the moon, the inner planets, and the outer planets.

- ☐ Identify some kinds of knowledge gained from exploration of other parts of the solar system.

- ☐ Describe two kinds of satellite orbits.

- ☐ Give examples of the ways satellites can be used to study the earth.

- ☐ Explain the importance of the space shuttle.

FIG. 5–1. A rocket lifts off the earth's surface when the thrust delivered by its engines is greater than the gravitational force on the rocket. (NASA)

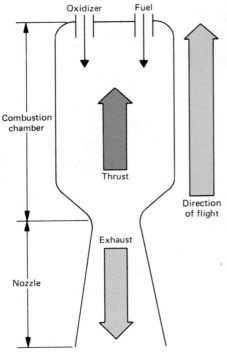

FIG. 5–2. The thrust of a rocket engine is produced by expansion of hot gases created when fuel and oxidizer are burned very rapidly. Thrust is an unbalanced force produced as a reaction to the exhaust. The size of the thrust is determined by the speed and mass of the gases leaving the exhaust.

ple, a rocket vehicle cannot start to move upward until the force delivered by its engine is greater than the gravitational force. Rocket engines are the only means now available that can furnish enough force to lift space vehicles away from the earth. In a rocket engine, hot gases are produced by very rapid burning of fuel and an oxidizer such as liquid oxygen. If the engine had no opening to the outside, the pressure of the hot, expanding gases would push in all directions against the walls of the engine. However, the hot gases do escape through the exhaust nozzle. As the gases flow out of the exhaust, the force on the front of the engine is no longer balanced by an opposite force to the rear. Unbalanced gas pressure pushing on the front of the engine thrusts the whole body of the rocket forward. This force is called the *thrust* of the rocket. See Figure 5–2. A rocket engine does not have to push against anything but itself. Thus it operates even in the near vacuum of space.

The force a rocket engine must produce to lift the rocket straight up is determined by the total mass of the vehicle. But, according to Newton's Law of Gravitation, the gravitational force pulling the rocket back toward the earth decreases as the rocket climbs farther away from the earth's center. For a rocket vehicle trying to leave the earth, overcoming the gravity force can be compared to trying to climb out of a deep hole or pit. The walls of the pit would be very steep near the bottom, but not nearly so steep at the top. See Figure 5–3. The slope of the pit's walls can be compared to the gravitational force pulling the rocket back to earth. At first, the gravity force is strong, as shown by the nearly vertical slope of the walls near the bottom of the gravity "pit." As the rocket climbs, the walls become less steep. Less force is needed to lift the rocket. Finally, at a certain altitude, the slope of the gravity "pit" becomes so small that almost no force is needed to keep the rocket from falling back. The shape of the gravity pit for a particular planet depends upon its mass. On the moon, for example, the smaller mass produces less gravity force on its surface. The gravity "pit" of the moon has less steep walls than the earth. As a result, a rocket of given mass needs less force to leave the moon than to leave the earth.

Once a rocket vehicle is out of the gravity "pit" of a planet, the engines can be turned off and the spaceship will continue to move in a straight line at nearly constant speed. Its path and speed will not change very much until its engines are turned on again, or it is affected by the gravity of another planet or other body.

Travel to the moon and planets. The moon is the easiest of all targets for space voyages. A vehicle bound for the moon is first lifted off the earth by powerful rockets. At the proper altitude, the rocket carrying the lunar vehicle goes into orbit around the earth. After several orbits around the earth, the rocket is aimed at the moon and the engines again started for a short time. The lunar spacecraft then separates from the main rocket and moves toward the moon, leaving the rocket behind.

About midway between the earth and moon, the moon's gravity begins to attract the spacecraft. Sixty hours after leaving earth orbit, the lunar vehicle is very near the moon. Its rocket engine is pointed toward the moon and fired briefly. This brakes the rapid descent of the spacecraft and allows it to ease into an orbit around the moon. A moon landing craft can then separate from the main vehicle. With its own small rockets it can guide itself to the lunar surface. Later the moon lander launches itself back into its original orbit to meet the lunar orbiter. The engines of the lunar orbiter will lift them both out of orbit and head them back toward the earth. The entire flight path for a lunar journey is shown in Figure 5–4.

FIG. 5–3. Escaping the earth's gravity can be compared to the energy needed to climb out of a pit shaped like the one shown here.

FIG. 5—4. Flights to the moon have been made using the plan shown in this diagram.

FIG. 5—5. Minimum energy flight paths to Venus and Mars. Left, to reach Venus the spacecraft is launched in a direction opposite to the earth's motion around the sun. Right, a flight path to Mars takes advantage of the earth's motion.

A flight to one of the inner planets can be made using the least amount of fuel if the spacecraft becomes a temporary satellite of the sun. Such a path to one of the inner planets can be called a "minimum energy orbit." For example, to reach Venus a spacecraft might be launched so that it achieves a velocity slower than the speed the earth moves in its orbit. The sun's gravity would then pull the craft in a direction toward the center of the solar system until it would finally cross the orbit of Venus. For a flight to Mars, a spacecraft is launched in the same direction as the earth moves around the sun. The earth's motion and the craft's velocity cause it to move in a path that will carry it outward to intersect the orbit of Mars. See Figure 5–5. These are minimum energy paths since the sun's gravity supplies part of the energy needed. As a result, flights between earth and its closest neighbors do not require much more power than flights to the moon.

However, flights to the planets involve difficult problems in guiding the spacecraft. An interplanetary spacecraft following a minimum energy path is a temporary satellite of the sun with its own orbit. The main rocket engines can be shut off except for brief firings to make small corrections in the path. This means that the craft's orbit must be carefully planned to bring it and the target planet together at some point in the planet's orbit.

An example of the kind of difficulty met in guiding an interplanetary flight is this: The orbit of Venus is tilted about 3 degrees from the plane of the earth's orbit. Although this is a small difference, it means that Venus is sometimes as much as 4,800,000 km (about 3,000,000 mi) above or below the plane of the earth's orbit. Unless this is taken into account, a vehicle launched in a minimum energy path toward Venus will miss its target. See Figure 5–6.

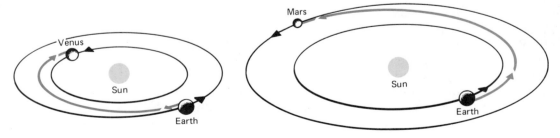

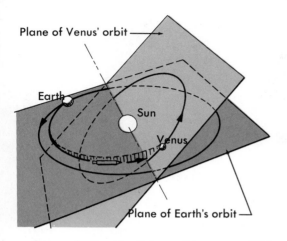

FIG. 5–6. Because the orbits of earth and Venus are not on the same plane, a Venus-bound space vehicle launched in a minimum energy path would miss its target. The difference between the planes of the orbits of earth and Venus is much less than shown in this diagram.

One solution to this problem would be to launch the vehicle when the earth crosses the plane of the orbit of Venus. At this time, there could be no error caused by differences between the orbital planes of earth and Venus. Thus the flight path would require no major corrections during the journey. This would also hold true if the vehicle were launched when Venus crosses the plane of the earth's orbit. This situation is shown in Figure 5–7. If the Venus-bound vehicle were launched at any other time, it would have to be aimed at a slight angle to the plane of the earth's orbit. See Figure 5–8. This type of flight path would require greater speed at launching and major mid-flight adjustments. It would never be a minimum energy flight path.

Like the orbit of Venus, the orbits of all other planets have some degree of tilt with respect to the plane of

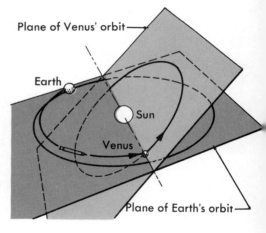

FIG. 5–7. If a vehicle is launched so that it meets Venus when that planet crosses the plane of the earth's orbit, it can follow a minimum energy path.

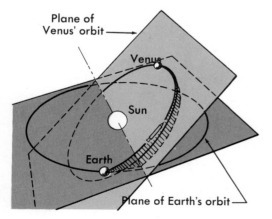

FIG. 5–8. A spacecraft arriving at Venus when it is not crossing the plane of the earth's orbit cannot follow a minimum energy path, since corrections are needed, as shown in this diagram.

FIG. 5–9. The orbits of all the other planets are tilted a certain number of degrees when compared to the plane of the earth's orbit.

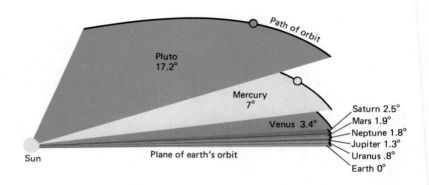

earth's orbit. See Figure 5–9. A minimum energy flight path to any planet requires arrival at the target planet when the planet is at or near the plane of the earth's orbit. Thus interplanetary spacecraft can be launched only during certain periods of time. These times are often called "launch windows." There is a different "launch window" for each planet that could be a destination.

In voyages to the outer planets, the gravity of the planets themselves can be used to direct the flights. For example, a spacecraft could be launched toward Jupiter following a minimum energy orbit. It would pass this giant planet in such a way that Jupiter's gravity would bend its path. It would then be flung toward Saturn as if hurled from a giant slingshot. See Figure 5–10. As the craft flew near Saturn, that planet would turn it toward Pluto. A similar plan could use the gravity force of Jupiter to guide a spacecraft to Uranus and Neptune.

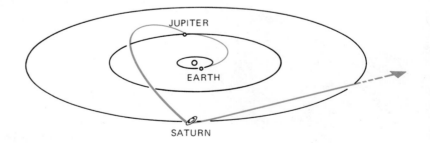

FIG. 5–10. A minimum energy flight path to Jupiter and Saturn uses the gravity of those planets to supply some of the needed energy.

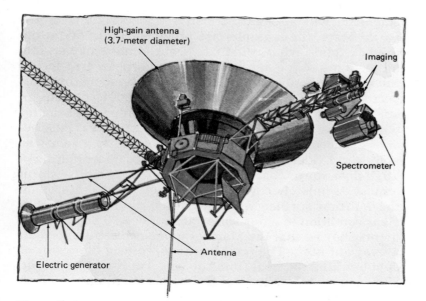

High-gain antenna
(3.7-meter diameter)

Imaging

Spectrometer

Antenna

Electric generator

Knowledge from the planets. The moon is the only other part of the solar system that has been visited by astronauts. All other exploration in space has been accomplished by pilotless spacecraft such as the one shown in Figure 5–11. These spacecraft are marvelous robots controlled from the earth. At the same time, they are able to act independently and solve many of their own problems. Their instruments collect an enormous amount of information that is transmitted back to the earth.

Why do we expend huge amounts of money and the efforts of thousands of people to send these spacecraft to the other planets? There are four main reasons. First, detailed knowledge of other planets and their satellites will help scientists to understand the earth's history. Each of these bodies is in its own stage of development. Thus they can show what the earth might have been like in the past, and how it might change in the future.

A second important reason for investigating the planets is to try to find evidence that life exists elsewhere. The discovery of life on other planets might mean that living things exist in many places in the universe.

Exploration of the other planets will also help prepare for the possibility that at some time the earth may become overpopulated. It may be necessary to establish colonies in other parts of the solar system to provide resources or living space.

Finally, by studying other planets at close range,

FIG. 5–11. The Voyager spacecraft shown in this diagram has flown past Jupiter and Saturn and may visit Uranus in 1986.

Investigate
Make a list of the food, water, and materials required to support a 50-person colony on a spaceship for one year while it moves to another planet. Include any recycling you wish, but only those types we know how to do at this time.

scientists will also gain a better understanding of the earth. Mars, for example, has provided information that might explain what could cause our climate to pass into an ice age. An ice age on earth would mean disaster for the world's population. Scientists have for some time suspected that smoke, dust, and other kinds of pollution in our atmosphere could trigger a critical drop in the earth's temperature. But there was no direct way to test this theory. A spacecraft that orbited Mars for nearly a year was able to provide the answer. It observed Mars during a huge dust storm that lasted for months. The dust storm almost completely covered the planet. Temperatures on the Martian surface were measured as it went from a clear condition to a dust-filled atmosphere. The measurements showed that the Martian temperatures dropped an average of 20°C. The temperature drop was the result of dust blocking the heat from the sun. This provides strong evidence that a pollution-filled atmosphere on the earth could set off an ice age.

Exploration of parts of the solar system has already provided a clearer picture of how the earth and other planets probably were formed and developed. The parts yet to be explored represent a giant scientific laboratory in which our knowledge of the earth can be expanded and tested.

IN ORBIT AROUND THE EARTH

Earth satellites. If a rocket leaves the earth with a speed too low to escape the gravity "pit," it will fall back toward the earth. However, such a rocket could be given a sideways push somewhere near the top of the flight path. This would be done by tilting the rocket and briefly firing its engine. The return path of the rocket as it falls back to earth would then be more curved. If the rocket is aimed carefully so that the curve of the return path is equal to the curvature of the earth's surface, the rocket will never actually reach the earth. It will fall continuously in an orbit around the earth. See Figure 5–12. The rocket becomes a satellite of the earth.

The altitude of the satellite determines the speed at which it must move to stay in orbit. For example, to remain in an orbit about 240 km (about 150 mi) above the earth's surface, a satellite must have a speed of 8 km/sec (5 mi/sec). At this speed the satellite will circle the earth in

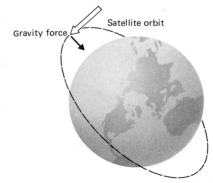

FIG. 5–12. A satellite will remain in a circular orbit around the earth as long as the curve of its flight path is equal to the curvature of the earth's surface.

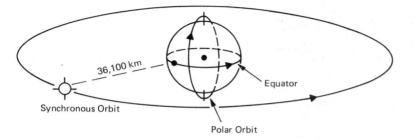

36,100 km

Synchronous Orbit

Equator

Polar Orbit

FIG. 5–13. A satellite in synchronous orbit above the equator always remains over the same point. A polar orbit passes over all parts of the earth, since the satellite always follows the same path as the earth turns beneath.

1½ hours. An orbit with an altitude of 6400 km (4000 mi) requires a speed of 6 km/sec (3.5 mi/sec), which allows the orbit to be completed in 4.25 hours. If the altitude is increased to 36,100 km (22,400 mi), the satellite completes one revolution in 24 hours. This is called a *synchronous* orbit because the satellite is synchronized, or in step, with the earth's rotation. A satellite directly above the earth's equator with a synchronous orbit will appear to be stationary in the sky. This is because it has a period of revolution equal to the period of rotation of the earth. Thus it always remains above the same point on the equator.

On the other hand, a satellite can be launched into an orbit that is at right angles to the equator. This kind of orbit is called a *polar* orbit because it carries the satellite over the earth's poles. A satellite in a polar orbit has characteristics opposite to one in a synchronous orbit. The orbit of any satellite stays nearly always the same in space. This means that each time a satellite in polar orbit crosses the equator, it passes west of its previous crossing because the earth has rotated toward the east. Figure 5–13 compares synchronous and polar orbits. A satellite in a polar orbit with the proper altitude will pass over every part of the earth's surface at least once each day. Such a satellite can be used to constantly observe the entire surface of the earth.

Once it has been put into orbit, a satellite will continue to move in that orbit forever if its speed never changes. However, a satellite may meet some kind of resistance that slows its speed. For example, a satellite at altitudes below about 320 km (about 200 mi) will be slowed by the atmosphere. The atmospheric gases are extremely thin at such a high altitude. Nevertheless, they are still able to cause a small drag. When a satellite in a circular orbit is slowed, its orbit becomes elliptical. See Figure 5–14. If the elliptical orbit comes close enough to the earth at

Discuss
The moon, like all earth-orbiting space vehicles, is an earth satellite. What would happen if something were to make it suddenly speed up? slow down? stop completely?

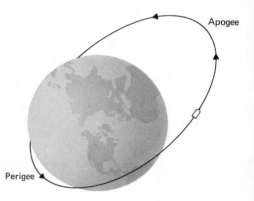

Apogee

Perigee

FIG. 5–14. An elliptical orbit may cause a satellite to come so close to the earth at perigee that it reenters the atmosphere and is destroyed.

activity

Suppose you are on the moon. You and two members of your crew are returning from a mission when your ship crashlands 300 km from the base ship. The crash is on the sunlit side of the moon. Most of your equipment has been destroyed in the crash. Your survival depends on reaching the base ship. In addition to space suits, your crew is able to remove the following items from the wreck: 4 packages of food concentrate, 20 m nylon rope, 1 portable heater, 1 magnetic compass, 1 box matches, 1 first aid kit, 2 50-kg tanks of oxygen, 20 L water, 1 star chart, 1 case dehydrated milk, 1 solar powered receiver-transmitter, 3 signal flares, 1 large piece nylon fabric, 1 flashlight, 2 knives.

1. Rate each item in the list according to how important it will be to your survival.

2. Make a list of the items, in the order of their importance.

3. After each item, write what it will be used for that makes it important or less important.

4. Compare and discuss your list with other members of the class.

5. Compare your list with one prepared by astronauts for NASA. Your teacher will give you this list.

6. List revisions you would make in the order of the items if the crash landing was on Mars instead of the moon.

perigee, friction with the denser part of the atmosphere will cause the satellite to burn up. Most satellites complete a number of elliptical orbits, with each orbit bringing them closer to the earth at perigee, before their final fiery plunge.

If every satellite now in orbit around the earth marked its path with a trail of smoke, the sky would be marked with thousands of trails. Most of these satellites will eventually fall back to earth. Most will burn up in the atmosphere with a brief meteor-like display. Parts of some larger satellites can survive the trip through the atmosphere and strike the earth's surface.

Exploring the earth from space. From space, it is possible to see the earth in new ways. Never before has so much of the earth's surface been "seen" at one time through the use of orbiting instruments. For example, a type of satellite called LANDSAT has been placed in a polar orbit at an altitude of about 1000 km (about 600 mi). Circling the earth about 14 times each day, the LANDSAT can pass over the most populated regions of the globe at least once every 18 days. Its instruments are able to examine 34,253 square km (13,225 square mi) each 25 seconds. See Figure 5–15.

Satellites such as LANDSAT do not use cameras to photograph the earth's surface. Instead, they use electronic sensors that can be made to respond to any of the wave-

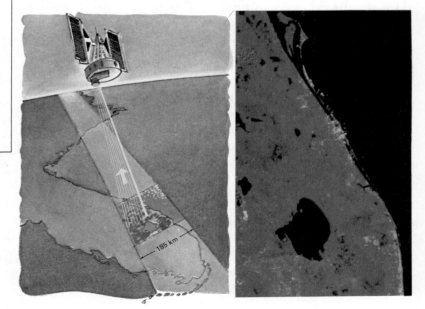

FIG. 5–15. LANDSAT instruments produce an image of a large area of the earth's surface.

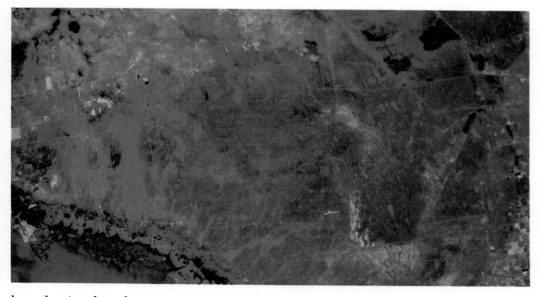

FIG. 5–16. An image made by LANDSAT of southern Florida showing the Everglades region and Miami. Plant growth is shown in red. (NASA)

lengths in the electromagnetic spectrum. Many of the wavelengths reflected or radiated from the earth are not included in visible light. For example, plants reflect infrared wavelengths very strongly. Electronic sensors responding to infrared are able to detect vegetation more accurately than can be done by using ordinary photographs. Information collected by sensors carried by the satellite is radioed back to computers in ground stations. The computers are able to process the signals into an image similar to a TV picture. The image shows surface features in false color. That is, the colors of the image are different from the true colors seen in ordinary light. See Figure 5–16. The computers can also produce images that show particular features, such as the damage caused by a forest fire.

LANDSAT images have provided information that revealed previously unknown oil and mineral deposits in Oklahoma, Alaska, the Rocky Mountains, and the Brazilian jungles. The growth of cities and other changes in the way land is being used can also be easily seen in the satellite images. One of the most important uses of LANDSAT involves its ability to survey the world's agriculture. In a matter of hours, the farmland of a large area can be examined and each kind of crop identified. This information can be used to predict food shortages and surpluses. Such advance warning greatly improves the distribution of food throughout the world.

FIG. 5–17. Well-defined cloud patterns over the United States are visible in this weather satellite photo. Note the rain and snow clouds over the Pacific states, the patchy cloudiness over Texas and the Great Lakes, and the snowy overcast covering parts of New England. (NASA)

No scientific tool has done more to make accurate weather forecasts possible than the special satellites designed to observe the earth's atmosphere. Some weather satellites circle the earth in polar orbits. They are timed to pass over the same areas at about the same time each day. Other weather satellites are placed in synchronous orbits. This kind of orbit allows a continuous view of a particular region of the earth's surface. During daylight hours, a weather satellite generally radios images made using the wavelengths of visible light to receiving stations. These images clearly show the arrangement of clouds. See Figure 5–17. At night, the satellite's sensors can use moonlight or infrared wavelengths that show cloud patterns from their temperatures. Infrared sensors on the satellite can also measure surface temperatures in the sea. This reveals the movements of ocean currents, as shown in Figure 5–18.

The space shuttle. Most satellites are designed to send information back to the earth. The satellite itself remains permanently in orbit, or, if it does return, it is destroyed when reentering the lower atmosphere. The space shuttle is an exception. It is designed to carry a cargo into a satellite orbit and then return to the earth's surface to land as an aircraft. The shuttle is the first space cargo ship designed to fly into space and return.

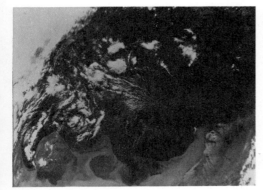

FIG. 5–18. An infrared image shows the warm water carried by the Gulf Stream along the eastern coast of North America. (NOAA)

FIG. 5–19. The shuttle orbiter is launched while attached to a large external fuel tank and two solid fuel rockets. (NASA)

When the space shuttle is launched, it has three main parts: an orbiter, a large liquid rocket-fuel tank, and two solid-fuel rocket boosters, all bolted together. The most important part is the orbiter, which resembles a small airliner. It can carry up to sixteen tons of cargo and has work and living space for as many as seven people. Three powerful main engines can develop enough power to light up the state of New York. The liquid fuel needed for launching is carried in the large external tank. Two additional solid-fuel rockets act as boosters to provide the extra thrust required to lift the entire load. Both the orbiter's main engines and the solid rockets are burning at liftoff. See Figure 5–19. After two minutes, the solid rockets are exhausted and dropped off. Just before reaching the correct altitude to enter orbit, the liquid-fuel tank is discarded. The solid rockets are recovered to be used again. But the liquid-fuel tank is burned up as it dips back to earth over a remote ocean area. Small maneuvering rockets on the orbiter allow it to make the precise movements necessary to enter the chosen orbit at a maximum altitude of about 1000 km (about 600 mi).

Once in orbit, the shuttle is able to carry out many kinds of missions. Satellites carried up can be released into their own orbits at the shuttle's altitude. Or the shuttle can act as a base for satellites launched into higher orbits under their own power. Satellites already in orbit can be picked up and serviced aboard the shuttle. The shuttle's cargo is

FIG. 5–20. Two orbiters shown as they might appear while ferrying supplies to a space station under construction.

often made up of scientific experiments. For example, the color of ocean water is measured as a guide to finding

FIG. 5–21. Spacelab is fitted into the cargo space of the orbiter. It includes a laboratory and sections where telescopes, sensors, and antennas have a clear view of space. A tunnel and airlock connect the laboratory to the crew section of the orbiter.

good fishing regions. A special type of radar useful for making maps identifying promising sites for oil and mineral exploration can be tested. Some experiments take advantage of the near vacuum of space and the microgravity of an orbiting spacecraft. Microgravity means that the gravity force of the earth is hardly felt when in orbit. This is a result of the continuous falling motion of a satellite in an orbit around the earth. In the future, space shuttle missions will also be concerned with the construction of permanent space stations.

When the orbiter has accomplished its various missions in space, it turns so that its engines face ahead. A brief firing of the rockets slows the ship, causing it to begin reentry into the atmosphere. The craft is turned nose up as it dips deeper into the atmosphere. This allows its specially coated underside to absorb the heat generated during reentry. Finally, it is guided to a landing area and glides to the runway. After careful inspection, the orbiter is again made ready for its next launch.

FIG. 5–22. When its space mission is completed, the orbiter returns to earth and lands like an aircraft. (NASA)

VOCABULARY REVIEW

Match the word or words in the column on the right with the correct phrase in the column on the left. *Do not write in this book.*

1. The first obstacle that must be overcome to get into space.
2. The force that makes a rocket move ahead.
3. A flight using the least amount of fuel.
4. The launch time when the least energy would get a rocket to a planet.
5. An orbit that keeps a satellite over one place on earth.
6. An orbit that shifts westward as the earth rotates.
7. A satellite that scans the earth's resources.
8. A space cargo ship designed to fly into space and return.
9. Plants reflect these wavelengths strongly.
10. The main part of the shuttle system.

a. infrared
b. launch window
c. polar
d. gravity barrier
e. LANDSAT
f. orbiter
g. thrust
h. microgravity
i. minimum energy orbit
j. ultraviolet
k. synchronous
l. space shuttle

QUESTIONS

Group A

Select the best term to complete the following statements. *Do not write in this book.*

1. Objects are pulled toward the earth by a force due to (a) pressure (b) gravity (c) atmosphere (d) vacuum.
2. At present, the only way we can escape the earth's gravity and travel into space is by use of (a) jet engines (b) nuclear engines (c) rocket engines (d) antigravity machines.
3. In a rocket engine hot gases expand, putting pressure on the front of the engine that is (a) not balanced toward the rear (b) always balanced toward the rear (c) the only force acting (d) not important.
4. A rocket engine (a) cannot work in empty space (b) can only work on earth (c) can work in empty space (d) must push against something to move.
5. As a rocket climbs farther away from the earth, the gravitational force pulling it back (a) increases (b) remains the same (c) decreases (d) increases then decreases.
6. Overcoming gravity can be compared to (a) climbing down a steep bank (b) reducing atmospheric pressure (c) climbing out of a pit (d) swimming.
7. Compared to leaving the earth, a rocket leaving the moon requires (a) less force (b) more force (c) the same force (d) an unpredictable force.
8. If a rocket is out of the gravity "pit" and it turns off its engines, the rocket will (a) travel in a straight line (b) fall back to earth (c) fall on the moon (d) travel in a circle.

9. To slow a rocket down, the engines are fired (a) backward (b) forward (c) sideward (d) upward.

10. For a spaceship in orbit around the earth to return to earth, its speed must (a) be increased (b) remain the same (c) be decreased (d) be changed according to where it is in orbit.

11. A flight to another planet that follows a path that uses the least amount of fuel is called a (a) conservation flight (b) no energy flight (c) full launch flight (d) minimum energy orbit.

12. In order to use the sun's gravity to help save fuel on a trip to Venus, a spacecraft would be launched so that it (a) has the same speed as the earth in its orbit (b) has less speed than the earth in its orbit (c) has greater speed than the earth in its orbit (d) is pointed directly at Venus.

13. Sometimes Venus is out of the plane of the earth's orbit by as much as (a) 480 km (b) 4800 km (c) 4,800,000 km (d) 4,800,000,000 km.

14. The importance of the answer to question 13 is that (a) Venus can't be reached (b) the launch time to Venus must be carefully calculated (c) a minimum energy orbit can't be used (d) the flight to Venus takes several years.

15. The times when interplanetary spacecraft can be launched in minimum energy orbits are often called (a) launch time slots (b) launch windows (c) starting times (d) activation times.

16. One of the main reasons for using money and the efforts of many people to explore other planets is to (a) make life interesting (b) use excess energy (c) discover how the earth formed (d) do things never done before.

17. A spacecraft to Mars showed that a dust storm can cause temperatures to (a) drop 20°C (b) rise 20°C (c) remain unchanged for long periods (d) change up or down by 100°C.

18. The answer to question 17 is evidence that a pollution-filled atmosphere could (a) scorch the earth (b) mean long periods of rain (c) set off a long hot spell (d) set off an ice age.

19. For a satellite to stay in orbit 240 km above the earth's surface, it must travel (a) 8 km/sec (b) 80 km/sec (c) 800 km/sec (d) 8000 km/sec.

20. The satellite in question 19 would circle the earth every (a) 1.5 months (b) 1.5 days (c) 1.5 hours (d) 1.5 minutes.

21. To stay in an almost circular orbit 6400 km above the earth's surface, a satellite must have a speed of (a) 6 km/sec (b) 60 km/sec (c) 600 km/sec (d) 6000 km/sec.

22. The satellite in question 21 would orbit the earth in (a) 2.55 min (b) 25.5 min (c) 255 min (d) 2555 min.

23. At what altitude will an orbiting satellite stay over the same spot on earth? (a) 10,720 km (b) 6700 km (c) 22,400 km (d) 36,100 km.

24. In question 23, the assumption is that the satellite (a) moves in a direction opposite to that of the earth's rotation (b) moves in the same direction as the earth's rotation (c) does not have any speed (d) changes its altitude constantly.

25. An orbiting satellite that stays over one spot on earth is in what is called (a) a synchronous orbit (b) a stationary orbit (c) a zero speed orbit (d) daytime orbit.

26. The satellite in question 25 can be (a) over any place on earth (b) in polar orbit (c) over the equator (d) over the equator or poles.

27. A satellite in polar orbit can make an excellent "spy in the sky" because it (a) stays in one position over the earth's surface (b) avoids most of the clouds in the earth's atmosphere (c) views the largest possible area of the earth's surface at any particular time (d) eventually passes over every part of the earth's surface.

28. On each cycle, an orbiting satellite that passes over the poles of the earth appears to move (a) northward (b) southward (c) eastward (d) westward.

29. The reason for the answer to question 28 is (a) the earth turns toward the west (b) the earth turns toward the east (c) the satellite always moves north (d) the satellite always moves south.

30. LANDSAT satellites circle the earth (a) around the equator (b) 14 times a day (c) at 36,100 km (d) all of these.

31. The images sent back to earth by LANDSAT-type satellites are (a) often in false colors (b) always in black and white (c) like an ordinary TV color picture (d) taken on photographic film.

32. LANDSAT satellites can give information on (a) crops (b) damage by forest fire (c) mineral deposits (d) all of these.

33. The important difference between satellites and the space shuttle is that the space shuttle (a) stays in one place in orbit (b) is larger than all other satellites (c) comes back to earth to be reused (d) all of these.

34. The space shuttle can be used for (a) taking satellites into orbit (b) servicing satellites in space (c) experiments with fishing regions and mineral deposits (d) all of these.

35. The shuttle is the first space cargo ship designed to (a) use both solid and liquid fuel (b) carry scientists into orbit (c) lift heavy payloads into space (d) fly into space and return.

Group B

1. Explain why the gravity barrier is a major obstacle to space study.
2. Explain the principle behind the operation of a rocket engine.
3. Describe what is meant by the gravity "pit."
4. What is the "minimum energy orbit" plan?
5. How should a spacecraft be launched relative to the earth's motion to reach Venus? How does this differ from launching to reach Mars?
6. What is one of the main difficulties in guiding a flight to Venus?
7. Explain what is meant by "launch windows."
8. How could Jupiter's gravity be utilized in a minimum energy flight to Uranus and Saturn?

9. Identify four reasons for sending spacecraft to other planets.

10. What have we learned from the huge dust storms on Mars?

11. How can a rocket "fall" continuously in an orbit around the earth?

12. Discuss what is meant by a synchronous orbit. Include altitude and times.

13. How are the speeds of satellites related to the altitudes of their orbits?

14. Describe the motion of a satellite in polar orbit.

15. State the functions of a LANDSAT satellite.

16. How are weather satellites of use to us?

17. Describe how the space shuttle orbiter is put into space.

18. State some of the functions of the space shuttle program.

19. Describe how a vehicle is sent to the moon.

20. Compare the characteristics of a satellite in synchronous orbit with one in polar orbit.

21. A satellite in orbit with an altitude less than 320 km will eventually fall back to earth. Explain why this is so.

22. Explain how weather satellites and satellites such as LANDSAT can send back images of the earth's surface during the night as well as during the day.

career paths in earth science

Do you like to draw? If you have some interest and talent in artistic work, you might think about a career as a mapmaker. Many kinds of maps are used in earth science. You can see numerous examples in this book. Each map must be carefully drawn by a mapmaker who uses information collected by other experts. Mapmakers may work for various branches of the government or companies in fields such as publishing, land development, and petroleum exploration. An interest and background in earth science is very helpful for this career. Many mapmakers have a college background, while others get all of their training on the job.

Mapmaking is only one example of a career that might follow from your study of earth science. Examples of other careers associated with earth science are shown on the opposite page. Other related careers are also described in special feature pages following each unit. Your teacher can provide additional information about these or similar careers.

Examples of careers that might be pursued after finishing high school are shown below and right.

Oil Drilling
Beginners working on an oil drilling rig usually handle the equipment needed to lower and raise the drilling pipe. After gaining practical experience with the machinery, workers might be promoted to supervision of the operation of one or more drilling rigs. Drilling supervisors must understand the nature of the rock layers and their effect on the drill.

Gemologist
An interest in minerals might open the way to a career in working with precious gems. Cutting and polishing of rough gemstones into valuable gems are learned by becoming an apprentice gem cutter. Other gemologists work only with the finished gems.

A career that requires some training after high school is illustrated below.

Water Treatment Plant Operator
Water treatment plants produce a supply of pure water for their communities. Operators of these plants usually have some education beyond high school. Some colleges have special two-year programs for training water and sewage treatment plant operators.

Many careers require a college degree. Persons in the two careers shown below and right would have spent at least four years in college.

Space Science and Engineering
Many people with college degrees in science and engineering are needed to design, build, and launch spacecraft.

Solar Energy Specialist
Design of systems to collect solar energy requires scientific training and skill. Specialists must apply their knowledge in developing economic and efficient methods of energy production.

unit
2
the planet
earth

6
models of the planet earth

objectives

- ☐ Compare ancient beliefs about the earth's place in the universe with what is believed today.

- ☐ List several reasons why the earth is believed to be a sphere.

- ☐ Describe a method of measuring the size of the earth.

- ☐ Describe how directions and locations can be determined on earth.

- ☐ Explain how map projections can be made.

- ☐ List the uses of maps.

- ☐ Describe how topographic maps can be interpreted.

Compared to the entire solar system, the earth is smaller than a speck of dust in your classroom. But to an insect crawling over the ground, the earth is a jungle of pebbles, fallen leaves, grass blades, twigs, and all sorts of giant barriers. However, if you walked over the earth's surface you would be in about the same position as the insect. Neither of you can see the earth as the ball-shaped planet it is, among a family of planets all moving around the sun. But unlike the insect, you are able to carry in your mind a picture or model that goes far beyond what your eyes and other senses tell you. Until recently, these models of the earth gave us the only clues to the earth's size, shape, and position in our solar system.

Humans probably began their search for an earth-model long before they learned to write. But their earliest recorded ideas of our planet did not correctly describe its size or shape. However, some early Greek teachers thought the earth was round. From these early beginnings, humans created many models to help describe the different features of the planet earth.

The most common models used to describe the surface features of the earth are called maps. A map is a visual model of the earth's surface features. Several of the ways maps can be made accurately, and then used to describe the earth's surface will be taken up in this chapter.

A CLOSER LOOK AT OUR PLANET

The earth's place in the universe. It was natural for ancient people to think of home as the center of all that was known to exist at that time. When wandering tribes

116

settled down to form communities and nations, these too were thought of as centers of the entire area. It was believed that the earth was central to the surrounding heavens which carried the sun, moon and stars around the sky. Each of the ancient civilizations developed a model of the earth to fit its own beliefs. See Figure 6–1. The one thing that all of these models seemed to have in common was that they placed the earth, or at least a part of it, at the center of the universe.

Overcoming these early beliefs required thousands of years of effort. Thoughtful men observed and compared their findings with these ancient models. It was not until men like Galileo and Kepler gathered enough evidence to show that the earth was just one of several minor bodies circling around the sun, that finally moved the earth from its central place in the universe.

The shape of the earth. Evidence for the earth's actual shape can be gathered from several familiar observations:

1. A large body of water shows a curved surface. Ships sailing over the horizon on the sea seem to sink out of sight, the masts or taller parts disappearing last. A more exact way of showing the curve of a water surface would be to set three long poles into a lake bottom. The poles should be set about one kilometer (.6 miles) apart and be of the same length above the water surface. If an observer sights along the tops of these poles, he sees that the middle one is several centimeters higher than the two at either end.

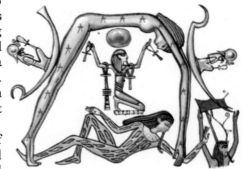

FIG. 6–1A. The universe as conceived by the Egyptians was a combination of gods. Keb, the earth-god, lay beneath the arching goddess of the heavens who was supported by the god of the atmosphere. Boats carrying the sun and moon gods sailed across the heavens.

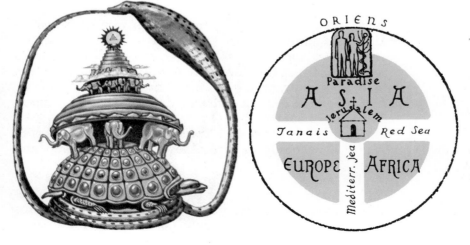

FIG. 6–1B. One Hindu idea of the earth had it supported on the backs of elephants who stood on a turtle. Around the entire arrangement was a cobra representing water.

FIG. 6–1C. One representation of the earth used during the Middle Ages was a disk. The continents were divided by the Red Sea, the Mediterranean, and the Don River. At the center was Jerusalem with the Garden of Eden in Asia.

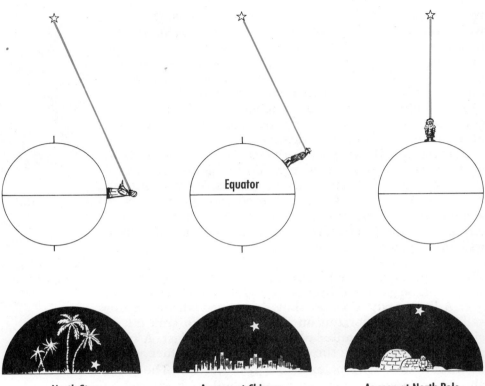

North Star
as seen at equator   **As seen at Chicago**   **As seen at North Pole**

FIG. 6–2. The elevation of the North Star (Polaris) above the horizon depends upon the position of the observer between the equator and North Pole.

2. The height or altitude of a particular star changes as an observer moves between the equator and poles. Figure 6–2 explains how the observer's viewing angle of the star is always changing as he moves over the rounded surface of the earth.

3. If the earth were flat, the horizon would always remain fixed along the edge of the earth. Then the horizon would never appear to move closer or farther away as an observer changes his viewpoint. Actually, the horizon does move away as an observer gains altitude and more of the earth's surface becomes visible. This could only happen on a rounded surface.

4. During a lunar eclipse the earth's shadow appears as the arc of a circle as it crosses between the sun and moon.

Other, more direct, evidence of the earth's shape is now available as a result of the exploration of space. Photographs taken by astronauts and from unmanned space vehicles reveal the earth's curved surface. See Figure 6–3.

Examine
How does the shape of the earth affect the number of miles in a degree of latitude as you move from the equator toward the poles?

The size of the earth. Determining the size of the earth has been a more difficult problem than finding its shape. However, more than two thousand years ago a Greek mathematician and astronomer named Eratosthenes (air uh TAHS thuh nees) measured the size of the earth with surprising accuracy. At the time he accomplished this, he was head of the library at Alexandria, the greatest institution of learning in the ancient world. He knew of a deep well in a city to the south of Alexandria where the sun's rays reached the bottom only once a year on about June 21. He reasoned that he could calculate the circumference of the earth if he knew the distance between

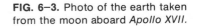
FIG. 6–3. Photo of the earth taken from the moon aboard *Apollo XVII.*

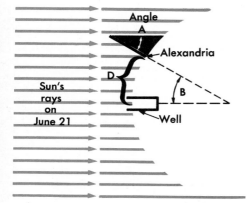

FIG. 6–4. Eratosthenes' method for measuring the size of the earth.

Discuss
How has our knowledge about the shape of the earth affected procedures for navigation over long distances?

Alexandria and the well, plus the angle of the noonday sun in Alexandria on June 21. His method is shown in Figure 6–4.

Angle A is equal to angle B. According to the principles of geometry, this angle has the same relation to 360° as the distance D does to the earth's circumference. Eratosthenes found that this angle was slightly more than 7 degrees. Since 7 degrees is about ¹⁄₅₀th of a complete circle, the distance between Alexandria and the well must be about ¹⁄₅₀th of the earth's circumference. Eratosthenes' calculation for the earth's circumference turned out to be about 46,250 kilometers (about 28,721 miles). This is not too far from the modern value of 40,000 km (25,000 miles). His final result was remarkably accurate, considering his crude measurements. Fortunately, the various errors almost canceled each other. Basically the same method used by Eratosthenes is still in use today.

During the seventeenth century, Sir Isaac Newton came to the conclusion that the earth could not be perfectly round. He reasoned that its rotation caused a slight flattening at the poles. His theory was proven correct in 1743 when scientists found that the earth's circumference at the equator was slightly larger than its circumference around the poles. This meant that the earth had the basic shape of an *oblate* (flattened) *spheroid*. That is, the outline of the earth is not a perfect circle.

However, the earth is so very close to a perfect sphere that it is rounder for its size than any bowling ball or basketball. Careful analysis of disturbances in the orbits of man-made earth satellites shows that the equatorial bulge is actually not at the equator as Newton had predicted. Instead it is a little south of the equator. This new location gives the earth an irregular shape that has been compared to the shape of a pear. However, this is not an accurate description. Since the earth's shape is so nearly perfectly round it is usually drawn as a regular circle. Figure 6–5 shows the generally accepted dimensions for the size and shape of the earth.

Directions and location on the earth. If the earth did not turn on its axis, it would be very difficult to describe any direction on its surface. Since the earth is very nearly a perfect sphere, it has no top, bottom, or sides to use in establishing direction. But the earth does rotate. Thus the ends of its axis of rotation, the north and south poles, provide the reference points needed. Based on the position of the poles, the four cardinal points of direction (N,E,S,W) can be located on the earth. The north-south direction runs along the axis line connecting the two poles. The east-west direction runs at right angles to the polar axis along a line parallel to the equator.

In describing a location on the earth's surface, we make use of imaginary lines. For example, to describe changes in location in a north-south direction, a system of imaginary lines drawn parallel to the equator is used. See Figure 6–6. The north-south location from the equator, of any place on the earth's surface, is called the *latitude* of that point. These imaginary lines are called *parallels of latitude.*

The latitude of a place is established in the following way: The distance from the equator to either of the poles is one-quarter of a full circle around the earth, or 90° of the full 360° circle. If the equator is taken as 0°, the location of any parallel of latitude can be described as a certain number of degrees north or south of the equator. For example, both the north and south poles have latitudes of 90°. Thus a point half way between the equator and one of the poles has a latitude of 45°. Of course, it is necessary to state whether the parallel lies north or south of the equator. Washington, D.C., for example, has a latitude of about 39°N. This fixes its position as 39° north of the equator.

To be more precise, each degree of latitude is divided into 60 equal parts called *minutes* (symbol ′). A more precise latitude for Washington, D.C. is 38°53′N. We can obtain even greater precision by dividing minutes into 60 equal parts called *seconds* (symbol ″). In actual distance over the earth's surface, a degree of latitude (60 minutes) is equal to about 111 km (69 mi). A minute of latitude is 1.85 km (1.15 miles, which is one *nautical* mile.

Given the latitude of a particular place we can only determine our north-south location. That is, our distance in degrees north or south of the equator. To locate where we are along this line of latitude, we must also know our east-west position. To do this, another set of imaginary lines, called *meridians*, are drawn extending from pole to pole. Each meridian outlines a complete circle when drawn around the entire earth. See Figure 6–7. The east-west position of a place, called *longitude,* is established by use of these meridian lines. This is done by using degrees, as with latitude, to establish the position of a certain meridian. However, there is no natural starting point that is called 0° longitude. To solve this problem, a particular meridian has been selected by agreement among all na-

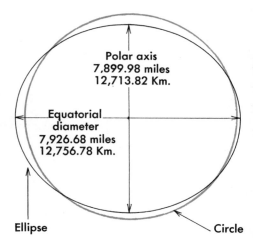

Ellipse   Circle

FIG. 6–5. For most purposes, the model of the earth with the shape shown in the diagram is satisfactory. The true shape of the earth is very slightly irregular, probably a little like a pear. However, the earth would appear as a perfect sphere if seen from a distance.

FIG. 6–6. Parallels of latitude are shown here at intervals of 15°. Note that each parallel forms a complete circle around the earth.

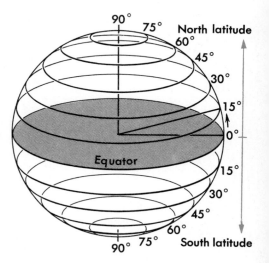

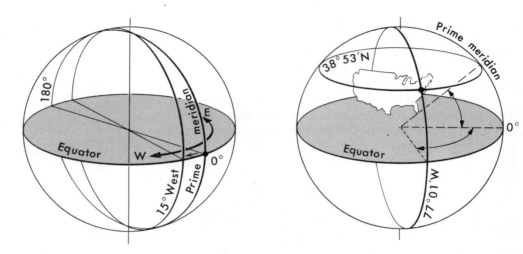

Meridians of longitude are shown here at intervals of 15°, measured east or west from the zero degree (0°) meridian.

FIG. 6-8. The exact location of Washington, D.C. is determined by measuring longitude west of the prime meridian, and latitude north of the equator.

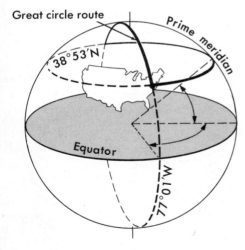

FIG. 6-9. A traveler flying a great-circle route from Washington, D.C. to Canton, China would have to pass over or near the North Pole.

tions. The meridian which passes through Greenwich, England, was selected as the 0° meridian. Greenwich is close to London and was originally the location of the Royal Observatory. The 0° meridian is often called the *prime meridian.*

Locations to the east of the prime meridian are said to have *east longitude;* those to the west have *west longitude.* The 180° meridian (directly opposite the prime meridian on the earth) also separates east and west longitude. Washington, D.C., is located west of the prime meridian; its longitude is about 77°W. to fix the position of Washington precisely, we say it is 38°53′N and 77°1′W. See Figure 6-8.

Another type of imaginary line often used on the earth is called a *great circle.* This is any line that completely circles the earth .and divides the globe into two equal halves. The equator is a great circle, and each circle of longitude is a great circle. Great circles are useful in aviation because they indicate the shortest distance between any two points. See Figure 6-9.

MAPPING THE EARTH'S SURFACE

How are maps made? A map is a model of the earth's surface, or a part of that surface taken from a globe and reproduced on a flat area such as a piece of paper. Maps can never be pictures of the earth's surface in the same way that photographs are. Since the earth's curved surface is drawn on a map as if it were flat, a part of the surface shown will always be distorted. For a map to give accurate information, certain systems must be followed in making and using a particular map.

It is possible to show how a map distorts the earth's surface by using the skin of an orange. If a large piece of orange skin is flattened, its shape will be changed by sketching and tearing. The larger the piece of skin, the more it must be distorted to flatten it. Similarly, the larger the part of the earth's surface that is being shown on a map, the greater the distortion will be. A map of a smaller area such as a city will have very little curvature and will show only slight distortion due to flattening.

Since the earth is almost perfectly round, its surface can be correctly shown only on another round body such as a globe. However, globes that are large enough to show small details are expensive and difficult to handle. Thus to represent the planet earth, we must rely mostly on maps drawn on flat surfaces.

The various ways that the curved surface of the earth can be transferred onto a flat map with the least amount of distortion are called *map projections*. Most map projections are the result of mathematical calculations. However, with just a little effort it is possible to picture the method used to produce them. Imagine a transparent globe, lighted from inside. Surface markings cast shadows on a piece of paper that is held against the globe. Differences in the way the flat paper is held against the lighted globe will produce a variety of patterns in the shadows cast on the paper.

A commonly used projection is a map made by wrapping a sheet of paper into a cylinder around a globe. If the cylinder is unrolled, the map projection would appear as shown, in Figure 6–10. The meridians of longitude are shown as equally spaced, straight, parallel lines. Since we know that meridians on a globe converge at the poles and are farthest apart where they cross the equator, this type of projection introduces a distortion of land areas that becomes greater near the poles. See Figure 6–11. Because the cylinder is in contact with the earth's surface only at the equator, this projection represents most accurately only the area near the equator.

However, the advantages of this type projection make it a most valuable tool for navigation. Some of these advantages are: All compass directions are shown as straight lines. The four cardinal points of the compass are located on the four sides of the map. All lines of latitude and

Discover
Look up the latitude of your school in an atlas or almanac. The latitude of a nearby city would be close enough. From your latitude in degrees, find the distance of your school from the equator.

FIG. 6–10. A cylindrical map projection can be visualized as the result of projecting the lines of latitude and longitude from the center of a globe onto a cylinder.

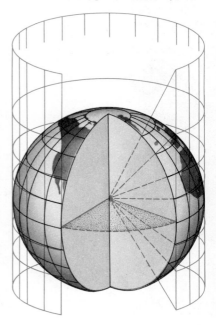

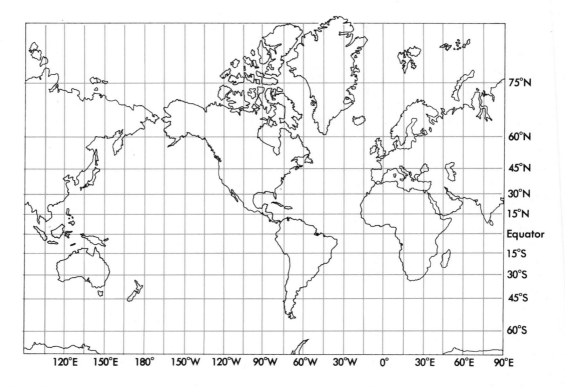

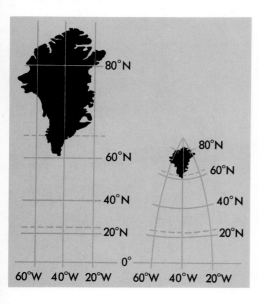

FIG. 6–11. This diagram illustrates how a cylindrical projection distorts high-latitude regions (above). The correct size is shown below.

longitude are clearly shown. You are probably familiar with this projection since it is most commonly used in your textbooks.

A second type of projection can be constructed by placing a flat sheet of paper on a globe so that it touches the surface of the earth at only one point. Distortion on this map is due to the unequal spacing of the parallels of latitude as you move farther from the point in contact. See Figure 6–12.

The advantage of this type of projection for navigation is that all great circle routes are shown as straight lines. This is extremely helpful for plotting the shortest route between any two points on the earth's surface. This kind of map projection is often used by navigators to plot polar routes for air travel.

A third type of projection uses a cone as the basis for a map projection. The cone is arranged so that the axis of the globe is in line with the axis of the cone. The parallel of latitude where the cone and the globe are in contact will show the least distortion on the completed map. See Figure 6–13. This type of projection is often used to accurately map relatively small areas. These areas can be taken from the part of the projection showing the least distortion. To map a number of neighboring small areas of the earth's surface, a series of cone projections may be used. This is called a *polyconic* projection. Each cone comes in contact with the earth at a slightly different latitude. The different latitudes where each cone touches the globe are fitted together to form a continuous map as shown in Figure 6–13. The advantage of this type of projection is that the shape and size of relatively small areas on the map are very nearly the same as those on the globe.

Using a map. People use maps to find out many kinds of information about the earth's surface. *Political* maps show national and local boundaries clearly. They often indicate the relative sizes of towns and cities along with their political importance. Political maps are often combined with *relief maps*. Relief maps show elevation of the land surface by using different color keys. *Navigation charts* provide navigators with routes and distances. *Hydrographic maps* are useful in showing depths of water and the shape of the sea floor. *Weather* and *climate maps* also give information about the general conditions of the atmosphere. *Geologic maps* show the arrangement of rock formations and are particularly useful in the mining and petroleum industries.

To understand any map, the first thing you must know is its relation to compass directions. The most common method of showing compass directions on a map is to make the top of the map north. Looking at the map, then, right is east, left is west and the bottom is south. Meridians of longitude are usually drawn as lines running from top to

FIG. 6–12. Left, the principle of a map projection on a plane surface. Right, an actual map produced by this method.

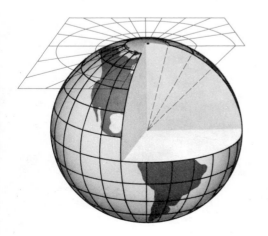
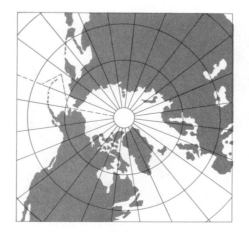

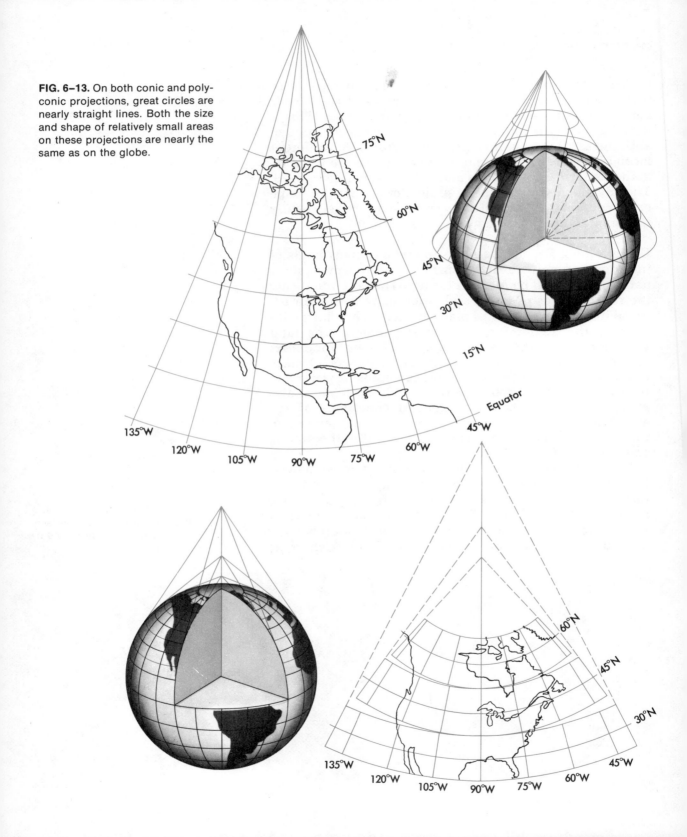

FIG. 6–13. On both conic and polyconic projections, great circles are nearly straight lines. Both the size and shape of relatively small areas on these projections are nearly the same as on the globe.

bottom. Parallels of latitude are lines running from side to side. Since these lines may either be straight or curved, depending upon the map projection, directions must always be read in relation to the parallels and meridians. This means that north is at the top of the map only if the meridians shown actually run from top to bottom.

In addition to knowing directions, the map user is usually interested in finding distances. All maps must indicate the relationship between actual distance on the earth and the same distance measured on the map. This relationship is called the *scale* of the map.

A map may be designed to show a large area of the earth. If this is the case, the scale selected will allow a short distance on the map to represent a large distance on the earth. For example, one inch on the map may represent 100 miles on earth. More detailed maps showing smaller areas use a larger scale, such as one inch to the mile. The scale of a map is commonly shown as a *graphic scale*. This is a line which is divided into parts marked to represent distances on the earth. Graphic scales are frequently used on maps covering smaller areas. Thus distances can be quickly found by directly comparing a measurement on the map to the number of divisions it covers on the graphic scale.

Another way of indicating the scale of a map is to use a fraction such as 1:62,500. This is called a *fractional scale*. It means that 1 unit of measurement on the map represents 62,500 of the same units on the earth. Occasionally the map scale may be given in a form such as "one inch equals one mile." This is known as a *verbal scale*.

TOPOGRAPHIC MAPS

A guide to the study of landforms. Some earth scientists use maps to study the landforms that give shape to the earth's surface. All the details that make up the surface features of the land are called its *topography*. A map made to show these details is known as a *topographic map*.

The governments of most countries make topographic maps of their territories for military, scientific and commercial use. In the United States, the Geological Survey, a branch of the Department of Interior, has mapped a large part of the country. The results are available in the form of detailed maps called *topographic sheets* or *quadrangles*. Most of these maps represent an area that covers 7.5 min-

Observe
Look at several road maps available at your local gasoline station. See if you can determine the type of projection used to draw these maps.

FIG. 6–14A. If the sea should rise by a number of equal increases, the new shorelines formed would correspond to contour lines showing the shape of the island.

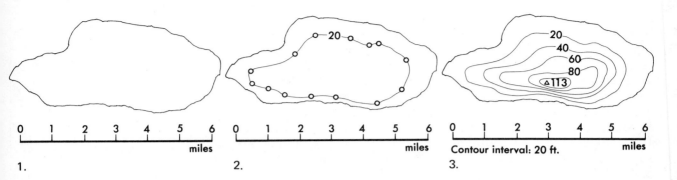

1.

2.

Contour interval: 20 ft.

3.

FIG. 6–14B. This sea island is 6 miles long, 3 miles wide, oval shaped, and 113 ft above sea level at its highest point. **1.** The map above shows only the shape of the land at sea level. The scale shows the length and width of the island. It does not show the *elevation*, the *steepness* of the slope, or the *shape* of the land above sea level.

2. A mapmaker surveys the island and proceeds to turn this map into a contour map. On the map he locates a series of points shown by his survey to be 20 ft above sea level. He then joins these points with a *contour line:* a line drawn through all points at the same height above sea level. Every point on this line is 20 ft higher (vertically) than any point at sea level, regardless of the uneven spacing between successive contour lines.

3. Now the mapmaker draws additional contour lines, showing where the island reaches the 40, 60, 80, and 100 ft elevations. Notice that the contour lines show the *elevation of the island, the steepness of the slopes* (which depends upon the closeness of the contour lines), and the *shape of the land above sea level*. Notice also that a symbol (△) called a bench mark is used to denote the elevation of any point which falls between the contour interval of 20 ft.

utes of latitude and 7.5 minutes of longitude. See pages 132–133. Topographic maps are also available for some areas which are 15 by 15 minutes; 30 by 30 minutes, and 1 degree by 1, 2, or 3 degrees.

Using topographic maps. Topographic maps almost always show the elevation and the shape of the land by using *contour lines.* A contour line is drawn on a map so that it connects all points on the ground of the same elevation. The shape of the contour lines gives us an idea of the shape of the land. The way these contour lines are able to show land features can be illustrated by a small island as shown in Figure 6–14A. Its shoreline could be one contour line connecting all places on the island at sea level elevation. Suppose the sea rose. Then a new shoreline with a new slope would be created. If the sea level should continue to rise, each time by the same amount, a series of new shorelines would be formed, each one outlining a new shape. If seen from above, each of the shorelines would represent a definite elevation above the original sea level.

The elevation of the land everywhere, no matter how far from the sea, is measured from *mean sea level.* This is a point midway between the highest and lowest tide. Mean sea level represents zero elevation or the starting point from which all elevations of the land are measured. Each new shoreline would at the same time also represent a contour line used to show the shape of the island. See Figure 6–14B. On an actual topographic map the contour lines are drawn from points of known elevation. The contour lines are drawn to connect points which are all at the same elevation above sea level.

The difference in elevation between any two contour lines is called the *contour interval.* A map maker chooses an interval suited to the size of the map and the topography of the region. The contour interval he chooses depends upon the *relief* of the land. Relief is defined as the difference in elevation between the highest and lowest points of the area being mapped. In maps of very mountainous areas where the relief is high, the contour interval may be as great as 50 or 100 feet to prevent crowding of the contour lines. In maps of low relief, some flat areas may have a contour interval of only one or two feet.

Interpreting a contour map. Just as printed words on a page transmit ideas, contour lines on a map are able to give a clear picture of the elevation, steepness and shape of the land surface. However, it takes some training and practice before contours on a topographic map can be interpreted. The following points should be kept in mind when working with topographic maps.

1. *A contour line connects all points having the same elevation.* The contour interval determines the elevation at which each contour line will be drawn. If a contour interval is 10 feet, contours will be shown for elevations of 10, 20, 30, 40, 50, 60 feet and so on. Usually, every fifth contour line is heavier and much darker. These are called *index contours* and are used to mark the actual elevation on the map.

Only the elevation of points directly on contour lines can be determined exactly. A point located between two contour lines has an elevation somewhere between that of the two lines. See Figure 6–15A.

Topographic maps almost always show a few points whose exact elevations are marked, although they are not located on contour lines. The elevations of these points are measured during the process of making the map. These known elevation points are marked with an X or BM (for "Bench Mark") printed in brown or black ink. Their exact elevation in feet is located near the mark.

2. The steepness of a given land surface can be easily determined from the contour lines. *Contours spaced widely apart show a gradual change in elevation and indicate nearly a flat surface. Closely spaced contours mean the elevation increases quickly and indicate a steep slope.* Contour lines that are almost touching indicate a very steep or vertical cliff. Where a map shows a vertical cliff grading into a more gentle slope the contours may appear to split. This, however, is only the effect of separate contour lines running together. See Figure 6–15B. A single contour line can never split. In the same way,

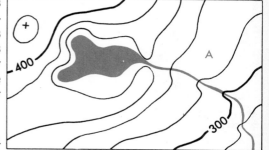

FIG. 6–15A. The elevation of point A is more than 320 feet but less than 340 feet. What is the contour interval?

FIG. 6–15B. A coastal valley shown in a side view or *profile* (top), and the same area represented by contour lines (bottom). The darker shading in the profile indicates the steepness of the slope also shown by the closeness of the contour lines. Notice that the 50-foot contour line intersects the stream channel at only one point. Explain.

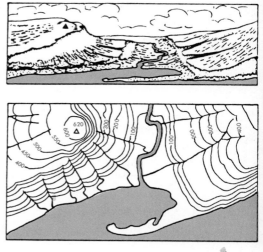

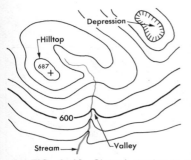

FIG. 6–16. Closed contours may indicate a hilltop as on left; with hachures they may signify a depression. Notice the heavier index contour and the contours bending in a V-shape with the point of the V directed up the stream valley.

Symbol	Color	Meaning
	Black	Buildings
	Black	Church and school
	Red or Black	Road or highway
	Black	Railroad
	Blue	Stream
	Blue	Intermittent stream
	Blue	Lake or pond
	Brown	Depression
	Blue or Green	Marsh or swamp

FIG. 6–17. The symbols shown here are some commonly used on topographic maps.

contour lines can never cross, since the point of crossing would have two elevations at the same place. The only exception would be a place where there is an overhanging cliff. But this is a very rare feature.

3. *When crossing a valley, contours bend to form a V shape.* The point of the V shows the upslope direction of the valley. If there is a stream flowing in the valley, the V in the contour lines will point upstream, or in the direction from which the water flows.

4. *Contour lines which form closed loops indicate a hilltop.* All contour lines join themselves again at some point. On a map of a small land area, however, most contours run off the map before closing on themselves.

5. A hole or depression is also indicated by using a closed contour loop. But, unless some indication is given that a depression is shown, the resulting contours might be mistaken for a hilltop. *A depression contour is indicated by use of hachure lines.* A hachure line is a short straight line pointing in the direction of the depression. See Figure 6–16.

Some of the colors and symbols used on topographic maps are shown in Figure 6–17.

Locating landforms. One of the most important uses of any map is to locate a particular point or place. With topographic maps the three methods commonly used are:

1. *Relation to easy-to-identify features.* The simplest and often most convenient method for locating a point is to give its distance and direction from any easily located feature on the map. The feature selected may be a mountain, lake, city, or any other feature that cannot be easily mistaken.

2. *By latitude and longitude.* On the topographic maps published by the United States Geological Survey, the eastern and western boundaries are marked as meridians of longitude. The longitude is marked at both the top and bottom of the maps in degrees (°) and minutes ('). Usually at least two other meridians are indicated on the maps between those at the edges.

The top and bottom boundaries on the map are marked as parallels of latitude. These parallels are marked at the left and right sides of the map. Usually at least two additional parallels are drawn or indicated by cross hairs (+) at 5 minute (') intervals.

3. *By township and range.* In the nineteenth century settlers began to move into the lands of the American mid-

west. At that time, a system of land boundaries was devised to establish ownership. The basic plan divided the land surface into squares six miles on a side. These areas were called *townships*. Each township was subdivided into thirty-six *sections*, each of which was usually one mile on each side. Each section could be further subdivided into half-sections, quarter-sections, or sixteenth sections.

On a map, townships are described by numbers and compass directions from some selected point of latitude and longitude. Vertical rows of a township square are called ranges and are numbered from east to west. Figure 6–18 shows how a point is located by means of township and range. The location of point X by this method would be described, for example, as: in the southeast quarter of the northeast one-fourth of Section 24, of the township which is second in the south horizonal row, and second in the west vertical row. States not included in this system of land division are all the eastern coastal states (except Florida), West Virginia, Kentucky, Tennessee, Texas, and some parts of Ohio.

activity

Studying a topographic map. Refer to the topographic map reproduced on pages 132–133. Answer the following questions that require the application of the rules for interpreting contour maps.

1. In what general direction does the Neversink River flow?

2. Determine the elevation to the closest contour of the bench mark (△) shown in the south–central part of the map. (Just below the G in FORESTBURG).

3. What is the latitude and longitude of Wolf Reservoir to the nearest minute? (At the cross hairs).

4. Is the railway line entering the city of Monticello ascending or descending as it moves north?

5. Which of the following terms would best describe the entire area lying between Route 209 and the railroad in the lower southeast corner of the map: plain, plateau, valley, peneplane, river basin? Explain your answer.

6. Using the scale found on the bottom of the map, determine to the nearest half mile the number of miles along Route 17 from Mastens Lake to Bridgeville.

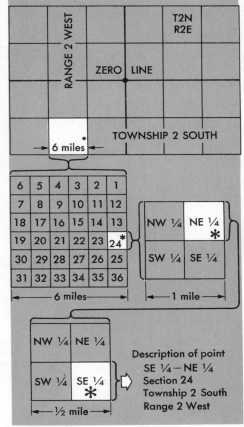

FIG. 6–18. The location of a point by the township and range method.

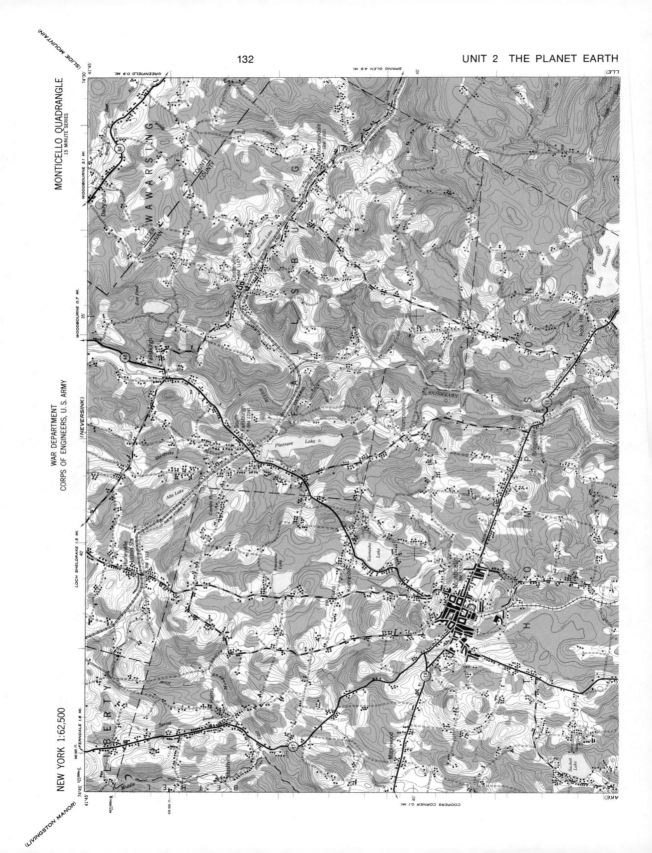

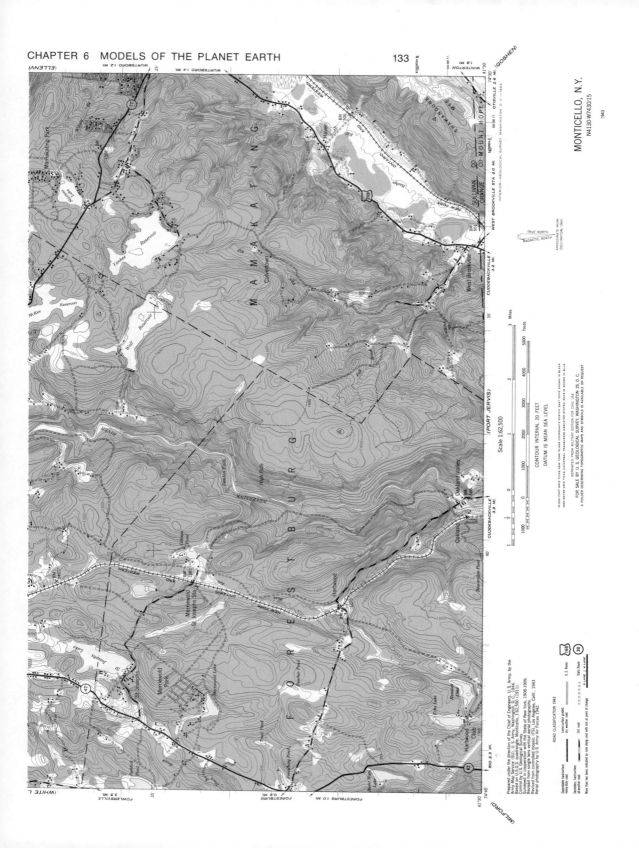

Scale 1:62,500

CONTOUR INTERVAL 20 FEET

DATUM IS MEAN SEA LEVEL

MONTICELLO, N.Y.

N4130-W7430/15

1943

7. Locate McKee Reservoir along Route 17. Just to the northwest of the reservoir is a small lake. Is the stream connected
   to the lake an inlet or an outlet from the reservoir? How do
   you know?

8. Locate Anawana Lake due north of the town of Monticello.
   A second lake, connected by a main stream to Lake
   Anawana, lies just to the southeast. Which lake has a higher
   elevation? How do you know?

VOCABULARY REVIEW

Match the word or words in the column on the right with the correct phrase in the
column on the left. *Do not write in this book.*

1. Measured the earth's circumference.
2. The distance around the earth.
3. The shape of the earth.
4. The most common models of the earth's surface.
5. Necessary to transfer curved surface of the earth to
   the flat surface of a map.
6. The north-south location of any place on the earth's
   surface.
7. 1/60th of a degree of latitude.
8. The east-west location of any place on the earth's
   surface.
9. The meridian which passes through Greenwich,
   England.
10. Any line drawn on the earth's surface which divides
    it into two equal parts.
11. Show national and local boundaries clearly.
12. Show depths of water and sea floor shape.
13. The relationship between a distance on the earth and
    the same distance measured on a map.
14. A map which shows the shape of the land surface.
15. A line drawn through all points with the same eleva-
    tion.

a. circumference
b. maps
c. longitude
d. prime
   meridian
e. Eratosthenes
f. oblate spheroid
g. minute
h. political maps
i. latitude
j. contour line
k. map projection
l. geologic map
m. great circle
n. topography
o. scale
p. hydrographic
   maps
q. topographic map

QUESTIONS

Group A

Select the best term to complete the following statements. *Do not write in this book.*

1. Early models of the earth gave clues to the earth's (a) size (b) shape (c) position in the solar system (d) all of these.

2. The most common models used in scientific descriptions of the earth are (a) layered (b) spheres (c) maps (d) full scale.

3. The earliest models of the universe placed the earth (a) on a crystal sphere (b) at a point which moved around the sun (c) at the center (d) at the top.

4. Which of the following statements is evidence for the curved surface of the earth? (a) The horizon is always at the edge of the earth. (b) The horizon remains fixed as an observer climbs to a higher viewpoint. (c) Ships suddenly disappear when they sail away. (d) The horizon moves away as an observer gains altitude.

5. Which of the following statements is *not* evidence for the spherical shape of the earth? (a) Ships sail out of sight over the horizon. (b) The elevation of a star changes from one place to another. (c) The earth's shadow is curved. (d) The stars move in circles around the pole star.

6. One of the first measurements of the size of the earth was made by (a) Alexander (b) Eratosthenes (c) Astronauts (d) Sir Isaac Newton.

7. The 7 degree angle which was part of the early measurements of the earth's size is nearest what part of a complete circle? (a) 1/50 (b) 1/360 (c) 1/7 (d) 1/5.

8. Which of the following statements is true of Eratosthenes' measurement of the earth's circumference? (a) his calculation of the earth's circumference was about 46,250 kilometers (b) his calculations were remarkably accurate (c) he used geometric principles (d) all of the above.

9. The earth's circumference is nearest (a) 46,250 km (b) 26,660 km (c) 40,000 km (d) 25,000 km.

10. Sir Isaac Newton decided that the earth could not be perfectly round because (a) it is rotating (b) it has ice caps (c) rivers wear it away (d) rocks are heavier than water.

11. The earth has a shape which is (a) more like a pear than a sphere (b) more like an orange than a sphere (c) more like a sphere than a basketball (d) more like a sphere than the best sphere that can be made.

12. Locations north or south of the equator are described by a system of lines called (a) parallels of latitude (b) lines of longitude (c) meridians (d) great circles.

13. Which of the following is *not* the latitude of a place on the earth (a) 90°N latitude (b) 120°S latitude (c) 30° latitude (d) 0° latitude.

14. Zero degrees latitude is the location of (a) the North Pole (b) the South Pole (c) the Equator (d) a point halfway between the Equator and the North Pole.

15. Washington, D.C. has a latitude nearest (a) 38°53′N (b) 38°S (c) 39°E (d) 53°38′W.

16. A minute of latitude is most nearly (a) 100 km (b) 60 km (c) 2 km (d) 1 km.

17. A minute of latitude is sometimes called (a) 1 mile (b) 1 second (c) 1 kilometer (d) 1 nautical mile.

18. Meridian lines are used to determine (a) latitude (b) north-south position (c) distance from the North Pole (d) longitude.

19. A degree of latitude is equal in distance to a degree of longitude (a) nowhere on the earth (b) at the equator (c) at the poles (d) only at the prime meridian.

20. California is located at about 120 degrees (a) north latitude (b) south latitude (c) east longitude (d) west longitude.

21. Which of the following is *not* a great circle? (a) the 45° parallel of latitude (b) the equator (c) the 180° meridian combined with the prime meridian (d) any circle on the earth's surface which passes through both the North and South Pole.

22. A map projection has the least distortion at the (a) equator (b) poles (c) point it touches the globe (d) points far from the point it touches the globe.

23. The only undistorted representation of the earth's surface is (a) a cylindrical map projection (b) a globe (c) a conical map projection (d) a flat surface map projection.

24. If north is at the top of a map, east would be (a) at the bottom (b) to the left (c) to the right (d) at the right or left.

25. The scale is most likely to be correct for a whole map if it is the map of (a) a large area (b) a small area (c) an ocean area (d) a land area.

26. Close spacing of contour lines on a topographic map indicates (a) a depression (b) a steep slope (c) a gentle slope (d) a flat area.

27. Contour lines which cross (a) never occur (b) occur only on flat areas (c) occur only on steep slopes (d) occur only where there is an overhanging cliff.

28. A township is a square whose area is (a) 6 square miles (b) 12 square miles (c) 24 square miles (d) 36 square miles.

29. Index contours are always (a) even numbers of feet (b) multiples of 10 (c) marked for the elevation represented (d) drawn as dashed lines.

Group B

1. What is an oblate spheroid?

2. What is the latitude and longitude where the prime meridian crosses the equator? What is the latitude and longitude of the North Pole?

3. What is meant by the relief of the land?

4. Give another name for a topographic sheet and tell what it is.

5. The capital of which state in the United States is at 38°35′ north latitude and 121°30′ west longitude? Which at 46°48′N and 100°47′W?

6. To the nearest degree, what is the latitude and longitude of your city?

7. Why does the distance measured by a degree of latitude always stay the same while the distance for a degree of longitude varies?

8. If the earth did not rotate, how would you establish reference points for directions?

9. Is a degree of longitude the same distance on the earth as a degree of latitude? Explain.

10. Can contour lines cross each other? Explain.

11. How many kilometers are equal to one degree of latitude?

12. How many kilometers are equal to one minute of latitude?

13. How many meters are equal to one second of latitude?

14. Some map projections are made by placing a cylinder around a globe of the earth. What effect would it have on the map projection if the diameter of the cylinder were twice the diameter of the globe?

15. A mile is about 1600 meters. Which is larger, a mile or a nautical mile? How many meters larger?

16. Two cities north of the equator are at the same latitude. They are connected on a map by a line which is a great circle route between them. Where will the great circle lie in relation to the parallel of latitude for the two cities?

17. The scale of a map is 1:24,000. What distance on the map would represent a distance of 2.4 km on the earth? Express your answer in centimeters.

18. Explain Eratosthenes' method for measuring the earth's circumference.

19. Suppose that Eratosthenes had measured the distance between the well and a city 550 kilometers to the north. Suppose also that he found the angle to be 5.0 degrees. What value would he have obtained for the earth's circumference?

7
earth chemistry

objectives

☐ List the three fundamental particles found in an atom.

☐ Describe the characteristics of these particles.

☐ Explain the role of atoms in the makeup of elements.

☐ Define the terms atomic number and atomic mass.

☐ Determine the number of particles in an atom, when given the atomic number and atomic mass.

☐ Describe two ways electrons are involved in chemical bonding.

☐ Explain how forces that hold molecules together form a gas, liquid, or solid.

Moon rocks are not very different from any rocks found on the earth. Rocks found on Mars also strongly resemble earth rocks. Scientists believe that pieces of the solid surface of any of the inner planets will not differ greatly from the many rocks on the earth's surface. If there are other systems of planets around distant stars, fragments from many of those planets will probably be very similar to some kinds of earth rocks. Why is there such a similarity between earth rocks and rocks from the moon, Mars, or even other solar systems?

The reason we make this conclusion is based on our knowledge of how all matter is put together. These ideas are part of the *atomic theory*. This theory assumes that every substance, whether a part of the earth or part of Mars, is made up of atoms. In addition, the atomic theory states that there are only a certain number of the kinds of atoms in the universe. Thus, in all forms of matter there are only different arrangements of these same kinds of atoms. Thus, rocks found on any planet must contain the same kinds of atoms as those found on earth.

Although the atomic theory allows us to make predictions about the kind of matter found on other planets, its greatest importance is in understanding the composition of the earth. The atomic theory provides us with a tool for studying the materials that make up this planet.

ATOMS

Structure of atoms. One of the most important things to keep in mind about the atom is its size. A single atom is so small that it is difficult to even imagine. In fact, no one has

138

actually seen a single atom! It would take more than a million average atoms, side by side, to equal the thickness of the paper on which this page is printed. Anything so small cannot be studied individually. Again, it was necessary to invent a scientific model to describe the structure of a single atom.

Experiments carried out over hundreds of years have shown that atoms must have some connection with electricity. Scientists of the eighteenth century, such as Benjamin Franklin, demonstrated that matter contains two kinds of electricity. It was Ben Franklin who first called the two kinds of electrical charge "negative" and "positive."

FIG. 7–1. An area on earth seen from five increasingly closer viewpoints. (A) An astronaut's view of the Salton Sea in southern California. The area circled is shown from a distant view at ground level (B) and at close range (C). A mineral sample taken from the rock of mountains is shown in (D) along with its atomic make-up in (E). (NASA–7–1A)

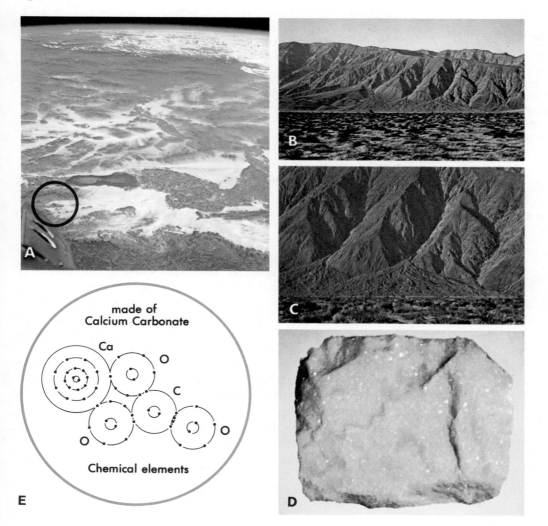

made of
Calcium Carbonate

Ca

O

C

O

O

O

Chemical elements

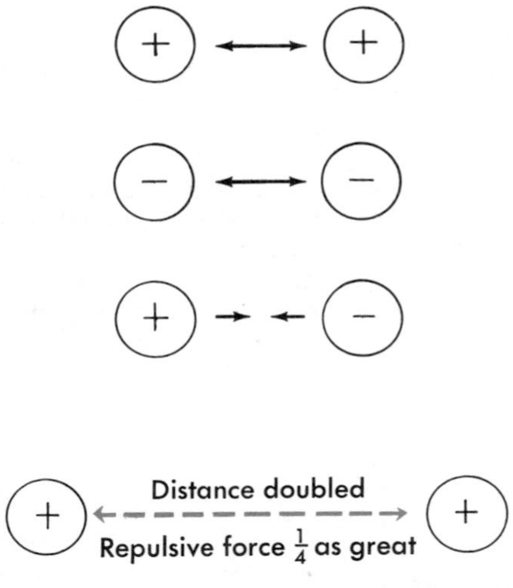

FIG. 7-2. The rules which explain attraction and repulsion of electrically charged bodies are illustrated. Notice in the lower diagram that increasing the distance separating the charged bodies greatly decreases the force acting between them.

Experiment
Use magnets to study attraction and repulsion. Use electrically charged objects for the same kind of investigation.

The model which best accounts for the various properties of individual atoms has three smaller parts to its structure. These fundamental atomic particles are *electrons, protons,* and *neutrons.* Models which describe the structure of atoms use different numbers and arrangements of these three basic atomic particles. One of the most important things to know about electrons and protons is that they carry electrical charges. Electrons carry a negative electrical charge; protons carry a positive charge. When any object has an electrical charge, it is able to influence another electrically charged body. If two objects have the same charge, such as two electrons, they will repel each other. Two positively charged bodies, such as two protons, will also repel each other. This rule is: *Like charged objects repel.*

On the other hand, charges that are not alike, attract one another. An electron and proton, for example, will be drawn toward each other because they carry opposite electrical charges. The rule here is: *Unlike charged objects attract.* The force of attraction or repulsion between charged bodies depends on the distance between them. The rules of electrical attraction and repulsion are shown in Figure 7-2. The third basic particle, the neutron, has no electrical charge. It is electrically neutral.

Besides the opposite electrical charges which make them attract each other, electrons, and protons in atoms have another important difference. All of the protons within a particular atom are packed together into a small region in the atom's center. This central core is called its *nucleus.* The atomic nucleus also contains any neutrons which the atom may have. The atom's electrons move in a certain region of space around the nucleus because they possess a certain amount of energy. Since the electrons are negatively charged, they are attracted toward the nucleus. Thus the electrons tend to remain relatively close to the nucleus of the atom to which they belong.

Electrons are not actually pulled into the body of the nucleus by the attraction of the protons. This is because the very rapid movements of the electrons around the nucleus give them enough energy to remain outside the nucleus. Keep in mind that neutrons in the nucleus play no part in the attraction of electrons since they are electrically neutral.

Kinds of atoms. The word "atom" comes from a Greek word which means "not able to be divided." This is an important feature of the atomic theory. An atom is the smallest complete part of any kind of matter. They are the smallest objects that can be recognized as ordinary matter. If any atom were separated into its parts, the result would be a collection of electrons, protons, and neutrons. These atomic particles bear no resemblance to ordinary matter, such as a piece of rock.

Careful investigation over a long period of time has shown that there are only a few kinds of matter that cannot be charged into a simpler form. This meant that these substances were made up of only one kind of atom. Since atoms cannot be changed into any simpler kind of matter, a substance composed of only one kind of atom cannot be simplified. Matter made up of only one kind of atom is called an *element*.

About ninety elements occur naturally on or in the earth. Around a dozen additional elements have been made artificially. This means that there is a total of about one hundred different kinds of atoms that are known to exist. Of the natural elements, a few very common ones make up most of the earth's crust. Table 7–1 shows the most abundant elements found in the earth's crust. Notice that in Table 7–1 a symbol is given for each element named. Each of the elements has a symbol of one or two letters. The symbol for an element is understood to represent a single atom of that element. Each kind of atom is also given an *atomic number*. This number is equal to the number of protons in the atom's nucleus. For example, the atomic number of an oxygen atom is eight. See Table 7–2. Since an atom normally contains the same number of protons and electrons, the atomic number also tells the number of electrons in that particular atom. Thus the oxygen atom has eight electrons moving around its nucleus along with the eight protons in the nucleus.

Thus the atom of each element has its own number of protons and electrons as a characteristic property that helps to identify it. But the atoms of different elements can be identified in another way. Each kind of atom has its own *atomic mass*. However, since atoms are so small, it is not practical to use ordinary units of mass such as grams. Instead, a special atomic mass scale is used on which the mass of a proton is assigned the value "1." Thus each of the protons in an atomic nucleus is said to have a mass of 1. Neutrons, which have nearly the same mass as protons, also have a mass of 1. Electrons are much lighter than either protons or neutrons. It takes

Table 7–1

Element	Symbol	Per cent by Weight in Crust
Oxygen	O	46.60
Silicon	Si	27.72
Aluminum	Al	8.13
Iron	Fe	5.00
Calcium	Ca	3.63
Sodium	Na	2.83
Potassium	K	2.59
Magnesium	Mg	2.09
Eighty other elements		1.41

Investigate
Examine Table 7–1. Compare the symbols used for the eight most abundant elements with their actual names. Find out why some symbols do not match the letters found in their names.

Table 7–2

Element	Atomic number (Protons)		Neutrons		Atomic Mass
Oxygen	8	+	8	=	16
Silicon	14	+	14	=	28
Iron	26	+	30	=	56
Lead	82	+	125	=	207
Uranium	92	+	146	=	238

Mass = 1   Mass = 2   Mass = 3

FIG. 7–3. The isotopes of hydrogen. Solid dots represent protons and open circles represent neutrons.

1,840 electrons to equal the mass of one proton. Since electrons add so little to the total mass of an atom, it is usually ignored. This means that the atomic mass for any atom is calculated by the number of protons and neutrons in its nucleus. The total atomic mass for an atom is the sum of the masses of its protons and neutrons. For example, the atomic mass of oxygen is 16 as shown in Table 7–2. This is the sum of its protons and neutrons. The atomic number for oxygen is 8 which means that it has 8 protons. Therefore, an oxygen atom must have 8 neutrons in its nucleus.

Isotopes. Not all the atoms of a given element contain the same number of neutrons. This can be illustrated by the 3 different atoms of hydrogen. The atomic number for hydrogen is 1 which indicates that it is the simplest kind of atom. With an atomic number of 1, a hydrogen atom has only a single proton in its nucleus with one electron moving around it. This is illustrated in Figure 7–3, left diagram. Almost all hydrogen atoms have this arrangement.

FIG. 7–4. A graph of the energy required to remove an electron from elements with atomic numbers 1–20.

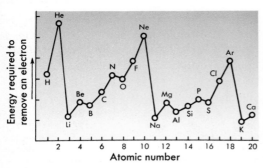

But a few hydrogen atoms also have a single neutron in their nuclei. These hydrogen atoms still have only one proton and one electron but the neutron adds 1 unit to the atomic mass. These hydrogen atoms have an atomic mass of 2. A very rare form of hydrogen has two neutrons in its nucleus. It has an atomic mass of 3. All three forms of hydrogen differ from each other in the number of neutrons found in the nucleus. Each additional neutron gives each type of hydrogen atom an additional unit of atomic mass. When atoms of the same element differ in mass because of the different number of neutrons in their nuclei,

they are isotopes of that element. Hydrogen has three isotopes as shown in Figure 7–3.

MOLECULES

Electron arrangement in atoms. In the earth's crust,
atoms of a particular element are not usually found alone.
Almost all the elements which occur in the earth's crust
are chemically joined with other kinds of elements. When
two or more elements are chemically united, the resulting
new substance is called a *compound*. Although the smallest complete unit of an element is an atom, the smallest
complete unit of a compound is called a *molecule*. Each
compound is made up of only one kind of molecule just as
an element is made up of only one kind of atom.

Different kinds of atoms join together to form molecules because of the way their electrons are arranged. Remember that the electrons in an atom move around its
nucleus. For example, the single electron present in hydrogen atoms moves in a sphere-shaped region around
the nucleus. The electron remains in the neighborhood
of the nucleus because it is in a state of balance. Although
its motion tends to carry it away, its negative charge attracts it toward the positively charged nucleus.

However, it is possible to remove electrons from their
atoms. If energy in the form of heat or light is added to
hydrogen atoms, they may lose electrons. The same is
true for almost all other kinds of atoms. The correct addition of energy can result in the loss of one or more of the
atom's electrons.

In experiments performed to determine the amount
of energy needed to remove an electron from various
kinds of atoms, an important discovery was made. The
graph shown in Figure 7–4 shows the results of this kind
of experiment with some of the lighter elements. We can
see that there is a pattern in this graph. Certain kinds
of atoms require high energies to lose an electron. These
atoms, found where there are peaks on the graph, are
helium (He), neon (Ne) and argon (Ar). If the graph were
extended to include all the elements, similar peaks for
the elements krypton (Kr), xenon (Xe) and radon (Rn)
would appear.

A conclusion that we can draw from this information
is that these are elements which do not easily lose electrons. Thus they have a more stable electron arrangement
than other atoms. The atomic numbers of these elements

activity

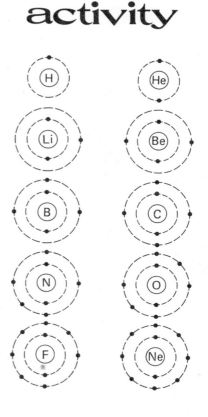

Use the electron structures above and
Fig. 7–4 to answer the following:

1. Do hydrogen (H) and lithium (Li)
   require less energy to remove an
   electron than helium (He)?

2. When an electron is removed from
   a hydrogen atom, is a pair of electrons broken up?

3. Is this also true for lithium?

4. Does boron (B) require less energy
   to remove an electron than beryl-
   lium (Be) and carbon (C)? Explain.

5. As electrons are added to an atom,
   what effect is there on the amount of
   energy needed to remove them?

6. Draw the electron structure of a
   sodium atom.

are: helium = 2, neon = 10, argon = 18, krypton = 36, xenon = 54 and radon = 86. We can further conclude that 2, 10, 18, 36, 54 and 86 electrons are favored numbers to be found around a nucleus.

Chemical bonds. For a molecule to be created, two or more different atoms must be held together by a *chemical bond.* Certain numbers of electrons indicate a stable atomic structure. This fact helps explain why some elements more readily form chemical bonds.

To illustrate the chemical bonding of elements into compounds, consider a common substance such as water. If we analyze water, we find that it is a compound made up of the elements hydrogen and oxygen. Hydrogen atoms (atomic number = 1) normally have only one electron. Since two electrons would be a more stable number, hydrogen will accept another electron if it is available. The element oxygen (atomic number = 8) holds eight electrons around its nucleus. Ten is the closest stable number of electrons. Therefore, we can predict that oxygen will accept two electrons to its structure.

If hydrogen atoms and oxygen are chemically combined, it is possible for both elements to acquire their stable number of electrons. Two hydrogen atoms can each share their single electrons with an oxygen atom, giving it a stable number of 10. At the same time, the oxygen atom can share two of its electrons; one with each hydrogen atom, resulting in a stable number of 2. The chemical bond which produces a water molecule is shown in Figure 7–5. Since the water molecule has a structure unlike either the hydrogen or oxygen atom, its properties are also unlike these elements.

Because of their electron arrangements, two hydrogen atoms will always join with a single oxygen atom to form a water molecule. This makes it possible to represent a water molecule by the formula, H_2O. A *chemical formula* represents one molecule of a certain compound. The formula for a compound tells what elements it contains and the number of atoms of each element present.

Some compounds are formed by chemical bonds that do not involve the sharing of electrons between atoms. An example is the compound sodium chloride, NaCl (common salt). Sodium atoms (atomic number = 11) have one more than the stable number 10 electrons. Chlorine atoms (atomic number = 17) have one less than the stable electron number of 18. Sodium atoms have a strong tendency

FIG. 7–5. A water molecule. Notice that the electrons of oxygen are arranged in two levels around its nucleus.

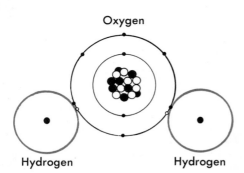

Oxygen

Hydrogen   Hydrogen

to lose one electron while chlorine atoms try to gain an electron. This means that one sodium atom will give up an electron to a chlorine atom as shown in Figure 7–6. Thus the chemical bond which is formed between the sodium and chlorine atom does not result from the sharing of electrons as in the water molecule.

When an electron is transferred from one atom to another, the atoms become electrically charged. Normally, atoms such as sodium and chlorine carry no overall electrical charge. This is because their charged particles, protons and electrons, are present in equal numbers. Their opposite electrical charges cancel each other exactly, thus neutralizing the atom. However, if an atom loses or gains electrons, it will take on either a positive or negative charge. Sodium, for example, becomes positively charged when it gives up an electron to chlorine. The loss of one electron still leaves a sodium atom with eleven protons in its nucleus. With only ten electrons circling the sodium atom's nucleus, there is then one excess proton not canceled by an electron. Thus, when the sodium atom loses an electron, it takes on a positive charge. It is then said to be a positively charged sodium *ion* whose symbol is Na⁺. An ion is an atom or group of atoms which carry an electrical charge.

In the same way that sodium becomes an ion by losing an electron, chlorine becomes an ion by gaining one. The addition of an electron to a chlorine atom gives it eighteen electrons but only seventeen protons in its nucleus. The extra electron changes the neutral state of a chlorine atom to a negatively charged chloride ion (Cl^-).

Forces which hold molecules together. Two or more kinds of atoms may combine to form a chemical compound. And, in some cases, it is easy to see why the resulting compounds are usually solids. For example, you know that the compound sodium chloride, NaCl, or common table salt, is normally solid. The reason for this can be found in the way this compound is formed.

When sodium and chlorine atoms combine chemically to form sodium chloride, each one of these atoms becomes an ion. The sodium takes on a positive charge (Na⁺) and the chlorine takes on a negative charge (Cl^-). When a number of these oppositely charged ions attract each other they arrange themselves in a definite pattern. Each positive (Na⁺) ion is surrounded by negative (Cl^-) ions. See Figure 7–7. Each ion is strongly attracted to its op-

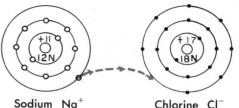

Sodium Na⁺   Chlorine Cl⁻
Ion   Ion

FIG. 7–6. The compound sodium chloride is formed by an electron transfer from the outer shell of sodium to chlorine, as indicated by the arrow.

FIG. 7–7. A crystal of sodium chloride is made of sodium ions (smaller spheres) and chloride ions held by their opposite electrical charges. Notice that each sodium ion is surrounded by oppositely charged chloride ions.

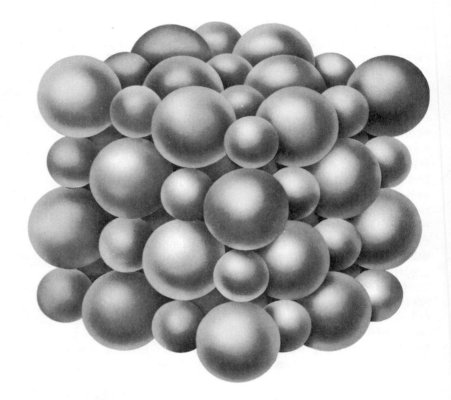

FIG. 7–8. Salt crystals have the shape of a cube because of the arrangement of their ions. (B.M. Shaub)

positely charged neighbor and can only fit into the pattern in a definite position. The sodium chloride structure that is formed will have a definite shape as shown in Figure 7–8.

Each piece of sodium chloride formed has the shape of a cube, because the ions from which it is made fit together in a cubical pattern. Any solid substance, such as sodium chloride, having a definite natural shape is called a *crystal*. Many compounds are able to form crystals, although the crystal arrangement is often difficult to see. The individual crystals are frequently so small that they can only be seen with a microscope. Also, the crystals often break after they are formed so that their original shape is not easily seen.

Other than its crystal shape, the ions which make up sodium chloride also give it another important property. It is hard to melt. To melt a solid material, heat energy must be added. This energy causes the particles (molecules or atoms) in the solid to move faster. When enough heat energy has been added, the individual particles move so rapidly that they break away from each other. In a solid, the particles are locked tightly together and the material holds its shape. But a liquid has no definite shape and can flow freely because its particles move around each other. If more heat is added to a liquid, some of its particles will move so fast that they will escape from the liquid. At this point the liquid begins to change to a gas. The relative amount of motion of particles in solids, liquids and gases is illustrated in Figure 7–9.

Sodium chloride is a difficult compound to melt because of its strong *ionic bond*. Ionic bonds are attractions of unlike charged particles. These bonds form substances that require a large amount of heat energy for them to melt and become liquid. This is not the case for all substances. Some substances may be composed of molecules which are held together less strongly than ions. An example of such a substance is water.

Unlike sodium chloride, water is made up of separate molecules. Each molecule has the formula H_2O representing the two hydrogen atoms sharing electrons with an oxygen atom. But these electrons are not shared by equal attractions. The oxygen has a larger number of protons in

FIG. 7–9. The relative motion of molecules in solids, liquids, and gases is represented in these diagrams.

its nucleus than does the hydrogen. Because of this, the oxygen attracts the electrons slightly more than the hydrogen does. This means that the water molecule has a slight negative charge around its oxygen atom. At the same time, there is a slight positive charge around the end of the molecule where the hydrogen atoms are located. Water molecules can then attract each other with their oppositely charged ends. See Figure 7–10.

If enough heat is lost, the attraction between water molecules can produce water in the solid form (ice). This form has a definite crystal pattern. If heat is then added to the water molecules, they begin moving farther apart. At a certain temperature the ice melts and becomes a liquid. Ice melts at a much lower temperature (0°C) than sodium chloride (801°C). This is because water molecules are held together by much weaker forces than the strongly attracted ions in salt. The relatively weak attraction between water molecules also allows them to become a gas at low temperature (100°C at ordinary atmospheric pressure).

Many of the materials found on the earth are liquids and gases. This is so because, as in the case of water, their molecules are not strongly attracted to each other. The substances which make up the gases of the atmosphere are another example of weakly attracted molecules. The most abundant gases in the atmosphere are nitrogen and oxygen. Each of these is made up of molecules which contain two identical atoms. Nitrogen gas has the formula N_2. Each of its molecules consists of two nitrogen atoms. The formula for oxygen gas is O_2. In both nitrogen and oxygen molecules, the electron arrangements of the atoms become stable by sharing electrons with an identical atom. The resulting molecules then have very little at-

FIG. 7–10. The arrangement of water molecules in (A) water vapor, (B) liquid water and (C) ice, is represented in these diagrams. Attraction of water molecules for each other causes them to form hollow rings in ice crystals.

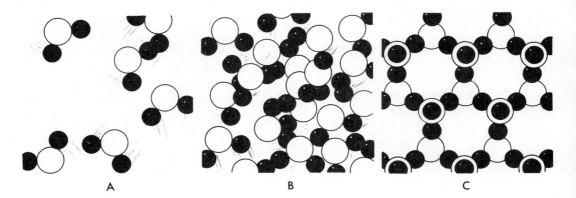

A   B   C

tractive force for each other. A similar situation exists
among the molecules of the other gases found in the
atmosphere.

The only liquid present in large amounts at or near
the earth's surface is water. Temperatures at the earth's
surface are such that water exists in the liquid state. If
the earth were much colder or warmer, this would not be
true. As already mentioned, water molecules are held to
each other by relatively weak forces. Thus there is enough
heat energy on most of the earth's surface to prevent
water from becoming solid ice. See Figure 7–11.

Most of the matter in the earth is in the solid form. This
means that most of the molecules of the earth are bound
together by relatively strong forces. These forces are due
to ion formation or electron sharing. These solid substances exist in a great many different forms. Just think
of the variety of rocks and minerals that can be found in
the crust.

FIG. 7–11. The sun's radiant energy allows water near the earth's
surface to remain in a liquid state.
(H.R.W. Photo by Russell Dian)

VOCABULARY REVIEW

Match the word or words in the column on the right with the correct phrase in the column on the left. *Do not write in this book.*

1. Assumes that every substance is made up of atoms.
2. Have only a negative electrical charge.
3. Part of the nucleus which carries a positive electrical charge.
4. Part of the nucleus which has no electrical charge.
5. Central core of an atom.
6. Means "not able to be divided."
7. Matter made up of only one kind of atom.
8. Equal to the number of protons in the nucleus of an atom.
9. Equals the sum of the number of protons and neutrons in the nucleus of an atom.
10. Atoms of the same element but with different numbers of neutrons in their nuclei.
11. The resulting substance when two or more kinds of atoms are chemically united.
12. Has the same relationship to a compound as an atom has to an element.
13. Holds a molecule together.
14. An atom or group of atoms which carries an electrical charge.
15. Any solid substance with a definite natural shape.

a. protons
b. nucleus
c. repulsion
d. atomic number
e. electrons
f. atom
g. neutrons
h. atomic theory
i. atomic mass
j. element
k. chemical bond
l. molecule
m. crystal
n. isotopes
o. chemical formula
p. compound
q. ion

QUESTIONS

Group A

Select the best term to complete the following statements. *Do not write in this book.*

1. The atomic theory states that (a) all atoms are identical (b) only light atoms are identical (c) there are only a certain number of kinds of atoms (d) only heavy atoms are identical.

2. Atoms cannot be studied individually because they (a) are too small (b) are not solid (c) move around too rapidly (d) are electrically charged.

3. Which of the following does *not* describe a kind of electrical charge? (a) positive (b) negative (c) neutral (d) nuclear.

4. Which of the following does *not* carry an electrical charge? (a) protons (b) neutrons (c) atomic nucleus (d) electrons.

5. Which of the following is *not* a part of every atom? (a) proton (b) neutron (c) nucleus (d) electron.

6. An atom which has lost an electron is brought near an atom which has gained an electron. The atoms will (a) attract (b) repel (c) sometimes attract and other times repel (d) have no electrical effect on each other.

7. Two atoms have charges which cause them to repel each other. If the atoms are moved so that their separation is twice as great, the repulsive force will be (a) zero (b) twice as great (c) one-half as large (d) one-fourth as large.

8. In an atom, neutrons will most likely be found (a) near the electrons (b) near the protons (c) in the outer regions of the atom (d) only if it is a heavy atom.

9. A substance which is composed of only one kind of atom is called (a) an element (b) a compound (c) a solid (d) an ion.

10. Of the elements known to exist, the number which occur naturally is (a) very few (b) about 12 (c) about 90 (d) about 100.

11. The most common element by weight in the earth's crust is identified by the symbol (a) Al (b) Fe (c) O (d) Si.

12. The atomic number of an atom is equal to the number of (a) electrons in the nucleus (b) neutrons in the nucleus (c) protons plus neutrons in the nucleus (d) protons in the nucleus.

13. The atomic number of oxygen is (a) 16 (b) 12 (c) 8 (d) 4.

14. The atomic number of uranium is (a) 238 (b) 146 (c) 92 (d) 82.

15. Which of the following elements has atoms whose atomic number and atomic mass are numerically equal? (a) oxygen (b) silicon (c) iron (d) hydrogen.

16. The atomic mass of oxygen is (a) 16 (b) 12 (c) 8 (d) 4.

17. The atomic mass of uranium is (a) 238 (b) 146 (c) 92 (d) 82.

18. The atomic number of argon is 18. Its atomic mass is 40. The number of neutrons in the nucleus of argon is (a) 18 (b) 22 (c) 40 (d) 58.

19. From the information in problem 18, the number of electrons in argon atoms is (a) 18 (b) 22 (c) 40 (d) 58.

20. How many electrons does it take to equal the mass of one proton? (a) 16 (b) 32 (c) 920 (d) 1840.

21. Hydrogen atoms have atomic masses of any of the following except (a) 1 (b) 2 (c) 3 (d) 4.

22. Most hydrogen atoms have an atomic mass of (a) 1 (b) 2 (c) 3 (d) 4.

23. Most argon atoms have an atomic mass of 40 but natural argon is given in most references with an atomic mass of 39.94. This means that (a) some argon atoms have an atomic mass greater than 40 (b) all argon atoms have an atomic mass of 39.94 (c) some argon atoms have an atomic mass less than 40 (d) some argon atoms have an atomic mass of 39.94.

24. The smallest complete unit of a compound is called (a) an atom (b) a molecule (c) a proton (d) a neutron.

25. A water molecule is made when two hydrogen atoms combine with (a) one oxygen atom (b) one oxygen molecule (c) two oxygen molecules (d) two oxygen atoms.

26. The chemical bond in a water molecule is produced by (a) hydrogen giving up electrons (b) oxygen giving up electrons (c) the sharing of electrons between hydrogen and oxygen (d) the taking on of electrons by oxygen.

27. Sodium atoms combine with chlorine atoms by (a) sodium atoms taking on electrons (b) sodium atoms giving up electrons (c) chlorine atoms giving up electrons (d) sodium atoms sharing electrons with chlorine atoms.

28. A crystal of sodium chloride is held together by forces (a) due to sharing of electrons (b) due to gravity (c) that are weak (d) caused when ions of unlike charge come near one another.

29. Water exists as a liquid at room temperature. If heated to 100°C, the forces holding it together may be broken converting it to (a) its elements (b) a solid (c) a gas (d) ions.

30. Most of the matter in the earth is in the solid form, which means that most of the molecules of the earth are (a) ionic (b) giving up electrons (c) held together by relatively strong bonds (d) held together by relatively weak bonds.

Group B

1. List the four most abundant elements in the earth's crust. What are their chemical symbols?

2. Where in atoms do scientists find protons? neutrons? electrons?

3. Why are electrons neglected in determining atomic mass?

4. What is an ion?

5. Define what is meant by isotope, atomic number and atomic mass.

6. Why aren't atomic masses expressed in grams?

7. Atoms of the element aluminum have 13 protons and 14 neutrons. What is the atomic number and the atomic mass of aluminum?

8. Atoms of the element potassium have 19 electrons and an atomic mass of 39. What is the atomic number of potassium? How many neutrons are in each atom?

9. Of what importance is it that the numbers of electrons and protons in atoms are equal?

10. What role does electricity play in the structure of atoms?

11. Draw a diagram of a hydrogen atom and an oxygen atom. Show locations for neutrons, protons and electrons in the diagram.

12. When the atomic mass for a naturally occurring element is given it is seldom a whole number. Why?

13. Why does hydrogen gas usually occur as molecules of two atoms?

14. What is the difference in the way atoms combine to form sodium chloride and water molecules?

15. Tritium is the name for an atom which has one proton and two neutrons in its nucleus. Deuterium has one proton and one neutron. Protium has just one proton. What is the common name for all three of these atoms?

16. Explain the following statement briefly. "The reason for different kinds of atoms joining together to form molecules is found in their electron arrangement."

17. The element silicon (Si) has atomic number 14 and oxygen (O) has atomic number 8. Suppose silicon and oxygen combine so that the oxygen is grouped around one silicon atom. What is the chemical formula that will give all atoms a stable number of electrons?

18. Water may exist as a solid, liquid or gas depending on its temperature. Explain why this is so.

objectives

☐ Describe several ways silicon and oxygen combine to form silicates.

☐ List several examples of silicate minerals.

☐ List and describe the composition of several non-silicate minerals.

☐ Describe several properties and tests that are used to identify a mineral.

☐ Explain how rocks can be classified according to how they were formed.

Almost any rock that you pick up off the ground was originally part of the earth's crust. The crust forms the outer layer of the earth's surface on which you live. Since it is an actual piece of the crust, the rock is very likely to be made up of about 50 percent of the element oxygen. The other 50 percent of the materials in the rocks of the crust could probably be accounted for by about eight other kinds of elements. See Figure 8–1A. The properties of the rock, and almost the entire earth's crust, depend upon the way in which these relatively few, but commonly found elements join together.

Each element has its own chemical properties. It can become part of the molecules in compounds that form the materials of the earth. How do these elements join together to produce the many different kinds of rocks and other materials that form the crust? This chapter will serve as a brief answer to this difficult question.

MINERALS

Basic mineral types. A *mineral* is a single chemical compound or element that is found naturally. *Rocks* are a combination of different minerals found in the earth's crust. Since oxygen and silicon are the most abundant elements in the earth's crust, the most common mineral compounds must include these elements. These minerals are known as *silicates*. There are five main groups of silicate minerals.

The key to identifying the silicate minerals is understanding the structure of silicon atoms. An atom of silicon (atomic number = 14) has a total of fourteen electrons.

153

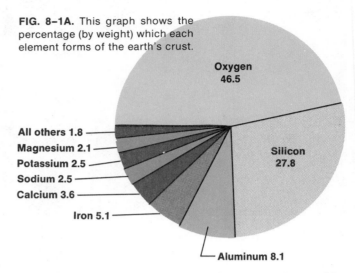

FIG. 8–1A. This graph shows the percentage (by weight) which each element forms of the earth's crust.

Oxygen 46.5
Silicon 27.8
All others 1.8
Magnesium 2.1
Potassium 2.5
Sodium 2.5
Calcium 3.6
Iron 5.1
Aluminum 8.1

FIG. 8–1B. The basic structure of many minerals is a tetrahedron made up of four oxygen atoms and one silicon atom.

FIG. 8–2. Quartz is made up of silicon and oxygen atoms arranged in a continuous network, as shown below. (American Museum of Natural History)

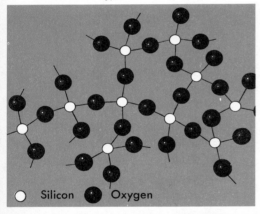

○ Silicon ● Oxygen

This number of electrons is mid-way between the stable electron numbers of ten and eighteen. One way a silicon atom can obtain a stable electron number is to share its four outer electrons with other atoms. A common arrangement is a silicon atom surrounded by four oxygen atoms. Each of these atoms shares an electron with the silicon. See Figure 8–1B. This arrangement forms a four-sided pyramid called a *tetrahedron*. But an oxygen atom (atomic number = 8) is able to accept two more electrons before reaching a stable electron number (10). By sharing only one of the four available electrons in a silicon atom, each oxygen atom is able to form a second chemical bond with still another silicon atom.

Thus each silicon atom is bonded to four oxygen atoms in a continuing pattern, to form a network as shown in Figure 8–2. The resulting mineral is *quartz*. Quartz has the chemical formula SiO_2. Quartz is one of the hardest of all minerals because all of its atoms are tightly joined together. See Figure 8–3.

A group of minerals called *feldspars* have a structure similar to quartz. In the feldspars, however, aluminum atoms replace silicon in some of the tetrahedrons. Sodium (Na), calcium (Ca), or potassium (K) atoms may also be present in the feldspar crystal. These atoms form weaker bonds than those between silicon and oxygen. Thus the feldspars are softer than quartz. Some kinds of feldspar are shown in Figure 8–4.

Another group of silicate minerals is produced when only one oxygen in the silicon-oxygen tetrahedron fails to

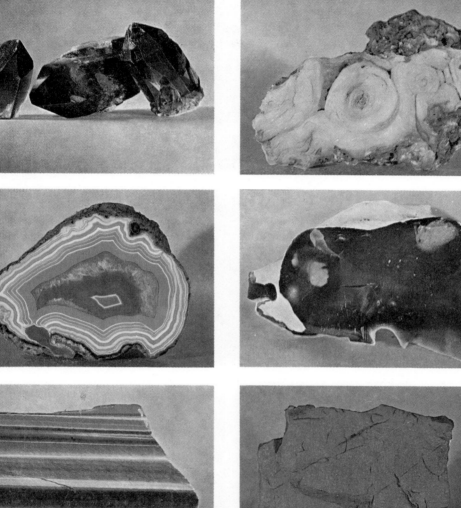

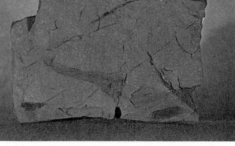

FIG. 8–3A. Top to bottom: Smoky quartz is a variety of colored quartz crystals. Agate is fine crystalline quartz. Tiger's Eye quartz is often cut into stones with a curved surface that reflects a band of light resembling a cat's eye.

FIG. 8–3B. Top to bottom: Chalcedony is a general name for fine-grained varieties of quartz having a waxy appearance. Flint is also a fine grained variety of quartz usually dark gray in color. Jasper is similar to agate, but has no color bands.

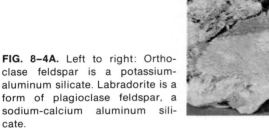

FIG. 8–4A. Left to right: Orthoclase feldspar is a potassium-aluminum silicate. Labradorite is a form of plagioclase feldspar, a sodium-calcium aluminum silicate.

FIG. 8–4B. Left to right: Microcline is a variety of orthoclase feldspar. Albite is a type of plagioclase feldspar which contains large amounts of sodium.

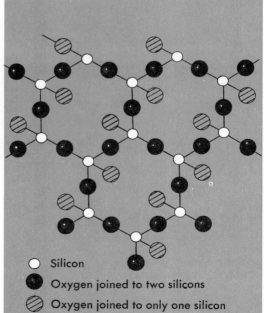

FIG. 8–5. Mica splits into thin sheets because its atomic arrangement is the one shown in the diagram at the left. Above is a photograph of the mineral "muscovite," a type of mica.

○ Silicon

● Oxygen joined to two silicons

◍ Oxygen joined to only one silicon

attach itself to another silicon atom. One oxygen may be joined to an atom of potassium or aluminum instead of silicon. See Figure 8–5. Minerals having this type of structure are called *micas*. The micas separate easily into thin layers because the silicon-oxygen bonds going in three directions are stronger than the fourth bond involving another kind of atom.

If only two oxygens in the silicon-oxygen tetrahedron are joined to other silicons, a fourth group of silicate minerals is produced. The oxygens which are not joined to silicon may be bonded to atoms such as magnesium (Mg), iron (Fe), calcium (Ca), and aluminum (Al). See Figure 8–6. Minerals with this structure are known as *hornblende* and *pyroxene*. Both of these minerals have a similar appearance and are difficult to tell apart.

If none of the oxygens in the tetrahedron are joined to other silicons the mineral *olivine* is formed. In olivine, the silicon-oxygen tetrahedrons are separate but are linked by magnesium and iron atoms.

In addition to the five basic groups of silicate minerals already mentioned, there are many kinds which are less common. Some are shown in Figure 8–7 to illustrate their variety in color and form.

Examine
Look at the clear window in an electric house fuse. Try to break out a part of the window. Observe how the material can be peeled into thin layers. What do you think this material is?

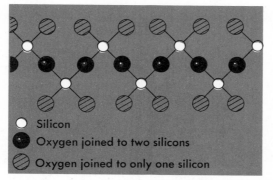

○ Silicon

● Oxygen joined to two silicons

◐ Oxygen joined to only one silicon

FIG. 8–6. The structure of pyroxene minerals is in a crystal pattern, based on the arrangement of atoms shown in the diagram at the right. The photographs are of two varieties of pyroxene minerals. Hornblende (left) and augite (right).

FIG. 8–7. All of these minerals are silicates. They are, top left, biotite mica; right, stilbite, a mineral belonging to a general group of silicates called "zeolites." Bottom left, chrysotile, a type of asbestos and right, serpentine, an ornamental stone often used in the same way as marble.

Other mineral types. Almost all common rocks are mixtures of the various kinds of silicate minerals. However, there are many other kinds of minerals which do not contain silicon and oxygen. These minerals can be put into three groups:

1. *Rock forming minerals other than silicates.* These include a large number of compounds as well as a few native elements such as sulfur. See Figure 8–8.

2. *Metal ore minerals.* An ore is a rock which is an important source of some useful metal. To be considered a valuable ore a rock must contain enough of a metal to make its removal cheap and practical. See Figure 8–9.

3. *Gem minerals.* Most gems are mineral crystals of unusual color and brilliance. They are usually cut to increase their brilliance and color. The value of a gem is determined by its size, lack of flaws, hardness, color and brilliance. See Figure 8–10.

Identification of minerals. Laboratory tests with special equipment are needed for the complete identification of a

FIG. 8–8A. Non-silicate minerals. Top, two varieties of calcite, $CaCO_3$. Bottom, two varieties of gypsum, $CaSO_4$.

FIG. 8–8B. Non-silicate minerals. Left, top to bottom, halite (NaCl) and fluorite. Right, top, kernite which yields borax and, bottom, apatite which is mainly calcium phosphate.

FIG. 8–8C. These two non-silicate minerals are elements. To the left is graphite, which is a form of carbon. At the right is sulfur.

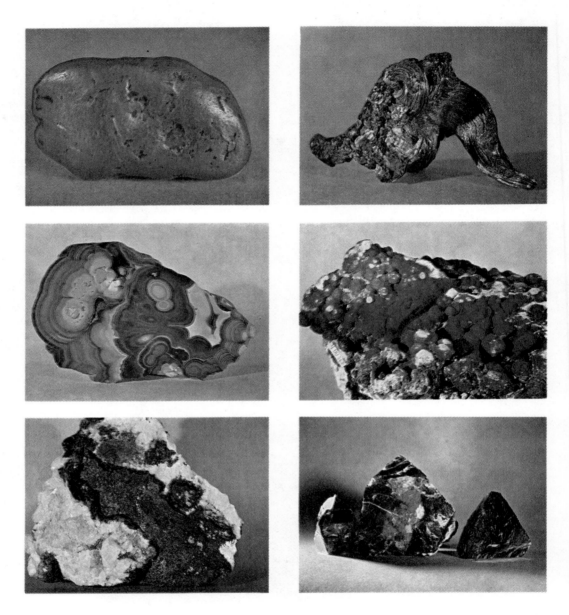

FIG. 8–9
NATIVE GOLD Bolivia
MALACHITE (COPPER) Urals, U.S.S.R.
CUPRITE (COPPER) Arizona

NATIVE SILVER Norway
AZURITE (COPPER) Arizona
SPHALERITE (ZINC) Spain

FIG. 8–9 (cont.)
HEMATITE (IRON) England
LIMONITE (IRON) New Hampshire
SIDERITE (IRON) Austria

CINNABAR (MERCURY) California
BAUXITE (ALUMINUM) Arkansas
URANINITE (Black) (URANIUM) India

FIG. 8–10
OPAL Australia
TOPAZ Burma and Brazil
TOURMALINE California

GARNET Alaska and Connecticut
BERYL, Aquamarine S.W. Africa
CORUNDUM, Ruby Madagascar

mineral. However, there are some simple tests by which many common minerals can be identified. Some of these tests are described below:

1. *Color.* One of the more easily seen characteristics of a mineral is its color. Look at the color photographs in the chapter and note the differences among the many minerals shown. The color of the surface tarnish of those minerals which look like metal should be considered. For example, the iron mineral pyrite is the color of gold on a fresh surface. But it is a much darker yellow when tarnished. Other minerals contain different elements which affect their color. The colorless mineral called corundum becomes a ruby when colored red by traces of chromium. Sapphires are produced from corundum colored bluish by traces of .iron and titanium. Such differences must be kept in mind when identifying unknown minerals according to color.

2. *Streak* is the color left by the mineral when rubbed against a streak plate. The back of a piece of building tile, or a piece of porcelain with a dull surface may be used for a *streak plate.* Again it must be kept in mind that the color of the streak may not be the same as the mineral. For example, the streak of gold-colored pyrite is black For most minerals, however, the streak is either colorless or very light.

3. *Luster* is the ability of the mineral to reflect, bend or absorb light. Many terms are used to describe luster: *dull, pearly, waxy, metallic, glassy, brilliant* (diamond-like).

4. *Crystal form.* The internal structure of a mineral can be understood by studying its crystal form. In most minerals, the atoms, ions and molecules are arranged in a particular pattern. This pattern gives the crystal its characteristic shape. Some mineral crystals are so perfectly shaped that they seem to have been cut and polished artificially rather than created naturally. The chemical properties of these atoms make it possible for them to fit together in a definite pattern.

There are six basic shapes of mineral crystals. See Figure 8–11. Think of these shapes as having imaginary internal lines called axes. The mineral molecules are arranged along the axes giving a crystal its shape which is different from that of other mineral crystals.

Other combined forms exist besides the basic forms shown in Figure 8–11. Most minerals are not found as separate large crystals. They are more commonly masses of small crystals which can be seen only with a micro-

activity

Classification keys are used to show relationships among a variety of similar things. To make and use such a key, do the following:

1. Print the capital letters of the alphabet across the top of a piece of paper.

2. Some letters are made from straight lines only, while others contain curved lines. Write "Straight Lines" on the left side of the page and "Curved Lines" on the right side.

3. List each capital letter under one of these headings.

4. In the curved line group, some letters have closed curves, such as B. Others have open curves such as C. Make headings "Open" and "Closed" below the curved group. List each letter in the curved group under one of these headings.

5. In the open group, the letters open to the right, left, both right and left, or upward. Make these four headings. List each letter under one of the headings. (See the key below.)

6. Now classify the closed group. Use headings such as: circles, vertical lines, slanted lines, etc.

7. Break the straight line group into smaller groups until you have 1 and 2 letters by themselves.

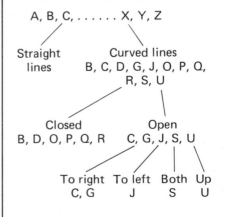

The Six Basic Crystal Systems

Sulfur

Isometric or Cubic System: Three axes of equal length intersect at 90° angles. Ex. galena, pyrite, halite.

Orthorhombic System: Three axes of different lengths intersect at 90° angles. Ex. olivine, topaz.

Zircon

Orthoclase   Amphibole

Tetragonal System: Three axes intersect at 90° angles. The two horizontal axes are of equal length. The vertical axis is longer or shorter than the horizontal axes. Ex. rutile, cassiterite, chalcopyrite.

Monoclinic System: Of three axes of different lengths, two intersect at 90° angles. The third axis is oblique to the others. Ex. gypsum, micas, augite, cryolite, kaolinite.

Tourmaline

Rhodonite   Chalcanthite

Hexagonal System: Of the four axes, the three horizontal axes intersect at 60° angles. The vertical axis is longer or shorter. Ex. quartz, calcite, apatite, hematite.

Triclinic System: The three axes are of unequal length and are oblique to one another. Ex. plagioclase feldspars, including albite and labradorite, turquoise.

FIG. 8–11

scope. Sometimes only x-ray will show it. A few minerals are not made up of crystals. There is no orderly arrangement of their molecules. These are called *amorphous* or *massive.*

5. *Cleavage and fracture.* These terms tell how a mineral splits or breaks. Some minerals split along certain directions leaving flat surfaces. The micas for example, split easily along one direction. This means that mica has one cleavage plane. The different kinds of cleavage are described according to the number and direction of the cleavage planes. See Figure 8–12.

Most minerals do not break along cleavage planes, but split unevenly in one of several ways. Some types of fracture are *conchoidal* (shell-like), *splintery, irregular* and *earthy.* See Figure 8–13.

6. *Hardness.* The hardness of a mineral is determined by how easily it can be scratched. Certain common minerals are used as standards of comparison. The scale of mineral hardness, called *Moh's Scale,* is given in Table 8–1. The range is from talc, one of the softest minerals, to diamond, the hardest of all minerals.

To test an unknown mineral for hardness find out which mineral on the scale it can scratch and which it cannot. For example, gypsum is harder than talc, but softer than calcite. Any mineral which scratches gypsum but not calcite has a hardness on Moh's Scale between 2 and 3.

The following standards of hardness are useful additions to the scale:

Fingernail	2.5
Copper penny	3.0
Glass or knife-blade	5.5
Steel file	6.5

Care must be taken in testing hardness. Talc, which is a soft mineral, may only appear to scratch an unknown mineral. A soft mineral can make a mark on a harder one that can be mistaken for a scratch. Such a mark can be rubbed off, a true scratch cannot. It is wise to test both ways. Try the known mineral on the unknown as well as the reverse.

7. *Specific gravity.* You have noticed that some minerals are heavier than others. The relative weights of two or more minerals can be compared. This can be estimated by weighing or by handling (hefting) equal sized pieces of different minerals.

FIG. 8–12. Two types of cleavage in three directions are shown in the top diagrams. Lower diagram illustrates cleavage in one direction.

FIG. 8–13. Fracture. Left, conchoidal fracture in obsidian. Right, earthy fracture in yellow ocher, an impure form of limonite. (American Museum of Natural History)

Table 8–1

Moh's Scale of Mineral Hardness

1.	Talc
2.	Gypsum
3.	Calcite
4.	Fluorite
5.	Apatite
6.	Orthoclase
7.	Quartz
8.	Topaz
9.	Corundum
10.	Diamond

Compare
What is the relationship between the terms "specific gravity" and "density?"

A much more exact method is to express the relative weight of a mineral by comparing its weight to the weight of an equal volume of water.

This can be determined by first weighing the mineral sample. It is then suspended by a fine thread and weighed while it is completely under water. Its weight in water is then subtracted from the dry weight. This difference is the weight of a volume of water equal to the volume of the mineral sample. Specific gravity can be obtained by dividing the weight of an equal volume of water into a dry weight.

For example, a sample of mineral is found to weigh 40 grams dry and 30 grams when submerged in water. Then:

$$\text{Specific gravity of the mineral} = \frac{\text{weight of sample in air}}{\text{weight of equal volume of water}}$$

or

$$\frac{40 \text{ grams}}{(40{-}30) \text{ grams}} = \frac{40}{10} = 4$$

Perhaps you already have some idea of the specific gravity of common minerals. You know from experience how a certain size of the most common minerals should feel (heft). When you pick up a pebble it will probably consist of quartz (Sp. gr. = 2.65), feldspars (Sp. gr. = 2.60 to 2.75) and calcite (Sp. gr. = 2.72). A pebble made of the mineral pyrite (Sp. gr. = 5.02) would feel too heavy for its size and you would probably consider it an unusual pebble. With a little practice it will be possible for you to handle a mineral specimen and judge whether its specific gravity is average, high or low.

Special properties of minerals. Following is a list of some other properties of certain minerals which will help you identify them.

1. *Magnetism.* Some minerals are attracted to a magnet. *Magnetite* and some samples of *pyrrhotite* (PIR-oh-tyt), both iron ores, are attracted to a magnet. Lodestone, a form of magnetite, acts as a magnet. Its magnetic properties have been known since ancient times.

2. *Fluorescence.* In some minerals certain atoms absorb ultraviolet light and give off visible light rays. These minerals glow under ultraviolet (black) light and are called fluorescent. See Figure 8–14. A few minerals continue to glow after the ultraviolet light is cut off. They are called *phosphorescent.*

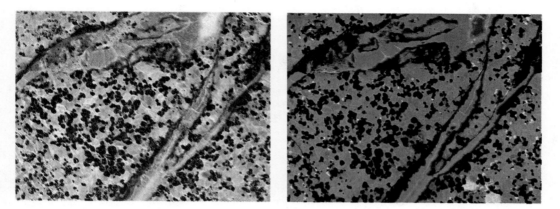

3. *Radioactivity.* Minerals which contain uranium and radium are radioactive, and are usually fluorescent. These atoms can be detected with a Geiger counter.

4. *Optical properties.* The way that light is changed as it passes through minerals is often useful in identifying them. For example, when looking at an object through calcite, a double image will be seen. See Figure 8–15. Some optical properties of minerals cannot be seen without the use of special equipment. A special kind of microscope is needed to measure the degree that light rays are bent as they pass through a mineral crystal. This is one of the most dependable methods used to identify a mineral.

Simple testing methods. Many tests have been developed for identifying minerals. Some that need only simple equipment are described below.

1. *Simple chemical test.* The common mineral calcite ($CaCO_3$) can be identified by an acid test. A drop of cold, dilute hydrochloric acid placed on calcite will cause bubbles of carbon dioxide (CO_2) gas to form. The same will happen to dolomite if hot acid is used, but will act more slowly if cold acid is used.

2. *Flame tests.* A Bunsen burner flame can be used to test minerals.

If you blow air through a metal tube (blowpipe) into the flame, the flame can be directed against a mineral sample. See Figure 8–16, top. When an oxidizing flame is used, oxygen reacts with the mineral to produce colors that indicate the presence of certain metals. In a reducing flame, hot gases remove oxygen from the mineral. This may cause a color change.

One way to use the blowpipe and flame is in bead tests. This is done by dipping a loop of platinum wire into

FIG. 8–14. Above, left, the minerals calcite (greenish white), willemite (light brown), and franklinite (dark brown), as they appear in ordinary light. Right, under ultraviolet light, calcite and willemite fluoresce red and green, respectively. Franklinite does not fluoresce.

FIG. 8–15. The mineral calcite, in a clear form, produces a double image of objects viewed through it.

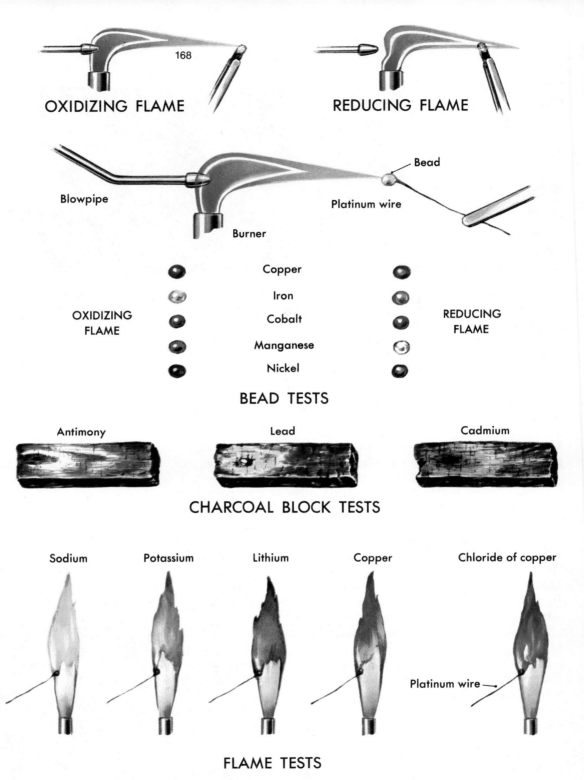

OXIDIZING FLAME

REDUCING FLAME

168

Bead

Blowpipe

Platinum wire

Burner

OXIDIZING
FLAME

Copper		
Iron		
Cobalt		
Manganese		
Nickel		

REDUCING
FLAME

BEAD TESTS

Antimony

Lead

Cadmium

CHARCOAL BLOCK TESTS

Sodium Potassium Lithium Copper Chloride of copper

Platinum wire

FLAME TESTS

FIG. 8–16. These color tests are frequently used in identifying minerals.

powdered borax. The loop is then held in the burner flame while the borax melts and forms a colorless bead. Then the bead is applied to the powdered mineral. When the bead is again held to the flame, the borax and metals in the mineral may react to produce a color. Different colors may be produced by using the oxidizing and reducing flames.

Another common test is to heat a small amount of the mineral on a block of charcoal. The appearance of the material left on the block after heating may help to identify some of the elements in the mineral. Characteristic fumes or a bead of metal may also be produced.

A bit of the powdered mineral may also be moistened with acid and held on a clean platinum wire. The wire is then put into a burner flame. A change in the color of the flame shows the presence of certain elements. See Figure 8–16, bottom.

ROCKS

Types of rocks. Suppose that you decide to start a rock collection. It is easy to collect rocks since they are found everywhere. They have many different forms and colors. But how will you organize your collection? Rocks can be grouped by their color, roughness or smoothness, or their mineral composition. But there is another way which will probably work better. It will include all rocks and is based on how they were formed.

Any rock that you would pick up for your collection is only a small piece broken off a larger rock. The earth's crust is made up of great bodies of rock which were formed in different ways. Scientists have discovered after careful study, that there are three ways that rocks are produced. See Figure 8–17.

One group is called *igneous*. The word "igneous" means "from fire." Igneous rocks are formed by the cooling and hardening of hot melted (molten) rock below the earth's crust. All rocks now in the earth's crust are believed to have first been igneous.

When igneous rocks are at the surface of the earth, conditions there help to break up the large masses. Small pieces are produced by the wearing away of solid rock. These pieces are called *sediments*. Sediments can be joined together again to form solid rock. Any rock made from sediments is called *sedimentary rock*.

Explore
Take some pictures along road cuts or of outcrops. If you live in the city examine the walls of buildings or park statues and try to identify the type of rock.

FIG. 8–17. These masses of rock in Zion National Park came into existence as layer upon layer of sediment was deposited. (Ramsey)

The third way that rocks are formed calls for the changing of igneous or sedimentary rocks. When these are buried very deep in the earth, they are under great pressure and become very hot. The heat and the pressure cause the rock to change into another form. This is called *metamorphic* rock. Only the conditions deep in the earth can produce metamorphic rock.

By examining a small piece of rock carefully we can often find signs that tell us something about the large rock from which it came. But to do this we need more detailed knowledge of the properties of the three types of rock.

Igneous rocks. The molten rock from which igneous rocks are formed is called *magma.* Magma may cool either at or below the earth's surface. When magma reaches the surface it is called lava. Igneous rocks can be divided into two groups according to where the magma cools and hardens.

How quickly or slowly the magma cools determines the appearance of the igneous rock being formed. When magma cools slowly mineral crystals are able to grow to a large size. The slow cooling of magma below the surface produces *intrusive* igneous rocks. The slow loss of heat allows atoms enough time to join together in well-organized patterns. Because of this the silicon-oxygen tetrahedrons form well-developed crystals. This results in intrusive igneous rocks having a coarse texture due to the larger crystals present. See Figure 8–18A. Granite is an example of a coarse-grained intrusive igneous rock. Such rocks are often called *plutonic* rocks. See Figure 8–18B.

Rapid cooling of lava at the surface produces *extrusive* igneous rocks. The rapid cooling may prevent large crystals from forming. This makes the rock look glassy like in *obsidian*. See Figure 8–19. Gases may be trapped in the lava as it hardens forming *pumice*. Slower cooling allows the growth of small mineral crystals. The resulting rocks such as *basalt* have a fine texture. See Figure 8–20.

Differences in the rate of cooling of magma affect the mineral composition of igneous rocks as well as their texture. Probably all magma begins with the same composition. It is thought to be dark, rich in iron and magnesium, very like basalt. If magma is forced out without any opportunity to cool before reaching the surface it will harden as *basaltic* rock. Large areas of the earth's surface are covered with such rocks formed by basaltic lava from volcanoes.

When magma cools below the surface, the first crystals to form will contain much of the iron and magnesium. These crystals tend to sink into the lower part of the still liquid magma. Thus the upper layers of magma will contain larger amounts of silicon and potassium than were originally there. The upper layers form a light-colored *granitic* rock. Granite type rocks are found here. The dark material below forms basaltic rock.

Sedimentary rocks. There are three types of sedimentary rock. One is called *fragmental*. The rock fragments are carried away from their source by water, wind, or ice. They are then left as deposits. The fragments are cemented into rocks by great pressure and the cementing action of other materials. One example of a fragmental sedimentary rock is *conglomerate*. This is a coarse-grained rock made up of gravel, pebbles or even boulders.

Cemented grains of quartz sand make up another group of sedimentary rocks. These are the *sandstones*. Gaps between the sand grains leave spaces which liquids move through easily.

QUARTZ   *HORNBLENDE*
FELDSPAR   *MICA*

FIG. 8–18A. A piece of granite is composed of minerals representing four mineral groups.

FIG. 8–18B. Plutonic rocks. Top, pink granite. Bottom, hornblende syenite.

FIG. 8–19. Massive obsidian formations. (Ramsey)

FIG. 8–20A. Natural salt formations found near the Sierra Nevada Mts. form evaporites of the mineral Halite. (Shostal)

Shale is a clay deposit that has become rock mainly as a result of pressure. The flaky clay particles are usually pressed into flat layers. Because of this shale tends to split into flat pieces.

A second group of sedimentary rocks is *chemical* in origin. These are rocks which are formed of materials that were dissolved in water. When the water evaporates it leaves materials behind, which become solid again. These materials are called *evaporites*. *Halite* and *gypsum* are part of this group. See Figure 8–20A.

The smallest group of sedimentary rocks are the *organic* rocks. These rocks are made from materials that once were living things. *Limestone* is a sedimentary rock made up of calcium carbonate ($CaCO_3$). Animals and plants take calcium carbonate from water. Their skeletons settle to form organic limestone. In *coquina*, an organic limestone, pieces of shells are often found. *Coal* is a sedimentary rock of *organic* origin. It is made up of plant material that is partly decayed, and then buried. The carbon from these plants forms most of the material in coal. See Figure 8–21A.

Metamorphic rocks. The third large division of rocks are those called *metamorphic*. These rocks are formed deep in the earth's crust. Great pressure and heat from the weight of the rock above, intrusion by magma, and movement of the crust cause the rocks to change. Minerals can be melted to produce new compounds. Pressure can mash

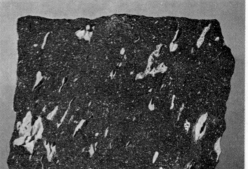

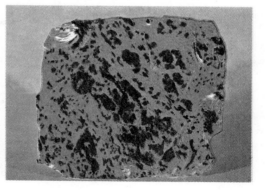

FIG. 8–20.

FINE-GRAINED IGNEOUS ROCKS
PYROXENITE Ontario
ANDESITE PORPHYRY Nevada
RHYOLITE BRECCIA Mexico

TRACHYTE TUFF West Germany
VESICULAR BASALT Colorado
RED OBSIDIAN Oregon

FIG 8–21A

CONGLOMERATE Colorado   **SANDSTONE** New York
SHALE Texas   **ROCK SALT** New York
LIMESTONE Massachusetts   **DOLOMITE** Minnesota

FIG. 8–21B

SLATE Pennsylvania
QUARTZITE South Dakota
MICA SCHIST Vermont

MARBLE Georgia
PHYLLITE Massachusetts
GNEISS Massachusetts

FIG. 8–22. Stratified sedimentary rocks. (Ramsey)

FIG. 8–23. A cliff formed of massive igneous rock in Yosemite National Park. (Ramsey)

rocks into new textures, and chemical changes can be brought about by gases and liquids escaping from magma. See Figure 8–21B.

All metamorphic rocks are made from sedimentary, igneous or other metamorphic rocks. For example, *slate* is a metamorphic form of the sedimentary rock shale. The tendency of shale to split into flat sheets can still be seen in slate.

Greater heat and pressure can change slate into another metamorphic stage called *phyllite*. And still greater heat and pressure can change phyllite into *schist*.

Marble is formed from limestone. The limestone crystals grow larger and many impurities are driven off. In general, marble is white, though it may have some dark streaks containing other minerals.

Quartzite is a metamorphic rock formed from sandstone. It is usually more compact and firmly cemented, so that the spaces between the sandstone particles disappear. Because quartzite is very hard and durable it forms the large part of many hills and mountains from which weaker rocks have been worn away.

Graphite is almost pure carbon, formed from the metamorphic changes of coal. It is a soft black solid widely used in pencils and as a lubricant for machinery.

Rock structures. Large masses of the three kinds of rocks are put together in different ways. Some are arranged in layers called *beds* or *strata*. The strata may be seen because of differences in color, texture or composition. See Figure 8–22. Sedimentary rocks are almost always *stratified*.

Those rocks which show no strata are said to be *massive* or unstratified. Igneous rocks usually are arranged in massive formations. See Figure 8–23. In some lava flows, however, strata is shown by beds of volcanic dust or fragments lying between layers of lava.

Metamorphic rocks can be either stratified or massive depending on their composition, and on the conditions under which they are formed. See Figure 8–24(A & B).

One thing that all three kinds of rock have in common are cracks or joints in fairly regular patterns. In igneous rocks the *joints* are probably formed during the cooling process. A column-like formation may result from jointing in the fine-grained igneous rock. See Figure 8–25.

In sedimentary rocks, jointing may be the result of the drying and cracking of the beds. It may also be caused by movements of the crust which may slowly bend and twist the beds. Joints in sedimentary rocks are usually at right angles to the direction of the bed. This often separates the rocks into large blocks. When these blocks fall, cliffs are formed. See Figure 8–26.

Jointing in metamorphic rocks is similar to that of the rock from which it originated. For example, in a rock such as slate, joints may be similar to those found in the sedimentary rock called shale.

FIG. 8–24A. A formation of massive metamorphic rock in Zion National Park. (Ramsey)

FIG. 8–24B. Stratified metamorphic rock formation. (Ramsey)

FIG. 8–25. Devil's Postpile National Monument is a large cliff of basaltic rocks with vertical joints which form separate columns. (Ramsey)

FIG. 8–26. These strange formations in Bryce Canyon National Park were formed from a cliff made of layers of weak sedimentary rocks. (Union Pacific Railroad)

VOCABULARY REVIEW

Match the word or words in the column on the right with the correct phrase in the column on the left. *Do not write in this book.*

1. A single chemical compound found naturally as part of the earth.
2. Silicon and oxygen joined to form a network of SiO_2 molecules.
3. Similar to quartz with aluminum atoms in place of some silicon atoms.
4. Separates easily into thin layers.
5. Only two oxygens in the silicon-oxygen tetrahedron are joined to other silicons.
6. The color of a thin layer of the finely powdered mineral.
7. Ability of a mineral to split with a flat surface along certain directions.
8. A volcanic rock light enough to float on water.
9. A form of magnetite.
10. Rocks formed directly from magma.
11. Formed from rock fragments produced by the wearing of solid rock at the surface.
12. Rocks formed from igneous or sedimentary rocks under great heat and pressure.
13. Igneous rock with a glassy appearance.
14. Clay that has become rock due mainly to pressure.
15. Formed from phyllite under great heat and pressure.

a. feldspars
b. streak
c. pumice
d. basaltic
e. mineral
f. micas
g. lodestone
h. quartz
i. hornblende
j. igneous
k. cleavage
l. metamorphic
m. shale
n. graphite
o. obsidian
p. schist
q. sedimentary rocks

QUESTIONS

Group A

Select the best term to complete the following statements. *Do not write in this book.*

1. A majority of the minerals in the earth's crust contain (a) oxygen and silicon (b) silicon and hydrogen (c) hydrogen and aluminum (d) aluminum and silicon.

2. Minerals which contain oxygen and silicon are known as (a) oxidation (b) silicates (c) silicons (d) hydrates.

3. A mineral which contains only silicon and oxygen is (a) hornblende (b) mica (c) feldspar (d) quartz.

4. Micas can be separated into thin layers because one oxygen in the silicon-oxygen tetrahedron has a (a) strong bond with a silicon (b) weak bond with a silicon (c) strong bond with another kind of atom (d) weak bond with another kind of atom.

5. A mineral whose structure is closest to that of hornblende is (a) mica (b) olivine (c) pyroxene (d) quartz.

6. Which of the following is a group of minerals, all of which contain silicon and oxygen? (a) rock forming minerals other than silicates (b) gem minerals (c) metal ore minerals (d) micas.

7. The most obvious characteristic used in the identification of minerals is (a) streak (b) luster (c) color (c) specific gravity.

8. Which of the following is *not* a term used to describe the luster of minerals? (a) bright (b) dull (c) silky (d) adamantine.

9. Which of the following is *not* a term used to describe fracture? (a) irregular (b) conchoidal (c) flat (d) earthy.

10. Moh's scale is used in describing the mineral characteristic of (a) cleavage (b) specific gravity (c) magnetism (d) hardness.

11. Any mineral which can be scratched by a fingernail has a hardness (a) less than calcite (b) greater than gypsum but less than calcite (c) greater than calcite (d) less than gypsum but greater than talc.

12. A common substance which could be used in place of calcite on Moh's scale is a (a) fingernail (b) copper penny (c) piece of glass (d) steel file.

13. The specific gravity of a mineral sample can be found by dividing the weight of the sample by (a) the weight of the dry sample (b) the volume of the sample (c) the volume of an equal weight of water (d) the weight of an equal volume of water.

14. Which of the following is not attracted by a magnet? (a) magnetite (b) pyrite (c) lodestone (d) pyrrhotite.

15. Fluorescence is the process in which substances absorb (a) visible light and also give off visible light (b) visible light and give off ultraviolet light (c) ultraviolet light and give off ultraviolet light (d) ultraviolet light and give off visible light.

16. Which of the following mineral tests may not require a flame? (a) borax bead (b) charcoal block (c) colored flame (d) all of these require flames.

17. To organize a rock collection, you may start by separating them by (a) color (b) texture (c) how they were formed (d) any method you wish.

18. Which of the following does *not* refer to one of the three basic ways in which rocks are formed? (a) igneous (b) evaporation (c) metamorphic (d) sedimentary.

19. Which one of the following rocks is not related to the others in the way it was formed? (a) basalt (b) obsidian (c) dolomite (d) pumice.

20. Granitic rocks belong to which of the following rock types? (a) igneous (b) metamorphic (c) sedimentary (d) basaltic.

21. All of the following are similar, in that they are made up of individual grains joined together except (a) shale (b) conglomerate (c) sandstone (d) halite.

22. Halite and gypsum are examples of (a) igneous rock (b) plutonic rocks (c) evaporite rocks (d) fragmental rocks.

23. Limestone and dolomite are similar except that dolomite (a) is not as old (b) is older (c) contains magnesium (d) has been exposed to the atmosphere longer.

24. Which of the processes listed below does not produce metamorphic rocks (a) precipitation (b) heat (c) pressure (d) chemical changes.

25. All metamorphic rocks are produced from existing (a) igneous rocks only (b) sedimentary rocks only (c) igneous or sedimentary rocks (d) plutonic rocks only.

26. Which rock listed first does not come from the rock listed second as a result of metamorphism? (a) marble from limestone (b) quartzite from sandstone (c) slate from shale (d) phyllite from schist.

27. If the following rocks had a common origin, which would be the last to form? (a) slate (b) schist (c) shale (d) phyllite.

28. Diamonds are a form of pure carbon. Which of the following is most closely related to diamonds? (a) graphite (b) coal (c) slate (d) shale.

29. Stratified rocks are rocks that are (a) massive (b) in layers (c) all the same color (d) all the same composition.

30. Joints in rocks are (a) only found in igneous rocks (b) places where rocks bend (c) cracks in the rocks (d) only found in sedimentary rocks.

Group B

1. How is a borax bead made and how is it used in identifying minerals?

2. Give a brief description of the three general ways in which rocks are formed.

3. Which mineral groups do not contain silicon and oxygen?

4. Explain why it is difficult to identify a mineral simply by its color.

5. What is the special optical property of calcite?

6. Name at least eight tests which can be used in identifying minerals.

7. What is the streak of a mineral and how is it obtained?

8. Explain how evaporites are formed and give at least one example.

9. A mineral sample has a dry weight of 24 grams and weighs 16 grams when submerged in water. What is the specific gravity of the mineral?

10. How do metamorphic rocks differ from the rocks from which they were formed?

11. Name at least three igneous rocks, three sedimentary rocks and three metamorphic rocks.

12. Describe the steps in the formation of a schist.

13. What is Moh's scale and how is it used?

14. Describe three types of mineral identification tests which require flames.

15. Explain the difference between cleavage and fracture.

16. Draw a diagram of a silicon-oxygen tetrahedron.

17. How do rocks produced by intrusive and extrusive volcanic activity differ?

18. How is olivine different from quartz, mica and hornblende?

19. What factors determine the properties of silicate minerals?

20. When an object is submerged in water, it appears to lose one gram for each cubic centimeter of its volume. The density of an object is the number of grams of mass for each cubic centimeter of its volume. Density has units of $\frac{grams}{cm^3}$. What is the specific gravity of a mineral whose density is $3.5 \frac{grams}{cm^3}$?

9
the movement of continents

objectives

- [ ] Describe the theory of plate tectonics.
- [ ] Explain how the earth's interior can be studied.
- [ ] Identify the main layers of the earth.
- [ ] Describe three kinds of crustal plate boundaries.
- [ ] Give evidence for sea-floor spreading.
- [ ] Relate plate tectonics and volcanism.
- [ ] Account for different kinds of magma.
- [ ] Identify volcanic features above and below the surface.

Have you ever noticed that parts of the continents seen on a world map seem to fit together like pieces of a jigsaw puzzle? South America and Africa, for example, look as if they could fit together across the Atlantic Ocean to make a single huge continent. Looking to the northern Atlantic, the eastern coast of the New World seems to generally fit the western coast of the Old World in a gentle S-shaped curve. Such fitting together of opposing coastlines became evident hundreds of years ago when exploration began to produce fairly accurate maps. In many places the shape of the continents seemed to suggest that all were once part of a single supercontinent. Somehow that original land mass broke apart. The fragments then moved away to form separate continents with the ocean basins between them. However, no convincing evidence could be found to support this belief. The idea of moving continents continued to be a subject for argument among scientists.

In 1915, Alfred Wegener, a German scientist, published a book in which he proposed a theory that was known as *continental drift.* Wegener claimed that there was once only a supercontinent, which he called *Pangaea* (Pan-*gee*-a), meaning "whole earth." His theory stated that the breaking up of Pangaea and the drifting apart of the smaller continents explained the growth of mountains. As evidence to support his theory, Wegener pointed out the similar rock formations and fossils found in land masses now separated by wide oceans. However, most scientists were not convinced by Wegener's evidence for continental drift. Furthermore, the theory did not account for the forces that might push the continents over the earth's rigid surface like rafts in a lake. Scientific controversy over the theory of continental drift continued until the 1960s. Then new discoveries caused a scientific revolution.

A new theory called *plate tectonics* has caused an explosion of knowledge about the forces that shape the earth. Plate tectonics includes the idea of continental drift, but from a new point of view. According to plate tectonics, the continents do not drift like individual rafts. The continents move because they are passengers on a number of rigid plates of rock that move along the earth's surface. Using the theory of plate tectonics, scientists have begun to understand the growth of mountains, occurrence of earthquakes and volcanoes, arrangement of rock layers, development of life forms as shown by fossils, and location of important ore deposits and oil fields. Plate tectonics has become the foundation for scientific exploration of our home planet.

THE BODY OF THE EARTH

Inside the earth. According to the theory of plate tectonics, the earth's surface is made up of a number of separate rigid plates. The plates are slowly moving relative to each other. This movement has been responsible for creating the familiar land and ocean features on the earth's surface. "Tectonics" refers to the forces that move and shape the plates. These forces come from deep inside the earth. The theory of plate tectonics is founded on knowledge of the earth's interior.

Scientists have only one way of studying the body of the earth. They use *seismic (size-mik) waves*. Seismic waves are vibrations that travel through the earth. When an earthquake or explosion occurs on or near the earth's surface, seismic waves are created. One kind of seismic wave is caused by the up-and-down vibrations in the matter of the earth. Another kind of wave results from back-and-forth vibrations. See Figure 9–1. Both kinds of seismic waves can be detected by instruments located far from the source. Seismic waves move at different speeds, depending on the kind of material through which they are traveling. As a result of their change of speed, the waves change direction when they pass from one layer to another. Measurement of the path of the seismic waves gives clues about the location and composition of the materials deep within the earth.

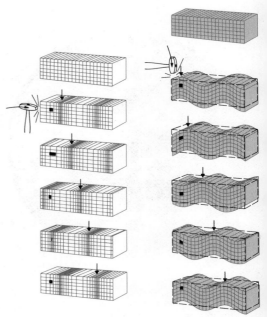

FIG. 9–1. These diagrams illustrate two kinds of seismic waves.

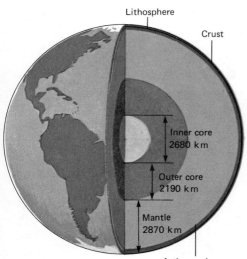

Lithosphere

Crust

Inner core
2680 km

Outer core
2190 km

Mantle
2870 km

Asthenosphere

FIG. 9–2. The earth is made up of layers, as shown in this diagram. The thickness of the crust is not actually as great as shown. The crust is about 10 km thick in the oceans and about 35 km thick in continental areas.

Examine
Use a fresh hard-boiled egg to help demonstrate the layered earth. Do not remove the shell. Cut the egg in half and compare what you see with Fig. 9–2.

Illustrate
Make a drawing that includes all of the following dimensions to scale. Let 1 cm = 500 km. Earth-radius = 6400 km, Core-radius = 1340 km, Mantle-radius = 6389 km; Satellite orbit = 250 km, Ocean deep = 11 km, Mt. Everest = 9 km. Explain any problems you may have.

In addition to causing seismic waves, a large earthquake can cause the entire earth to vibrate like a bell struck with a hammer. But the earth's vibrations are too low to be heard by the human ear. Instruments, however, can detect these vibrations, which may continue for several weeks after a severe earthquake.

Measurement of the earth's vibrations along with changes in seismic waves provide important information that scientists use to construct a model of the earth's interior. This is done by using computers to test many possible models of the earth against the observed vibrations and seismic-wave travel times. Models that fit the measurements are then considered to be possible, correct descriptions of the earth. At the present time, there is no single model that fits all the observations. In the future, more precise measurements may give a single model. However, all the various models now proposed have certain features in common that may be used to provide a general description of the body of the earth.

A model of the earth. There are three major divisions in the modern scientific model of the earth. The *crust* is, by comparison, a very thin outer layer; beneath the crust is the very thick layer called the *mantle;* and at the earth's center is a dense *core,* which is divided into an inner and outer part. See Figure 9–2. How did the earth become separated into distinct crust, mantle, and core layers? The answer to this question is hidden in the first billion years of the earth's history. Scientists have no direct evidence of the events of these early times. However, by comparing the amounts of different chemical elements now found in earth materials with recent findings of space probes and other explorations, many researchers believe that a general picture of the earth's earliest stages can be worked out. According to one widely held theory, the earth was formed by the coming together of smaller bodies in the young solar system. The new planet was probably made up of a uniform mixture of chemical compounds of silicon, iron, magnesium, and oxygen, along with smaller amounts of all the other chemical elements. Initially the young planet was cold. Its temperature began to rise as a result of three processes that produced heat. First, the smaller bodies that had combined as the earth grew added their energy of motion. Much of that energy was changed to heat as each body collided with the growing planet. In turn, most of that heat was lost back into space. Part of it, however, was trapped within the interior of the planet as layer upon layer of new material was added. This trapped heat began to increase the temperature inside the growing earth. A second source of inner heat was the squeezing together of the planet's interior by the increasing weight of the growing outer layers. All materials become warmer if compressed. For example, the air in the barrel of a bicycle pump becomes hotter as it is compressed when a tire is pumped up. As the earth grew, much of the heat produced by compression of the interior rocks was trapped and accumulated inside the planet. A third source of heat was the presence of *radioactive* atoms. Such atoms throw off atomic particles. As these particles are absorbed by the surrounding rock, their energy of motion is changed into heat. Although radioactive atoms are not common, the small amounts of heat they produce can become significant over many millions of years.

FIG. 9–3. The shadow zone is a region on the earth's surface where waves from a given earthquake are rarely felt. It results from the bending of some waves and the blocking of others by the earth's core. The location of the shadow zone is different for each earthquake.

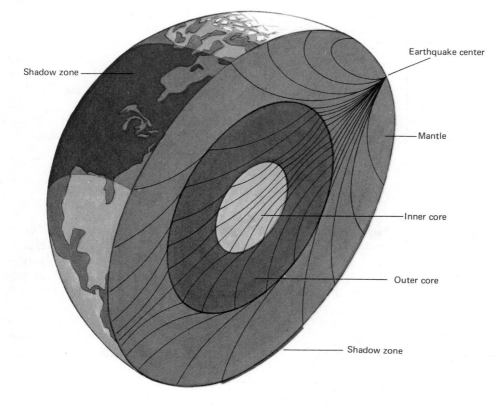

Shadow zone

Earthquake center

Mantle

Inner core

Outer core

Shadow zone

activity

Most volcanic activity occurs in certain regions of the earth's surface. In this exercise you will locate several volcanoes on a map of the world.

Obtain a map of the world that has the crustal plates outlined. Your teacher may provide you with one.

Volcano	Name	Longitude	Latitude
A	Aconcagua	70W	35S
B	Tungurahua	80W	0N
C	Pelée	61W	15N
D	Tajumulco	90W	15N
E	Popocatepetl	100W	20N
F	Lassen	122W	40N
G	Rainier	122W	47N
H	Katmai	155W	60N
I	Fujiyama	139E	35N
J	Tambora	120E	10S
K	Krakatoa	108E	5S
L	Mauna Loa	155W	20N
M	Kilimanjaro	37E	3S
N	Etna	15E	38N
O	Vesuvius	14E	41N
P	Teide	16W	28N
Q	Laki	20W	65N

Find the location of Volcano "A" on the map by reading the longitude across the top or bottom of the map and the latitude along the side. Write the letter "A" at that location. Do the same thing for each of the other volcanoes. Show each location by writing the proper letter.

When finished locating the volcanoes, answer the following questions:

1. Which ocean has a ring of volcanoes around it?

2. Are most of the volcanoes located on the edge of crustal plates?

3. Which volcanoes are not located on the edge of a crustal plate?

At some time after its formation, probably between a few hundred million and a billion years, the interior of earth had become hot enough to melt iron. Since iron is the most dense of all the common elements in the earth, gravity would cause iron to move toward the center. Thus iron probably moved inward to form the earth's dense core. The existence of this dense core is confirmed by a *shadow zone* for seismic waves. This is a region on the earth's surface where the waves from a particular location cannot be detected. The dense core bends and absorbs the waves, as shown in Figure 9–3, to produce a shadow zone. The behavior of the waves indicates that the core consists of an outer liquid region and a solid inner part. Its density suggests that iron is the most abundant material. About 15 to 20 percent of the core is lighter elements such as silicon.

Migration of iron to produce the core caused a great change in the outer layers of the developing earth. Heavy iron moving toward the center would release a huge amount of energy in the same way that falling water releases energy, which may be used to generate electricity. Most of the energy produced by the growth of the core was changed into heat. This heating raised the temperature of the entire planet to the melting point. The lightest molten material floated up to the surface to make the outermost layer. Later cooling of this light rock formed the earth's ancient crust. Denser rock material remained as the thick mantle between the crust and core. With this final separation of the crust and mantle, the earth became the three-layered planet that we find today.

CRUSTAL PLATES

Moving plates. In 1909 the Yugoslavian scientist A. Mohorovicic discovered that some seismic waves show a sharp change in speed a short distance beneath the earth's surface. This behavior of the waves marks the boundary between the crust and mantle. This dividing line is called the *Mohorovicic discontinuity*. It is usually simply called the "Moho." The way the waves change at the Moho indicates that the mantle is more dense than the crust and must be made of a different kind of rock. The depth of the Moho at different places on the earth's surface shows that the crust varies in thickness. It is thicker under the continents than under the oceans. Its average thickness beneath the oceans is 10 km (6.21 mi). Below the continents, the crust averages about 35 km (21.7 mi) in thickness. Beneath high mountains, the crust is always thickest and may be as much as 65 km (40.4 mi) in depth. See Figure 9–4.

In the crust, the lightest rocks are found in the continents. The most common kind of rock in continental crust is granite-type rock. The sea floor is made of basaltic rock. The lighter rock that makes up the continents seems to float on the mantle like an iceberg floating on the ocean. An iceberg floats because the less dense ice can be supported by the more dense water. However, a large part of an iceberg is always submerged beneath the surface of the water. In the same way, the weight of mountains causes them to sink deeper into the mantle than the surrounding continental crust does.

At a depth of about 70 km (43.5 mi) there is another boundary within the mantle. At this depth, the hard, rigid rock of the upper mantle appears to change into a partly molten condition. This boundary separates the hard outer shell of the earth from the rest of the mantle. This hard

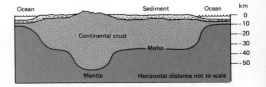

FIG. 9–4. The crust is thickest under mountains.

FIG. 9–5. General outlines of the separate plates that make up the lithosphere are shown on these global maps.

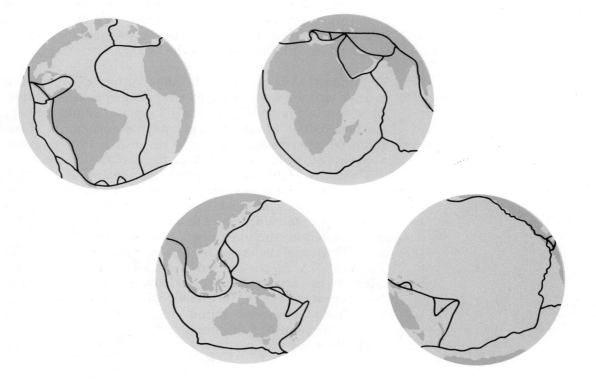

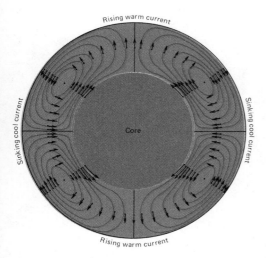

FIG. 9–6. Motion of the lithospheric plates may be caused by slow currents in the mantle resulting from uneven heating. The actual pattern of the currents would be much more complicated than the simple pattern shown in this diagram.

outer shell is called the *lithosphere*. The crust makes up the uppermost part of the lithosphere. The softer, partly melted layer of the mantle below the lithosphere is known as the *asthenosphere*. See Figure 9–2. About 250 km (155 mi) beneath the surface, the asthenosphere ends as the great pressure prevents any melting and softening of the rigid mantle rock.

According to the theory of plate tectonics, the lithosphere is broken into about a dozen major plates. See Figure 9–5. Each of these plates moves by sliding slowly over the partly melted asthenosphere, carrying the continents with it like logs frozen in moving ice in a river. But what forces drive the plates? Scientists have not yet been able to provide a complete answer to this question. There is evidence that the rock of the mantle is not heated evenly. The warmer rock is less dense than the cooler material. Thus it slowly rises in the mantle in the same way a balloon filled with hot air rises in the atmosphere. Near the surface, cooling causes the rock to sink back into the mantle, where it will be reheated. Such a very slow but continuous rising and sinking of mantle rock could form giant circular flow patterns in the mantle, as shown in Figure 9–6. Much research is now under way to prove the existence of such movements in the mantle and their exact connection with the motion of the lithospheric plates.

Action at the plate edges. The most active regions on the earth's surface are the boundaries of the separate plates. Most of the changes in the earth's crust take place along the plate edges. Study of the events that occur along these plate boundaries has caused a revolution in our understanding of the forces that shape the earth's surface.

There are three kinds of plate boundaries. They differ according to the way each plate moves in relation to the other. For example, one kind of plate boundary occurs where both plates are moving away from each other. The gap created as the plates separate is filled with molten rock. The liquid rock rises up from the asthenosphere. See Figure 9–7A. As the molten rock cools, it is added to the edges of the separating plates. Thus new crust is always being created as the plates slowly move apart. This type of plate boundary is now found almost entirely on the sea floor. The locations of these spreading boundaries are marked by the *mid-ocean ridges*. These mid-ocean ridges are mountain chains running around the entire globe along the floor of every ocean of the world. The mountains have been made by the molten rock rising from below as the plates on each side of the ridge move away. The separation of the plates along the mid-ocean ridges is called *sea-floor spreading*. The speed at which the plates move apart in sea-floor spreading is not the same along all parts of the mid-ocean ridges. A typical rate of motion is about 5 cm per year. Your fingernails grow at about this speed.

Pushing apart of the plates by sea-floor spreading means that the plates must have other boundaries where they are pushed together. The direct collision of one plate with another makes the second type of plate boundary. Two types of collisions can occur. The first kind of collision involves one plate made of dense rock, such as the basaltic rock of the sea floor, colliding with another plate made of lighter granitic rocks, typical of the continents. The denser sea floor is pushed down while the continent rides up. See Figure 9–7B. The region along plate boundaries where one plate is forced under the other is called a *subduction zone*. To "subduct" means to take away. In a subduction zone, the part of the plate that is pushed down is thought to be incorporated back into the mantle. The second type of collision occurs when parts of two plates carrying the lighter continental rocks come together. Neither plate is forced downward. Instead, the colliding edges are crumpled and uplifted to produce large mountain ranges like the Himalayas.

Plates can also collide and only slip past each other. This kind of collision forms the third type of plate boundary. See Figure 9–7C. The plates' edges usually do not slide smoothly past each other. Instead, they scrape together and move with sudden slips separated by periods of little or no motion.

Some of the plate boundaries, such as the mid-ocean ridges, are not too difficult to locate on the earth's surface. Other boundaries are much more difficult to trace. As a result, scientists are not certain about exactly how many separate crustal plates exist. There may be as many as twenty individual plates making up the lithosphere. Some plates cover large areas of the earth's surface. Others are small. The boundaries run down the middle of oceans, around the edges of continents, and through continents. Each plate has some combination of three kinds of plate

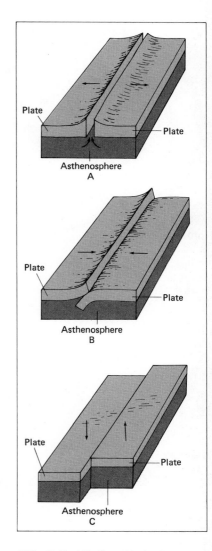

FIG. 9–7. (A) One kind of plate boundary is formed when molten rock rises up from the asthenosphere into the gap between separating plates. (B) When two plates collide, one may be pushed down while the other rides up. (C) In some cases when plates collide, they slide past each other.

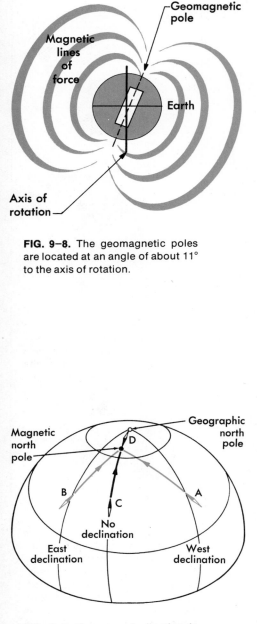

FIG. 9–8. The geomagnetic poles are located at an angle of about 11° to the axis of rotation.

FIG. 9–9. Compass declination is the error caused by the compass needle pointing toward the magnetic pole rather than the geographic pole.

boundaries. Some plates are almost all land. Others are partly land and partly covered with ocean. Some plates seem to be made entirely of sea floor. The familiar outlines of the continents and oceans on maps do not closely resemble the outlines made by the plate boundaries. The continents are only the high places on the plates; the oceans are the low areas.

Testing the theory. Like any scientific theory, plate tectonics must be supported by observations. Most scientists believe that there is considerable evidence to support the belief that the earth's outer layer is made of solid, moving plates. Examples of observations that can be explained by plate tectonics are found in the study of the earth's magnetism.

Anyone who has used a compass to determine direction knows that the earth is like a giant magnet. One way to describe the earth's magnetism is to imagine a powerful bar-shaped magnet buried inside the earth. This imaginary magnet would be tilted at an angle to the geographic poles or axis of rotation. See Figure 9–8. The points on the surface just above the poles of the imaginary magnet are called the *geomagnetic poles*. The tilt of the imaginary magnet means that the geomagnetic poles and the geographic poles are found in different locations. Since a compass needle always points toward the geomagnetic poles, a compass does not show the direction of the north and south geographic poles. The angle between the direction of the geographic pole and the direction in which the compass needle points is called *magnetic declination*. In the northern hemisphere, declination is measured in degrees east or west of the geographic north pole. In Figure 9–9, a compass needle at point A points west of the geographic pole. At point C, the compass needle lines up with the geographic north pole so that there is no declination. At point B, the compass indicates a declination east of geographic north. For example, the magnetic declination in the region of Houston, Texas, is 10°E. This means that a magnetic compass used near Houston will point 10° east of the direction of the geographic north pole. Because of its importance in navigation, magnetic declination has been determined for points all over the earth. Charts and tables that give the declinations for most regions of the world are available. The pattern of magnetic declination for most of the United States is given in Figure 9–10.

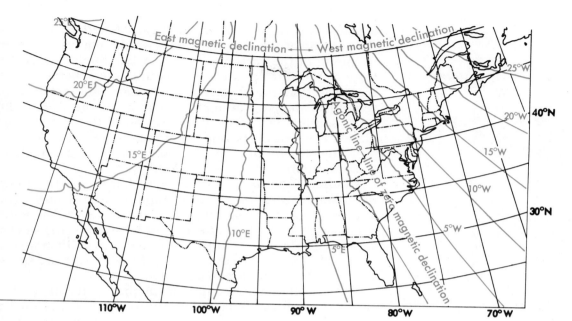

FIG. 9–10. The lines on this map connect points having the same magnetic declination. Unevenness of the lines shows that the earth's magnetic field does not have the same strength everywhere.

Assuming that the earth has a permanent magnet buried inside provides a useful way of describing its magnetism. However, this cannot be the true explanation for the earth's magnetism. No permanent magnet could exist at the high temperatures found inside the earth. Magnetism is created, however, by electric currents. Scientists believe that the liquid parts of the earth's iron core may produce electric currents. Uneven heating of the core's molten iron by radioactive elements could cause motions independent of the earth's rotation. These movements could generate electrical currents. Although final proof for this theory is lacking, it does seem to provide the best explanation for the earth's magnetism.

Like a compass, some kinds of minerals are affected by the earth's magnetism. When a piece of rock containing some of these minerals is heated, the individual mineral crystals line up in the direction of the earth's magnetic poles. When the rock cools, these crystals are frozen in the direction of the magnetic poles. Study of rocks that were heated at some time in the past shows a record of the direction of the earth's magnetic field at that time. See Figure 9–11.

Ancient magnetic records found in rocks have revealed two surprising facts. First, the earth's magnetic poles seem to have shifted position many times. However, experiments suggest that the magnetic poles and geograph-

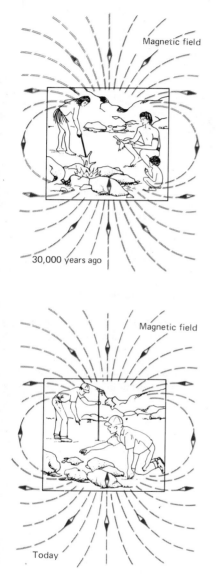

Magnetic field

30,000 years ago

Magnetic field

Today

FIG. 9–11. Rocks heated in the fireplace of an ancient campsite (above) would have become magnetized in the direction of the earth's magnetic field at that time. Later (below) the rocks can be studied and compared with the direction of the present magnetic field.

ic poles will always remain fairly close together. A shift in the position of the magnetic poles would require a change in the tilt of the earth's axis of rotation. Such changes in the way the earth rotates are very unlikely. A better explanation is that the magnetic and geographic poles have remained almost stationary. Their apparent wandering must actually be the result of motion of the lithospheric plates. Thus ancient evidence of magnetism in rocks supports the theory of plate tectonics and helps scientists trace the motion of the plates in the past.

Another surprising discovery was that the earth's magnetic field reverses itself from time to time. That is, the magnetic north and south poles change places. Magnetic evidence in rocks indicates that at least nine such reversals have taken place during the past 3.5 million years. Reversals are also used as a way of measuring the speed with which the plates move apart.

When molten rock comes up in the center of a mid-ocean ridge, it hardens and is added to the edges of the plates that are moving away on either side of the ridge. See Figure 9–12. As a result, the molten rock is split into two narrow stripes as it hardens. One stripe lies on each side of the ridge. The rocks in these stripes carry a record of the direction of the earth's magnetic poles at the time they were found. This means that each reversal of the magnetic field will be recorded as a magnetic pattern running parallel to the ridge along each side. See Figure 9–13. During millions of years, the magnetically marked stripes of the ocean floor move away from the mid-ocean ridge. Almost the entire floor of the modern ocean shows this pattern of magnetic bands.

The discovery of these magnetic stripes on modern sea floors means that sea-floor spreading almost certainly does occur over millions of years. The rate at which new sea floor is spreading can be determined from the times of the magnetic reversals and the distance of the separate stripes from the mid-ocean ridges. The magnetic patterns found on the sea floor thus supply strong evidence to support the theory of moving lithospheric plates.

VOLCANISM

Plate tectonics and volcanism. We know that the earth's interior is hot. In deep holes, such as the diamond mines of South Africa, the temperature has been found to increase at an average of about 2 or 3°C for every 100 m (300 ft) of

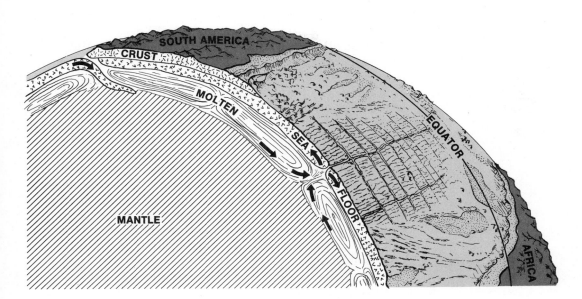

depth. There is no direct way to measure temperatures very deep in the earth. However, it seems very unlikely that the temperature keeps increasing at the rate of 2 or 3° per 100 m all the way to the earth's center. The temperature probably increases very slowly below about 100 km. Scientists believe that they can estimate the temperature of the interior from the behavior of seismic waves. A

FIG. 9–12. Material rising from the mantle at the mid-ocean ridges forces the plates making up the sea floor to move apart. Where the spreading plate meets another, as along the west coast of South America (left), one of the plates is pushed beneath the other.

FIG. 9–13A. Reversals of the earth's magnetic field leave a record in the form of magnetic stripes in the rock flowing out along the mid-ocean ridges.

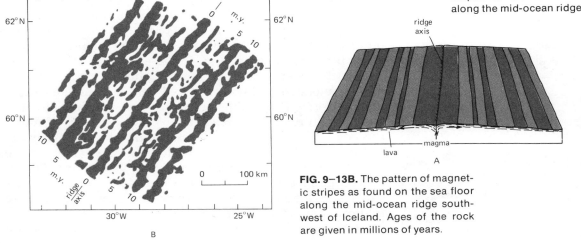

FIG. 9–13B. The pattern of magnetic stripes as found on the sea floor along the mid-ocean ridge southwest of Iceland. Ages of the rock are given in millions of years.

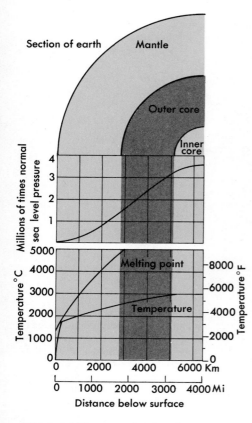

Section of earth   Mantle

Outer core

Inner core

FIG. 9–14. These graphs show how temperature and pressure increase beneath the earth's surface.

FIG. 9–15. This volcano appeared out of the sea along the mid-Atlantic ridge south of Iceland. (John V. Christiansen/Earth Image)

graph showing the temperature and pressure of the earth's interior is shown in Figure 9–14.

Almost all of the rock in the lithosphere is solid, since the temperature in most parts is too low to cause melting. However, parts of the lithosphere may be melted, mainly as a result of changes taking place where two plates collide. In the upper mantle, temperatures are high enough to create the asthenosphere layer of rock. In some places, the more mobile rock of the asthenosphere begins to migrate toward the surface. As it moves upward, the pressure is lowered and the rock becomes so mobile that it behaves much like a liquid. Such very fluid rock close to the earth's surface is called *magma*. When magma reaches the earth's surface, it is known as *lava*. Activities caused by the movement of magma are described as *volcanism*. Any opening on the earth's surface that gives off lava is a *volcano*.

If all of the about 600 active volcanoes known to exist on the earth are located on a map, it can be seen that they are not scattered in a random way. A pattern appears in their locations. Most active volcanoes are found at locations that trace boundaries of the lithospheric plates. For example, one great system of volcanoes is found along the mid-ocean ridges. The fracture between the separating plates reaches all the way down to the asthenosphere. Volcanism occurs when magma rises within the fracture to produce the ridges and new sea floor. Most of the volcanic eruptions along the mid-ocean ridges are not seen, since they take place beneath the sea. Occasionally an eruption beneath the ocean will create a volcanic island that will rise above the sea level. See Figure 9–15.

Volcanism also commonly occurs along subduction zones. This kind of volcanism, produced along colliding plate boundaries, is much more evident, since it takes place primarily on the land. If the edge of a plate that is part of the sea floor meets another plate carrying a continent, the oceanic plate is pushed beneath the continent. See Figure 9–16A. As the oceanic plate is pushed down, it bends and produces a deep *trench* on the sea floor along the edge of the continent. A mountain chain usually develops along the edge of the continent as the oceanic plate squeezes against the land. As the oceanic plate is subducted and pushed deep toward the mantle, increasing pressures and temperatures act upon the plate. Less dense material separates from denser material and eventually rises to produce volcanoes among the mountains along

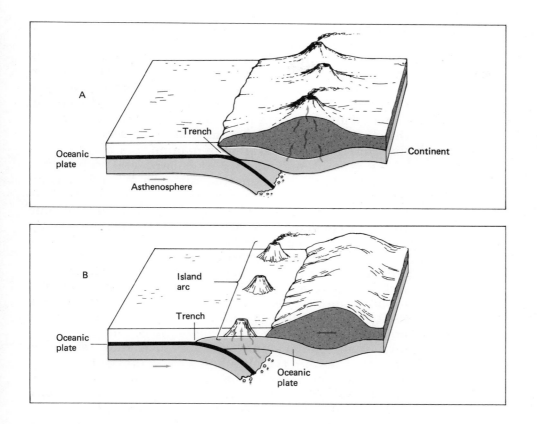

the continental edge. Because this magma does not come entirely from the asthenosphere, it is usually made of different rock than the magma connected with the mid-ocean ridges.

When the colliding plates are both part of the sea floor, a different arrangement of volcanoes results. A chain of volcanoes called an *island arc* grows on the sea floor along the trench marking the boundary of the plates. See Figure 9–16B. The earliest stages of an island arc produce a series of small volcanic islands such as the Aleutian Islands stretching across the North Pacific Ocean. Older island arcs grow larger, like the group of volcanic islands that make up Japan.

Not all volcanoes are found along plate boundaries. There are some places within the interiors of plates where magma is able to work its way upward from the asthenosphere. These locations of volcanism within plates are called *hot spots*. The reason for the occurrence of hot spots is not well understood.

FIG. 9–16. (A) When part of a plate made up of sea floor meets part of a plate carrying a continent, the oceanic plate is subducted, melts, and forms magma that rises to produce volcanoes along the edge of the continent. (B) When one oceanic plate meets another, a trench is produced on the sea floor where one plate is subducted. Volcanoes form an island arc along the trench.

Interpret
From the graph in Figure 9–14, determine the temperature and pressure at the boundary between the mantle and the core.

Volcanoes. Volcanoes are a kind of window to the deep interior of the earth. Lava coming out of volcanic openings provides the only opportunity we have to study the composition of the deeper parts of the earth. Examination of lava shows that its composition varies from one volcano to another. Thus the parent magmas that supply the volcanoes cannot all have the same composition. Analysis of the minerals found in different kinds of lava suggests that there are three general kinds of magma.

One type of magma is dark-colored and rich in the chemical elements magnesium and iron. When lava coming from this type of magma cools, igneous rock such as basalt develops. This type of magma is believed to come from the asthenosphere. Almost all magma is probably of this type as it begins its journey from the upper mantle toward the surface. Later it may come into contact with other materials from the earth's upper layers. These materials melt and mix with the magma. One of the main materials usually added is the chemical substance silica (SiO_2). The addition of silica yields the second type of magma. This magma has a lighter color and produces a lava that solidifies into igneous rock that resembles granite. The third type of magma has a range of compositions that falls between the dark, basaltic and the light, granitic varieties. The many different kinds of igneous rocks found in the earth's crust are produced from lavas having different compositions and by the way in which the minerals separate from the molten rock as it cools.

The way in which a particular volcano erupts is largely controlled by the type of parent magma that forms the lava. For example, basaltic-type magmas are usually very fluid while still molten. They flow easily, like water. The red-hot liquid lava may flow out like a river from the volcanic opening. See Figure 9–17. Volcanoes in Hawaii

FIG. 9–17. Lava, magma that reaches the earth's surface, flows in river-like streams from the eruption of Kilauea volcano, Hawaii. (Werner Stoy/FPG)

usually erupt with this type of lava flow. Cooling of the very thin liquid lava may cause it to harden with a wrinkled surface. This type of solidified lava is known by its Hawaiian name of *pahoehoe* (pah-*hoy*-ay-*hoy*-ay), which means "ropey." Rapid cooling on the surface of a lava flow can form a crust that breaks into jagged chunks as the liquid below flows on. This produces a lava deposit described as *block lava* or, in Hawaiian, *aa* (*ah*-ah). See Figure 9–18. Sometimes the outer parts of a lava flow cool rapidly, while the interior remains liquid. This liquid material may flow out later, leaving tunnels in the solid lava. See Figure 9–19.

Magmas rich in silica tend to produce thick lavas. This is a result of the linking together of the silica molecules. Gases such as water vapor, which are normally released when magma cools, do not easily escape from the thick lavas. Pressure from the trapped gases builds up until there is an explosion that first releases these gases and later may eject lava out of the volcano. Whether a volcano erupts with quiet lava flows or explosively depends largely upon the amount of gases trapped in the magma and how easily they can escape.

When a volcano erupts explosively, the solid fragments thrown out are called *pyroclastics*, which means "fire fragments." Some pyroclastics come from cooling magma that is broken into fragments by the rapidly expanding gases. Other pyroclastics are formed by clots of liquid lava

FIG. 9–18. Pahoehoe type lava is seen near the person. Aa type lava is visible in the lower left of this photograph. (Ramsey)

FIG. 9–19. This tunnel was formed when liquid lava drained away from the solidified lava overhead. (Bob and Ira Spring)

FIG. 9—20. This large volcanic block was thrown out during the eruption of Mount Lassen in 1917. (Ramsey)

FIG. 9—21. These volcanic bombs were given their shape during their flight through the air. (Ramsey)

FIG. 9—22. (Top) A shield cone. (Bottom) A cinder cone. (Ramsey)

that cool during their flight through the air. Some pyroclastics come from rock surrounding the volcanic opening.

The smallest pyroclastic rocks are particles as tiny as grains of flour that make up volcanic *dust*. Particles the size of rice grains are called volcanic *ash*. These smallest products of a volcanic eruption may be carried away from the volcano by the wind. They settle mainly on the surrounding land. However, some volcanic dust may be carried completely around the world in the upper atmosphere. Larger particles called *cinders*, about the size of golf balls, are deposited close to the volcanic opening. Very large chunks of rock, some as big as a house, thrown out in an eruption are known as *blocks*. See Figure 9–20. Larger fragments, still red-hot when thrown out of an erupting volcano, are called *volcanic bombs*. They often take on a streamlined shape as they fly through the air. See Figure 9–21.

Volcanic features formed on the earth's surface. The lava and solid materials that are produced during volcanic eruptions can create massive rock piles around the volcanic opening. These rock piles are called *cones* and are classified into three main types. *Shield cones* are broad at the base with gentle slopes. They are made from layer upon layer of lava flowing out around the volcanic opening. The lava flow is fluid enough to cover a wide area. Shield cones are generally formed as a result of quiet eruptions. The Hawaiian Islands are a cluster of shield cones built up from the sea floor.

Explosive eruptions will form *cinder cones*. These have fairly narrow bases and steep sides. A cinder cone is a pile of solid fragments thrown out during an explosive eruption. Because their materials are loosely arranged, cinder cones are seldom very high. Figure 9–22 illustrates a shield and a cinder cone.

Many volcanoes have quiet eruptions during which a cone is built mainly by lava flows. Then an explosive eruption occurs that throws out large amounts of pyroclastics. This is followed again by quiet lava flows. The resulting cone is built of alternating layers of lava flows and pyroclastic materials. Such volcanic mountains are known as *composite cones*. They may also be called *stratovolcanoes*, since the cone is often quite high. Some of the world's best known volcanic mountains, such as Fujiyama in Japan, and Mounts Rainier, Hood, Shasta, and Mt. St. Helens in the United States, are stratovolcanoes.

FIG. 9-23. The waters of Crater Lake, Oregon, fill the caldera of the ancient volcano called Mount Mazama. A small cinder cone forms an island in the lake. (HRW Photo by Richard Weiss)

Openings commonly develop in cones some distance below the main opening. Materials flowing from these openings can build up smaller volcanoes called *parasitic cones*. Very small cones that form in the lava fields away from the main opening are called *spatter cones*. As the lava cools, a crust forms. Hot gases then force lava out in spattering fountains through openings in the crust.

The funnel-shaped pit at the top of a volcanic cone is known as the *crater*. The crater usually widens as the very hot magma melts and breaks down the crater walls. The rim of the volcanic cone can also collapse when explosions occur within the crater. Often a smaller cone is built within the crater by materials erupting from the vent.

Occasionally a large, basin-shaped depression can be formed when a volcanic cone collapses. A cone can collapse into a basin created when magma pours out or drains downward again. Explosions can also completely destroy the upper part of the cone. When the cone collapses or is blown away, the immense pit or depression formed is called a *caldera*. See Figure 9–23.

When a volcano is active, it continues to build its cone much faster than weather and erosion can wear it away. When it stops erupting for a long period, valleys and gullies are cut deeply into its slopes by erosion. Eventually, the softer parts of the dormant, or inactive, volcano are worn away, and only the hard rock formed by the solidified

FIG. 9–24. This volcanic neck has been exposed as the surrounding land was eroded away. The rock is called Shiprock and is located in New Mexico. (Shostal)

FIG. 9–25. Most lava reaches the earth's surface through fissures, or cracks, in the crust. Accumulation of the lava from fissure flows can cover large areas of the land.

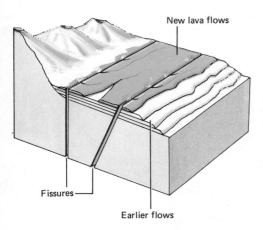

New lava flows

Fissures

Earlier flows

magma remains. Often the hard, central neck, or plug, is left standing as an isolated shaft of rock. Narrow outgrowths of solidified magma extending from the central plug can also be exposed. See Figure 9–24.

The best known volcanoes are those whose eruptions come from a central opening. These volcanoes usually build the familiar volcanic cones and mountains. However, the largest amounts of lava always come out of long, narrow cracks in the crust that are called *fissures*. Lava coming from fissures does not build a cone, but spreads out over a large area. See Figure 9–25. An example of a fissure flow covering a very large area is found in the northwestern United States. A region known as the Columbia Plateau was formed millions of years ago by fissure flows of basaltic lava that flooded out over about 260,000 sq km (about 100,000 sq mi). These ancient lava flows have covered the original land surface of this region with a lava plain that is nearly 2 km (about 1 mi) thick in some places. Although some parts of continents have been covered by fissure flows, most such eruptions occur beneath the ocean. Lava flowing out along the mid-ocean ridges erupts almost entirely from fissures.

Volcanism beneath the surface. Magma always tends to move upward because expanding rock is less dense than an equal volume of solid rock. It works its way upward through the cracks and weaker parts of the rocks above it. Sometimes magma cools and becomes solid before it reaches the surface. A body of magma that solidifies underground forms a mass of rock called a *pluton*. When plutons are later exposed by erosion or studied by digging or drilling holes into the crust, they are found to be of different sizes and shapes. See Figure 9–26. The principal types of plutons are:

1. **Batholiths.** These are the largest of all the structures formed by bodies of magma. *Batholiths* are very large masses of solidified magma covering hundreds of square kilometers. The name "batholith" means "deep" rock. Batholiths extend downward to unknown depths. They form the cores of many major mountain ranges. A small batholith, less than 100 square kilometers, is usually called a *stock*.

2. **Laccoliths.** Magma can move between existing rock layers and spread outward toward the surface. The magma pushes the rock layers above it into a dome-shaped mass. This results in the formation of small mountains on the surface of the earth. These dome-shaped structures are called *laccoliths*. "Laccolith" means "lake" of rock. Laccoliths, unlike either batholiths or stocks, have a definite floor. They are frequently found in groups.

3. **Sills.** A sheet of fluid magma sometimes flows between the layers of existing rock. This raises the rocks just enough to make room for the intruding magma. The

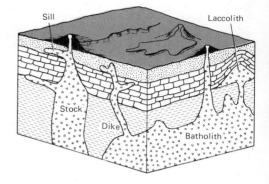

FIG. 9–26. This diagram shows the various types of plutons formed when magma invades the rocks beneath the surface.

FIG. 9–27. A sill can be seen near the top of this hill in Yellowstone National Park. (Ramsey)

FIG. 9–28. A dark-colored dike cuts through the rock layers exposed in this road cut. (Ramsey)

FIG. 9–29. A laser ranger such as this is used to measure changes in distance related to volcanic deformation. The time it takes a laser beam to return from a reflector station accurately measures the distance between two marked points to a few millimeters. The changes in distance are the result of slight bulges in the surface of a volcano. (U.S. Geological Survey/H. Glicken)

magma then hardens to form a *sill*. The walls of a sill are roughly parallel to the surrounding rock layers. See Figure 9–27.

4. **Dikes.** A mass of magma can fill and then solidify in fractures that generally cut across the surrounding rock layers. This structure is called a *dike*. See Figure 9–28. Dikes are often outgrowths of the large pockets of magma that form batholiths.

Predicting volcanic eruptions. At least 200,000 people have been killed as a result of volcanic eruptions during the past 500 years. Recently, in the United States, 65 people lost their lives when Mount St. Helens erupted in 1980. The sudden eruption of a volcano near a large city or other heavily populated area could cause thousands of deaths, as well as the destruction of valuable property and resources. If it were possible to accurately predict the time of eruption, many lives could be saved and property damage prevented. Scientists believe that it may soon be possible to predict eruptions of volcanoes that are active or temporarily inactive.

By monitoring active volcanoes with sensitive instruments, certain signs can be detected that signal the beginning of an eruption. One of the most important signals is the occurrence of small earthquakes. These earthquakes seem to be caused by the growing weight of the magma on the surrounding rocks as it works its way upward. Temperature changes, the motion of the magma, and gas explosions may also help to create the earthquake. As the time for an eruption approaches, the number of earthquakes often increases until they occur almost continuously. An increase in the strength of the earthquakes may also be a signal that an eruption is near.

The upward movement of magma beneath the surface before an eruption can cause the volcano's surface to bulge out very slightly. These gentle bulges cause the distance between two marked points to change slightly, like spots on a balloon would move apart as the balloon is inflated. Instruments can measure those small changes in distance, as well as the changing tilt in the ground surface. See Figure 9–29. Such measurements help to warn of a possible eruption. Changes in the composition of the gases given off by volcanoes are being investigated as another possible signal of an increase in volcanic activity.

Knowledge of previous eruptions at a particular volcano is a very important part of forecasting an eruption. The

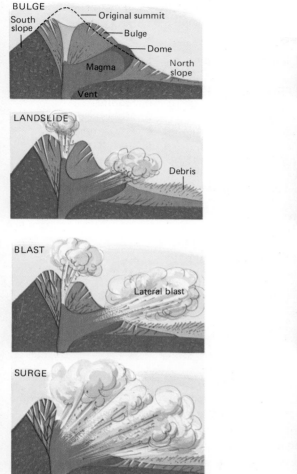

past behavior of the particular volcano can be compared with the daily measurements of earthquakes, deformation of the surface, and changes in gases to provide the best forecast. Unfortunately, only a small number of the world's active volcanoes have been under study long enough to discover any pattern in their activity. Volcanoes that have been extinct for long periods of time may, with little warning, suddenly become active. Predicting the eruption of these dangerous volcanoes will require a solution to the difficult problem of detecting the presence of magma while it is still deep inside the crust.

FIG. 9–30. In May, 1980, scientists warned that Mt. St. Helens in Washington was expanding and might erupt. However, the explosive violence of the eruption was a surprise. These diagrams and photographs show the first few minutes of the eruption, which threw 400 million tons of dust and ash into the atmosphere. (Gary Rosenquist/Earth Image)

VOCABULARY

Match the word or words in the column on the right with the correct phrase in the column on the left. *Do not write in this book.*

1. Very large body of solidified magma extending to unknown depths.
2. Sent out when an earthquake occurs on or near the earth's surface.
3. Activities caused by movement of magma.
4. The modern theory stating that the earth's crust is made up of several rigid parts.
5. The location of the spreading boundaries of the crustal plates.
6. The boundary between the crust and the mantle.
7. All underground molten rock.
8. The thin outer layer of the earth.
9. Massive rock piles created around volcanoes by eruptions.
10. The angle between the direction of the geographic pole and the direction the compass needle points.

a. plate tectonics
b. seismic waves
c. crust
d. mantle
e. Moho
f. lithosphere
g. mid-ocean ridges
h. magnetic declination
i. magma
j. volcanism
k. cones
l. batholith

QUESTIONS

Group A

Select the best term to complete the following statements. *Do not write in this book.*

1. The theory of continental drift is supported by the (a) size of continents (b) equal spacing of mountains (c) shapes of the continents (d) thickness of the mantle.
2. In his theory on continental drift, Wegener claimed there once was a supercontinent called (a) Valhalla (b) Pangaea (c) Bali Hai (d) Pandora.
3. The theory of continental drift is (a) entirely acceptable (b) partly acceptable (c) completely unacceptable.
4. According to plate tectonics, the continents (a) drift like rafts (b) are fixed and do not move (c) move only northward (d) are passengers on a number of moving plates.
5. According to the theory of plate tectonics, the earth is made up of (a) several rubbery plates (b) thousands of small pieces (c) a single flexible plate (d) several rigid plates.
6. The term that refers to forces that move and shape the crust is (a) plates (b) regulators (c) tectonics (d) volcanism.
7. To study the interior of the earth, scientists use (a) waves caused by explosions (b) waves caused by earthquakes (c) seismic waves (d) all of these.
8. Starting at the surface, the order of the layers of the earth is (a) core, mantle, crust (b) core, crust, mantle (c) crust, core, mantle (d) crust, mantle, core.

9. The layer of the earth that we know best is the (a) inner core (b) crust (c) outer core (d) mantle.

10. Together, the crust and upper mantle make up the (a) outer core (b) asthenosphere (c) atmosphere (d) lithosphere.

11. The high temperature inside the earth is due to heat trapped during formation, heat due to radioactive atoms, and (a) heat due to pressure (b) heat due to evaporation (c) heat from space (d) heat from people.

12. The ring-shaped region where waves from a particular earthquake are not felt is called the (a) thermal zone (b) density zone (c) shadow zone (d) pressure zone.

13. The behavior of seismic waves indicates that compared to the earth's surface, the earth's core is (a) more dense (b) about the same density (c) less dense (d) indistinguishable.

14. The Mohorovicic discontinuity apparently separates the earth's (a) mantle and crust (b) core and mantle (c) inner and outer cores (d) crust and core.

15. The lightest materials in the earth's crust occur (a) under the oceans (b) under lakes (c) in the continents (d) high on mountains.

16. The thickest part of the earth's crust is about (a) 10 km (b) 35 km (c) 65 km (d) 250 km.

17. The hard outer shell of the earth is called the (a) asthenosphere (b) mantle (c) Moho (d) lithosphere.

18. Where plates move apart is a region of (a) mid-ocean ridges (b) subduction (c) no activity (d) all of these.

19. Where one plate is forced beneath another is a region of (a) mid-ocean ridges (b) subduction (c) no activity (d) all of these.

20. Compared to the geographic poles of the earth, the magnetic poles are (a) near the equator (b) at the same location (c) located in different places (d) all of these.

21. The angle between the directions of true north and magnetic north is called magnetic (a) inclination (b) declination (c) geographic (d) azimuth.

22. The magnetism of the earth is often compared to that of a (a) permanent magnet (b) coiled spring (c) donut (d) horseshoe.

23. Proof of sea-floor spreading is found in a study of the earth's (a) life forms (b) earthquakes (c) magnetism (d) mountains.

24. Magnetic records on the sea floor show that the earth's magnetic field (a) has always been in the same direction (b) has reversed (c) has changed only slightly (d) none of these.

25. Temperatures taken deep in wells and mines show that with depth temperature (a) decreases (b) remains about the same (c) increases (d) changes unpredictably.

26. Even though much of the rock found deep in the crust and mantle is very hot, it is kept solid because of (a) low pressures (b) high pressures (c) radioactivity (d) solar energy.

27. Other than hot spots, most volcanoes are found (a) along boundaries of crustal plates (b) in the middle of crustal plates (c) in a line across crustal plates (d) in Europe.

28. Molten rock found below the surface of the earth is called (a) lava (b) a batholith (c) a laccolith (d) magma.

29. Explosive eruptions of volcanoes form (a) blocks (b) bombs (c) cinder cones (d) all of these.

30. When magma solidifies in vertical cracks, it forms (a) sills (b) block lava (c) dikes (d) pahoehoe lava.

Group B

1. Briefly describe the theory of continental drift.

2. How is the theory of plate tectonics similar to the theory of continental drift?

3. How does the theory of plate tectonics differ from the theory of continental drift?

4. Draw a diagram showing the main layers of the earth. Label the layers using their correct names.

5. Describe the Moho.

6. What evidence do we have to support the existence of the core of the earth?

7. Describe the layers of the earth.

8. Explain the relationship between mid-ocean ridges and sea-floor spreading.

9. Describe the three kinds of crustal plate boundaries.

10. How is the earth's magnetic field like the field of a permanent magnet?

11. Explain the relationship between the geographic and geomagnetic poles.

12. What is magnetic declination?

13. What do scientists think causes the earth's magnetic field?

14. How has the earth's magnetic field changed since the earth formed?

15. Describe the nature of the evidence that supports sea-floor spreading.

16. Describe how the temperature changes as you move from the earth's surface toward the center.

17. Write a short paragraph correctly using the terms magma, lava, volcanism, and volcano.

18. Name and briefly describe three types of cones associated with volcanoes.

19. How is volcanism related to plate tectonics?

20. How are island arcs and ocean trenches related?

movement of the earth's crust

Climbers reaching the top of Mount Everest plant their victory flags over the remains of animals that once lived in the sea. Rocks at the top of the highest mountain in the world are made of limestone from ancient sea-floor sediments. Material that was once far below sea level has been thrust up to an altitude of 9524 m (29,028 ft). There can be no more convincing evidence than this for the existence of huge forces that squeeze and stretch the solid rocks of the earth's crust. Beds of rock are tilted, bent, and broken into complicated patterns that shape the crustal surface into mountains and valleys. The scientific study of the earth is called *geology*. When geologists study the arrangement of rocks in the crust, they use knowledge of events far back in time to trace the changes that have created the observed features.

The modern science of geology can be traced to the late 1700s and the work of James Hutton, a Scottish medical doctor and gentleman farmer. Hutton thought about aspects of the earth that he observed. He watched the streams on his farm carrying the soil into rivers. He knew that the burden carried by the streams would eventually be deposited in the sea, where it would build up layer upon layer. But Hutton realized that there must be some force that caused the rock layers formed on the sea floor to be lifted and shaped into the mountains and lowlands that are found on the land. If not, the earth's surface would have been leveled off long ago. Hutton concluded that all the forces seen changing the rocks in the crust at the present time have also worked to change the earth's surface in the past. His theory was that *the present is the key to the past*. By studying the rocks as they are now, it is possible to go back in time and reconstruct ancient landscapes.

objectives

☐ Explain the theory of uniformitarianism.

☐ Describe how rocks may become folded.

☐ Distinguish between folding and faulting.

☐ Identify three main sources of stress in rocks of the crust.

☐ Explain earthquakes and how they are studied.

☐ Describe how most earthquakes are believed to be related to crustal plates.

☐ Compare different kinds of mountains.

FIG. 10–1. Solid rock is plastic under high pressure. The limestone cylinder at the left was distorted into the form at the right by a vertical pressure of 125,100 pounds per square inch and a pressure of 22,100 pounds per square inch applied from the sides. (U.S. Department of the Interior)

FIG. 10–2. Horizontal beds of rock can respond to pressure by folding to form anticlines and synclines, as shown in the diagram. Strongly folded metamorphic rock layers near Calico, California, are shown in the photograph. (Ramsey)

Geologists call this theory *uniformitarianism*. It means that the same forces that changed the rocks in the past are still operating today, causing the same kinds of changes, although not necessarily at the same rate. Before Hutton, it was commonly believed that the earth was shaped by a single great event in the past. The landscape we see today was formed. Very little change took place as time went by.

Most forces that work today to change the crust appear to operate very slowly. It is therefore reasonable to assume that it must have taken many millions of years for those forces to create the complicated rock structures that are now found in the earth's crust. Because of this, Hutton's theory of uniformitarianism implied that the earth is very old.

CHANGING THE SHAPE OF THE CRUST

Deforming rocks. When force is applied to some kinds of materials, such as an iron bar, they may respond by bending. Other materials, like glass, are more likely to break when force is applied. Rocks can respond to forces in both of these ways. Forces act within the crust to squeeze rocks together or pull them apart. When these forces are applied slowly, rocks usually deform like a piece of rubber first. Then they return to their original shape when the force is removed. However, all rocks have a certain limit beyond which they cannot be deformed without permanently changing their shape. When this limit is reached, the rock either continues to bend, or it breaks. Generally, rocks that are under high pressure and temperature deep within the crust bend and are deformed a great deal. See Figure 10–1. Rocks near the surface also bend, but they reach their limit sooner and then break.

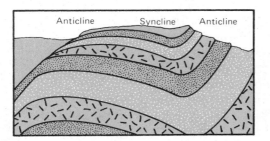

Folding. When flat beds of rock are squeezed together, they often move into new positions without breaking up. This process is called *folding*. When rocks are folded, the layers are often pushed into shapes that resemble waves on water. Sometimes cracks or breaks may appear in the rocks, but the layers remain complete. Geologists believe that of the forces that produce folding, some are caused by the collision of crustal plates of the lithosphere. Such collisions would be able to squeeze large areas of continental rock. Giant folds pushed up along the edges of colliding continents have formed mountain ranges such as the Appalachians. Rock layers may also be folded by the force of gravity. This happens when large sheets of sedimentary rocks are lifted up and tilted. The layers may then slide downslope as a whole, causing them to be squeezed together and folded. This kind of folding often results when magma pushes up from below and lifts the older sedimentary beds near the surface.

Some folds in rock layers are large enough to raise mountains. Other folds may affect only a small area. Whatever its size, a fold usually has an upturned and downturned section. The section that is raised is called an *anticline*. The corresponding lowered section is a *syncline*. See Figure 10-2.

If the earth's surface was not constantly worn down, large folds in rocks would usually form ridges from the anticlines. Valleys would form in the synclines. See Figure 10-3A. However, the folds are often worn off as they develop. This action keeps the surface almost flat. See Figure 10-3B. When this happens, the only evidence of folding is the tilt of the rock layers beneath the surface. The anticline areas can also be worn down more rapidly because they are the first to be exposed. More rapid wearing down of the anticlines leaves the original syncline as a mountain or ridge. See Figure 10-3C and D. Ridges formed from synclines and valleys that were originally anticlines are common in the Appalachian Mountains.

Rocks exposed at the earth's surface are constantly changing. As a result, it is not always easy to find evidence of folding and other features of the rock structure in a region. Useful clues that help identify the rock structure of an area are found in sedimentary layers. All sedimentary rock layers are nearly flat or horizontal when formed. This is because most sedimentary rocks are formed from material deposited on the level surfaces at the bottom of bodies of water. A sedimentary rock layer that is not level was

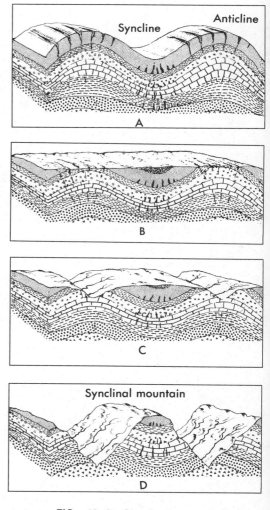

FIG. 10-3. Synclinal mountains develop as weaker beds are worn down. (A) Leveling the initial anticlines and synclines. (B) Weak beds in the anticlines are exposed. (C) Uplift brings about further wearing away. (D) The resistant rocks of the original syncline remain as a ridge.

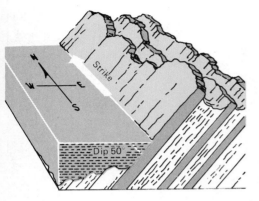

FIG. 10-4. The dip and strike of some beds of sedimentary rock are shown along the shore of a lake. What is the angle of dip and the direction of strike shown?

FIG. 10-5. In a pitching fold, the axis of the fold is tilted, compared to the horizontal plane of the surface.

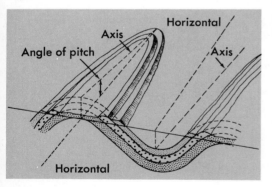

probably tilted by movements of the rocks in the crust. The angle at which a rock layer tilts from the horizontal is called its *dip*. For example, a certain bed of rock might have a dip of 50°. The direction of dip is also usually given, using compass directions. Thus the dip of a rock layer might be 50° west. It is useful to indicate whether the rock layer runs in a general north-south or east-west direction. This is called the *strike* of the rock layer. The strike is always measured at right angles to the direction of its dip. A rock layer might be described as having a dip of 50° west and a north-south strike. See Figure 10-4.

Once the dip and strike of the rock layers in an area are known, the rock structure beneath the surface can usually be determined. For example, the dip of a group of layers may show that they are part of a fold whose *axis* is not horizontal. The axis of a fold is an imaginary plane that cuts an anticline or syncline in half. A fold whose axis is not horizontal is said to be a *pitching fold*. See Figure 10-5. The surface appearance of a typical pitching fold is clearly illustrated by Sheep Mountain, Wyoming, shown in Figure 10-6.

Faulting. Rocks do not always respond to changes by folding. Sudden or abrupt crustal movements can fracture, or crack, rocks. This is most likely to occur on rocks near the surface, since they are under the least amount of pressure. If there is no movement between the sides of a frac-

FIG. 10-6. Sheep Mountain, Wyoming, shows how the anticline of a pitching fold can appear after the surrounding layers have been worn away. (Barnum Brown)

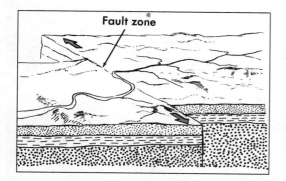

FIG. 10–7. A strike-slip fault is produced when blocks on either side of the fault move in opposite directions. Notice the stream channel that has been offset by movement along the fault.

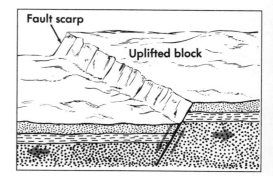

FIG. 10–8. When a fault slants away from the uplifted block, a "normal" fault is produced. This type of faulting often forms a steep cliff, or "fault scarp." Arrows indicate the direction of movement that caused the fault.

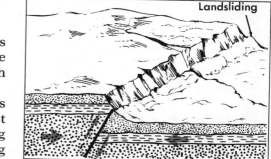

FIG. 10–9. A fault slanted toward the uplifted block produces a "reverse" fault. The fault scarp then becomes an overhanging cliff that usually collapses.

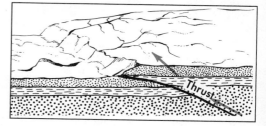

FIG. 10–10. If a fault is slanted very slightly, one block may be thrust or pushed up over the other to form a "thrust" fault. This action can produce a tightening of the crust.

ture, it is called a *joint*. Joints may also form in igneous rocks when cooling takes place. (See page 178.) These joints are always at right angles to the surface at which cooling occurs.

Often the same pressures that produce rock fractures and joints also force the fracture surfaces to slip against each other. If there is such movement of the rock along the sides of the fracture, it is called a *fault*. When faulting occurs, the rocks can slip against each other along the fracture in several ways. The rocks along one side of the fault can move horizontally. See Figure 10–7. This is called a *strike-slip fault*. A more common kind of movement along faults is an upward or downward movement of the rocks. This type of movement produces *vertical faults*, since the movement of the rocks is mainly vertical in direction. There are several kinds of vertical faults. They differ in the way rock blocks on each side of the fault move in relation to each other. Figures 10–8, 9, and 10 show several ways that vertical faulting can occur. Many faults show evidence of both horizontal and vertical motion.

In a few places on the earth, vertical faults have caused large blocks of rock to drop downward. This movement forms broad valleys with steep cliffs on both sides. This kind of valley is called a *rift valley* or *graben*. Graben is a German word meaning "trench." Huge rift valleys have

FIG. 10–11. East Africa and the Arabian Peninsula are shown in this photograph taken from a spacecraft. The Red Sea on the left and the Gulf of Aden on the right occupy large rift valleys that were formed when part of the African plate separated to form the Arabian Peninsula. (NASA)

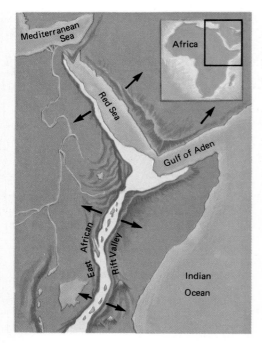

FIG. 10–12. The African continent is separating at the Red Sea as the Arabian Peninsula moves away at the rate of several centimeters per year. The Red Sea is a young ocean that will grow larger in time. Toward the south, the separation line in East Africa has formed the Great Rift Valley.

been produced where the crustal plates are spreading apart. The Red Sea occupies a large rift valley formed by separation of a part of the African plate. See Figures 10–11 and 12. Smaller rift valleys, or grabens, are also found within the continents.

Origin of forces that deform the crust. According to the theory of plate tectonics, the earth's solid outer crust is made of a series of rigid moving plates. As the plates rub together, collide, or spread apart, a *stress* is created in the rocks of the lithosphere. Stress is a force that tends to change the shape or size of the rock. Up to a certain point, deformed rocks can return to their original shapes when the stress is removed. This is like a stretched rubber band. However, stress that builds up quickly can cause the rock to fracture. These breaks commonly occur along existing or old plate boundaries. Stresses can build up more slowly over many thousands of years. When this occurs, the rocks are more likely to be folded without breaking.

Not all changes of the rocks in the crust are caused by movement of the crustal plates. Stress can also be created because the solid lithosphere floats on the more mobile asthenosphere below. Just as a ship floats high or low on water, depending on its cargo, the crust can move up or down if its weight changes. The principle that states that the solid crust floats in a state of balance is called *isostasy* (eye-*sos*-tuh-see). The word isostasy derives from Greek and means "equal standing." Isostasy means that the heavier parts of the crust sink deeper toward the mantle. Mountains, for example, have deep roots that make them heavy and cause them to sink. They could be compared to a heavy piece of wood that floats deeper in water than a lighter piece. As the top of a mountain is worn off, the rest of the mountain will continue to rise until its roots are level with the crust. The material that is worn off the mountain will be deposited as sediment in a nearby body of water. The added weight of the sediment will cause that region to sink. See Figure 10–13. When the crust rises and falls due to the effects of isostasy, the rocks are exposed to stress that may result in folding and faulting.

Isostasy also explains what happens when very thick deposits accumulate on the floors of large bodies of water. Sediments from the higher parts of the land wash into the ocean basins near the shore. The weight of this sediment, pressing down on the floor of the basins, forces them to sink and more sediments can be deposited. As a result of the sinking, the basin remains at a lower level than the surrounding land. Accumulation of a great thickness of sediment formed this way produces an area called a *geosyncline*.

Studies of the effects glaciers have on the crust also seem to show the principle of isostasy in operation. During the earth's past, glaciers covered much of the continent's surface. The weight of the ice pressed down on the land. Now that the glaciers have almost disappeared, the land that was covered with ice is slowly rising. See the map in Figure 10–14.

Gravity also creates stress in rocks. Large sheets of sediment can slide downhill, folding and breaking as they move. This happens when a large amount of magma in a batholith pushes up from below. Pushing up the crust causes thick layers of sediment to slide down the gentle slopes that are created. Parts of the Appalachian Mountains may have been folded in this way.

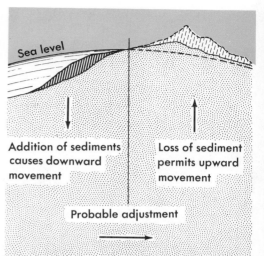

FIG. 10–13. Diagram illustrating the theory of isostasy.

FIG. 10–14. The darker areas on the map are rising, while the lighter areas are sinking. The numbers indicate the rising (+) or sinking (−) in millimeters per year. The lifting is probably the result of weight being removed after the glaciers melted. Sinking areas can be caused by the formation of geosynclines.

EARTHQUAKES

Earthquakes. Most earthquakes occur when there is movement of the masses of rock on the sides of a fault. The walls of a fault are usually pressed very closely together. Because they are close, it means that the rocks along a fault do not usually slide smoothly against each other. Instead, pressure slowly builds up until the rocks bend and then finally break. Due to an abrupt release of this pressure, there is a sudden slipping of great blocks

Pre-existing fault plane

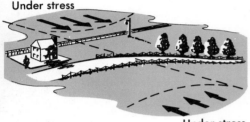

Under stress

Under stress

After earthquake

FIG. 10–15. According to the elastic rebound theory of earthquakes, stress builds up slowly and is then suddenly released.

of rock along the fault zone. The result is the trembling and vibration of solid rock called an *earthquake*.

This explanation of the way that earthquakes are produced is usually called the *elastic rebound theory*. It is commonly accepted as the explanation for the occurrence of most earthquakes. Measurements made along faults over many years show that very slow movement of rock on both sides of large faults does take place. This movement builds up strain in the rocks that causes them to bend and stretch like rubber bands. When the strain becomes too great, the rocks break and spring back (rebound). As the rocks along the fault snap into new positions, they also release energy in the form of earthquake waves. The actual point along a fault where the slippage occurs and causes the earthquake is called the *focus*. The way that elastic rebound actually works along a fault is shown in Figure 10–15. Small earthquakes can also be created by volcanic eruptions, or by land slides.

Some earthquakes are caused by faults very deep beneath the earth's surface. These earthquakes can probably not be explained by the elastic rebound theory. The temperatures and pressures at those depths are much different from the conditions near the surface. However, it is the sudden release of energy stored in rocks under great strain that is responsible for all large earthquakes.

Detecting earthquakes. When an earthquake occurs, vibrations in the form of waves move out in all directions through the surrounding rock. The effect of these waves is felt most strongly at the surface, where there are loose or water-soaked sediments instead of solid rock. Most earthquake damage to buildings is likely to occur where the structures are not built on solid rock. Observe what happens to a bowl of gelatin when the bowl is struck quickly. The bowl shakes very little but the gelatin vibrates in a series of waves.

Since most earthquakes occur along fault lines, the focus is usually a line rather than a point. The point, or line, on the earth's surface directly above the focus is called the *epicenter*. During an earthquake shock, the waves travel out from the focus in widening circles. Their movement is similar to the ripples made when a rock is thrown into quiet water.

Earthquake waves can be detected by an instrument called a *seismograph* (*syz*-muh-graf). The principle behind the working of a seismograph is illustrated in Figure

10–16. Analysis of seismograph records shows that there are three types of earthquake waves. The seismograph record shown in Figure 10–17A, indicates the three kinds of waves. The straight line beginning at the left is the record shown in Figure 10–17A indicates the three kinds line suddenly takes on a wavy appearance at the letter P. At the letter S the wave also changes and continues to letter L, where it changes again. The first disturbance that appears on the seismograph is the *primary wave,* the second is the *secondary wave,* the third is the *surface-long wave.*

The primary and secondary waves travel from the focus. See Figure 10–17B. They are often called *body waves* because they travel through the body of the earth. Primary waves are the result of a back-and-forth vibration of rock. Secondary waves are caused by an up-and-down (or side-to-side) motion of the rock. Notice in Figure 10–17B that secondary waves are not detected on the side of the earth opposite an earthquake. Surface waves are created when either type of body wave reaches the surface.

Another important difference between the types of earthquake waves is the speed at which they move. Primary waves are the fastest, secondary waves are slower, and surface waves are the slowest of the three. Although primary and secondary waves travel at different speeds,

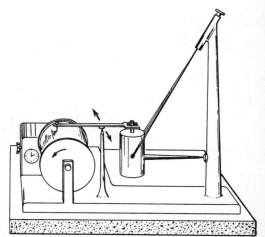

FIG. 10–16. A simple seismograph like the one shown in the diagram can be constructed. When the earth vibrates, the suspended weight remains stationary, allowing the attached pen to make recordings on the revolving drum. This kind of seismograph is best able to record side-to-side waves.

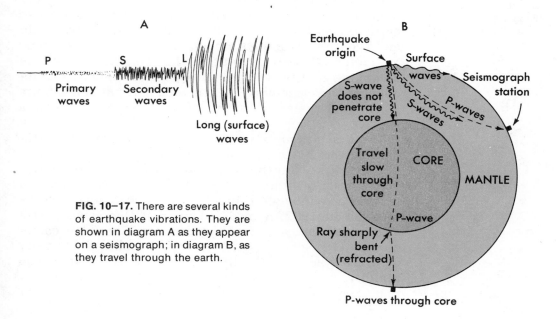

FIG. 10–17. There are several kinds of earthquake vibrations. They are shown in diagram A as they appear on a seismograph; in diagram B, as they travel through the earth.

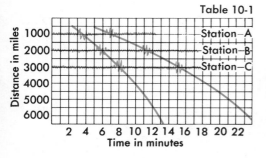

Table 10-1

Table 10-2

Distance miles	Time of Travel					
	P wave		S wave		S-P interval	
	M	S	M	S	M	S
1000	3	22	6	03	2	41
2000	5	56	10	48	4	52
3000	8	01	14	28	6	27
4000	9	50	17	50	8	00
5000	11	26	20	51	9	25
6000	12	43	23	27	10	44
7000	13	50	25	39	11	49

FIG. 10–18. The epicenter of an earthquake can be located by means of seismograph recordings taken by at least three different stations.

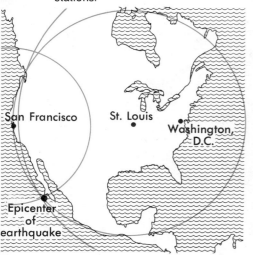

the differences in speed between them always remains the same. Primary waves travel about 1.7 times faster than secondary waves. Knowing this makes it possible to use a seismograph record to find the distance to an earthquake focus and its epicenter. Secondary waves will fall farther behind as the distance from the focus increases. Think of it as similar to a race between a fast horse and a slow one. The slow horse will fall farther and farther behind as the distance away from the starting gate increases. The time lag between the arrival time of the two kinds of waves can be plotted on a graph similar to the one in Table 10–1. Data from the graph can then be translated into a distance in Table 10–2.

For example, if secondary waves arrive 2 minutes and 41 seconds after the primary waves at seismograph station A, the earthquake focus is 1610 km away. If several kinds of seismograph records are made at one station, the direction from which the waves came can also be determined. It is possible then to locate the general position of the epicenter of the earthquake. However, a more accurate way of locating the epicenter makes use of the records from several stations. In this method, circles are drawn for at least three stations. Each station is used as a center, and the distance from it to the focus is the radius of a circle. The epicenter is very near the point where the three circles meet. See Figure 10–18.

The thousands of earthquakes that occur every year have many different intensities. On the average, only two earthquakes each year actually cause destruction to life and property. A limited amount of damage is caused by an additional twelve. About one hundred more would be damaging if they occurred in populated regions. The intensity of an earthquake is measured in terms of its ability to cause damage to structures and change the earth's surface.

The first attempts to measure the strength of earthquakes were based on descriptions of what happened during the quake. But the people who experienced the earthquake described their observations differently. In order to provide a reliable description, a scale was developed by Mercalli in 1902. A modified form of this scale is still often used today to describe the intensity of an earthquake. The scale starts with the number I intensity described as so slight that it can barely be felt. The numbers go as high as XII, which indicates total destruction. However, many things influence the amount of damage done by an earthquake. The Mercalli scale is not useful in comparing the intensity of earthquakes. A scale is needed to measure the *magnitude* or total energy released by an earthquake.

The scale that measures earthquake magnitude is called the *Richter scale*. Earthquake magnitude is determined on this scale by using seismograph records. Numbers 1 to 8.6 are used to describe the magnitude. Each number indicates a greater release of energy. Each higher number indicates an energy about 30 times greater than the preceding number. Thus, an earthquake with a magnitude of 5.5 releases about 30 times more energy than one of magnitude 4.5. The largest earthquakes measured have magnitudes near 8.6. Those with magnitudes less than 2.5 are usually not felt by people in the area. See Table 10–3.

Locations of earthquakes. We know that the depth of the focus of an earthquake can be measured. Most earthquakes occur at shallow depths within 55 km of the surface. However, some have a focus with intermediate depth that is between 55 to 240 km. A few deep-focus earthquakes come from a depth of 300 to 650 km. At a depth greater than this, the rock is probably too soft to accumulate the strain needed to produce an earthquake.

Locating the epicenters of shallow, intermediate, and deep-focus earthquakes on a map can begin to help show the cause of many severe earthquakes. Often, the epicenters of the three types of earthquakes are found to be

Table 10–3

Richter Magnitudes	Earthquake effects
0	Smallest detectable quake.
2.5–3	Generally not felt but recorded. About 100,000 such earthquakes of this magnitude occur each year.
4.5	Can cause local damage.
6.0	Destructive in a populated region.
7.0	Called a major earthquake.
7.8	San Francisco earthquake of 1906.
8.0 or greater	Great earthquakes. Cause total destruction to close population centers.

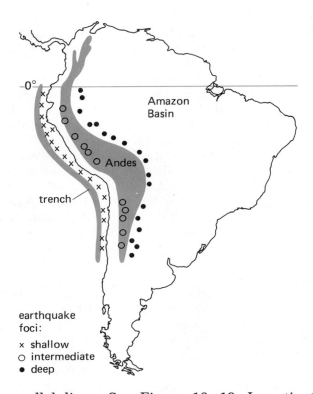

FIG. 10–19. Epicenters of earthquakes of various depths along the western edge of South America.

nearly parallel lines. See Figure 10–19. Investigation shows that these earthquakes mark the location where one crustal plate is diving beneath another. The source of the earthquake follows the edge of the descending plate. See Figure 10–20. It appears that earthquakes are one of the phenomena resulting from movement of boundaries of the crustal plates.

FIG. 10–20. A sloping fault zone along the west coast of South America accounts for the various depths of earthquakes.

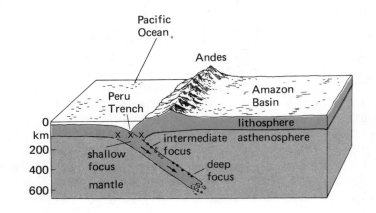

FIG. 10–21. Aerial photo of San Andreas fault near San Francisco. (U.S. Geological Survey)

An example of earthquake activity that occurs along plate boundaries is found on the San Andreas fault in California. See Figure 10–21. Movement along this fault was the cause of the San Francisco earthquake of 1906. Many other less severe earthquakes have occurred in the area since then. The San Andreas fault is the boundary where the north moving Pacific plate brushes against the west moving North American plate.

Earthquakes can occur anywhere, although they are rare in regions not near plate boundaries. One of the strongest earthquakes known to have occurred in North America happened near New Madrid, Missouri in 1812. A major earthquake also occurred in Charleston, South Carolina in 1886. These earthquakes are thought to be the result of isostatic adjustments to the extra weight of accumulated sediments in the areas. Occasional earthquakes occur in New England. These earthquakes may be the result of adjustments of the crust to weight lost after the ice of the last glacial age melted.

Predicting and controlling earthquakes. In the future, earthquake warnings may be given just as serious storms are forecast now. Scientists are discovering how to detect changes in the crust that can signal an approaching earthquake. Most faults that are near centers of population have been located and mapped. Instruments are placed along these faults to measure small changes in the zone around the fault. Built-up strain can be detected. The slight tilting of the ground that precedes many earthquakes can also be observed. Small changes in the earth's magnetic field can also be useful information for predicting a coming earthquake. A severe earthquake is often preceded by a decrease in the speed of P-waves generated by milder disturbances. Just before the severe earthquake, the P-wave speed suddenly increases. Studying the waves produced by mild earthquakes can then give advance warning of the more severe quakes.

Scientists are also concerned about the problem of controlling earthquakes. It is known that earthquakes have been artificially set off when water was forced into deep wells. The weight of water stored in reservoirs and the explosion of nuclear bombs underground have also set off earthquakes. When the processes involved in the formation of earthquakes are better understood, it may be possible to develop a system for earthquake control. This could be done by releasing the strains along a fault and producing a slow, steady motion. Rapid movements that cause the elastic rebound characteristic of earthquakes could thus be avoided.

MOUNTAIN BUILDING

Kinds of mountains. Mountains are more than simple elevated parts of the earth's crust. They are complicated features whose rock structures give evidence of their formation. Mountains are grouped according to the most characteristic process that created them. There are four main groups. For example, the largest mountain ranges of the world are made by *folding*. The Alps, Rockies, Himalayas, Appalachians, and Urals are all basically made up of very large and complex folds. However, these giant mountain systems also show much faulting and intrusive igneous activity.

Other mountains have been formed by faults that broke parts of the crust into large blocks. These blocks were then tilted to create *fault-block* mountains. See Figure 10–22. The Sierra Nevadas of California and the Teton Moun-

Investigate
Would alluvial fans be more likely to form on the sides of folded mountains or block mountains?

FIG. 10–22. The basic structure of a typical fault-block mountain.

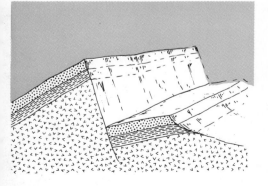

tains of Wyoming are examples of fault-block mountains. Much of Nevada, Arizona, western Utah, southern Oregon, northern New Mexico, and southeastern California are covered by fault-block mountains. The blocks form nearly parallel mountain ranges averaging 80 km in length. The Grand Canyon has been cut into one of these upturned blocks. See Figure 10–24.

Volcanic mountains are formed from piles of lava and other volcanic materials that have been forced from volcanic openings. The Cascade Mountains of Washington and Oregon are examples of mountains made up of a series of volcanic cones. Most volcanic activity occurs on the sea floor along the mid-ocean ridges. The sea covers some of the largest volcanic mountains on the earth.

An unusual kind of mountain can be produced from old igneous or metamorphic rocks that have been covered with layers of sediments. Some force, such as a batholith or laccolith rising from below, raises a circular dome like a giant blister on the earth's surface. See Figure 10–25A on page 222. As the sedimentary layers are worn off the dome, *domed* mountains are formed. See Figure 10–25B on page 222. The Black Hills of South Dakota and the

FIG. 10–23. The Sierra Nevadas are fault-block mountains. (Ramsey)

FIG. 10–24. The Grand Canyon has been cut into one of the tilted blocks that is part of the system of fault-block mountains of southern Utah and northern Arizona.

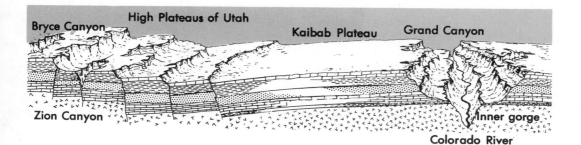

FIG. 10–25. *A)* Domed mountains are formed when bedrock with a sedimentary covering is uplifted. *B)* Later the surface is worn away, creating a more or less circular mountainous area.

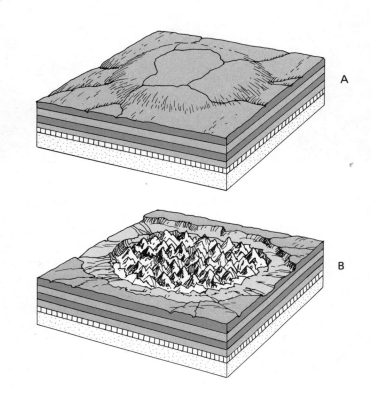

activity

In this activity, you will locate some of the world's major earthquakes and mountain ranges. You will then see how they are related to other features on the earth. To do this, you will need a map of the earth that shows the crustal plate boundaries and volcanoes.

Each of the positions listed in the chart is a place where a major earthquake has occurred. On your map, write the letter identifying each earthquake at the longitude and latitude stated.

The position stated for each of the major mountain ranges is near the center of the range. On your map, write the name of each major mountain range, as near as you can, to the proper longitude and latitude. Then answer these questions:

1. Describe the general relationship between the earthquake regions and the crustal plates.

2. Describe the general relationship between the mountain ranges listed and the crustal plates.

3. How are the positions of earthquakes, volcanoes, and mountain ranges related?

4. Are all the earthquakes listed found along the edges of crustal plates?

5. Are all the mountain ranges listed found along the edges of crustal plates?

Adirondack Mountains of New York are examples of domed mountains.

Plate tectonics and mountains. One kind of boundary between crustal plates forms when a part of the sea floor is pushed under the edge of a continent. Such a boundary is found where the western edge of South America meets the Pacific sea floor. This boundary was probably formed when South America separated from Africa, moved westward and was thrust up over the Pacific Ocean floor. The action at this kind of plate boundary is believed to be responsible for creating great mountain systems. It may also have built the continents themselves. Plate tectonics and mountain building are not completely understood. However, scientists believe that the following series of events happen repeatedly during the earth's long history. First, materials wash off the land and accumulate as sediment on the sea floor near the edge of a continent. Because of the weight of sediment, the lithosphere breaks near the edge of the continent. A plate boundary is then formed. The lighter continental plate is pushed up over the oceanic plate. The oceanic plate plunges down where it, and the sediments, become molten. Melting of the oceanic plate results in volcanic activity. This activity creates an island arc of volcanoes. Sediments that continue to accumulate between the island arc and the edge of the continent form a geosyncline. See Figure 10–26A. Continued movement of the plates pushes the old island arc toward the continental edge. The sediments of the geosyncline are crumpled and pushed upward, forming folded layers. Magma from below joins with these folded layers. The magma becomes the core of a growing mountain range that is forming along the margin of the continent. See Figure 10–26B. Continuing uplift and volcanic activity finally build a large mountain system. The Andes Mountains of South America are examples of the kind of mountain range that is still growing.

The continents might have grown to their present size by repeated episodes of mountain building. Each time a thin band of mountains grew up, they were added to where the continent's edge met the sea floor. If this theory is correct, the continents were once made up of mountains that have since worn away. Evidence of these ancient

EARTHQUAKE	LONGITUDE	LATITUDE
A–China	110E	35N
B–India	88E	22N
C–Pakistan	65E	25N
D–Syria	36E	34N
E–Italy	16E	38N
F–Portugal	9W	38N
G–Chile	72W	33S
H–Chile	75W	50S
I–Equador	78W	0
J–Nicaragua	85W	13N
K–Guatemala	91W	15N
L–California	118W	34N
M–California	122W	37N
N–Alaska	150W	61N
O–Japan	139E	36N
P–Japan	143E	43N

MOUNTAIN RANGE	LONGITUDE	LATITUDE
Himalayas	75E	30N
Alps	10E	45N
Atlas	0	30N
Appalachian	80W	40N
Andes	70W	30S
Coast	120W	40N

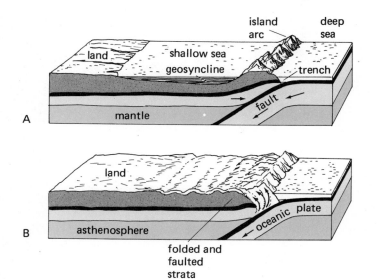

FIG. 10–26. These diagrams show how mountains can be formed along the edges of continents and increase the size of the land area.

mountains is found in rock structures of the land that remains.

Some mountains also formed when two continents collided head-on. When this happened, one continent was not pushed beneath the other because both were made up of light rock. Rather, the colliding plates were crumpled and deformed to create a mountain range. An example of a mountain system formed this way is the Himalayas. These mountains were created about 45 million years ago, when India collided with Asia.

The Appalachian Mountains also exhibit evidence of continental collision. More than 200 million years ago, North America collided with North Africa. The two continents have since separated. But evidence of that collision still remains within the Appalachians. Continental collision has also played a part in the formation of the Alps in Europe.

VOCABULARY REVIEW

Match the word or words in the column on the right with the correct phrase in the column on the left. *Do not write in this book.*

1. Scientific study of the earth.
2. Theory that the present is the key to the past.
3. Rocks respond to forces by moving into new positions without breaking.
4. Raised section of a fold.
5. Movement of rock along the side of a fracture.
6. Present in the center of a mid-ocean ridge.
7. Principle that the solid crust floats in a state of balance.
8. Accumulation of great thickness of sediments in the ocean near shore.
9. Results from abrupt release of pressure along a fault zone.
10. Actual point along a fault where slippage occurs.
11. Point on the earth's surface directly above the focus.
12. Instrument used to detect earthquake waves.
13. Causes the first disturbance that appears on the seismograph.
14. Scale used to measure earthquake magnitude.
15. Results from a force such as a laccolith rising from below.

a. folding
b. isostasy
c. anticline
d. rift valley
e. geology
f. earthquake
g. uniformitarianism
h. focus
i. fault
j. seismograph
k. secondary wave
l. geosyncline
m. stress
n. epicenter
o. dome mountain
p. strike
q. Richter scale
r. primary wave

QUESTIONS

Group A

Select the best term to complete the following statements. *Do not write in this book.*

1. Evidence that the earth's crust has been lifted in places is found in (a) remains of sea animals high on mountains (b) heavy rocks in the deep ocean (c) remains of sea animals deep in the ocean (d) heavy rocks on mountain tops.
2. The theory that "the present is the key to the past" was first stated by (a) Albert Einstein (b) Johanne Kepler (c) Nicolaus Copernicus (d) James Hutton.
3. Forces can change the features of the earth because (a) the forces are abrupt (b) the forces are small (c) the forces act over a long period of time (d) rocks of the crust were molten at one time.
4. Which of the following is not closely related to the others? (a) syncline (b) anticline (c) geosyncline (d) folding.
5. Sedimentary layers that are tilted show movements of the rock, because sedimentary layers are originally deposited in (a) a horizontal position (b) a vertical position (c) a slanted position (d) dry weather only.
6. The tilt of a rock layer from the horizontal is called its (a) axis (b) strike (c) slant (d) dip.

7. Movement of rock along the sides of a fracture is called a (a) joint (b) strike (c) slip (d) fault.

8. Stress in rocks might be due to (a) motion of crustal plates (b) joints (c) a dip (d) a strike.

9. The center of a mid-ocean ridge is a (a) high mountain (b) fold (c) rift valley (d) riverbed.

10. The principle that says the heavier parts of the crust sink deeper toward the mantle is called (a) float (b) isostasy (c) uplift (d) thrust.

11. Sediments washed into the ocean near shore can result in the accumulation of a great thickness that produces an area called (a) a geosyncline (b) an anticline (c) a syncline (d) a plate.

12. The actual point along a fault where slippage occurs and causes the earthquake is called that earthquake's (a) circumcenter (b) epicenter (c) orthocenter (d) focus.

13. The point on the earth's surface directly above the source or center of an earthquake is called the (a) circumcenter (b) epicenter (c) orthocenter (d) focus.

14. Body waves are made up of (a) primary waves only (b) secondary waves only (c) primary and secondary waves (d) long waves.

15. The earthquake waves that travel fastest through the earth are (a) primary waves (b) secondary waves (c) surface waves (d) long waves.

16. Using the Richter scale of earthquake magnitudes, each higher number represents how many more times as much energy than the preceding number? (a) 5 (b) 10 (c) 15 (d) 30.

17. Most earthquakes seem to be caused by movements (a) in the earth's core (b) deep in the earth's mantle (c) at shallow depths in the earth's crust (d) on the earth's surface.

18. Most earthquakes occur (a) near the edges of crustal plates (b) in the centers of crustal plates (c) in the middle of an ocean (d) at the earth's poles.

19. In addition to predicting earthquakes, it is hoped that we can control them by (a) increasing their magnitudes and having fewer of them (b) decreasing their magnitudes and having fewer of them (c) decreasing their magnitudes and having more of them (d) increasing their magnitudes and having more of them.

20. Mountains formed when the crust breaks into large blocks that are then tilted are called (a) folded mountains (b) fault-block mountains (c) volcanic mountains (d) dome mountains.

21. The greatest volcanic mountains of the earth are (a) in southern Europe (b) in North America (c) in South America (d) in the ocean.

22. When part of the sea floor is pushed under the edge of a continent (a) the continent rises (b) the ocean rises (c) the continent sinks (d) a desert forms.

Group B

1. Describe the principle of uniformitarianism.

2. What is the difference between a syncline and an anticline?

3. Describe the formation of a rift valley. What is another name for a rift valley?

4. In Figure 10-12 on page 000, what evidence shows that these are faults and not joints in the rocks?

5. How does the theory of plate tectonics account for stress in the earth's crust?

6. Give a brief description of the principle of isostasy.

7. What is the elastic rebound theory for earthquakes?

8. What is believed to be the principle cause of most earthquakes?

9. How are scientists able to tell how far away an earthquake has occurred?

10. How are the focus and the epicenter of an earthquake related?

11. How do scientists plan to control earthquakes in the future?

12. What relation exists between the location of most earthquakes and most volcanoes?

13. What do the origins of the Alps, Rocky, Himalaya, Appalachian, and Ural mountains have in common?

14. How are fault-block mountains formed?

15. Briefly explain how mountain building is related to crustal plates.

our dynamic planet

Hot Spots

Narrow columns of hot material called "plumes" flow upward from deep in the mantle and create hot spots. Volcanic activity occurs where the plume pushes upward through the crust. While most of the earth's volcanoes are located along the edges of the crustal plates, there are about 100 places around the globe with volcanic activity that are not plate boundaries. These are hot spots.

Two hot spots are known to be currently active in the continental United States. One is located in the area of Yellowstone National Park. At the left is a picture of Yellowstone Park's Old Faithful, a geyser that erupts every 66 minutes. It is evidence of the molten plumes forcing their way up to the crust and heating the water trapped below the surface. Here steam builds up and erupts at regular intervals, relieving the pressure. Another hot spot is in the northeastern corner of New Mexico.

A much larger and more active hot spot on the floor of the Pacific Ocean has formed the Hawaiian Islands. Pictured here is a volcanic eruption on the Hawaii island. Bursting forth from the hot interior, molten lava pours down the side of the mountain, destroying everything in its path.

The cause of plumes within the mantle is not known. They must be part of the activity within the earth that is responsible for the motion of the plates that make up the lithosphere. Further study of hot spots may lead to a better understanding of the forces that have repeatedly broken apart the continents and created ocean basins during the earth's long history.

Can Animals Predict Earthquakes?

For many years observers all over the world have reported unusual behavior by animals before severe earthquakes. Some of the strange behavior has taken place weeks before the earthquake occurred. In other cases, the animals became disturbed only hours or minutes before. There have been reports of horses trembling, neighing, and running wildly, of dogs howling, of cattle bellowing and refusing to graze, of rats leaving their holes, and even of fish leaping out of the water. All reports of strange animal behavior before an earthquake agree on restless and noisy responses. In China, the observations of such animal behavior have been used as part of a system to make a prediction of major earthquakes.

How animals are able to anticipate earthquakes is not known. It may be that they are able to sense slight changes in earth conditions that are known to precede large earthquakes. Among these slight changes are shifts in the earth's magnetic field and tilting of the land as the crust is lifted slightly. Animals may also be able to detect the very small shocks that occur before many large earthquakes. Additional research on the way animals seem to be able to predict damaging earthquakes may lead to development of instruments that can duplicate their apparent abilities. Such instruments would be a valuable part of an earthquake warning system.

People Who Study the Earth in Change
A surveyor pictured at left measures the topography of the land. From these measurements come various maps that give us a detailed picture of the earth. Because of changes constantly occurring on our dynamic planet, a surveyor's work is never done.

A geologist examines cores from the crust of the earth. Each layer tells a story of hundreds of years in earth's history. The earth builds up and wears away in ever-changing cycles. Through the cores, the geologist interprets this story.

unit
3
forces that sculpture the earth

11
weathering and erosion

objectives

☐ Name and describe two ways that weathering occurs.

☐ Explain how the nature of rock, climate, and topographic conditions affect the rate of weathering.

☐ List and describe the layers of a mature soil.

☐ Explain how soil is produced.

☐ Describe the roles gravity and wind play in erosion.

☐ Describe several methods currently being used to conserve soil.

☐ Describe the life cycle of mountains, plains, and plateaus.

Rocks appear to us to be solid and permanent. But they are almost always undergoing change. When rocks are exposed to air and water at the earth's surface, they meet different conditions from those under which they are formed. Gases of the atmosphere, moisture, and temperature changes bring about changes in the appearance and composition of the rock. Most rocks will slowly crumble into small pieces as they react to conditions at the surface. All changes that destroy the original rock structure are called *weathering*. When rocks are exposed to weathering processes, the formation of soil is one important result. *Soil* is a mixture of loose rock fragments and materials produced by living things growing in the soil.

Weathering of large masses of rocks allows movement of resulting smaller fragments. Before the breaking up of rocks by the processes of weathering, little movement can take place. Movement of rock materials over the earth's surface produced by various natural agents is called *erosion*. The smaller fragments produced by weathering can fall down a slope as a direct result of gravity. They can also be easily carried off by erosional agents, such as wind and running water. Once weathering has taken place, the agents of erosion work to move the products. It is the combined action of weathering and erosion that accounts for much of the present shape of the land surface.

WEATHERING

Types of weathering. The effects of weathering may be seen wherever rock structures are exposed. Small loose

232

chips are easily pulled off, and some of the fragments may even crumble at a touch. When the rock is broken open, the inner parts are found to be firm and often different in color from their surface appearance. Air and water have been able to penetrate into cracks and openings near the surface. Here, too, the exposed rock is often stained and softened. See Figure 11–1.

The weathering process that produces these changes takes place in two ways. The rock may be broken up without any change in its mineral composition. This is called *mechanical weathering*. At the same time, processes are at work which change the individual mineral crystals. This is called *chemical weathering*.

An example of mechanical weathering is the effect of freezing water on exposed rock surfaces. When water freezes, it expands by about ten percent of its volume. If water seeping into cracks in rocks freezes, pressure is created by its expansion. This can force pieces of the rock to break off from its original structure. The breaking off of great blocks of rock when water freezes and expands is called *frost action*. See Figure 11–2. However, the greatest effect of frost action is the constant chipping away of small grains of rock. Frost action is a common cause of rock weathering at high altitudes and other places where temperatures often fall below freezing.

FIG. 11–2. Frost action is breaking the rock on this mountainside into smaller pieces. (Ramsey)

FIG. 11–1. The solid granite rock on this hillside is being slowly reduced to fragments by the processes of weathering. (Ramsey)

FIG. 11-3. Mud cracks form when wet soil containing large amounts of clay or silt particles is dried. (Ramsey)

A similar process takes place in drier climates. Over long periods during which water evaporates from rock surfaces, salt crystals are likely to grow in small openings. These salt crystals originate from minerals dissolved by the water. Their growth as the water evaporates can produce enough pressure to break off fragments of the rock.

Rocks which contain clay particles can absorb water at their surface. This absorption of water can create a swelling in the rocks. Later, drying and shrinking will then produce cracks. This process is also responsible for the familiar pattern of cracks found in dried mud rich in clay particles. See Figure 11-3.

Working together with mechanical weathering processes are the chemical changes that cause rock to crumble. In chemical weathering, reactions take place between minerals in the rock, carbon dioxide, oxygen, and water. Dry carbon dioxide (CO_2) gas, which is always present in the atmosphere, has no effect whatever on rock. But many minerals are affected when carbon dioxide from the air dissolves in water. Carbon dioxide gas dissolves in water to produce a weak solution of carbonic acid. Carbonic acid readily reacts with certain minerals, such as the feldspars and calcite, which are commonly found in rocks. This action of carbonic acid is an example of chemical weathering. The rocks containing minerals which react with carbonic acid, weather very rapidly in moist climates. This process is called *carbonation*.

Like carbon dioxide, oxygen gas in the atmosphere can also affect certain rocks when dissolved in water. This kind of weathering is called *oxidation*. Iron-bearing minerals will quickly combine with dissolved oxygen to form reddish-brown iron oxide. The red color of certain rocks and soils is almost always due to chemical weathering, which accounts for the presence of iron oxide.

Besides aiding the action of carbon dioxide and oxygen, water also changes some minerals in other ways. Water can dissolve or combine directly with some minerals. This is known as the process of *hydration*. Most of the minerals affected by hydration dissolve in water. In this way, they are removed from the rocks. Often water seeping down from the surface dissolves minerals as it passes through the upper rock layers. These dissolved minerals are then carried down to greater depths and finally deposited. This is called *leaching*. In leaching, minerals are transported from the upper layers of soil and rock to the lower parts. Valuable mineral ore deposits may be created when leaching concentrates minerals in a small region beneath the surface.

In addition to mechanical and chemical weathering, plants and animals also play a part in the decomposition of rocks. Roots of trees and shrubs can work their way into the cracks of rock and wedge it apart. See Figure 11–4. Burrowing animals, such as gophers and prairie dogs, are also a factor in weathering. Their digging activities constantly expose new rock surfaces to the process of weathering. Even such a small creature as the earthworm plays a role in weathering. Earthworms bring fine rock particles to the surface where they are exposed to the weathering action of the atmosphere. Their burrows also allow space in the ground for water and air to penetrate the upper layers more easily.

The results of weathering. The various processes of mechanical and chemical weathering that have been described are relatively slow. But they have been at work for millions of years. The combined effects of these processes have broken up and decomposed almost all the rocks on or near the surface of the crust. For this reason, the land surfaces of the earth have accumulated a layer of rock fragments of various types. The solid unchanged

activity

Punch about 10 small holes in the bottom of a paper cup. Cover these holes, on the inside, with a piece of facial tissue. Half fill the cup with sand. Add one teaspoonful of salt. Now place the cup inside a cup that has no holes. Support the sides in several places with pieces of paper, so that the inner cup is about one centimeter off the bottom of the outer cup. Add 5 teaspoons of water, one at a time. Answer these questions:

1. Describe the appearance of the sand and salt before the water was added.

2. Describe the appearance of the sand and salt after the water drained through it.

3. Where has the salt gone?

4. What simple test can you use to check your answer to 3?

FIG. 11–4. Trees growing in cracks can force rocks apart. (Ramsey)

FIG. 11–5. Weathering takes place most rapidly along existing cracks in the rock. This has resulted in the deep grooves following vertical cracks in this resistant sandstone. (Ramsey)

FIG. 11–6. This California mountain is covered with residual boulders. (Doug Wilson-Black Star)

rock beneath the layers of looser material is often called *bedrock.*

The appearance of weathered rock fragments depends upon the nature of the original rock and the way in which weathering has taken place. Coarse-grained igneous and some metamorphic rocks usually separate into their individual mineral grains. The result is a coarse sand or gravel.

Usually joints or cracks are already present in bedrock. Weathering can then take place fairly rapidly. As a result, rocks often tend to separate into a series of blocks as weathering takes place along parallel sets of joints. See Figure 11–5. An unbroken pattern of weathering along a series of joints can eventually leave separate blocks with rounded off edges. Smaller fragments are usually carried away, leaving large *residual boulders* in their place, similar to those shown in Figure 11–6.

Some rocks tend to break apart in shells or plates. This is called *exfoliation,* and gives the weathered rock surface a rounded appearance. See Figure 11–7. Exfoliation seems to be the result of a combination of both mechanical and chemical weathering. It may also be the result of pressure released when rocks buried under a great weight are brought to the surface.

The various shapes and forms created by weathering in the bedrock mean that new rock is being exposed. As the surface rock is being reduced to small fragments, erosion continues removing the old material. Fresh bedrock then becomes exposed to the forces of weathering. The speed with which rock is weathered away depends upon three main factors:

1. *The nature of the rock.* The rate at which rocks weather varies according to their composition. Igneous and metamorphic rocks in general, react slowly to weathering by mechanical processes. They are affected largely by chemical processes. Although their decomposition is slow, they eventually crumble as the minerals change and allow the grains to separate. Quartz is the least affected mineral commonly present in igneous and metamorphic rocks. It remains as individual grains; quartz makes up most of the mineral matter in ordinary sand.

Among the sedimentary rocks, limestone and some rocks containing calcite are the most rapidly weathered. They are affected by carbonation and decay rapidly in a moist climate. See Figure 11–8. Many other sedimentary rocks are attacked mainly by mechanical weathering processes. The rate of weathering of most sedimentary rocks depends upon the material that holds the fragments of the sediment together. Shales and sandstones that are not firmly cemented together gradually return to their original condition of separate particles of clay and sand. On the other hand, conglomerates and sandstones that are strongly cemented by silicates, last even longer than most igneous rocks.

2. *Climate.* In dry climates, weathering takes place slowly. The lack of water slows the rate of the chemical weathering processes. The same is true of cold climates where extremely low temperatures reduce chemical weathering. In humid, warm climates, weathering is

FIG. 11–8. These old tombstones on Cape Cod show how various kinds of rock weather at different rates. The marble slab on the right is dated 1854 while the slate marker on the left is from 1835. Which shows the greater evidence of weathering? (Ramsey)

FIG. 11–7. Exfoliation results in the peeling and flaking of rock surfaces, as shown in these granite outcrops. The picture was taken at the summit of the Sierra Nevadas in California. (U.S. Geological Survey)

FIG. 11–9. These two photographs show how the weathering processes in dry and humid regions produce different landscapes. (Ramsey)

Discuss

A farmer would be very concerned about the type of soil on his land. Explain why a soil having either a high sand content or high clay content would be very poor farming soil.

fairly rapid. In a changeable climate, weathering is generally more rapid than in a more constant one. Temperature changes affect mechanical weathering. The difference in landforms between parts of the earth is mainly due to the effects of weathering in different climates. See Figure 11–9.

3. *Topographic conditions.* Altitude, the slope of the land, and exposure to sun and rain all influence weathering. With greater altitude, temperatures decrease and rainfall increases. Steep slopes allow quick removal of weathered rock. New surfaces are thus exposed more quickly to the action of weathering.

Soil. Once a layer of loose rock fragments has been created by weathering, the upper exposed rock material continues to decompose. The lower parts are partly protected and therefore do not weather as rapidly. Eventually, a layer of very fine particles is created at the surface. This layer of small rock fragments forms the basis for the production of soil.

When soil is first formed, it consists of grains of mineral matter. The minerals present in the original soil are those of the rocks which have been weathered. A variety of feldspar minerals are the most plentiful of the rock forming minerals. When the feldspars are subjected to weathering, fine grains of a silicate mineral containing aluminum and water are formed. This material makes up *clay*. Other minerals containing aluminum may also form clay when they weather.

Granite and other rocks containing a large percentage of quartz can form *sandy soils*. Such soils contain large amounts of the relatively unaffected quartz grains. These are left after the other minerals in the rocks have decomposed. In addition to clay and sand, soils may also contain a number of other rock particles. Their sizes range from the microscopic clay particles to the larger sand grains. These particles are classed as *silt*.

The weathered mineral and rock grains that form soil very often do not remain in the same location as the bedrock of their origin. Erosion transports them to other locations. Thus soil may have a composition different from the rock on which it rests.

The weathering of rock is only the first step in the formation of soil. The second step begins when plants grow in the young soil and start to add organic matter. Plants help the growth of animals. As the plants and animals die, their remains form *humus,* a dark organic material which is a part of fully developed soils.

Several well formed layers make up the *soil profile* of a mature soil. A typical soil profile is shown in Figure 11-10. The layers themselves are known as *horizons.* Soils generally consist of three principal horizons labeled by the letters A, B, and C. See Figure 11-11.

The top, or A horizon, of a mature soil is the topsoil. The longest period of the soil's life occurs in this layer. During this time it nourishes the growth of plants, bacteria, fungi, and small burrowing animals. It is also the zone from which surface water *leaches* the most soluble minerals. Leaching is the process by which minerals are removed from the soil. Immediately below the A horizon is the subsoil or B horizon. In most soils the B horizon contains the dissolved materials leached from the surface layer. In dry climates, the B horizon may also contain minerals brought up from below by the rapid evaporation of water on the soil surface. The lowermost or C horizon consists of partly weathered bedrock. This bottom layer is in the first stages of mechanical and chemical change. These changes eventually produce the B and A horizons.

The type of soil found at a particular place is determined mostly by the climate of the region. Where it is humid, large amounts of water pass down through the upper soil horizons. Much of the mineral matter in the A horizon is leached and then concentrated in the B horizon. The presence of these substances gives the B horizon a dense and hard composition. Where rainfall is abundant, the B horizon is often very thick.

In drier climates the soil horizons are frequently distinct. This is due to the lack of water passing downward through the top layers to leach minerals in the lower soil layers. The soils of drier regions are often very fertile because there has been little opportunity for leaching to remove needed minerals.

A mature soil reacts to the particular climate of its location. Therefore, there are hundreds of different sets of conditions that influenced its weathering and the slow organic processes that brought it to maturity.

EROSION

Gravity and erosion. When rock fragments are separated from a mass of solid rock by weathering, they fall to the

FIG. 11-10. A photograph of a mature soil showing the separate layers which have been formed. (Ramsey)

FIG. 11-11. A diagram of a typical mature soil with each horizon labeled.

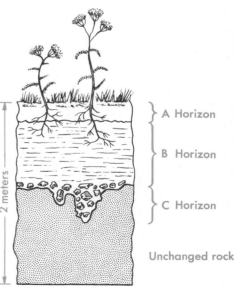

A Horizon

B Horizon

C Horizon

Unchanged rock

2 meters

FIG. 11-12. Rock material falling to the base of this cliff has formed a talus slope. (Ramsey)

FIG. 11-13. The soil on this hillside is moving down as shown by its appearance on the lower slope. Near the top, movement is in the form of soil creep which is not so obvious. (Ramsey)

lowest part of the surrounding surface. If the weathered rock lies on a steep slope or on the face of a cliff, the rock fragments tumble down until they reach the base of the slope. A pile of rock fragments of various sizes begins to accumulate at the base of most cliffs and steeper slopes. Such accumulation of broken rock is called *talus*. See Figure 11-12. On gentle slopes the rock fragments move down so slowly that the talus weathers sufficiently to form soil. In this case, the talus slope becomes covered with vegetation.

The formation of talus is only one of the more obvious results of gravity-produced erosion. Less obvious is the slow downhill movement of loose, weathered rock material pulled along by its own weight. This process is called *soil creep*. Usually it is not noticed unless buildings, fences or other objects on the surface are moved along with it. See Figure 11-13. Soil creep is generally most rapid when the ground is wet. Water lubricates the rock particles and allows them to move more freely. Absorption of water by the soil also increases its weight.

The movement of large amounts of loose material down slopes may be much more sudden and dramatic than soil creep. In a landslide, masses of loose rock and soil on a slope abruptly break loose and come tearing down. At the beginning a landslide may be only a small amount of loose rock breaking loose from the top of a slope. As the loose rock moves down the slope, more material breaks loose. The moving rock mass gains momentum until it ends as a great avalanche of boulders, rocks and soil. Such a landslide destroys everything in its path until it reaches ground-level and comes to a halt.

Landslides are common in mountainous regions where steep rock cliffs have been carved by glaciers. Landslides are most likely to happen without warning. It is not obvious what sets them off. Heavy rainfall or melting snow may be responsible in some cases. Earthquakes also have been known to trigger large landslides.

Usually landslides take place in stages, with a small amount of land slipping, first in one part of the slope, then in another. These minor landslides are said to be *slumping* and give many slopes a hilly, bumpy appearance. See Figure 11–14. Slumping may take place on a large scale along some cliffs which have a weak rock base. Large blocks may slump down the cliff-side as erosion removes rock from its base.

In mountainous regions that are ordinarily dry, heavy rains may produce a flowing movement of a mixture of fine rock particles and water called a *mudflow*. When a mudflow occurs, masses of mud move through valleys, frequently spreading out in a sheet at the base of the mountains.

The kinds of erosion already described are those which are caused directly by gravity acting upon individual fragments of rock. All such processes of erosion directly controlled by gravity are called *mass-wasting*. However, gravity also acts indirectly to cause erosion. The most important of all the causes of erosion is water falling on, and running over the earth's surface. Running water moves because gravity pulls it down to a lower level. Ice, mainly in the form of great glaciers, is also a powerful

FIG. 11–14. A large block of this slope has slumped, probably as a result of material removed when the highway was built. (Ramsey)

FIG. 11–15A. An area of desert pavement in the Colorado desert of California. (Ramsey)

FIG. 11–15B. The unusual rock formation was the result of the sand-blast action of wind-blown mineral grains of quartz. (Black Star-Shulthess)

agent of erosion. A more detailed study of the effects of running water and glacial ice as agents of erosion will be taken up in Chapters 12 and 13. However, a third agent of erosion, the wind, is not as closely controlled by gravity.

Wind erosion. Wind which blows with enough speed is able to pick up loose particles of rock. However, winds are usually able to move only the smallest rock fragments, such as silt, dust, and sand grains. In regions where rainfall is abundant, there are usually enough plants to protect the ground against the full force of wind. But in dry areas the plant cover is spotty and much of the surface of weathered rock is exposed. Under these conditions, wind becomes an important agent of erosion.

The most common form of wind erosion is *deflation* (from the Latin, *deflare,* meaning to blow away). This refers to the lifting and movement of silt and dust particles. Usually deflation tends to remove a layer of the finer soil particles, leaving behind fragments too large to be lifted. The remaining pebbles and gravel often form a sheet which protects the materials below from further erosion. Such a layer of closely fitted small stones is called *desert pavement.* See Figure 11–15A. Deflation may also cause the formation of shallow depressions. These are produced when bare soil is exposed to wind in only a limited area. This can happen where, for some reason, the natural plant cover has disappeared. Once the protective plants are gone, the wind strips off a layer of material to form a *blowout* or *deflation hollow.* Running water carries fine sediments into this depression. Later, wind action blows these sediments away. A deflation hollow tends to expand by this method. They may grow to several kilometers wide and 5 to 20 meters deep.

Sand grains carried by the wind increase its ability to carry on erosion. In this case, the erosion is caused by the hard sand grains, which are usually quartz, striking softer rock. Sand grains are seldom lifted more than a short distance off the ground. This means that erosion takes place only close to the ground surface. See Figure 11–15B.

Various rock shapes such as natural bridges, rock pinnacles, rocks perched on pedestals, and even large desert basins are believed to have been caused by the eroding effects of wind-driven sand. However, it is very doubtful that such large features could be produced by wind action. Erosion of large masses of rock by wind-blown particles takes place very slowly. It is effective only close to the ground where heavier sand grains can be lifted. Only in the few locations that have strong and steady winds, large amounts of loose sand and relatively soft rocks, can there be much erosion by wind-blown sand.

An exception is where pebbles and larger stones have been exposed to the wind in deserts and along beaches. Stones that develop one or more smooth polished faces are called *ventifacts* (from the Latin, *ventus*, meaning wind). See Figure 11–15C.

Deposits made by wind. All the material moved by wind eventually settles and is deposited. The ability of wind to move material diminishes as its speed decreases. Thus deposits are formed where a decrease in wind speed drops a part of the load. Such deposits are often temporary and are carried away by the next strong wind. Just as often, however, the material becomes covered over. Pressure from the overlying sediments then forces the fragments together, and perhaps substances are deposited between the grains to cement them together. The deposit of wind-blown material then becomes sedimentary rock, a relatively permanent part of the earth's crust.

The most common of all wind deposits are *dunes*, or hills of wind-blown sand. Dunes are formed where there is a supply of dry, unprotected soil and winds strong enough to move it. Large areas of dunes are common in deserts. But sand dunes are also common along the shorelines of the sea and larger lakes. A dune is started when an obstruction breaks the speed of the wind. With the reduction

FIG. 11–15C. Ventifacts when developed to perfection have three curved surfaces intersecting in three sharp edges. (U.S. Geological Survey)

FIG. 11–16. Movement of sand across the surface of a dune creates a gentle slope facing the wind. Notice the size relationship in the photo between the dune and the man. (Photo Researchers–Fran Hall)

FIG. 11–17. This deposit of loess in Linn County, Iowa, is about 11 meters thick. (U.S. Geological Survey)

in wind speed, a small deposit of material accumulates on the sheltered or lee side of the obstacle. As the small mound of material grows, it acts as a larger windbreak. More material is then deposited so the pile grows larger.

The gentlest slope of a dune is typically on the side facing the wind. It is in this location because the force of the wind tends to flatten the side of the dune against which it blows. The sand that the wind pushes over the crest tumbles down on the lee side, giving it a steeper slope. The wind sweeping around the sides often builds two long pointed extensions which give the dune a crescent shape. See Figure 11–16. These crescent-shaped dunes are called *barchans* (bahr-kans). They are most common in areas where there is a limited supply of sand. Most dunes have a complicated shape, a result of one dune being piled on top of the other.

The wind, blowing continuously, moves sand grains up the windward slope and over the crest of dunes to the leeward side. Generally if the wind blows from the same direction, it will force the dunes to move in the direction of the wind. In fairly level areas this dune migration or movement continues unless there are enough plants to hold the sand in place. Sometimes it is necessary to plant grasses, trees, or shrubs to prevent dunes from drifting over highways, railroads, farmland, or buildings. Protective fences are often used for the same purpose.

Dunes are formed of the heavier sand particles carried by the wind for fairly short distances. The finer particles of dust are carried to greater heights by the wind and travel much farther. The fine wind-blown material is probably finally deposited in such thin layers that it is not noticed. However, in certain places throughout the world, thick, unlayered deposits of a yellowish, fine-grained sediment have been formed by the accumulation of fine wind-blown dust. This material is known by its German name of *loess* (luhs). Although *loess* is soft and easily eroded, it sometimes forms steep bluffs because of its tendency to break into vertical slabs. See Figure 11–17.

A large area in northern China is covered entirely with loess. Apparently the deposited material was carried by the wind from the interior deserts of Asia. Larger deposits of loess are also found in Central Europe. In North America loess appears in the north central states along the eastern border of the Mississippi valley, and in eastern Oregon and Washington. These deposits were probably built up by dust from the dried beds of lakes and streams after the last ice age.

Soil erosion and conservation. Erosion is a natural process. It is part of the chain of natural events that constantly change the shape of the earth's solid surface. Erosion of the soil which covers bedrock is a part of this wearing down of the crust. It would occur even if people were not present on the planet. However, natural soil erosion is a slow process. It is usually kept in balance by the accumulation of new soil. If human activities did not disturb this balance, new soil would be formed about as fast as the existing soil was eroded. However, we use the land to grow crops and to pasture their animals. The natural balance between soil erosion and soil formation can be upset by the unwise use of the land.

Rapid soil erosion is encouraged when the natural plant protection is removed. This happens when land is cleared of trees and small plant cover in preparation for farming. Grazing animals can also destroy the natural grasses and low-level plants. These activities expose the upper soil layers to the full effect of the erosion processes. Plowed land becomes cut through with deep gullies. Furrows in plowed land, particularly in rows running up and down slopes, allow water to run swiftly over the bare soil. Each furrow becomes a small gulley which expands with each rain. Finally the land is filled with miniature canyons. See Figure 11–18. This type of destructive soil erosion is called *gullying*.

Less obvious but equally destructive is *sheet erosion*. This type of erosion occurs when water strips away ex-

FIG. 11–18. Gullying has ruined great areas of valuable farmland. This field is practically useless for farming because of heavy erosion. (Soil Conservation Service)

FIG. 11–19. Dust storms cause tremendous damage, as is illustrated by this Colorado farmyard after repeated storms. (Wide World Photo)

FIG. 11–20. Contour plowing helps to prevent gullying by causing water to flow along the plowed furrows. Another soil conservation method is shown in this photograph. Can you find it? (U.S.D.A. Photo)

FIG. 11–21A. Strip cropping helps to hold the topsoil in place on gentle slopes. Runoff is reduced by the strips of cover crops. (U.S.D.A. Photo)

FIG. 11–21B. Terracing builds up low ridges that slow down runoff water, resulting in greater absorption by the soil, thus causing relatively little erosion. Photo Researchers (Charbonnier/ Realities)

posed topsoil slowly and evenly. Finally only a smooth surface of subsoil or rock is left in a once fertile field. In dry periods wind replaces water as the means of rapid erosion. The loose topsoil is blown away as clouds of dust and drifting sand. See Figure 11–19.

Continuing erosion by water and wind reduces the fertility of the soil by removing the upper layers, thus exposing the lower horizons which are lacking in organic matter and are difficult to cultivate. The non-fertile subsoil prevents plants from growing which might otherwise protect it from further erosion. The cycle ends with complete removal of all the soil layers during a few years' time.

Rapid and destructive soil erosion can be controlled. Cover plants can be laid out on the bare soil as protection. But the most important methods of conservation are those which allow the soil to be cultivated and crops grown. Some of the methods used are:

1. *Contour plowing:* plowing in such a way that the furrows follow the contours of the land. This prevents direct flow of water down slopes which might cause gullying. See Figure 11–20.

2. *Strip cropping:* arrangement of crops in alternate bands, such as corn, and cover crops which help to hold the soil. See Figure 11–21A.

3. *Terracing:* construction of step-like ridges following the contours of the field. These hold or slow down the run-

off water to prevent rapid erosion on the slopes. See Figure 11–21B.

4. *Crop rotation:* alternation of row crops which expose the soil one year with cover crops which protect the soil the next year. This stops erosion in the early stages and allows small gullies to fill with soil.

Much of the soil over the land surface of the world has been damaged by erosion. To protect the soil while using it is a never-ending struggle that required knowledge and skill. Once soil has been swept away by erosion, no amount of knowledge or work can repair the damage quickly. Thus loss of soil resources through careless and unwise use is a problem for everyone. Care of the soil is a national and international problem, for all of our lives depend upon its fertility.

EROSION SHAPES THE LAND

Origin of the minor landforms. All land forms, regardless of the forces that shaped them, have one thing in common. They are always temporary results of the action of two opposite influences on the earth's crust. One of these forces is diastrophism, the movements that bend, break, and lift the crust into elevated landforms such as mountains. Associated with diastrophism in raising parts of the crust are the processes of volcanism. Opposing the building actions of diastrophism and volcanism are the powerful agents of weathering and erosion. The wearing action of these two forces tends to level off the land surface.

However, the landscape that exists at any particular place always has a more ancient chapter in its history. A mountain exists because it was built up by diastrophism or volcanic action that raised it above the surrounding land. But a mountain is always being worn down by weathering and erosion. Its present shape is partly a product of these forces, which are at work now and have been for millions of years.

The rocks which make up a mountain have a much older history than the mountain itself. They were created long before any landform of which they are now a part. The building and wearing forces which constantly carve the land surface are also controlled by the structure of the ancient rocks on which they must work. Thus all landforms have their present appearance as an outgrowth of

FIG. 11–22. These landscapes illustrate the three factors that control land forms. Yosemite Valley (A) is a land form which was formed by a combination of glacial processes and the resistant rock structure of the region. The landscape in (B) is entirely the result of its basic rock structure. The mountain shown in (C) is in a mature stage in its life history and has a form typical of its age. (Ramsey)

FIG. 11–23. Closest in this photograph are the Alabama Hills which are mature mountains that contrast with the youthful Sierra Nevadas in the background. (Ramsey)

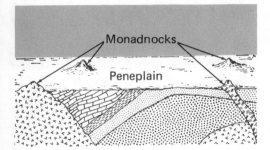

FIG. 11–24. Monadnocks in a peneplain developed on a mountain region are remnants of folded and faulted underlying rock formations.

FIG. 11–25. The Catskills of New York are the remains of a mature plateau. The ridges are relatively flat-topped and have horizontal terraces along their slopes. (Litton Industries, Aero Service Division)

a complicated, ancient, and fairly recent history. This must be taken into account when thinking about the way in which any landscape has come into being.

The weathering and erosion of the major landforms (mountains, plains, and plateaus) result in the formation of a variety of minor landforms. All of the influences which give these *minor landforms* their various shapes can be grouped into one of three categories:

1. *Structure* is a term which refers to the rocks from which a minor land form may be carved. Two different features of rocks are included in their structure. One feature is the *type* of rock, such as sedimentary, igneous or metamorphic. This is important and will determine how well the rock is able to resist the forces acting to wear it away. A second feature of rock structure is *formation*. This is the way they have been faulted, folded, or lifted by crustal movements. Formation also controls how the landform will be carved by erosion, since bending and breaking generally make rock less resistant.

2. *Processes* are agents of weathering and erosion which attack the rock structure. Streams, glaciers, wind, and waves are the most important forces which produce landforms with characteristic shapes. Usually one process is most important at a particular place. For example, waves are the most powerful influence on the shape of most shorelines.

3. *Stage* is a term which describes how long a particular process has been at work and how effective its work has been. A stream, for example, in its early stages may barely affect the surface over which it flows. In time the stream can cut a deep channel and create a major landform. However, the main factors in producing landforms are the rock structure and the various processes which shape it. Stages in the development of a landform are often interrupted and changed. An old river can be made youthful if the underlying strata undergo uplift. Such changes in

rock structure often make it difficult to establish the stage of a given landform.

The three factors that control the origin of minor landforms are illustrated in Figure 11–22.

Life cycle of mountains. No matter how they have come into existence, mountains usually follow a cycle of development as they are exposed to the forces of erosion. Mountains are created by the forces of uplift. As long as these forces continue, the mountain is usually raised faster than it is worn away. These mountains are said to be *youthful.* They are rugged with sharp peaks and deep narrow valleys. However, mountain growth eventually stops. Then the rugged scenery becomes worn down to form *mature* mountains. These are recognized by their rounded tops and gentler slopes. See Figure 11–23. As erosion continues over a very long period of time, mountains may become worn down until the area is almost level. This period of the mountain cycle is called *old age.* The final appearance of the surface at this stage is called a *peneplain*, which means "almost a plain."

A peneplain usually has low rolling hills with occasional resistant knobs called *monadnocks.* See Figure 11–24. A peneplain might be mistaken for a true plain. However, it hides under its surface the folded, twisted or titled rocks that are unlike the horizontal layers of true plains. Southern New England is a raised peneplain with typical monadnocks in New Hampshire and Massachusetts.

Some scientists believe that continents do not remain in a stable condition long enough for streams to form peneplains by erosion. If this is true, the way these wide level surfaces are formed is still an unsolved problem.

Life cycle of plains and plateaus. Both plains and plateaus usually pass through a series of stages as erosion wears away their surfaces. In the case of a plain, the changes are not great because the streams have little of the land surface to cut away before they reach base level. A mature or old plain is different from one in youth mainly because it possesses well developed streams. A plateau, on the other hand, is exposed to much erosion because it is at a higher elevation. The effect of the streams in attacking the rock of a plateau depends upon the climate and the type of rock. A young plateau usually has deep stream valleys with broad, flat regions separating them. Mature

FIG. 11–26. Note the structures of the mesa above and the buttes below. (Santa Fe Railway Photo; Union Pacific Railway Photo)

plateaus, such as the Catskill region in New York State, are rugged areas and are generally called mountains. See Figure 11–25. Many streams have cut wide valleys through the broad surfaces of the original plateau. In old age, plateaus are worn almost level, with traces of the original plateau left here and there. In dry regions these remnants have steep walls with flat tops. If they have broad tops they are called *mesas*. The smaller ones with narrow tops are called *buttes*. See Figure 11–26. In humid regions the last remnants of old plateaus are more rounded.

VOCABULARY REVIEW

Match the word or words in the column on the right with the correct phrase in the column on the left. *Do not write in this book.*

1. Changes that destroy the original rock structure.
2. The action of CO_2 when dissolved in water.
3. Water combining with minerals in rock.
4. Breaking apart in sheets or shells.
5. Dark organic material in soil.
6. Accumulation of broken rock at the base of a cliff.
7. The layers which make up soil.
8. All types of erosion caused directly by gravity.
9. Stones or pebbles which have been smoothed by wind erosion.
10. Large accumulations of fine wind-blown dust.
11. Furrows that follow the contour of the land to prevent erosion.
12. Movement of rock materials over the earth's surface by various natural agents.
13. Refers to a period of time during which a particular process has been at work.
14. The surface in the mountain cycle when it is almost a plain.
15. An occasional resistant knob on a peneplain.
16. Remnants of plateaus.

a. loess
b. mesa
c. humus
d. monadnocks
e. exfoliation
f. stage
g. deflation
h. peneplain
i. contour plowing
j. soil profile
k. blowout
l. ventifacts
m. weathering
n. erosion
o. carbonation
p. terracing
q. hydration
r. mass wasting
s. talus

QUESTIONS

Group A

Select the best term to complete the following statements. *Do not write in this book.*

1. The most important result of the weathering of rock as far as living things are concerned is the (a) release of oxygen by frost action (b) formation of soil (c) erosion which occurs (d) leaching which occurs.

2. Rock in which a change in mineral content is taking place is said to be undergoing (a) erosion (b) mechanical weathering (c) frost action (d) chemical weathering.

3. Which one of the following involves mechanical weathering? (a) exfoliation (b) hydration (c) carbonation (d) oxidation.

4. The breaking up of rock by freezing water is called (a) hydration (b) leaching (c) frost action (d) mass wasting.

5. Which one of the following ingredients of air is *not* involved in weathering? (a) nitrogen (b) oxygen (c) carbon dioxide (d) water.

6. A process similar to frost action which causes rocks to break apart but which is found in drier climates is (a) exfoliation (b) growth of salt crystals (c) blowout (d) leaching.

7. Rocks containing clay undergo a weathering very similar to (a) hydration (b) oxidation (c) frost action (d) carbonation.

8. Large caves in limestone rock are formed when carbon dioxide dissolves in water to produce (a) quartz (b) oxides (c) hydrates (d) carbonic acid.

9. Sand or gravel is formed from the weathering of rocks such as (a) limestone (b) shale (c) granite (d) calcite.

10. Large unweathered rocks left behind after weathering of bedrock are called (a) ventifacts (b) talus (c) residual boulders (d) barchans.

11. Weathering takes place slowly in climates that are (a) dry or very cold (b) humid (c) moderate (d) warm.

12. Soil composed mostly of a mixture of particles ranging from microscopic clay particles to larger sand grains is called (a) humus (b) talus (c) loess (d) silt.

13. The fine grains often formed from weathered feldspars are called (a) sand (b) clay (c) humus (d) talus.

14. A mature soil commonly has a soil profile consisting of (a) one layer (b) two layers (c) three layers (d) four layers.

15. Subsoil is the name given to the layer of the soil profile which is generally labeled (a) A (b) B (c) C (d) D.

16. The layers of the soil profile are called (a) horizons (b) ventifacts (c) barchans (d) sublayers.

17. The layer of soil where most of the animal and plant life is found is labeled (a) A (b) B (c) C (d) D.

18. The dissolved materials of the soil generally leach into the (a) A horizon (b) B horizon (c) C horizon (d) bedrock.

19. Soils in drier climates are often very fertile, mainly because (a) no plants can grow to use up the minerals (b) they have well defined horizons (c) much leaching occurs (d) very little leaching occurs.

20. Directly or indirectly, the cause of most erosion is (a) water (b) wind (c) gravity (d) the earth's rotation.

21. Slow, downhill movement of weathered rock is called (a) a landslide (b) slumping (c) soil creep (d) deflation.

22. Slumping is considered an early stage of (a) soil creep (b) landslide (c) mudflow (d) deflation.

23. The most important agent of erosion is (a) water (b) wind (c) heat (d) man.

24. The most common form of wind erosion is (a) oxidation (b) exfoliation (c) mass wasting (d) deflation.

26. Humans can increase the weathering and erosion of soil by (a) grazing animals on it (b) clearing the plant life (c) plowing the soil (d) all of these.

27. Once soil has been swept away by erosion (a) no amount of work can quickly repair the damage (b) crop rotation should be started (c) contour plowing should be done (d) terracing will repair the damage.

28. The life cycles of plains and plateaus are very similar except that plateaus generally pass through the series of erosion stages faster since they (a) are at a higher elevation (b) have more streams (c) are flatter on top (d) are made of more durable materials.

Group B

1. Why does weathering generally occur before erosion?

2. Describe how you can tell whether a given rock sample has undergone weathering.

3. Contrast mechanical weathering and chemical weathering.

4. Give three different examples of how water produces mechanical weathering of rock.

5. Why are limestone and calcite caves found only in areas with an ample water supply?

6. How can you tell whether iron-bearing minerals have been weathered?

7. Give at least three examples of how living things cause weathering.

8. What three main factors determine the speed of rock weathering?

9. Describe the two steps involved in the formation of soil.

10. Describe a typical soil profile.

11. What relationship exists between soil creep, slumping and landslides?

12. Why is gravity the indirect cause of all erosion regardless of the eroding agent?

13. What fact seems to make it doubtful that such large features as natural bridges, rock pinnacles, rocks perched on pedestals, and the like were formed by wind erosion?

14. Describe the process by which a sand dune moves along the ground.

15. How are sand dunes and loess related?

16. Although natural erosion is constantly changing the earth's solid surface, it would not generally cause much concern if it were not for the activities of man. Why is this so?

17. What is the principal means by which igneous, sedimentary, and metamorphic rocks weather?

18. Describe how the four methods used by farmers aid in controlling erosion.

19. Explain the formation of monadnocks.

20. What do buttes and mesas have in common? How do they differ?

21. What features of a peneplain are like those of a true plain? What features are different?

objectives

☐ Explain the operation and importance of the hydrologic cycle.

☐ Describe the importance of water conservation.

☐ Trace the steps in the development of a river system.

☐ Describe erosion caused by moving water.

☐ Explain how the water table is related to the water in the ground.

☐ Describe the formation of springs and geysers.

☐ Describe the formation of sinks and caverns.

Water is the most powerful of all erosion agents. It is more effective in moving weathered rock fragments than all other forces of erosion combined. Each drop of water that falls on the land does its share in wearing away and moving the rocks that make up the earth's crust.

The earth has not always had its present supply of liquid water. There were no oceans, lakes, or rivers during the earth's earliest history. These bodies of water could form only after the earth's crust cooled. It is likely that much of the earth's water appeared originally in liquid form beneath the earth's surface. It may have been given off as a gas in the violent volcanic eruptions that were common in the earth's early stages.

Heavy layers of clouds probably surrounded the young planet. Eventually as the cloud cover began to drop its accumulated moisture, rain fell and filled the ocean basins. At this time, an endless water cycle was started. Solar energy absorbed by water in the sea caused evaporation. This water vapor again formed clouds, then rain. Raindrops fell to become running streams which used the energy from gravity to carve and move rock materials. It is this continuous cycle of water movement which gives water its power as an agent of erosion.

THE WATER CYCLE

The hydrologic cycle. Part of the earth's water supply is bound up as water molecules in certain sedimentary rocks. The remaining water on earth follows a never-ending path, leading from evaporation to water vapor, then condensing back to clouds to produce precipitation in the form of rain or snow. Then run-off flows over the surface or sinks into the ground. Finally, it is stored until evaporation again takes place. This series of events is known as the *hydrologic cycle,* or water cycle.

Each year about 95,000 cubic miles of water enters the air as water vapor. Most of this, about 80,000 cubic miles, comes from the sea. This water evaporates when the sea absorbs the sun's heat energy. However, an additional 15,000 cubic miles of water evaporates each year from other sources. An important source of this moisture is the process of *transpiration,* a means by which plants give off water vapor. Water is also evaporated from lakes, streams and from the upper layers of the soil.

Most of the water that evaporates into the air falls back again into the oceans. About 71,000 cubic miles of water is returned annually to the oceans as rainfall. This leaves 24,000 cubic miles of water to fall on the land surfaces. Of this, 9,000 cubic miles is run-off in streams and rivers. This run-off water moves over the land surface and returns to the sea within a few weeks. But the remaining 15,000 cubic miles of water soaks into the land to become *ground water.* Some of this ground water fills the spaces between soil particles, providing an available water supply to plants for use in their vital life processes. A great deal of ground water sinks far into the pores and cracks of the deeper bedrock. This water slowly meanders by various routes back to the oceans. The complete hydrologic cycle is illustrated in Figure 12–1.

We can say that this continuous hydrologic cycle gives the earth a *water budget.* The income part of the budget is the water received by precipitation, rain, and snow. Outgo is represented by the ways that water evaporates and enters the air. The water budget is probably in balance for the whole earth. For example, the amount of water lost by evaporation over the complete surface of the earth equals the amount of moisture which falls to the earth as precipitation.

The water budget, however, much like the heat budget, is not in balance at every particular place on the earth's surface. Precipitation is affected by the patterns of weather and the shape of the earth's surface. One side of a mountain range may receive large amounts of precipitation while the other side remains a dry desert. Evaporation also differs greatly from one location to another. Far more solar energy is received in regions near the equator than in areas further away from it. Evaporation is also affected by winds which sweep up a great deal of moisture.

The amount of evaporation and precipitation at a particular place also changes with the seasons of the year. Some locations receive more moisture every month of the year than is lost by evaporation. In these places, surplus water constantly runs off in streams or becomes ground water. Other very dry locations consistently receive less water than is lost by evaporation. Under these conditions, plant life is scarce and run-off occurs only after occasional, heavy rains.

In places which are neither close to the equator nor to the poles, the situation varies throughout the year. During the cooler months, temperatures remain low enough for evaporation to use less water than the amount that is received by precipitation. During these months a surplus of moisture exists. The soil becomes saturated and excess water runs off in streams. During the warmer months, however, an increase in the rate of evaporation causes an overall loss of moisture. In the warmer period, the amount of moisture in the soil drops to a low level. Streams can continue to flow only if they are supplied from distant sources or from ground water.

Investigate

Blow up three balloons to nearly the same size. Place one balloon outside in sunlight; the second balloon in a shady spot; and the third balloon on some crushed ice. After 15 minutes observe the size of the three balloons. In which balloon would evaporation take place fastest?

FIG. 12–1. A summary of the hydrologic cycle.

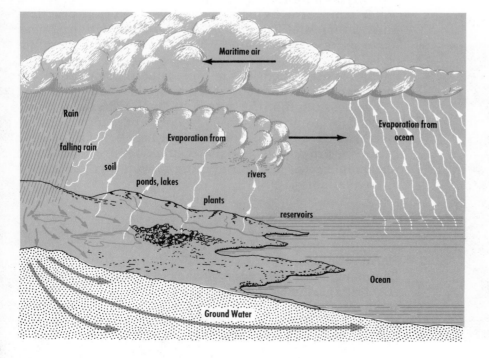

Table 12.1 – Water in Industry

Quantity	How Consumed
5 gallons	To process 1 gallon of milk.
10 gallons	To produce 1 gallon of gasoline.
80 gallons	To generate 1 kilowatt-hour of electricity.
300 gallons	To manufacture 1 pound of synthetic rubber.
65,000 gallons	To produce 1 ton of steel.

FIG. 12–2. The surface of the ground shown in this photograph has been eroded by falling raindrops, except where it was protected by stones. Notice the coin at the left for scale. (Ramsey)

Water conservation. It is estimated that each person in the United States drinks an average of 200 gallons of water per year. Each individual uses up another 15,000 gallons each year for washing, laundry, cooking, and the operation of heating and air conditioning equipment. This personal use of water is unimportant, however, when compared with 160,000 gallons per person which are used each year by industry. Table 12–1 will give you some idea of the vast amounts of water consumed in America's industrial processes.

To these figures must be added the billions of gallons of water needed to irrigate crops. All of this vital water must come from the water which runs off in streams or enters ground water supplies. This amounts to about 1300 billion gallons daily in the United States. It represents the entire supply available which can be obtained at reasonable cost. As the population of the country grows, the demand for water also grows. Conservation of this limited resource becomes essential.

On the average 90 percent of the water used by cities and industries is returned to rivers or the sea as waste. The untreated sewage of many cities is disposed of by dumping. Much waste water discharged into rivers by industries contains harmful materials. In some cases, the water can never be used again even after all practical purification methods have been tried. Such practices leave parts of our limited water supply unfit for human consumption. To insure a future supply of water large enough to meet the needs of an expanding population, we must achieve better control of water pollution.

It is possible that the supply of fresh water can be increased by treatment of sea water. This subject is discussed in Chapter 16. However, our best hope in the near future to provide an adequate supply of fresh water at reasonable cost is by conserving the water which is presently available.

RUN-OFF OF WATER OVER THE LAND SURFACE

Formation of a river system. Erosion by running water begins when each raindrop strikes the earth. Energy from the falling rain is delivered against the surface at the moment of impact. See Figure 12–2. Even after reaching the solid surface, however, a raindrop still does not lose all its energy until it reaches sea level. Gravity forces water to move or try to move down slopes until it reaches the level of the oceans. Thus the water that falls and remains on land surfaces must run off in established streams or else erode new channels. When a new path is carved, the first step in the formation of a river system begins as the water tumbles in broad sheets down the slopes.

The run-off from several slopes collects in low places. The energy from the moving water is then concentrated on a smaller surface and the rate of erosion increases. Soon a gully is formed. The gully serves as a collection channel for water from all nearby slopes. The increasing volume of water further enlarges the gully. These are the earliest stages in the development of a river. A beginning stream is not likely to flow except immediately following a rain. Even so, each time the water flows, it enlarges the gully and collects more water. Eventually this process results in a fully developed river valley containing a permanent stream.

In addition to the growth of the beginning river and its valley, branch gullies are produced by streams flowing in from the sides. These streams become the *tributaries* for the growing river system. See Figure 12–3. Erosion at the head of the gullies, where the water enters the stream, lengthens them and continues to cut back the existing slopes. This is called *headward erosion*. Chiefly it is the process by which a beginning river system extends its main branches back towards its source to collect the run-off water from a larger area of land.

Compare
Determine the conditions necessary for the formation of each of the following drainage patterns: deranged, dendrite, trellis, rectangular, and radial.

FIG. 12–3. Stages in the development of rivers and streams. (Top left) Gullies appear on the steeper slopes. (Top right) Gullies expand by downcutting and headward erosion. (Bottom right) Streams establish separate drainage basins. (Bottom left)

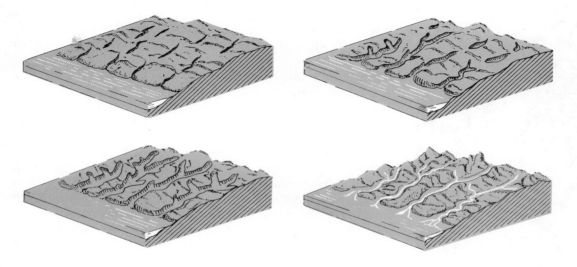

FIG. 12–4. Sharp ridges shown in this photograph are divides separating desert streams which flow during periods of heavy rainfall. Can you find an example of headward erosion which will probably cause stream piracy in the future? (Ramsey)

FIG. 12–5. The valley of the Yellowstone River shows the characteristics of a young stream. (Ramsey)

With headward erosion, the main stream and its tributaries eventually form a branching system of channels. These channels drain a series of slopes that make up the *drainage basin* or *watershed* of the river system. The ridges or regions of high ground that separate the various watershed areas from each other are called *divides*. See Figure 12–3.

Divides between neighboring streams gradually become lower as erosion continues its work from both sides. The rate of erosion is generally not equal on both sides of the divides. This is due to differences in the kind of material through which the stream flows or the speed of stream flow. Occasionally, a speed-up of erosion of one stream cuts through a divide. When this happens, all the water then flows in one of the stream channels. This process is called stream capture or *stream piracy*. See Figure 12–4. An established river system is often enlarged by stream piracy.

A stream's ability to cut down and widen its channel by erosion depends largely upon the swiftness of its water flow. In turn, the velocity of a stream depends upon the amount of water it carries and on its slope or *gradient*. A stream's gradient is a term expressing the difference in elevation between its head and mouth. Rivers with a steep gradient generally have a high velocity, allowing the river to rapidly erode its channel. However, as the river grows older, its gradient becomes less steep and the channel is cut down toward the level of its mouth.

In its early stages, when the gradient is still steep, a river usually deepens its channel more rapidly than it can

cut into the sides. This produces a *V-shaped* valley such as that shown in Figure 12–5. A river in this stage of development is said to be in its youth. Besides a V-shaped valley, *waterfalls* and *rapids* are very common features of youthful streams. Both waterfalls and rapids frequently appear where there is unusually hard rock in the stream channel. Hardened rock resists erosion by the stream and tends to remain as a barrier. However, such barriers are usually only temporary, and both waterfalls and rapids disappear as the stream grows older. See Figure 12–6.

Young rivers usually have relatively few tributaries. This is because there has not been time for a large system of feeder streams to develop. For this reason, a young river usually carries a small volume of water. Much of the water falling in the watershed of a young river system does not reach the main stream. It remains in the higher lands to form lakes and swampy areas.

On the other hand, a *mature* river has well-established tributaries. It effectively drains off the water falling into its drainage system. Because of its many tributaries and good drainage, a river carries its largest volume of water at the mature stage. In its later stages, a stream has less

FIG. 12–6. This aerial photograph of Niagara Falls shows how it has been worn back over the past few hundred years. Like all waterfalls, it will eventually disappear from the river channel. (Niagara Frontier State Commission)

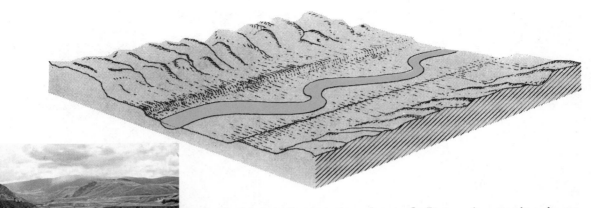

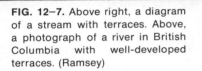

FIG. 12–7. Above right, a diagram of a stream with terraces. Above, a photograph of a river in British Columbia with well-developed terraces. (Ramsey)

FIG. 12–8. The process by which a water gap may become a wind gap following stream capture.

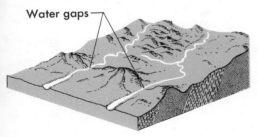

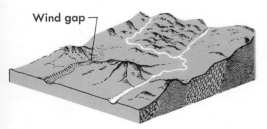

tendency to deepen its channel. Its main erosion is on the sides of its valley walls. Thus an *older* river usually loses the V-shape of its youthful channel and creates a wider valley with a relatively flat floor. The waterfalls and rapids of the youthful phase start to disappear and the channel becomes flat. Then the river channel normally occupies only a small part of the valley floor.

Movements of the crust can raise or lower the surface of the land. Any movement that increases the slope of the land will change the gradients of existing streams. A stream whose gradient has been increased in this way is said to be *rejuvenated*.

The steeper gradient of a rejuvenated stream allows it to cut deeper into the valley floor. Rejuvenation often results in the formation of step-like *terraces* in stream valleys. See Figure 12–7. When the land surface rises very slowly, the existing streams are cut down only as fast as they are elevated. When this happens, the mountains raised across the channel of an existing river will be deeply notched. This notch is called a *water gap*. The river that cuts the water gap may later be captured by a stream running parallel to the mountains. Thus the water gap is abandoned and it becomes a *wind gap*. See Figure 12–8.

Erosion by moving water. Running water carries on its work of carving the land surface by a series of processes. The first event that produces a movement of rock fragments is the splash of a raindrop. The force of raindrops on bare soil tends to seal off the surface. The impact of the drops shifts the particles around on the surface and the finer soil grains are thus fitted between larger ones. The pores and channels through which water might enter the soil are now plugged up. Soil particles are also pressed together by energy from the falling drops, closing the spaces between the particles. The effects of these actions waterproof the soil surface so that most water falling on it will be lost as run off.

At first, the run-off appears as shallow layers of water running over the surface. This brings about *sheet erosion*. This type of erosion removes the lighter soil particles and substances which dissolve. When the run-off water enters lower places on the surface, its energy becomes more concentrated. At the same time, its power to erode is directed against smaller areas of the surface, allowing the larger rock fragments more freedom of movement. Small particles are carried along, suspended in the water. Heavier fragments of rock are rolled or bounced along the stream bed. Each contact between the loose rock fragments carried by the water and the rocks of the stream bed wears off a small amount of both. The moving rocks become rounded and smooth, while the stream bed is worn away. Rocks are often swirled around for some time in a whirlpool motion over the same area on the stream bed. This creates a bowl-shaped cavity in the rock of the stream bed. Such a cavity is called a *pothole*. See Figure 12–9.

As the load carried by the stream increases, its rate of flow slows down and there is a tendency for bends to develop. Once a slight bend develops in a stream bed, the channel curve in that place usually grows larger. The reason for this is that water usually flows fastest around the outside edge of the curve. The faster flowing water erodes the outside bank more readily than it does the inner bank. This results in an ever-increasing enlargement of the curve, as shown in Figure 12–10. Slower moving water on the inside of the curve allows sediments to settle and form a bar. Older streams often develop a series of looping curves called *meanders* (mee-*an*-durz). See Figure 12–11. Frequently these meanders become so curved that the river cuts across the narrow neck of land

FIG. 12–9. A pothole formed in the solid rock of a stream channel. The continuous motion of the stones swirled along in the water helped to grind and enlarge the hole. (Ramsey)

Explain
Part of the Rio Grande river along its older stream bed, acts as a boundary between Texas and Mexico. Why is this a poor physical boundary line?

FIG. 12–10. A stream tends to develop larger curves because the outside bank of a small curve is eroded faster than the inner bank.

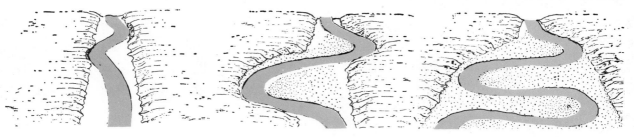

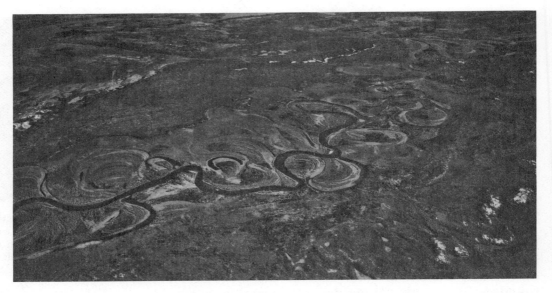

FIG. 12–11. A meandering river showing deposits which form bars along the inside curve of each meander. (National Air Photo Library)

FIG. 12–12. A stage in the formation of an oxbow lake; the cutting off of a meander loop. The river cuts deeply into the bank along the outside of the meander's curve.

at both ends of the curve, isolating the meander from the river. If the water remains in the isolated meander, it will form an *oxbow lake*. The process by which a meander becomes an oxbow lake is shown in Figure 12–12.

The combination of dissolved substances, suspended particles, and rock fragments rolled along by the water make up the *load* of a stream. Under certain conditions, a stream will decrease its load, generally when for some reason the water slows down. Because stream velocities are constantly changing, the loads of streams are also increasing and decreasing. As a result, deposits of solid material carried by the stream are formed, then picked up again by moving water. In some cases, stream deposits remain in place only a short time. In others, they become a more or less permanent feature of the land.

Most of the load carried by a stream is deposited when the stream reaches a larger body of water. As a stream empties into the relatively quiet water of an ocean, gulf, or into a lake, all of the remaining load is deposited at its

1   2   3   4

FIG. 12–13A. Delta of the Nile River as seen during an orbit of Gemini 4. The small diagram shows the area covered by the photograph. (NASA)

mouth. The sediment deposited usually takes the form of a triangle with its apex turned directly upstream. These fan-shaped deposits at the mouth of a stream are called *deltas*. See Figure 12–13A.

When streams flow down a steep slope from high land onto a level surface, a deposit resembling a delta is formed. As the stream flows swiftly down the steep mountain slope, it gathers a heavy load. When the stream reaches the level surface, it deposits its load in a delta-shaped form called an *alluvial fan*. The word "alluvial" refers to any deposit laid down by running water. See Figure 12–13B.

Floods. A stream adjusts itself to carry the amount of water flowing within its channel. If a stream always flowed with the same volume, its channel would change very little. But the flow of water (in nearly all streams) changes constantly. Most noticeable are the times when the stream overflows its banks. When this happens, the

FIG. 12–13B. The major difference between an alluvial fan and a delta deposit is their place of deposition. A fan is an alluvial *land* deposit. (Photo Researchers-Russ Kinne)

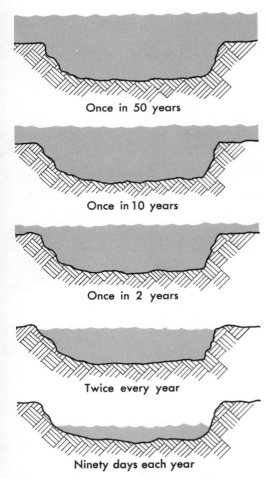

Once in 50 years

Once in 10 years

Once in 2 years

Twice every year

Ninety days each year

FIG. 12–14. The water-flow in a typical river varies greatly over a period of time.

stream is said to be in *flood stage*. Changes in water flow which produce floods are illustrated in Figure 12–14.

Spring floods are common in regions where the winters are harsh. Since rainfall and water released by melting snow cannot be absorbed by the frozen ground, it appears mostly as surface run-off in streams. Ice jams also increase the chance of spring flooding by blocking stream channels.

In many areas human activity has increased the size and number of floods. The natural ground cover of plants which helps protect the surface from heavy run-off has been removed. Forest fires and destructive logging operations, or the clearing of land for cultivation in watershed regions increase the run-off load in streams.

When a stream overflows its banks and spreads out over the nearby land, it deposits part of its load along the borders of the channel. Accumulation of these deposits along the edges of the stream eventually produces raised banks. Such elevated banks are called natural *levees*. These long, raised banks may be quite high and prominent along mature river channels.

However, not all the load deposited by a stream in flood stage goes into the formation of levees. Large amounts of sediment are left behind in fairly even layers after the water level has gone down. A series of floods produces a thick layer of deposited material on the level areas on both sides of the stream channel. Since this accumulation is generally very even, the part of the valley floor usually covered in the flood becomes a relatively flat flood plain. A *flood plain* with levees is shown in Figure 12–15. Swampy areas are common on flood plains because drainage is usually poor in the area between the levees and the outer walls of the valley.

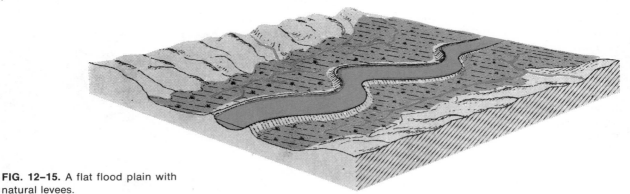

FIG. 12–15. A flat flood plain with natural levees.

Floods are a natural part of the development of streams and will continue to occur. They are a problem because many cities and farms are located on the fertile flood plains of rivers. To prevent loss of life and property in these areas, some control over floods becomes necessary. There are direct and indirect ways of controlling floods.

Indirect methods of flood control include forest and soil conservation measures which prevent heavy run-off during periods of high rainfall. The most common example of direct flood control methods is a dam. Dams help to control floods by creating artificial lakes which absorb excess run-off. The stored water can then be used for generating electrical power and for irrigating cropland during dry periods. Artificial levees are also built to directly control floods. However, these levees offer only temporary protection. As a river deposits sediment along its bed, the height of the levees must constantly be raised. They also require protection against erosion by the river. In some cases, a permanent overflow channel or floodway can be an effective means of *flood control*. During flood stage, a floodway helps carry excess water to prevent the main stream from overflowing.

WATER BENEATH THE EARTH'S SURFACE

The water table. Water can seep into the earth's crust because much of the land surface is covered by loose, weathered material. In many places, the outer layer of fragments lies over porous rock, through which water can penetrate. Below this is rock which may be completely solid. As water seeps down from the surface, it first enters the zone of aeration, a region where the spaces between the rocks contain both water and air. Most of the water in this zone is found as a thin film clinging to the surface of the rock and soil particles. Some of the water in the *zone of aeration* moves down to lower levels; some of it evaporates and some is absorbed by the roots of plants. The depth of the zone of aeration and the amount of water it contains varies greatly from one time or place to another. During a dry period this zone would increase in depth. In mountainous or hilly regions, it may be hundreds of feet thick. At other locations where the soil is saturated with water, the zone of aeration will be only a few inches thick or else completely absent.

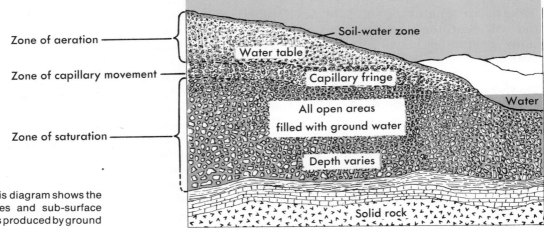

Zone of aeration

Zone of capillary movement

Zone of saturation

Soil-water zone

Water table

Capillary fringe

All open areas filled with ground water

Depth varies

Water

Solid rock

FIG. 12–16. This diagram shows the various features and sub-surface moisture zones produced by ground water.

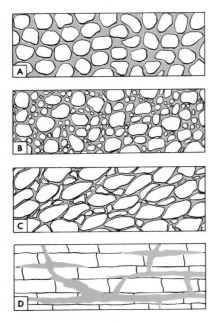

FIG. 12–17. (A) Rock composed of particles of nearly the same size has high permeability because of many pores or open spaces; (B) Rock with particles of many different sizes has lower permeability; (C) Mineral deposits between the rock particles reduce permeability to nearly zero; (D) Fine-grained rock which dissolves in water has high permeability because water dissolves cavities in it.

At the bottom of the zone of aeration is a region where water is drawn up from below by capillary action. Capillary water is water drawn upward against the force of gravity by a force known as *capillary tension*. In a soil profile, the top of the *zone of capillary movement* is called the *capillary fringe*. The depth of this zone depends on the size of the rock particles and is seldom greater than a few feet. In gravels and coarse sands the rise of capillary water is only a fraction of an inch, but in the smaller soil particles the grains are closer together and the rise may be several feet.

Below the zone of capillary movement lies the *zone of saturation* in which all spaces are filled with water. Water found in the zone of saturation is referred to as *ground water*. The top of the zone of saturation that marks the depth needed to reach ground water is usually called the *water table*. See Figure 12–16.

The depth of the water table below the ground surface at any particular location depends upon several things. One important influence is the amount of rainfall in the recent past. Another factor is how easily water can penetrate into the rock. The ability of rock to allow entrance of liquids is called *permeability* (per-mee-uh-*bil*-i-tee). The most permeable rocks are those composed of very coarse grains. These rocks usually have large pores between the grains. An example of this type of rock is sandstone. Very fine grained rocks, such as limestone, may become permeable if cracks are produced after the rock is formed. Various ways in which rocks become permeable are shown in Figure 12–17. Ground water ordinarily sinks deeper until it strikes a completely solid rock layer. The rocks above this layer then become completely saturated up to the water table.

Generally speaking, the water table follows the contours of the land surface. It slopes down where the surface slopes and rises in higher ground. However, the general slope of the water table is usually not as sharp as that of the land surface. This is true because ground water flows very slowly as it works its way through small cavities in the rock.

In many places depressions on the land surface dip below the water table. When this happens, ground water may flow out onto the surface to join with surface streams. Water may also collect on the surface to form lakes or swamps. See Figure 12–18. Many streams continue to flow in periods of low rainfall because they are fed by ground water.

Surface streams flow because they have a gradient which is determined by the slope of their beds. Ground water also flows because the gradient of its bed corresponds in a general way to the slope of the surface

activity

Obtain two dry-glass or plastic tubes (about 1 cm dia. and 15 cm long). Place a wad of cotton in one end of each tube and cover with cotton gauze. Through the open end, add about 10 cm of dry, fine sand. Add about 10 cm of dry, coarse sand to the other tube. Place the tube with the fine sand, holding a ruler next to it, above a container of water. Your partner will tell you every time 5 sec. passes. Place the gauze covered end of the tube about 1 mm into the water. Every 5 sec. thereafter, record the height of the water inside the tube. Continue for 2 min.

Do the same thing for the tube with the dry coarse sand.

1. What caused the water to rise in the tubes?

2. In what tube did the water rise the fastest? The slowest?

3. Is there a limit to the height to which water will rise?

4. Would capillary action have an effect on evaporation of water from the soil? Explain.

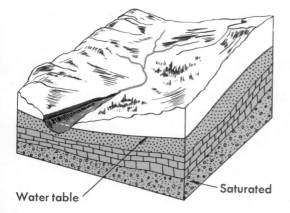

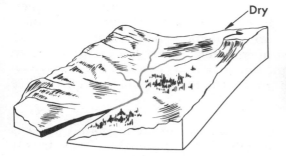

Water table   Saturated

Dry

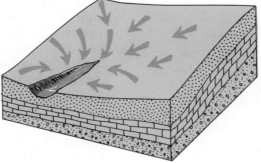

FIG. 12–18. Above, a small stream channel permanently fills with water where it first cuts below the water table. Below, the dry upper zone has been lifted to show the surface of the water table.

FIG. 12–19. The type of rock formations that result in artesian wells. At A, the artesian well does not flow. At B, the well flows because it is below the origin of the water, where the aquifer is exposed at the surface. Well C collects only ground water from the local area.

above. The speed with which ground water moves through the rock depends upon the gradient and the permeability of the rock structure.

Any hole below the water table will fill with water and form a well. The rate of water flow into a well depends upon the amount of ground water available at that particular place. An *artesian well* obtains water from a porous rock layer exposed at one end and sandwiched between an upper and lower layer of solid rock. The porous layer, called an *aquifer* (*ak*-wi-fer), slants downward from its exposed surface. See Figure 12–19.

FIG. 12–20. Springs often appear where the water table intersects the surface above a layer of impermeable rock.

Springs. A *spring* is found where ground water comes naturally to the surface. There are many types of springs; any exposed surface between the water table and the ground may result in the formation of a spring. Hillsides are common locations for springs because a slope on the surface often drops below the level of the water table. Such springs usually do not flow continuously, since the water table may drop below the slope in dry periods. However, one type of hillside spring is likely to flow continuously. This is a spring formed at the zone of contact between permeable and impermeable rocks. Water filters down through the permeable rock and comes to the surface on a hillside when it meets the impermeable rock layer. See Figure 12–20.

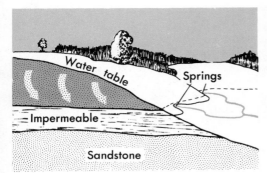

Hot springs and geysers. In some areas ground water is heated by volcanic activity. The water then finds its way

to the surface before cooling, and produces a *hot spring*.
The water from a hot spring is heated by passing close to
a body of magma. It may also be heated by mixing with
steam and hot gases escaping from bodies of magma deep
in the crust. Hot water is better able to dissolve minerals
than cold water. Thus the water of hot springs is likely to
contain unusually large amounts of mineral matter. Much
of this dissolved mineral material is deposited when the
water cools after reaching the surface. It often accumulates around the mouths of hot springs and builds up
layers or terraces. See Figure 12–21. The chief mineral
in these deposits is usually *travertine*, a form of calcite.
Travertine is mostly white when freshly deposited but
turns gray upon weathering. It may be colored red, brown,
or yellow by impurities, such as iron compounds.

Geysers are a type of hot spring that, at times, throw
steam and water into the air. Some geysers erupt from
open pools and throw up sheets of water and steam.
Others erupt through a small surface opening sending a
column of water and steam high in the air. Both types
tend to build deposits of a silicate mineral called *geyserite*.

FIG. 12–21. A hot spring in Yellowstone National Park. The water
has deposited layers of travertine
around the opening of the spring.
(Ramsey)

FIG. 12–22. (A) a geyser in Yellowstone National Park. (B) a deposit
of geyserite built around an extinct geyser in Yellowstone National Park. (C) a diagram showing
the underground structure of a
typical geyser. (Shostal Photos—A)
(Ramsey–B)

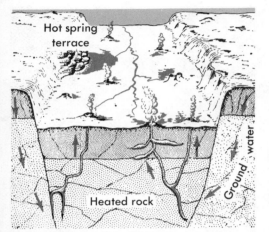

FIG. 12–23. Geysers and hot springs are generally found only in trenches formed by parallel cracks. Typical features and their origins are shown in this diagram.

FIG. 12–24. A limestone sink in Weston County, Wyoming. Sinks are a feature of erosion by ground water. (U.S. Geological Survey)

The underground structure of a geyser consists of a crooked tube leading to the surface. The tube usually connects several underground chambers, as shown in Figure 12–22. Heated ground water fills both the tube and chambers. Steam trapped in the rock cavities creates pressure that forces some water out onto the surface. This produces a small surge of water which signals all geyser eruptions. The water of a geyser boils explosively when the escaping water reduces pressure on the very hot water below. A roaring column of steam and boiling water is driven up and erupts on the surface. The eruption continues until most of the water and steam are emptied from the tube and storage chambers.

Afterwards, the ground water begins to collect again for a repetition of the process. The period between every geyser's eruption is determined by its supply of ground water, heat, and the shape of its tube and chambers. Geysers are found in special valleys called *geyser basins*. See Figure 12–23. Almost all of the geysers of the world are found in Yellowstone National Park, New Zealand, Iceland and Japan.

Ground water and rock. Pure water does not easily dissolve any but a few minerals. If ground water were pure, it would not have much effect on the rocks through which it moves. However, ground water does contain carbon dioxide, oxygen, and other dissolved materials. These are picked up during its movement on the surface, before moving down into the rock. Thus rocks beneath the ground can be affected chemically by ground water. An example is the process of *carbonation*. In this process carbon dioxide dissolved in ground water forms an acid which attacks certain minerals. This reaction is identical to the one that takes place in chemical weathering on the surface.

Rocks composed mainly of the mineral calcite, such as limestone, dolomite, and marble, are especially vulnerable to attack by carbonation. They dissolve slowly in water containing carbon dioxide. Thus ground water penetrating into openings in these rocks enlarges the rock cavities. If this dissolving action enlarges a number of connecting cracks, a *cavern* will be formed in the rock. Often the cavern will grow quite large with many large and small connecting chambers. This is particularly true in areas where large deposits of limestone rock exist beneath the surface. Examples of such caverns formed in

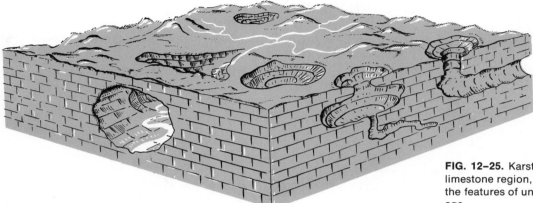

FIG. 12–25. Karst topography in a limestone region, showing some of the features of underground drainage.

limestone are the Carlsbad Caverns in New Mexico and Mammoth Cave in Kentucky. Many other places in North America have limestone regions with caverns.

When part of the roof of a cavern collapses, it often creates a roughly circular hole on the ground surface. Such holes are called *sinks*. See Figure 12–24. Sinks frequently become filled with water, forming ponds or lakes.

Cavern growth in limestone regions often produces a particular type of landscape where sink holes and similar depressions on the surface are very common. Ponds and lakes are also common, but there are almost no surface streams. The absence of streams means that there are no valleys in the region. This type of landscape is called a *karst plain*. See Figure 12–25. Karst plains are found in the limestone regions of Kentucky, Tennessee, and southern Indiana.

Caverns that lie above the water table cannot fill with water. Their appearance is marked by stone formations made of calcite deposited by water dripping into the spaces inside the cavern. Water dripping from the same spot on a cavern roof builds a hanging deposit called a *stalactite* (stuh-*lak*-tyt). At the point on the cavern floor where drops of water fall, another deposit of calcite called a *stalagmite* (stuh-*lag*-myt), is formed. Frequently the two formations grow until they meet forming a continuous *column*. See Figure 12–26.

FIG. 12–26. The limestone deposit that forms stalactites and stalagmites is known as *travertine* or *calcareous tufa*. (Luray Cavern)

VOCABULARY REVIEW

Match the word or words in the column on the right with the correct phrase in the column on the left. *Do not write in this book.*

1. The series of events in which water is evaporated and finally returned to storage.
2. Water that soaks into land to fill available space.
3. The probable balance between evaporation and precipitation.
4. Streams flowing into a river from the sides.
5. The series of slopes which make up the drainage basin.
6. Ridges or regions which separate the various watershed areas.
7. The notch cut in a mountain raised across a river channel.
8. Formed when water remains in an isolated meander.
9. The part of the load which is deposited at the mouth of a stream.
10. The region in which the spaces between the rock contain both water and air.
11. The name usually given to the top of the zone of saturation.
12. Obtains water from an aquifer located between two impermeable layers.
13. Occurs where the ground water comes naturally to the surface.
14. A hot spring which throws steam and water into the air.
15. Sometimes formed when the roof of a cavern collapses.

a. watershed
b. oxbow lake
c. hydrologic cycle
d. water gap
e. tributaries
f. meanders
g. ground water
h. delta
i. divides
j. water budget
k. spring
l. zone of aeration
m. cavern
n. geyser
o. artesian well
p. sink
q. water table

QUESTIONS

Group A

Select the best term to complete the following statements. *Do not write in this book.*

1. The earth's present supply of liquid water (a) has always been present in the oceans (b) probably first appeared as liquid water beneath the earth's surface (c) is the result of solar winds (d) was first present as ice.
2. Of the water which enters the atmosphere as vapor each year, what part comes from the sea? (a) all of it (b) about 15% (c) about 80% (d) about 95%.
3. The process by which plants give off water vapor is called (a) dew (b) precipitation (c) evaporation (d) transpiration.
4. Of the water that falls on the land surface each year, the part which becomes ground water is about (a) 50% (b) 40% (c) 20% (d) all of it.

5. Which of the following would play the smallest part in balancing the water budget for a particular area? (a) evaporation (b) winds (c) run-off (d) precipitation.

6. Which industrial process requires the greatest amount of water? To produce or process (a) a gallon of milk (b) a pound of synthetic rubber (c) a gallon of gasoline (d) a pound of steel.

7. Which process uses the greatest amount of water per person in the United States? (a) washing and laundry (b) cooking (c) drinking water (d) industry.

8. The amount of water which runs off in streams or enters the ground water supplies in the United States is about (a) 1300 gallons daily (b) 1,300,000,000,000 gallons yearly (c) 1,300,000,000,000 gallons daily (d) 1300 gallons yearly.

9. Water loses all its energy due to gravity only after it (a) falls as rain (b) reaches a river (c) is in a reservoir (d) reaches sea level.

10. The wearing of the land at the head of gullies which causes them to lengthen and reach up the slopes is called (a) headward erosion (b) tributaries (c) watershed (d) dividing.

11. Stream piracy involves (a) one county stealing another's streams (b) conservation (c) erosion of the divide between two streams (d) building dams.

12. Which of the following factors least affects the rate of stream flow? (a) slope of a stream bed (b) gradient of a stream bed (c) load of a stream (d) steepness of a stream bed.

13. As a stream grows older its gradient usually (a) becomes greater (b) becomes less (c) remains the same (d) becomes a channel.

14. Common features of a youthful stream usually *do not* include (a) large amounts of water (b) a V-shaped valley (c) waterfalls (d) many tributaries.

15. A stream whose gradient has been increased due to movements of the earth's crust is said to be (a) youthful (b) mature (c) rejuvenated (d) terraced.

16. When the stream running in a water gap is captured by another stream running parallel to the mountains, the water gap becomes a (a) terrace (b) wind gap (c) youthful gap (d) mature gap.

17. A bowl-shaped cavity worn in a stream bed by rocks swirling at that location for some time is called a (a) pothole (b) sink hole (c) rock hole (d) pithole.

18. Which of the following is *not* primarily related to looping curves in streams? (a) oxbow lakes (b) meanders (c) load (d) bars.

19. The deposit built when a stream reaches a level surface at the foot of a mountain is called (a) a stream delta (b) an alluvial delta (c) an alluvial fan (d) a stream fan.

20. Which of these is built in an attempt to control flooding by rivers? (a) flood stages (b) flood plains (c) deltas (d) artificial levees.

21. The region in the ground where all of the spaces between rock particles are filled with water is called the (a) zone of saturation (b) zone of aeration (c) capillary fringe (d) flood zone.

22. The ability of rock to allow liquids to enter is called (a) aeration (b) permeability (c) capillarity (d) saturation.

23. A porous water layer from which supplies of underground water may be drawn is called (a) a spring (b) a well zone (c) an artesian zone (d) an aquifer.

24. A form of calcite deposited by hot springs is (a) geyserite (b) travertine (c) pyrite (d) quartz.

25. A silicate mineral deposited by hot springs is (a) geyserite (b) travertine (c) pyrite (d) quartz.

26. Which of the following is *not* commonly found on a karst plain? (a) ponds (b) sink holes (c) surface streams (d) lakes.

27. A deposit of calcite hanging from the roof of a cavern is called a (a) stalagmite (b) stalactite (c) column (d) limestone icicle.

Group B

1. Describe the hydrologic cycle.

2. Name two factors which determine how fast water flows into a well.

3. What might cause the rejuvenation of a stream?

4. Explain how stalactites and stalagmites are formed.

5. What are several characteristics of a mature river?

6. Why do people inhabit flood plains when there is so much danger from floods?

7. At what time of year would you drill a well to a depth that would most likely provide a year-round supply of water?

8. What is meant by the water budget? Why might it not be balanced for a particular place on the earth's surface?

9. Explain how meanders are formed.

10. What are hot springs and how are geysers related to them?

11. Name several features of a youthful stream and describe how each changes with maturity.

12. How has man increased the possibility of floods? How has he decreased the possibility of floods?

13. Name several factors which affect the rate at which a stream channel is widened and deepened.

14. Would it be possible for a tributary to form a delta? Explain.

15. Explain briefly the relationship among the zone of aeration, the zone of saturation, the capillary fringe, and the water table.

16. How are oxbow lakes formed?

17. Why is long term prediction of stream flow difficult?

18. Would alluvial fans or deltas be composed of coarser materials? Explain.

19. How could you as an individual conserve water? How much water would be conserved by 3 billion people (the earth's population) if they all practiced this conservation idea? Do not suggest anything that would damage your health.

About ten thousand years ago early humans witnessed the end of an important chapter in Earth's history. Great sheets of ice that had covered about one-third of the earth's surface began to melt. Land that had been buried under thick ice layers for thousands of years began to be exposed. The blankets of ice had covered much of the continents of the Northern Hemisphere. In North America an ice sheet had reached as far south as what is now New Jersey, Ohio, Illinois, Kansas, Montana, and Washington.

Today the landscape throughout much of the world, and particularly in the northern United States and Canada, shows the effects of these great bodies of ice. Evidence contained in older rocks indicates that the earth was partly covered by ice at least several times in its ancient past. The ice sheets that retreated only a few thousand years ago were the last of a series of ice ages the earth has experienced. This chapter is concerned with the way the ice left its marks on the rocks.

ORIGIN AND MOVEMENT OF GLACIERS

Formation of glaciers. Water which remains on the earth's surface as a liquid quickly runs off into lakes, streams, and rivers, and finally back to the sea. However, if the average temperature is near or below the freezing point, the water falling on the surface freezes and remains as solid ice. Water accumulates from one year to the next in thick layers of snow. The great weight of the snow presses against the layers beneath. This pressure, along with some melting and freezing again, changes the snow into small grains of ice. The grainy ice thus produced is

objectives

☐ Explain how glaciers form.

☐ Describe the two kinds of glaciers presently found on earth.

☐ Describe how glaciers erode the land.

☐ List the two general kinds of glacial deposits.

☐ List the conditions necessary for a lake to form.

☐ Describe the life history of a lake.

☐ Describe the conditions that existed during the ice ages.

☐ List several possible explanations for the cause of the ice ages.

275

Interpret

Does the elevation of the snowline vary directly or indirectly with the latitude of the place in question? Explain.

FIG. 13–1. A part of the Columbia ice field in the Canadian Rockies is visible as the layer of snow and ice across the distant mountains. (Ramsey)

called *firn* or *névé* (nay-vay). In the lowest layers of the accumulated snow, the pressure is so great that the firn becomes a solid mass of ice. Each year more snow is added to the top of the ice layers. Eventually the weight becomes great enough to cause parts of the body of ice to move slowly over the earth's surface. The moving ice is called a *glacier.*

Today relatively few places have low enough average temperatures along with sufficient snowfall for glaciers to form. Only in the polar regions can ice survive all year at sea level. Another area is above the snow line in high mountains near equatorial, middle, and northern latitudes. The *snow line,* the elevation above sea level where snow remains all year, is higher near the equator and lower near the poles. If the amount of snow and ice that accumulates during a year is greater than the amount that melts during the warm season, an *ice field* (or snow field) will be formed. Ice fields cover all lands very near the poles and are found in mountains at lower latitudes. See Figure 13–1.

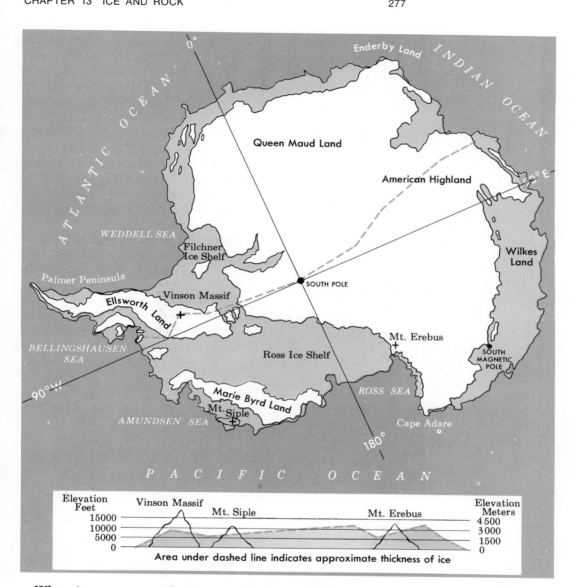

FIG. 13-2. Antarctica. The darker color represents the oceans surrounding the continent. The lighter color shows sea areas covered by ice. The white areas are land masses, also covered by ice. The dotted line in the drawing above shows the thickness of ice at varying elevations.

When ice moves out from a mountain ice field, an *alpine* or *valley* glacier is produced. These can be small glaciers or huge rivers of ice that fill an entire valley.

A second type of glacier found only in the polar regions is a continuous sheet of ice called a *continental glacier*. This type of glacier covers the entire land surface except for the tops of the highest mountains. Antarctica is covered by the largest of all existing continental glaciers. It is one and a half times as large as the United States (not including Alaska and Hawaii). Over much of its area, the

Antarctic ice cap is more than 3000 meters (about 10,000 feet) thick. See Figure 13–2. Greenland, the largest land body near the North Pole, is also buried under a continental glacier. It covers about 80 percent of the entire surface of this large island. Only small fringes of the land around the coast are exposed. At its thickest point the Greenland glacier is about 3000 meters in depth. If all the ice in the two great glaciers of Greenland and Antarctica were melted, it is estimated that the water released would raise the level of the sea by more than 60 meters (about 200 feet). This would be high enough to submerge completely all the major cities along every coastline in the world!

Movement of glaciers. Glacial motion has been a subject of much scientific study. The exact way in which glaciers move is still not completely understood; however, some principles are clear.

The advancing front wall of a glacier is called the *ice front*. In alpine glaciers the ice front moves as a result of pressure from above and from being on a sloping surface.

FIG. 13–3A. Movement of ice in an alpine glacier.

FIG. 13–3B. Relative speeds of flowage in various parts of a glacier.

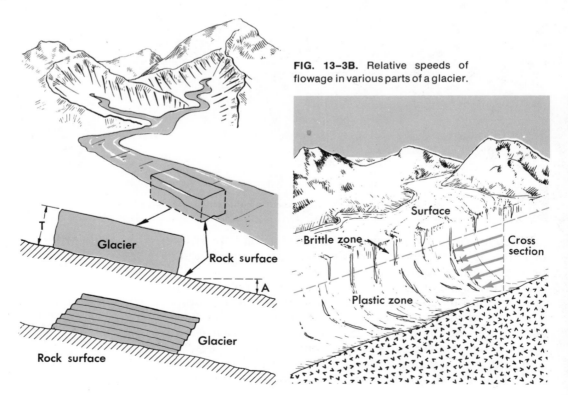

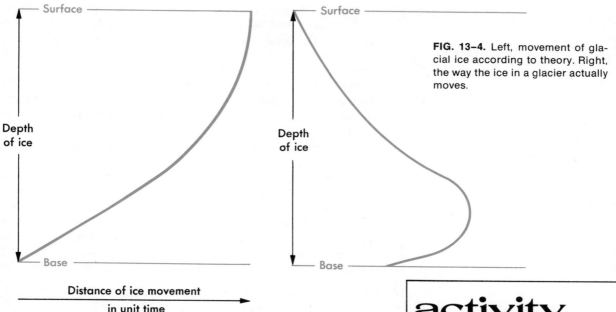

FIG. 13–4. Left, movement of glacial ice according to theory. Right, the way the ice in a glacier actually moves.

Think of a block of ice frozen to a sloping surface. The block of ice represents a glacier and the sloping surface, a valley floor. If the angle of the slope in Figure 13–3A is large enough (angle A) and the ice is thick enough (T), the ice should begin to move down the slope. The force that acts to pull it down is regulated by the angle of the slope. In glaciers, motion may also result from layers of ice sliding over each other.

Another explanation suggests that the bottom ice melts under pressure from the overlying weight, rolls forward on a thin film of water, and then refreezes. Still another possible way to explain glacial movement is that ice in the upper zone of the glacier is quite brittle. Below this brittle zone it is suggested that the ice may flow like melted plastic. See Figure 13–3B.

In the type of movements just described, the rate of ice flow plotted against depth would ordinarily produce a graph like that shown in the left part of Figure 13–4. But the actual measurement of the way glacial ice moves results in a different picture. See the right graph in Figure 13–4. The most rapid movement occurs at two-thirds of the distance down from the top of the glacier.

We can see then that a glacier does not just slide down a slope. Its movement is a combination of a sliding of upper

activity

The motion of silicone putty is very much like that of a glacier. Prepare a V-shaped trough by folding a 4 x 6 index card lengthwise and taping it to another index card. See the diagram below. Press the putty into the top of the trough and against the end card. Mold the putty so that it resembles a glacier. Press 5 small beads in a straight line across the putty as shown in the diagram. Prepare a data table to record how far each of the beads moved each hour for 5 or 6 hours. Compare your results to Fig. 13–5, page 280.

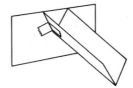

FIG. 13-5. The movement of different parts of an alpine glacier can be measured if stakes (dots) are driven into the ice. The crosses show the positions of the stakes at a later time.

ice layers and a plastic-like flow of deep layers. How rapidly any particular part of the glacier moves is determined by the slope, ice thickness, and temperature at that point. Thus not all parts of a glacier move at the same speed. By driving rows of stakes across the surface of an alpine glacier, geologists have determined that surface motion is most rapid near the center, decreasing toward the sides. The rate of movement is often one meter or more a day. See Figure 13-5.

The uneven surface movements of alpine glaciers cause *pressure ridges* to form in ice near the surface. This uneven pressure results in large cracks, called *crevasses* (kreh-*vas*-ez), appearing on the top and sides of the glacier. See Figure 13-6. Crevasses generally form across the width of the glacier. However, near the lower end they may develop in a lengthwise direction. See Figure 13-7A. Crevasses often extend more than 30 meters (100 ft) below the ice surface. The actual surface opening may be

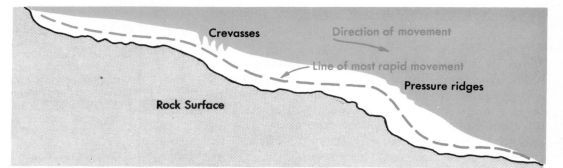

FIG. 13-6. Ridges and cracks (crevasses) appear on the surface of a glacier as a result of uneven movement.

FIG. 13-7A. A small glacier with many crevasses and pressure ridges. This type of glacier on a steep slope is often called a "hanging" glacier. (Ramsey)

hidden by a thin crust of snow which breaks at the slightest weight. Thus traveling over the top of a glacier can be very dangerous, requiring experience and great caution. See Figure 13–7B. Continental glaciers move as a result of the continuous pressure of additional snow which is added to their surface.

In an ice sheet there are usually centers from which glaciers move outward in all directions. The ice sheets which now cover Antarctica and Greenland, for instance, have a general movement outward toward the shoreline. Along the coast of Antarctica, the glacier has moved out over the sea in places to form wide ice shelves. The rise and fall of the tides are able to break off large sections of this leading front of ice. These large blocks of ice then float away as *icebergs*.

Glacial movement depends upon the balance between ice being added by snowfall and ice lost by melting and evaporation. As long as new snow is added faster than it melts, the glacier will continue to advance. The glacier will finally stop its forward movement and become stationary at the level where the ice melts as quickly as the snow is added. Very slight differences in average yearly temperatures and snowfall may upset this balance. The glacier will then advance or retreat accordingly. Thus the movement of a glacier is a sensitive indicator of climatic temperature changes.

FIG. 13–7B. A mountain climber shown scaling a treacherous glacial crevasse. (Springs)

THE WORK OF GLACIERS

FIG. 13–7C. A small round lake or *tarn* is commonly found in the bowl of a cirque. (U.S. Geological Survey)

Glacial erosion. The formation of alpine glaciers cause the landscape to become even more rugged. The glacial processes which change the shape of the mountains begin at the heads of the valleys in which the glaciers form. Frost action breaks off rock from the valley walls, causing them to become steeper. The glacier pulls loose blocks of rock from the floor of the upper valley as it moves. These actions create a bowl-shaped depression called a *cirque* (surk). This French word, meaning "circus," refers to the resemblance of the depression to a circus theater. See Figure 13–7C.

When a number of cirques are formed close together, the dividing ridges between become very sharp and jagged. These ridges are called *aretes* (a-rayts) which means "spines." See Figure 13–8. Several aretes may join to form a sharp peak called a *horn*.

FIG. 13–8. The cycle of glacier formation and the development of glacial features: cirques, aretes and horn peaks.

FIG. 13–9. An outcrop of glacially polished rock photographed north of Lake Superior. A layer of till still covers the rock at the right. (Ramsey)

As the mass of ice grows in the cirque, the glacier slowly begins to move to lower levels along the general path of the existing valley. Large amounts of rock become mixed in the ice. The rock may range in size from microscopic pieces, up to large blocks broken off by frost action or by the force of the moving ice. These rock fragments are evidence that help explain how:

1. A glacier can erode even very hard rock over which it moves.

2. The rocks buried in the ice, in turn, act like the teeth of a file as they are dragged over the rock floor.

Solid rock over which a glacier has moved often shows a polished effect from the action of the tiny rock particles in the ice. At the same time deep scratches or grooves are left by the larger rocks. See Figure 13–9. A large rock projection may be smoothed into a glaciated rock *knob*. Such a knob usually has a smoothly sloping side in the direction from which the glacier came. The other side is steeper as a result of rock being plucked away by the ice. See Figure 13–10.

As a glacier moves through a valley, the valley walls and floor are ground away. This changes the original V-shape of the valley into a U-shape. See Figure 13–11. U-shaped valleys are considered direct evidence of glacial erosion. Glaciers are the only means by which a valley can acquire a U-shape. The movement of a glacier through a valley also straightens its sharp bends.

Glaciers in smaller, adjacent valleys may often join the main glacier. As the glaciers melt, these side valleys end in mid-air high above the main valley floor. They are called *hanging valleys*. See Figure 13–8. Streams flowing through them form steep waterfalls that drop to the main valley below.

Because of their great thickness and mass, continental glaciers completely override all existing land features except for the largest mountains. The landscape that results from erosion by continental glaciers is the opposite of the sharp and rugged features produced by alpine glaciers. Continental glaciers produce relatively smooth landscapes by grinding down mountains and all other relief features. See Figure 13–12. In a few cases existing valleys may be gouged out and deepened, especially if the direction of ice movement is parallel to the valley. In most places, however, the land surface is smoothed.

Glacial deposits. The ice of a glacier may melt as a result of higher temperatures. This can happen when an alpine glacier reaches lower altitudes or when a change in climate melts the ice sheets of continental glaciers. When a glacier melts, all of the material accumulated by the ice is deposited. This forms various kinds of features which fall under the general name of *glacial drift*. It is found in two completely different deposits:

FIG. 13–10. Lembert Dome is a granite glacial knob in Yosemite National Park. The glacier flowed from the right, producing a smoothly rounded surface, then plucked off the rock at the left to create a steeper slope. (National Park Service)

FIG. 13–11. A glaciated valley in the Sierra Nevada Mountains. (Ramsey)

FIG. 13–12. These two photographs show the difference in the kind of landscapes produced by alpine and continental glaciers. One is a view of the Canadian Rockies; the other was taken in northern New York. (Ramsey)

FIG. 13–13A. Glacial till exposed along a road in New England. (Ramsey)

FIG. 13–13B. A large glacial erratic boulder deposited on top of Mt. Mansfield, Vermont. (Mary S. Shaub)

FIG. 13–14. The joining of lateral moraines to form medial moraines is clearly seen in this view of an Alaskan glacier. (Rapho-Guillumette, Lowry)

1. *Till* is the unsorted material deposited directly from the ice. It is either scraped from the bottom of the glacier or left behind when melting takes place. See Figure 13–13A.

2. *Stratified drift* is material that has been sorted and deposited in layers by the action of melt-water flowing out from subglacial streams.

Usually it is easy to spot glacial deposits. The visible rock surfaces are polished and scratched. Large boulders, called *erratics,* may be present. See Figure 13–13B. These are carried by the ice and have an entirely different composition from the bedrock.

To distinguish between till and stratified drift, it is necessary to dig into a deposit. Till is characterized by small rocks, gravel, sand, and clay all mixed together. Stratified drift is characterized by layers of rounded rocks, pebbles, and clay. When highways are cut through hills in glaciated regions, the nature of the deposits is often exposed.

Glacial deposits of till and stratified drift are responsible for many characteristic landscape features. Some of the most common of these features are described in the following sections.

Features formed by till. The material carried along by the glacier often forms ridges or mounds on the ground or on the glacier itself. These deposits are called *moraines.* Accumulations of rocky debris along the edges of a valley glacier in a string of long ridges are called *lateral moraines.* Joining of two or more valley glaciers brings the

lateral moraines together to form *medial moraines*. See Figure 13–14.

The bottom of a glacier may become so loaded with rock material that all of it can no longer be carried. When this happens, part of this load is deposited below the ice. It is then passed over as the ice moves along. All the material left beneath the glacier, along with that deposited directly from the ice when melting takes place, makes up the *ground moraine*. After the glacier has disappeared, the surface of the ground moraine is left with a gentle, hilly appearance. Much of the landscape from eastern Ohio west to the Rockies and north into Canada was formed by ground moraine.

At the melting edge of the glacier, a large moraine is usually formed. It is produced by material pushed along ahead of the moving ice or dumped by melting ice at the glacier's edge. This end deposit is called a *terminal moraine*. See Figure 13–15. In glaciated regions old terminal moraines are often seen as belts of small hills with many hollows that may contain lakes or ponds. Large terminal moraines are found in Minnesota, the Dakotas, Wisconsin, northern Illinois, Indiana, and Ohio.

A common feature of areas covered with ground moraine are *drumlins*. These are long, low mounds of till. They are rounded in shape and are often found in clusters lying with their long axis parallel to the direction of glacial movement. See Figure 13–16.

Illustrate

Draw a diagram which shows the location of each type of moraine as they would be found in a valley glacier.

FIG. 13–16. A series of drumlins near Rochester, New York. (E. R. Degginger)

Investigate
Check your local library for books on the geology of your area. See if you can find any evidence of local glacial activity. Look for examples of glacial deposits that you will recognize from your text illustrations.

Features formed by glacial meltwater. The melting of a glacier is an almost continuous process. Streams of meltwater flow from the edges and surface of the glacier, particularly during the warmer parts of the year. The glacial meltwater usually has a milky appearance, caused by the presence of very fine rock particles. This powdered rock is produced by the grinding action of rock against rock as the glacier moves. Along with the small rock particles, the meltwater carries enough drift to deposit a large *outwash plain* in front of the glacier. The plain is usually crossed by many streams carrying the meltwater.

The outwash plain is often pitted with many depressions called *kettles*. These are formed when pieces of ice break off from the glacier and are covered with drift. As the ice melts, the drift sinks and produces a depression. Kettles often fill with water to make ponds.

When the continental glaciers disappear, long, winding ridges of gravels and coarse sand may be left behind. These are called *eskers*. They are ridges of stratified drift deposited by a stream of meltwater flowing in tunnels within or beneath the glaciers. Eskers may extend for some distance over flat glaciated country, looking like a winding raised roadway. Glacial features such as eskers, produced by meltwater, are especially common where continental glaciers were melting and shrinking at the same time.

LAKES

Origin of Lakes. Any erosion process or other event that causes a depression in the earth's surface can produce a lake. The only other requirement is a plentiful supply of water.

Glaciers, a common source of basins, created lakes in several ways. As they moved over the surface, they gouged out depressions in low level areas. These filled with water and became lakes.

Hundreds of the lake basins in New England and New York were dug from solid rock by the continental glacier which lay over the region. The Finger Lakes of central New York were formed from existing stream valleys. These valleys were deepened by the ice, then dammed by glacial deposits. Almost every glacial cirque which is now exposed is occupied by a small lake called a *tarn*.

However, most of the lake basins were formed by deposits from glaciers rather than by glacial erosion. Many such lake basins were left in the uneven surface of the ground moraine that covers much of northern Europe and upper North America. The lakes were filled with water from melting ice. If they were below the level of the water table, rainfall and ground water helped to fill them.

Other glacial lakes came from terminal and lateral moraines damming existing streams. The belts of terminal moraines across Minnesota and the Dakotas produced many such lakes. Similar belts of moraines and associated lakes are found in Wisconsin, Indiana, Ohio and northern Illinois.

The Great Lakes are the direct result of a combination of glacial actions. These lakes were formed from existing broad river valleys covered by ice. Glacial erosion widened and deepened them. As the glacier melted, the meltwater flowed into these basins. The water, trapped by moraines to the south, formed dams. In their early stages, the lakes had outlets to the south through the Wabash and Illinois rivers, then to the Mississippi. However, with further melting of ice, the lakes grew in size. This caused them to drain to the Atlantic through the Susquehanna River of Pennsylvania and western New York. Later the lakes also drained through the Mohawk and Hudson valleys.

The Great Lakes grew in size until they were slightly larger than they are now. Then, as the glaciers retreated and the weight of ice lessened, the land surface was pushed up as the pressure decreased. The lake-beds were

FIG. 13–17. Steps in the formation of the Great Lakes. Note how the drainage pattern changed as the glacier retreated.

FIG. 13–18. The former glacial Lake Agassiz, the extent of which is outlined in light blue, once covered a very large area in Canada, North Dakota, and Minnesota.

Investigate
Our earliest evidence concerning Ice Age Man, has been found in bog sediment of ancient glacial lakes. Refer to the book "Bog People: Ice Age Man Preserved" by R. L. Bruse-Milford.

also uplifted, reducing them to their present size. This land shift also established final drainage to the north through the St. Lawrence River. This brief summary of the complicated history of the Great Lakes is shown in Figure 13–17.

A large glacial lake that formed at about the same time as the Great Lakes, covered parts of North Dakota, Minnesota, and the Canadian province of Manitoba. This was called Lake Agassiz (*Ah*-guh-see). It developed when the meltwaters from the continental glacier were blocked from their northward flow by an ice dam. When the ice dam melted, the lake disappeared. This body of water once covered an area greater than all the Great Lakes combined. See Figure 13–18. Lake Winnipeg in Canada is the chief remnant of the former giant Lake Agassiz.

The life history of lakes. Most lakes are relatively temporary features of a landscape. Even the largest is likely to exist for a much shorter time than many other features. There are two reasons for the rapid disappearance of lakes. First, the lake may lose its water, leaving the basin empty. Second, the lake basin itself may be destroyed.

Lakes often disappear because their water drains away. The most common cause is erosion by an outlet stream. If the outlet stream cuts its bed down below the level of the lake floor, all of the water in the lake will drain out. A lake may also lose its water because a change in climate brings dry conditions to the region. Evaporation of the water then causes the lake to disappear.

Lake basins may be destroyed by an overload of sediment. This sediment is carried in by the streams feeding the lake. The inflowing streams build deltas which grow toward the center of the lake. See Figure 13–19. Some of the sediment is washed down from neighboring slopes or by the action of waves cutting into the lake basins. Wind blows dust into the water and landslides add their share of soil. Plants growing around the lake's edges advance farther and farther into the lake. Then the lake becomes a bog or marsh. Finally the basin fills completely and becomes a meadow. See Figure 13–20.

If a lake has no outlet stream, it may become a *salt lake*. Salt lakes are similar to the sea. Since water can leave only by evaporation, the dissolved substances are left behind, making the water increasingly salty. Any lake will become salty if it is naturally without an outlet or is deprived of its outlet. If it is also located in a dry region

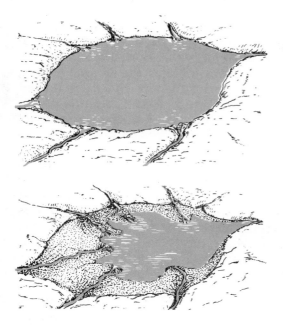

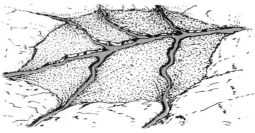

FIG. 13–19. Three stages in the filling of a lake by deposits of stream sediments.

FIG. 13–20. A small lake or pond may fill slowly or be converted into a bog by the growth of vegetation around its edges.

where evaporation is rapid, it has more of a chance of becoming salty. The composition of the water in salt lakes depends upon dissolved minerals brought in by streams and by the chemical and biological processes which take place in the lake. Substances commonly found dissolved in the waters of salt lakes include sodium chloride, and the chlorides of calcium, magnesium and potassium as well as the sulfates of sodium, calcium, and magnesium. The salt deposits left in the beds of extinct salt lakes often contain valuable minerals. For example, borax is mined from such deposits.

THE ICE AGES

Glacial periods. For unknown reasons, the earth passes through periods when the average temperature everywhere on the surface becomes lower. These periods are the ice ages or glacial periods when great continental ice sheets spread outward from the poles. There is evidence that there have been at least several ice ages during the earth's long history. An ice age seems to have taken place about 600 million years ago and another about 275 million years ago. But the most recent ice age occurred during the last one million years. Its great ice sheets retreated only during the last ten to fifteen thousand years. The areas of the Northern Hemisphere covered by glaciers during this period are shown in Figure 13–21. All of Canada and the mountainous regions of Alaska were buried under the ice sheets. In the western mountains of the United States, alpine glaciers gathered into a single ice sheet that flowed west to the Pacific and eastward to the foothills of the Rocky Mountains. A great continental glacier with its center in the Hudson Bay region spread as far south as the Missouri and Ohio rivers.

Europe was almost covered by an ice sheet centered over the Baltic Sea. It spread south to Germany and the Low Countries, and west to the British Isles. In the east the glaciers reached Poland and Russia. Alpine glaciers grew in the Alps in southern Europe and in the mountains and highlands of Siberia. Large glaciers were also formed in the Southern Hemisphere. The Andes Mountains of South America supported a large ice sheet as did the South Island of New Zealand.

The features built on the land surface by the glaciers of the most recent ice age are still fresh, and easily found. Studies of the landforms show four separate cycles of growth and retreat of glaciers during this period. These cycles seem to have been the result of brief periods of extreme cold when the glaciers advanced. Relatively long periods of warmer climate followed when the ice sheets retreated.

Possible causes of the ice ages. There is no clear explanation of the climate changes which cause ice ages. For glaciers to grow, temperatures must drop, or the amount of moisture falling as precipitation must increase, or both must happen at the same time. A number of theories have been proposed to account for the conditions necessary to produce an ice age. A brief description of some of these theories will show the general thought and the problems involved:

1. *Changes in the amount of energy from the sun.* The amount of solar energy received by the earth might change due to several factors. One would be a small change in the tilt of the earth's axis and a slight difference in the earth's orbit around the sun. Changes in the earth's motions in space could cause a slight decrease in average temperature on the earth's surface. Such changes in the tilt of the earth's axis and shape of its orbit are known to occur in 21,000- to 90,000-year cycles. However, the glacial periods have come at irregular intervals and are

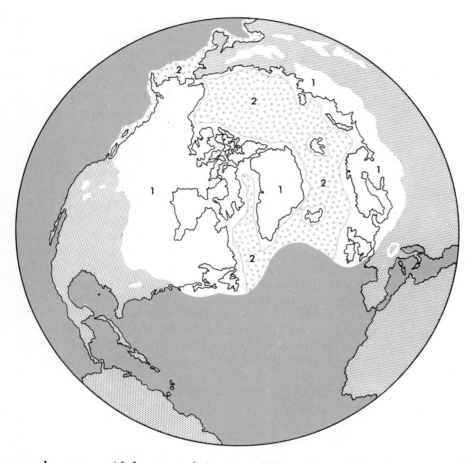

much more widely spaced in time. More than 300,000 years have gone by between some stages of the last ice age.

The amount of energy given off by the sun might vary. While the evidence is not certain, there is reason to think that the amount of solar energy produced does change slightly from time to time. However, a great deal more information about the behavior of the sun needs to be collected over a long period of time. A change in the sun's energy output, combined with other factors, might cause periods of glaciation on the earth.

2. *Topographical factors.* It may be that the ice ages are caused entirely by events which take place on the earth itself. For example, at least two of the glacial periods have occurred during times of mountain building. High mountains above the freezing level (snow line) act as snow-collecting areas and that might gradually help to reduce

FIG. 13–21. At its peak the last great glacial period covered about 30 per cent of the earth's land area. The glaciers (1) and ice pack covering the sea (2) are shown at their height. The solid lines mark present day sea coasts and lake shores.

the earth's temperature. This in turn would produce more snow and ice eventually leading to a glacial period. In addition, volcanic activity connected with mountain building might help to bring about cooler temperatures. Volcanic dust, thrown into the air, would shade the earth's surface and reflect solar radiation. Elevation of the earth's crust, particularly near the poles, might also help a glacier to form.

However, all of these topographic changes in the land surface require long periods of time. The glaciers seem to form and disappear relatively quickly. Further, it is difficult in this theory to account for the increased snowfall needed to create the large continental glaciers.

3. *Melting of the arctic ice.* Part of the Arctic Ocean is now covered by a thick layer of ice. Under present climatic conditions, this ice remains frozen. However, calculations show that if a large amount of warmer water from the North Atlantic were to enter the Arctic Ocean, the ice would melt. This does not happen now because a ridge on the sea floor keeps the two oceans from completely mixing. However, if the level of the Atlantic were to rise, water could pass over this barrier. Theoretically this is exactly what would be needed to produce an ice age.

If the Arctic Ocean were free of ice, more water would evaporate into the polar atmosphere. This would cause heavier snowfall on the continents bordering the north polar regions. The increased snowfall could possibly result in new glacier formations. But the trapped water in the glaciers would eventually cause the ocean levels to drop again. This would again result in separating the Arctic and Atlantic oceans. The Arctic Ocean would then freeze over again. The glaciers would be cut off from their supply of moisture and would begin to disappear. So the cycle of advance and retreat of the ice sheets would be complete.

The advantage of this theory is that there is no need to explain any basic change in the earth's position or motion to account for the ice ages. However, there is no evidence that removal of the ice in the Arctic Ocean would cause formation of the great glaciers. Also, the circulation of water in the oceans is not well understood. It would be difficult to establish whether mixing of Arctic and Atlantic water could actually take place on a large scale.

Many more questions need to be answered before a satisfactory explanation of the advances and retreats of the glaciers can be found.

VOCABULARY REVIEW

Match the word or words in the column on the right with the correct phrase in the column on the left. *Do not write in this book.*

1. Grainy ice formed from snow.
2. A large body of moving ice.
3. Large cracks in a glacier.
4. A chunk from a glacier "floating" in the sea.
5. Bowl shaped depression in which a glacier grows.
6. Material deposited directly from a glacier as it melts.
7. Large boulders left by a melting glacier.
8. Ridges or mounds formed on a glacier by the material it carries.
9. An elongated mound usually formed by continental glaciers.
10. Depressions found in an outwash plain.
11. A cirque filled with water.
12. Formed by continuous evaporation from a stationary body of water.

a. drumlin
b. salt lake
c. till
d. firn
e. tarn
f. cirque
g. eskers
h. kettles
i. crevasses
j. aretes
k. glaciers
l. moraines
m. iceberg
n. erratics

QUESTIONS

Group A

Select the best term to complete the following statement. *Do not write in this book.*

1. For water falling on the land to remain for long periods of time very near where it falls, it needs to (a) fall as snow (b) fall as hail (c) fall where the temperature on the ground is at or below freezing (d) be absorbed by a plant.

2. Portions of ice fields become a glacier when (a) movement of the ice begins (b) the spring thaws occur (c) snow continues to fall (d) the temperature remains below freezing.

3. The elevation in mountains above which snow remains all year is called the (a) frost line (b) ice line (c) snow line (d) glacier line.

4. An ice field forms when (a) snow falls at least once each year (b) more snow falls than melts each year (c) an equal amount of snow falls and melts each year (d) a crust of ice forms on deep snow.

5. When ice moves out from a mountain ice field. (a) a continental glacier is formed (b) an iceberg is formed (c) an esker is formed (d) an alpine glacier is formed.

6. At the present continental glaciers are found only (a) on the North American continent (b) in the Southern Hemisphere (c) in the Antarctic (d) in the polar regions.

7. The movement of an alpine glacier is best described as (a) a combination of sliding motion for top layers with a flowing motion for lower layers (b) only a sliding motion (c) only a flowing motion (d) a combination of flowing of upper layers with sliding of lower layers.

8. A direct evidence of the uneven movement of glaciers is the formation of crevasses and (a) icebergs (b) pressure ridges (c) terminal moraines (d) drift.

9. Glaciers are a sensitive indicator of climate changes since they are (a) found where most weather occurs (b) found high in the mountains (c) in delicate balance between addition of ice by snowfall and loss of ice by melting and evaporation (d) easily checked for average depth which is an indicator of temperature change.

10. A sharp peak formed by the joining of several aretes is called a (a) cirque (b) horn (c) till (d) tarn.

11. Hanging valleys, such as those found surrounding Yosemite Valley in Yosemite National Park in California, were formed (a) by water erosion (b) by wind erosion (c) by meltwater from a continental glacier (d) by several small glaciers joining one larger glacier.

12. The erosional features caused by alpine and continental glaciers (a) are different in many ways (b) are very similar (c) occur only while melting (d) are caused by the ice itself.

13. The material deposited by a glacier is called (a) drift (b) arete (c) cirque (d) kettle.

14. Material deposited in layers by meltwater is called (a) erratic (b) moraine (c) stratified drift (d) till.

15. Glacial deposits can frequently be recognized since visible rocks are (a) jagged and sharp (b) weathered and dark in color (c) often polished and scratched (d) rounded and smooth.

16. An accumulation of rocky debris along the edges of a valley glacier often forms long ridges called (a) lateral moraines (b) medial moraines (c) ground moraines (d) terminal moraines.

17. Material deposited beneath a glacier is called a (a) lateral moraine (b) medial moraine (c) ground moraine (d) terminal moraine.

18. Long ridges made of gravels and coarse sand exposed when a continental glacier melts are called (a) kettles (b. lateral moraines (c) medial moraines (d) eskers.

19. Glacial features produced by meltwater are especially common (a) where valley glaciers are melting (b) where continental glaciers melted (c) near where valley glaciers formed (d) in the polar regions.

20. Which of the following is a feature most commonly produced by glaciers? (a) waterfalls (b) V-shaped valleys (c) lakes (d) plateaus.

21. Any erosion process or other event that causes a depression can produce a lake if (a) rainfall is plentiful (b) a glacier melts to give it water (c) the water table of the surrounding land is higher than the depression (d) any of these may produce the lake.

22. Almost every glacial cirque that is now exposed is occupied by a small lake called a (a) tarn (b) till (c) finger lake (d) basin.

23. Final drainage of the Great Lakes to the north happened when (a) dams were erected on the Wabash and Illinois Rivers (b) the continental glacier eroded a new path (c) the continental glacier melted, allowing the surface to bulge upward (d) the Great Lakes took on extra water.

24. Lake Winnipeg in Canada is the remnant of the former giant (a) Lake Superior (b) Lake Agassiz (c) Lake Erie (d) Lake Iroquois.

25. Most lakes are relatively temporary features of a landscape, they generally disappear when (a) an outlet river cuts a channel draining the lake (b) the lake basin is destroyed by becoming filled with sediment from incoming streams (c) landslides fill the basin (d) all of these are ways a lake is destroyed.

26. A salt lake forms when (a) water drains from mountains rich in minerals (b) not enough rain falls (c) the lake has no outlet stream (d) the lake has no inlet streams.

27. The most recent ice age occurred about (a) 600 million years ago (b) 275 million years ago (c) 1 million years ago (d) 1 thousand years ago.

28. The last ice age produced a polar ice sheet that (a) covered all of Canada and Alaska (b) extended south to Germany (c) covered the British Isles (d) all of these are correct.

29. In order for glaciers to grow (a) temperatures must drop (b) more precipitation must form (c) temperatures must drop or more moisture must precipitate (d) any one of these will cause a glacier to grow.

30. The ice ages were caused by (a) a small change in the tilt of the earth (b) excessive mountain building (c) there is no correct explanation at this time (d) melting of the Arctic ice pack.

Group B

1. Describe how a glacier is formed.
2. In what general locations would you expect to find glaciers at present?
3. What are the two main types of glaciers?
4. Describe the kinds of motion shown by the parts of a typical glacier.
5. What causes crevasses and pressure ridges in glaciers?
6. How do rocks become mixed with glacial ice?
7. How does the movement of a glacier through a valley change the shape of the valley?
8. How does the erosion caused by continental glaciers differ from that caused by alpine glaciers?
9. How can one tell the difference between till and stratified drift?
10. Name the kinds of moraines and describe how each is formed.
11. Describe three features formed by glacial meltwater.
12. In what ways are lakes formed by glaciers?
13. What reason can you give for the main drainage of the Great Lakes being north through the St. Lawrence River?
14. In what ways might a lake lose its water?
15. What processes aid in the filling of a lake bed with sediment?
16. What areas were covered by glaciers in North America during the last ice age?
17. What are three popular theories about the causes of ice ages?
18. Why is it said that no present theory can really explain what causes ice ages?

14
shorelines

objectives

- ☐ Explain why shorelines constantly change.

- ☐ Describe the features that may be seen along shorelines with marine cliffs and beaches.

- ☐ Trace the movements of materials along a shoreline.

- ☐ Relate the general development of a coastline to changes in sea level.

- ☐ Describe the appearance of a shoreline with barrier islands.

- ☐ List the kinds of coral reefs.

- ☐ Explain how shorelines are affected by human use.

A surfboard and its rider cannot ride a wave without using energy. That energy is supplied by the wave. Sometimes some of the energy contained in the wave is spent moving the surfer and the surfboard in different directions, as shown in the photograph above. A surfer who is upset and tumbled in the surf is probably very aware of the power in a wave. More common evidence of the energy in waves can be found anywhere the land and the sea come together.

Almost all of the energy stored in waves comes from the sun. Solar heat added to the atmosphere causes the winds to blow over the surface of the sea. Energy contained in the moving air is transferred to the water by processes described in Chapter 17. Waves are the result of this shift of solar energy from the atmosphere to the sea. Energy accumulated by waves far out at sea is released when the waves meet the edges of the continents. As a result, the boundaries between the land and the sea become a *shoreline*. A shoreline is a place where the oceans, land, and atmosphere meet to create one of the most changeable parts of the earth's surface. Shorelines are a very temporary feature of the earth's surface. They result from the ability of waves to erode the rock and to transport and rearrange the fragments.

WAVES AND ROCK

Wave action. The power of waves striking the rocks along a shoreline can actually shake the ground like a small earthquake. Seismographs near the shore often record the vibrations made by waves hitting solid rock at the shore.

The direct force of the waves is great enough to break off pieces of the rock. These fragments are, in turn, thrown back against the shore, where they cut into the solid rock. As the waves cut back into the shore, the rock fragments grind together in the tumbling water. This action eventually reduces most of the fallen rock to small pebbles and sand grains. In addition to this mechanical erosion along the shoreline, chemical weathering attacks the rock. Wave action forces sea water and air into every small crack in the rocks. The chemical action of the water and air enlarges the cracks and openings. This allows the mechanical erosion of the waves to become more effective in breaking up the rock.

Much of the erosion along a shoreline takes place during storms. Large waves like those shown in Figure 14–1 can deliver tremendous amounts of energy to the shore. Blocks of rock weighing several tons can be broken off and moved around by such waves. A severe storm may completely change the appearance of a stretch of shoreline in a day's time.

Sea cliffs. When waves strike directly against the rock of the land, erosion usually produces a *marine cliff*, such as the one shown in Figure 14–2. At first only a small cliff is formed as the waves erode the rock. Waves continue to attack the base of the cliff until a notch is finally cut. Undercutting of the cliff causes the overhanging rock to fall. Rock debris accumulates at the base of the cliff. This material is slowly ground up by wave action. As the process continues, the cliff is gradually worn back and becomes steeper. Many shorelines are made up almost entirely of high and nearly vertical cliffs.

How rapidly a sea cliff is worn away by wave erosion depends upon the nature of the exposed land. Soft materials are worn away very rapidly. For example, the cliffs made up of loose glacial deposits along the shore of Cape Cod are retreating by about one meter each year. Old maps show that cliffs made of soft rocks along parts of the shoreline of England have been worn back by several kilometers during the past two thousand years. On the other hand, shorelines made of hard rock show little change over hundreds of years.

As a sea cliff is worn back, a nearly level platform is usually left extending out beneath the water at the base of the cliff. This platform is called a *wave-cut terrace*. As the waves cause the cliff to retreat and the terrace to

FIG. 14–1. Most of the erosion along shorelines takes place when large storm waves hit the shore. (L. R. Fairbanks)

FIG. 14–2. Sea cliffs develop where waves directly meet the land. (Ramsey)

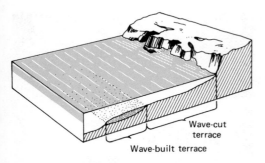

FIG. 14–3. This diagram shows the positions of a wave-cut terrace and a wave-built terrace at the base of a sea cliff.

widen, much of the ground-up rock taken from the cliff is carried away from the shore. This material may then be dropped some distance from shore, building an extension to the wave-cut terrace called a *wave-built terrace*. Figure 14–3 shows these two types of terraces.

In many places along shorelines made up of cliffs, the terraces have become very wide. As a result, waves lose most of their energy in the shallow water some distance out from the base of the cliff. In such places, the rate of wave erosion of the cliff is greatly reduced. Wave erosion may be halted unless waves or currents erode the terraces and again expose the cliff to the force of the waves. An increase in sea level can also allow wave erosion to begin again.

Wave erosion of a cliff seldom wears it away evenly. Fingers of harder rock may be left projecting out from the cliff. Erosion along the sides of these projections can cut through to produce a *sea arch*. See Figure 14–4. Continued erosion of a sea arch may cause it to collapse, leaving an isolated column of rock called a *stack*. See Figure 14–5. In time, the arches and stacks are eroded away, leaving only the wave-cut terrace. Along many marine cliffs, erosion penetrates deeply into the face of the cliff where there is a joint. The result is often the formation of a large hole or *sea cave*.

FIG. 14–4. A sea arch is made by waves cutting through the sides of a column of rock. (Josip Ciganovic/ Alpha)

FIG. 14–5. Sea stacks have been left standing on the wave-cut terrace as waves created cliffs on this part of the Oregon coast. (Ramsey)

Beaches. Waves work in two ways to create various features found along a shoreline. Waves erode the land. At the same time, wave action moves and deposits the rock fragments produced by wave erosion. *Beaches* are examples of a shoreline feature formed by the ability of waves to transport and deposit materials. A beach is a deposit of rock fragments along a seashore or the edge of a lake. A beach begins to grow when wave action moves small rock fragments forward as the wave washes up on the shore. As the water retreats, some rock fragments are moved again away from the shore. Beaches are formed at any place along the shoreline where conditions cause more rock fragments to be moved toward the shore than away.

Most people think of beaches as being made of light-colored sand. Actually, the size and kind of materials found on beaches can vary widely. Many beaches have no sand, but are covered with pebbles or larger rock fragments. See Figure 14–6. The various sizes of rock fragments found along beaches are given in Table 14–1.

The composition of beach materials depends upon the rock source. For example, granitic rock yields light-colored sand fragments composed mostly of quartz and feldspar. Beaches made of this type of sand are common along the North American shorelines because granitic rocks are abundant. On Hawaii and other volcanic islands, beaches may be made of black sand. See Figure 14–7. This sand comes from the dark-colored basaltic rocks that make up

Table 14–1 Beach Materials

Description	Size (millimeters)
Boulders	More than 200 (over 8 in)
Cobbles	76 to 200 (3 to 8 in)
Gravel	
Coarse	19 to 76
Fine	5 to 19
Sand	
Coarse	2 to 5
Medium	0.4 to 2
Fine	0.07 to 0.4
Silt	less than 0.07

FIG. 14–6. The beach at the base of these cliffs is made of pebbles and cobbles. (Ramsey)

FIG. 14–7. This beach in Hawaii is made of black sand from volcanic rocks seen in the foreground. (Ramsey)

Investigate and Interpret
Obtain samples of beach
materials from a nearby lake or
ocean beach. Study the
materials, using magnification
to determine their identity.
Suggest the origin of the
particles and how they ended
up on the beach.

these islands. Beaches along the Oregon-Washington shoreline also are often made of dark, volcanic sand. In many locations, sand is carried to the shore by rivers. Thus much of the sand found along a beach may not have come from rocks along the shore. Its source may be far up the channel of a nearby river. In some warmer regions, the beaches can be made of fragments of sea shells and coral. Beaches along the Florida shoreline are made of this type of material.

Although beaches may differ a great deal in the size and composition of the fragments, their materials are always moved in the same general way. All sizes of fragments are moved toward and away from the shore by wave action. However, the smaller and lighter fragments, such as sand grains, are moved most easily. In the following discussion of the changes in beaches, only the motion of sand grains will be considered.

Each succeeding wave reaching the shore moves the individual sand grains slightly forward. Although each advance is small, several thousand waves per day can move sand grains a considerable distance. The sand piles up on the shore, producing a sloping platform. During high tides or when large waves occur, sand is deposited at the back of a sloping beach. This causes most beaches to have an inner raised part called the *berm*. The berm is the part of the beach most often used by people when they go to the beach.

Most beaches change their appearance between winter and summer. Winter storms create large waves that carry a heavy load of sand up over the berm and deposit it on top. Thus large waves often increase the height of the

FIG. 14–8. Features that might be found along a beach are shown in this diagram.

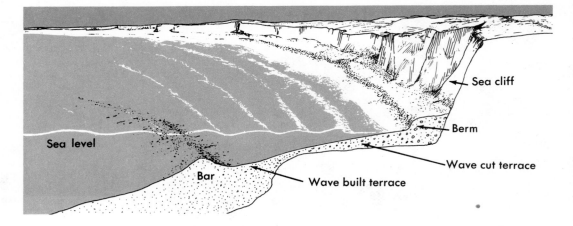

Sea level

Bar

Wave built terrace

Wave cut terrace

Berm

Sea cliff

berm during the winter. But the same waves also pick up sand and move it seaward. So the winter berm is usually higher than the summer berm, but not as wide. The sand carried away from the shore is deposited in deeper water just offshore. This creates a long underwater ridge called a *bar*. Because bars are usually underwater, they are not seen except at very low tides. Figure 14–8 shows these features of a beach.

Longshore movements. Waves almost never move straight onto a shoreline. They generally approach the shore at an angle. This motion affects the way sand grains and other materials on a beach are moved by the waves. A wave that rolls up on the beach and then washes back at an angle will not pick up and move sand grains with a straight in-and-out motion. The sand grains will be moved a short distance to one side. See Figure 14–9. This results in a zigzag motion of the sand grains along the beach in a direction determined by the way the waves approach the shore. Another effect of waves moving at an angle to the shoreline is the *longshore current*. The longshore current is a movement of water parallel with and close to the shoreline. See Figure 14–10. It is caused by the slight push given to the water in a direction along the shore by waves hitting at an angle. The longshore current also tends to move beach materials along the shore. As a result of these two effects of waves striking the shore at an angle, there is usually a general movement of sand parallel to the shoreline in one direction or another.

Sand that is moved along the shore will keep moving

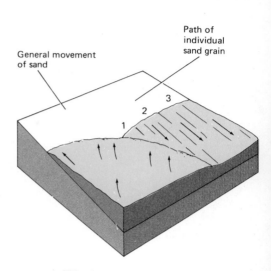

FIG. 14–9. Waves washing up on the beach at an angle tend to move each sand grain in a zigzag path parallel to the beach. This causes a general movement of sand along the beach.

FIG. 14–10. Longshore currents flow parallel to the shore (black arrows). They are the result of waves striking the shore at an angle.

FIG. 14–11. A spit has formed from the sand moved along the beach toward the bottom of the photograph. (John S. Shelton)

until the shoreline changes direction. A projection of the shore (a headland) or the opening to a bay slows down the longshore movement and builds up deposits of sand. A deposit of sand in which the outer end is surrounded by water is called a *spit*. See Figure 14–11. Often the end of a spit is curved inward or outward by tidal currents into the shape of a hook. A chain of spits frequently develops between islands, thus connecting them. These connecting ridges of sand, called *tombolos*, may also link an island to the mainland.

The beach budget. Beaches, if they are not disturbed, are usually in a state of balance. The balance exists between the processes that remove sand from the beach and those that add it. Sand is removed from beaches in several ways. It may be moved down the shoreline by wave motion. Large waves can remove sand from the beach and carry it too far offshore to be returned. On some beaches, the wind blows sand toward the land, forming ridges of dunes. Most sand added to beaches comes from rivers. The sediment carried to the sea by rivers can be deposited as deltas along the shore. In many locations, however, the longshore current carries the sediment along the shore and creates beach deposits rather than deltas. Erosion of sea cliffs may also add sand to beaches. This is particularly true if the cliffs are made of soft materials. Over a long period of time, the removal and addition of sand to a natural beach are nearly equal. The beach does not permanently grow wider or shrink. Its sand budget is balanced.

Many beaches have not been allowed to remain in their natural state of balance. As the population of the United States has increased, more houses and other structures have been built on beaches. Unless a careful study of sand budget of the shoreline has been made, the presence of the structures can change the beaches. For example, a pier that extends out from the beach can block the longshore current. Sand normally moved by the current will be dropped on one side of the pier. The beach will become wider on that side. On the opposite side, the beach will not receive its normal supply of sand. The beach will gradually disappear from that side. See Figure 14–12. Any structures built near a beach may disturb the balance of its sand budget in some way. If this fact is not carefully considered, the destruction of the beaches can result.

FIG. 14–12. Any structure built out from a beach can interfere with the movement of sand along the beach. (John S. Shelton)

COASTLINES

Changing sea level. A shoreline is a narrow boundary where the land and sea meet. Features formed along a shoreline, such as cliffs and beaches, are mainly the result of the work of waves. A *coast* is a larger region that often has a width of several kilometers inland and several hundred kilometers along the shoreline. Coasts are parts of the continents that are affected by changes in the level of the sea relative to the land.

Scientists believe that the earth has passed through several ice ages during its history. During each ice age, the accumulation of snow and ice in the glaciers represented a large share of the earth's water supply. The most recent ice age occurred during the last one million years. Based on the area covered by the ice sheets and their estimated thickness, scientists conclude that the glaciers must have held about 70 million cubic kilometers of ice. The total amount of ice contained in glaciers at the present time is about 25 million cubic kilometers. Thus enough water must have been taken from the oceans during the last ice age to make up the extra 45 million cubic kilometers of ice. This would lower the sea level by about 130 m (426 ft). That lowest sea level was reached about 15,000 years ago. Since then, the glaciers have been melting and the sea level gradually rising. See Figure 14–13. Sea level is now rising at an average rate of about 1 mm per year. This is equal to about 1 foot per century. If all the glaciers were to melt, the oceans would rise about 65 m (213 ft). Many of the largest cities in the world, such as New York, Tokyo, and London, would then be submerged. This would also be the fate of other coastal regions, where a large part of the world's population lives.

Coastlines may become covered by water or exposed when a change in the amount of water in the oceans causes sea level to rise or fall. However, the shore may also be flooded or uncovered if the land sinks or is lifted. For example, in some parts of Europe, particularly in Scandinavia, the shoreline is being lifted. Docks at the water's edge have actually been raised completely out of the water during the past several centuries. Apparently the heavy ice sheet that covered this region during the last ice age pushed the land down. As the ice sheet retreated, this weight was removed and the land began to come up. Parts of North America near Hudson's Bay are also rising, prob-

Thousands of years ago

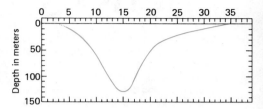

FIG. 14–13. This graph shows how sea level has changed over the entire earth during the past 35,000 years.

Interpret
The presence or absence of a continental shelf can influence the type of shoreline that develops. Describe the shoreline features that could be expected to develop where a continental shelf exists.

activity

The shorelines are temporary features whose constant changes are the result of many forces at work at the same time. Some of these forces result in land rising or sinking a measurable distance. The amount of this rise or sinking for the North American continent is shown in Fig. 10–14, page 213. Fig. 14–14, p. 305, shows how erosion causes a change in a shoreline that is sinking.

In the table below, five figures showing widely different locations of shorelines are listed. Copy and complete the table. You will need to look at each figure, one at a time, and at Fig. 10–14 to determine the rate of rising or sinking. Write or draw a description of what the shoreline might look like in several thousand years if the rising or sinking continues at the same rate. *Do not write in this book.*

Figure	Shoreline Location	Rate of Rising or Sinking	Description after 5,000 yrs
14–15	NW U.S.		
14–6	NE U.S.		
14–11	Central East Coast of U.S.		
14–16	SW U.S.		
14–18	Gulf Coast of U.S.		

ably for the same reason. See Figure 10–14, page 213.

Another cause of the lifting or sinking of a coast is the large-scale movements of the earth's crust discussed in Chapter 10. For example, much of the eastern coast of the United States is sinking. Some scientists believe that the edge of a continent sinks if it is on the trailing side of a moving crustal plate. This describes the eastern side of North America, which is carried on a plate moving in a westerly direction. Also, coastlines along the edges of continents near a plate boundary may be exposed to a variety of forces created as the plates come together. Some parts of such coasts may be lifted, while other parts sink. This situation seems to exist on the Pacific coast of the United States, which shows evidence of rising in some locations and sinking in others. For example, much of the coast of Southern California has been lifted during the past few million years. On the other hand, the coast from northern Oregon through Washington has been mostly submerged.

Origin of coastline features. Like all other parts of the land, coastlines are always changing. What we see at a particular place on the coast is like a photograph taken of a parade. The picture will show only one view of a series of events. The features found along a coastline are also part of a constantly changing series of events. Rise and fall of sea level, lifting and sinking of the land, erosion caused by waves, and deposition of sediments all work to make coastlines one of the most rapidly changing parts of the earth's surface. Scientists studying a coastline must try to guess what happened in the past and what developments might occur in the future. This is a difficult task, since most coastlines are the result of many different processes. The conditions responsible for the features of many coastlines are not easily determined. However, many coastline features can be described as though they are caused by a change in sea level relative to the land.

Think of a region near the coast where streams have cut valleys leading down to the sea. Then sea level rises or the land sinks. The sea moves in and down the river valleys to create a *submerged* coastline. It has many bays and inlets. Divides between neighboring valleys become headlands and points jutting into the sea. Beaches are usually short, narrow, and rocky. The highest parts of the submerged land may remain above water as offshore islands. See Figure 14–14A. Irregular coastlines in many parts of

FIG. 14–14. The cycle of erosion of a coastline, with features resulting from submergence. A—initial stage; B—youth; C—maturity; D—old age.

the world seem to have been formed by submergence of existing hills and valleys. Much of the northeast coast of the United States is a submerged coast. In time, such submerged coastlines are changed by wave erosion. The headlands are worn back and the shoreline becomes straighter, with longer and wider beaches. See Figure 14–14B, C, D. The relative sea level usually changes again before a submerged coastline reaches its later stages of development.

When a large river valley is flooded, it may produce a wide, shallow bay that extends far inland. Fresh water from the river mixes with the sea water in the upper parts of the bay. Such a bay is called an *estuary* (*es*-chu-ehr-ee). Many of the world's best harbors, such as San Francisco Bay, are estuaries. Estuaries may be an important source of seafood. Shellfish often are found in large numbers in the slightly salty water of estuaries. Any serious pollution of the river flowing into an estuary can destroy this important source of food.

The coasts of Alaska, Norway, Greenland, Chile, and Western Canada show many examples of an unusual type of submerged valley. Extending down to the sea along these coasts are many deep, U-shaped, drowned glacial valleys. These valleys create coastline with very narrow, deep, steep-walled bays called *fiords*. See Figure 14–15.

FIG. 14–15. A fiord in Norway. (Springs)

FIG. 14–16. The step-like features in this photograph are old wave-cut terraces exposed by uplift or drop in sea level. (John S. Shelton)

Not all coastlines are the result of a relative rise in sea level. If the land rises or there is a general drop in sea level, an *emergent* coastline can be produced. An emergent coast that has a steep slope and is exposed rapidly will usually develop a jagged shape dotted with sea cliffs, narrow inlets, and bays. In some cases a series of wave-cut terraces will be exposed, as shown in Figure 14–16. Some emergent coastlines can have a gentle slope. For

example, a part of the sea floor may be lifted and exposed to produce a relatively smooth coastal plain. This type of coastline has few bays or headlands and many long wide beaches. Much of the coast of Florida is this type of emergent coastline.

Barrier islands. The Atlantic and Gulf coastline of the United States, from New England to Texas, is mostly a relatively flat coastal plain. As sea level has risen during the past 15,000 years, the shoreline has moved inland as much as 150 km across this gently sloping coast. Large amounts of sand were moved with the advancing shoreline. Sediments in deltas built by coastal rivers were also exposed to wave action as the shoreline moved toward the land. When the rise in sea level slowed, waves, currents, and winds moved the sand shoreward to form *barrier islands*. These are long ridges of sand running nearly parallel to the shoreline 3 to 30 km (about 2 to 20 mi) offshore. Barrier islands are typically 2 to 5 km (about 1 to 3 mi) wide and 10 to 100 km (about 6 to 62 mi) long. Winds blowing toward the land usually create a line of dunes on the shoreward side of the islands. The dunes are usually 3 to 6 m (10 to 20 ft) high, and on a few islands may reach a maximum height of 30 m (98 ft). The eastern and southern coastline of the United States between Cape Cod, Massachusetts, and Padre Island, Texas, has nearly 300 barrier islands of various sizes. See Figure 14–17.

Between most barrier islands and the mainland is a narrow region of shallow water called a *lagoon*. Lagoons are often nearly filled with mud that comes from sedi-

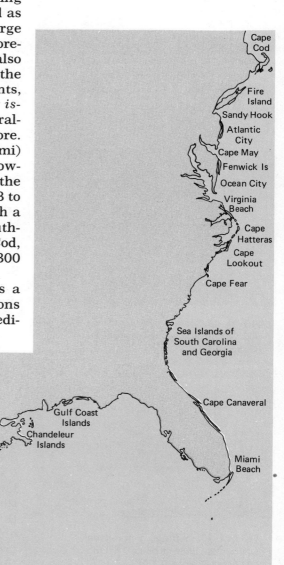

FIG. 14–17. This map shows the main groups of barrier islands along the Atlantic and Gulf coastlines of the United States.

FIG. 14–18. A barrier island is off the shoreline at the right. (J. R. Williams/FPG)

FIG. 14–19. Storm waves destroyed most of the houses on this barrier island. (U.P.I.)

FIG. 14–20. A fringing coral reef surrounds this tropical island. (Nicholas DeVore III/DPI)

ments brought in by streams. At high tide, water may cut channels through a barrier island to form tidal inlets into the lagoon. As a result, lagoons often have large areas that are *tidal flats*. A tidal flat is a muddy or sandy part of the shoreline that is usually above water but is flooded at high tides. See Figure 14–18. Most tidal flats are covered with plants that are able to grow in the salty water.

The low elevation of barrier islands allows them to be severely eroded by very high tides or large storm waves. During a storm, water may spill across the island and flow toward the lagoon. Sand is removed from the ocean side and moved toward the inland side of the island. This action is causing most barrier islands to migrate toward the mainland. Thus all barrier islands are very unstable, rapidly changing parts of the coastline.

Barrier islands have become popular recreation and resort areas. Houses and other buildings have been constructed on many of the islands, particularly during the past two decades. However, barrier islands can be dangerous places to live. Structures on these islands are often built on beaches that are exposed to very rapid erosion. The beaches on some barrier islands are eroding back at an average rate of 20 m (66 ft) per year. Many houses built along barrier island beaches have collapsed as the beach has disappeared. Storms can cause severe damage to structures on barrier islands in a short period of time. See Figure 14–19. A very large storm, such as a hurricane, could cause widespread loss of property and life among the growing populations of these fragile coastal islands.

Coral reefs. Corals are animals that are found in large numbers in warm, shallow waters. They are very small, soft creatures with a tube-shaped body form known as a *polyp*. A coral polyp is able to extract calcium carbonate from sea water to build a hard skeleton. Polyps grow rapidly and are usually attached to each other, forming large colonies. A coral reef is a coastal feature built mostly of great numbers of the accumulated skeletons of these colonies.

One kind of coral reef is found around many tropical islands. See Figure 14–20. It is believed that these coral reefs form around a volcanic island that grows up from the sea floor. See Figure 14–21A. As the volcano grows older and stops erupting, the corals in the shallow water around the shoreline build a *fringing reef*. See Figure 14–

21B. Spreading movement of the sea floor away from the elevated mid-ocean ridges carries the volcanic island into deeper water. Slow sinking of the island allows the coral reef to build up. It becomes a *barrier reef* surrounding the remnant of the volcanic island. See Figure 14–21C. Finally the island disappears completely, leaving a nearly circular coral reef, now called an *atoll*, surrounding an open-water lagoon. See Figure 14–21D.

Large amounts of coral are also found in shallow tropical waters near the continental coasts. For example, much of the floor of the shallow ocean around the Bahama Islands is covered by thick coral reefs.

Where people and oceans meet. The shorelines are not only a narrow margin of land where the oceans meet the continents. They are also a valuable resource to the people who live on the land. Among the most important uses of the coastal lands are ports, shipbuilding, industrial and residential development, recreation, commercial fishing, and dumping ground for wastes. Any shoreline is a mixture of cliffs and beaches, bays and harbors, lagoons and estuaries, islands and spits. All of these are combined in different ways to produce the individual shorelines of the world. But these coastal resources are fragile, and can be easily damaged or destroyed by human use.

As world population grows, the shoreline resources will be put to even heavier use. It is estimated that by the year 2000, half of the total population of the United States will live on 5 percent of the country along the coasts of the Atlantic and Pacific oceans and the Great Lakes. Dredging and filling operations, as shown in Figure 14–22, will increase. Pollution in all forms, which has already destroyed one-tenth of the shellfish-producing waters of the country, will continue to increase at a rapid rate. One particularly destructive kind of coastal pollution can occur when petroleum is accidentally spilled into the sea. The oil floats on the surface and may be carried onshore by winds and currents. Much of the shoreline life is killed by the oil. Although the affected area on the shore will eventually cleanse itself, its ability to support plant and animal life can be destroyed for years. Oil spills will become a constant threat as more tankers move near the coasts and oil wells are drilled offshore. If future use of the shoreline resources parallels the past, much of the existing coastal land of the nation will be destroyed.

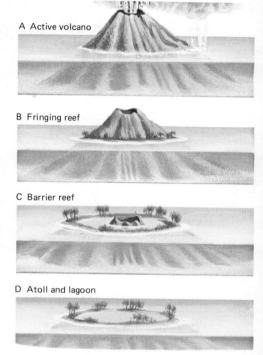

A Active volcano

B Fringing reef

C Barrier reef

D Atoll and lagoon

FIG. 14–21. Stages in the growth of a coral reef around a volcanic island.

Construct
Study the diagrams in Fig. 14–21. Use modeling clay or plaster of Paris to produce a class model of each of the four stages in the formation of an atoll.

FIG. 14–22. These two photographs of a portion of the Florida shoreline were taken twenty years apart.

Investigate
Locate a lake or a portion of a river near where you live. Find out if any of it is used for drinking or recreation, and if any part of it is polluted. See what you can find out about any actions taken to correct or control possible future problems.

Wise use of the coastal zone depends upon an understanding of the natural forces that created the shoreline features. With such an understanding, it should be possible to predict and control harmful changes in the natural course of events. It may even be possible to correct past mistakes. But knowledge that some changes in shorelines might be destructive does not always solve the problems. Control over development of coastal regions is divided among many separate groups. Some of the shore is privately owned. The remainder is managed by towns, cities, counties, states, and the federal government. There is no single group that can provide the leadership needed for an organized development of the coastal resources.

VOCABULARY REVIEW

Match the word or words in the column on the right with the correct phrase in the column on the left. *Do not write in this book.*

1. The raised portion of the beach on which people rest and play.
2. A separate column of rock found along oceanside cliffs.
3. Causes the movement of beach sand parallel to the shoreline.
4. Sand piled up in the quiet water protected from long-shore currents.
5. Ridges of sand connecting two land bodies.
6. A bay located at the mouth of a river.
7. The result of submergence of a glacial valley.
8. A bar that is above average sea level.
9. A body of water found between a barrier island and the mainland.
10. A circular reef enclosing a lagoon.

a. fiord
b. lagoon
c. atoll
d. tombolos
e. berm
f. estuary
g. barrier island
h. sea stack
i. spit
j. sea cave
k. sea arch
l. longshore current

QUESTIONS

Group A

Select the best term to complete the following statements. *Do not write in this book.*

1. The place where the oceans, land, and atmosphere all meet to create one of the most changeable of all parts of the earth is called a (a) beach (b) volcano (c) shoreline (d) fault zone.
2. Which one of the following events is not considered a part of mechanical wave erosion? (a) air and water pressure breaking the rock (b) pieces of rock thrown against the shore (c) small fragments of rock rubbing each other in the tumbling water (d) water and air forced into cracks dissolve some rock, enlarging the cracks.
3. When waves act directly against the rock of the land, erosion forms a (a) marine cliff (b) bar (c) spit (d) hook.
4. Waves attacking the shoreline of Cape Cod caused some of the shoreline to re-treat (a) about 10 meters since the time of the Romans (b) one meter each year (c) several kilometers each year (d) about 1 meter in the past century.
5. In the formation of a wave-cut terrace, some of the rock may be transported sea-ward to form an extension called a (a) bar (b) spit (c) hook (d) wave-built terrace.
6. Rock terraces along the shore can reduce wave erosion. Wave erosion may be halted entirely unless (a) the sea level is unchanged (b) the sea level is raised (c) a bar forms (d) a spit forms.
7. Sea caves result when wave erosion (a) occurs along a deep joint (b) forms a headland (c) forms sea arches (d) causes a sea arch to collapse.

8. A deposit of rock fragments along a seashore or the edge of a lake as a result of wave erosion and currents is called a (a) wave-cut terrace (b) wave-built terrace (c) beach (d) shoreline.

9. Analysis of some beach sand shows it is composed of rock fragments that are foreign to the shoreline. This sand most likely came from (a) the ocean floor (b) a volcano (c) sea caves (d) rivers.

10. If a beach is composed of coarse sand containing quartz and feldspar, it most likely is the result of wave action on (a) basalt rock (b) granite rock (c) limestone (d) clay.

11. Although beaches may be composed of various sized rock fragments, their materials always (a) make the same type of movements (b) have the same composition (c) have a white color (d) have the same shape.

12. Waves reaching the shore move sand particles (a) toward the shore (b) away from the shore (c) in circles (d) up and down in the same spot.

13. During the winter months, the large waves carry sand from the beach, depositing it just offshore to form a (a) spit (b) berm (c) bar (d) terrace.

14. Winter storms produce larger waves that can form a (a) low, wide berm (b) berm with a gentle slope (c) high, steep berm (d) beach made up of smaller rock fragments.

15. Longshore currents are usually the direct result of (a) waves striking the shore at an angle (b) deep sea currents (c) surface currents coming close to shore (d) very wide beaches.

16. Longshore currents result in the movement of sand (a) in and out from shore (b) higher on the berm (c) out to deeper water (d) parallel to the shoreline.

17. If not disturbed by human activities, the sand budget of a beach (a) is in an unbalanced state (b) is in a balanced state (c) causes it to shrink in size (d) causes it to increase in size.

18. The Pacific coast of the United States is rising in some places, sinking in others. This is due to (a) accumulation of glacial ice (b) melting of glacial ice (c) erosion of the coastline (d) collision of crustal plates.

19. A shoreline resulting from the fall of sea level is said to be (a) emergent (b) submergent (c) divergent (d) ancient.

20. The sinking of land or a rise in sea level can form (a) tombolos (b) arches (c) bays and inlets (d) sea cliffs and narrow inlets.

21. A shoreline feature often resulting from emergence is (a) a fiord (b) an estuary (c) a sea cliff (d) a headland.

22. Barrier islands that develop tidal inlets to a lagoon often have large marshy areas called (a) tombolos (b) headlands (c) bays (d) tidal flats.

23. Coral reefs are produced by living organisms that have the ability to (a) accumulate large quantities of sand (b) swim about, causing a deflection of ocean currents (c) form very hard skeletons by extracting minerals from sea water (d) attach themselves to sand bars.

24. A circular reef enclosing an open water lagoon is called (a) a barrier reef (b) a fringing reef (c) an atoll (d) an estuary.

25. A particularly destructive type of coastal pollution is (a) sewage from ships and coastal cities (b) oil spills from oil tankers and offshore oil wells (c) trash from ships at sea (d) dredging and filling operations.

26. If the shoreline resources are used in the future as they have been used in the past, (a) we have no need to worry (b) coastal land will be greatly improved (c) there will be little damage done that can't be repaired (d) existing coastal land of the nation will be destroyed.

Group B

1. Describe the four ways in which waves attack the shoreline rock.

2. What process other than wave action causes deterioration of the shoreline?

3. Compare the formation of a wave-cut terrace with that of a wave-built terrace.

4. What effect does the development of a terrace have on the erosion of the shoreline? What condition may alter this effect?

5. Name three prominent shoreline formations that often result from uneven erosion of sea cliffs.

6. How do beaches get their sand?

7. Describe the general pattern of movement of small rock fragments near the shoreline.

8. How can you tell by observations of the berm whether the recent waves have been larger than usual?

9. What causes sand to be moved parallel to the shore?

10. Describe two causes of the lifting or sinking of a coast. Give an example of each.

11. What features are typical of an emergent shoreline? of a submergent shoreline?

12. Classify the following shoreline features as due to emergence or submergence. (a) San Francisco Bay (b) Coast of the Gulf of Mexico (c) The fiords of Alaska.

13. Compare the manner in which the following were formed: (a) the island on which Miami Beach is located (b) the island off the south coast of Long Island, New York (c) an atoll.

14. Discuss the source of rock fragments that make up the following beaches: (a) coarse, light-colored sand composed of quartz and feldspar (b) coarse, dark rock fragments (c) fine sand composed of materials not found along the shoreline.

15. What effects are produced when great continental glaciers form and melt that indicate the crust of the earth rests upon a somewhat plastic (soft) base?

16. What geologic evidence is there for believing that the level of the sea has changed?

17. Describe the geologic results on a shoreline of a drop in sea level and a rise in sea level.

18. For each of the six uses of coastal lands listed in the text, give one specific example of how the coastline can be harmed if used carelessly.

19. Describe what must be done to save the coastal land from eventual destruction.

15
the unknown sea

objectives

- [ ] Name the branch of science that deals with the oceans.

- [ ] Explain why the earth might be called the "water planet."

- [ ] Name the seven seas.

- [ ] Describe three general ways measurements are made in the oceans.

- [ ] Identify the main features of the deep sea floor and the continental margins.

- [ ] Describe three kinds of sea floor sediments.

In the days when ships were moved by sails, the sea was often thought to be full of mystery and danger. There were tales of monsters that could seize and sink a ship. The tales probably originated from greatly exaggerated experiences with real forms of sea life. The true mysteries of the sea could be solved only by exploring beneath its surface. That part of the earth was as unfamiliar as the surface of the moon. During the past hundred years, however, there has been considerable progress in scientific exploration and understanding of the ocean depths. Modern sailors no longer fear sea monsters.

Scientific exploration of the oceans began with the voyage of the British Navy Ship, H.M.S. *Challenger*, in 1873–76. The team aboard this small ship made thousands of observations as the vessel crossed the Atlantic, Antarctic, and South Pacific oceans. Their findings marked the beginning of the new science of *oceanography*, the scientific study of the oceans. Oceanography is a blend of the four basic sciences of physics, chemistry, geology, and biology.

Since the voyage of the *Challenger*, oceanography has grown rapidly into one of the most important branches of science. Many ships are now equipped to perform oceanographic research. An example of a research ship is the *Glomar Challenger*, shown in Figure 15–1. This ship is able to drill into the rock of the sea floor that lies as much as 4 km (2.5 mi) beneath the ocean's surface. Samples of the sea floor from all over the world have been taken by the *Glomar Challenger*. These samples provide scientists with a foundation for understanding how the oceans formed. Other modern methods of oceanographic research are used to study the basic processes that control everchanging sea conditions. The once mysterious ocean depths are slowly becoming as familiar as the earth's surface that lies above the sea.

EXPLORING THE SEA

The water planet. The oceans are only a small part of the earth by mass and volume. The water in the oceans weighs 1560 million billion tons. Yet this enormous mass is only 240 millionths of the total mass of the earth. The volume of the solid earth is about 800 times greater than the volume of the water in the oceans. However, the oceans are a more prominent feature of the earth than the overall quantity of water suggests. Ocean water is spread in a thin layer over about 362 million square kilometers, covering about 70 percent of the earth's surface. As far as we know, no other planet has such an extensive covering of liquid water. Among the planets, the earth can be called the "water planet."

Since ancient times, the oceans covering most of the earth's surface have been incorrectly referred to as the "Seven Seas." The seven included the North Atlantic, South Atlantic, North Pacific, South Pacific, Indian, Antarctic, and Arctic oceans. The name "seas" is now applied to small areas of the oceans that are partly surrounded by land. See Figure 15–2. Actually, all oceans are parts of

FIG. 15–1. The *Glomar Challenger* is an oceanographic research ship that is able to drill into the deep sea floor to obtain samples of sediments and rocks. (Deep Sea Diving)

FIG. 15–2. This map shows the oceans of the world with their smaller areas called "seas."

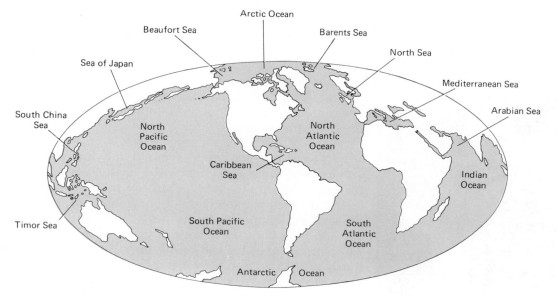

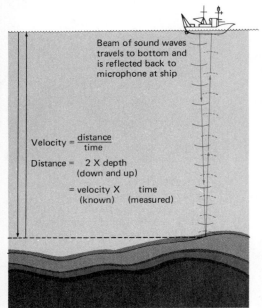

Beam of sound waves travels to bottom and is reflected back to microphone at ship

$$\text{Velocity} = \frac{\text{distance}}{\text{time}}$$

Distance =　2 X depth
(down and up)

=　velocity X　time
(known)　(measured)

FIG. 15–3. Sound waves reflected from the sea floor allow a continuous measurement of depth.

FIG. 15–4. This instrument is called a CTD (Conductivity, Temperature, Depth). It transmits these measurements back to a shipboard computer. (Jerry Dean/Woods Hole Oceanographic Institute)

one continuous body of water.

Nevertheless, the division of the sea into oceans is not totally artificial. Each ocean has its own characteristics. For example, water from the North Atlantic is saltier than water from the other oceans. This would not be unusual except for the fact that the entire Atlantic receives more fresh water from rain and run-off than the other oceans do. Thus the salt content should be more diluted. The Arctic Ocean is completely surrounded by land. The Antarctic Ocean completely surrounds a continent. Both of these polar oceans are less salty than the others. Melting ice constantly dilutes the water, and no rivers empty into them to supply additional salts.

The Pacific is the largest of the three principal oceans. With an average depth of 3.9 km (2.4 mi), it is also the deepest. It contains more than half of all the water found in the sea. Next in size is the Atlantic Ocean. It includes the Mediterranean, Caribbean, and Baltic seas, and the Gulf of Mexico. Because of some shallow parts, the Atlantic is the shallowest of the major oceans. Its average depth is 3.3 km (2.0 mi). The Indian Ocean ranks third in size and has an average depth of 3.8 km (2.4 mi). The oceans are not evenly divided between the earth's two hemispheres. About 39 percent of the Northern Hemisphere is land; the Southern Hemisphere is only about 19 percent land.

Exploring the depths. Scientific study of the sea is based on measurement. For centuries, sailors measured depth in shallow water with a weighted rope line. From this simple beginning, a great number of special instruments have been developed. These instruments allow modern oceanographers to measure many properties of the sea. The following is a description of the properties of the sea and the instruments and techniques commonly used for measuring these properties.

1. **Measurements made from a moving ship.** Sound waves, which pass through water easily, make it possible for a moving ship to draw a profile of the sea floor. A transmitter on the ship sends out a continuous series of sound pulses. The sound waves are reflected from the bottom of the sea floor back up to receiving equipment on the ship. The time it takes for the sound impulse to return is automatically translated into a depth reading. A graph that shows the profile of the sea floor along the ship's course is produced. See Figure 15–3. The kind of sound waves sent

out and the technique for recording them can be adjusted to show various features on the sea bottom. For example, one type of sound signal penetrates the sediment layers and shallow rock structures on the bottom. The returning signal indicates the thickness of these sediments and the nature of the underlying rocks in that part of the ocean.

One of the most significant properties of sea water is its temperature. Because the temperature of water changes slowly, it can remain constant in a certain mass of water. Thus temperature readings are often used to identify and trace moving masses of sea water. Temperature can be measured as a ship moves by towing a cable that has electronic temperature-sensing devices attached at certain depths. This gives a continuous record of the temperatures at the selected depths. Another type of temperature-measuring device is shown in Figure 15–4. As it is lowered through the water, it produces a record of the temperature and the electrical conductivity of the water.

Meters lowered from a ship are able to measure the speed and direction of currents. See Figure 15–5. Nets are also often towed behind a moving ship to collect samples of sea life.

FIG. 15–5. Speed and direction of currents measured by this instrument are recorded on tape cassettes, which are later recovered and analyzed. (John Hadley/Woods Hole Oceanographic Institute)

FIG. 15–6. This Nansen bottle is first lowered to the desired depth. A weight sent down the cable then opens the bottle so it can collect a water sample. Attached thermometers record the temperature at the time the sample is taken. (Ramsey)

2. **Measurements made at fixed locations in the ocean.** Sea water is a very complex solution that contains many dissolved substances. Chemical analysis of water samples taken from certain locations and depths yields some of the most important information used by oceanographers. Water samples for chemical analysis are usually collected with a special sampling container attached to a line. See Figure 15–6. When lowered to the desired depth, the container is closed by a weight dropped down the line. The trapped water is taken on ship and is either analyzed or preserved for future detailed study. Some chemical properties of sea water can be measured directly by instruments lowered into the water.

Measurements of the movement of sea water are also made from fixed locations. To detect movement of sea water, two basic measurements have to be made. First, the up-and-down movements are measured. These are waves and tidal movements. One common kind of wave detector uses electronic methods to record water motions. The electronic wave sensor can be aboard a ship, attached to a floating buoy, or placed on the sea bottom near the shore. A continuous record that shows wave activity at a particular location is thus available.

Currents, or horizontal movements of the water, also

have to be measured. The speed, direction, and location of the current are the most important information needed. The velocity, or speed, of the current can be measured with several different kinds of current meters. One type of electronic current meter can be installed in the water at a particular depth and left unattended. The meter will produce a record of the current velocity at that location. Floats are often used to track the direction of a current. The floats are followed as the current carries them along. Current direction can also be measured by adding certain "tracer" materials to the water. The movements of the tracer are followed by identifying its presence in water samples taken at different locations.

Valuable information can also be obtained from samples of the sediments that cover the sea floor. These samples are usually obtained through various devices lowered from ships and anchored or held at fixed locations. Some sampling devices have jaws that take a "bite" of exposed sediments on the bottom. Deeper sediments are sampled with long tubes that are driven into the sediment layers. When brought back aboard ship, the *core samples* are removed as long cylinders. These samples show the vertical arrangement of the layers. Core samples of this kind can be as long as 30 meters and can be taken from the deepest parts of the ocean. Other sampling devices drill into rock beneath the sediment layers to obtain samples of solid rock.

Cameras are used to photograph the surface of the sea floor. The photographs are useful in preparing maps and in locating areas that might yield valuable resources. Probes that are able to measure temperature can be driven into the ocean floor. These temperatures provide clues about the nature of the rock layers and the deeper parts of the crust beneath the oceans.

Humans beneath the sea. Instruments are vital to studying the sea. However, nothing can replace the first-hand knowledge that comes from direct observation of conditions beneath the sea. No matter how fascinating the sea is, it is still a foreign environment for humans. To penetrate this vast world of "inner space" beneath the sea, special equipment and great skill are needed.

A basic requirement for survival beneath the sea is a continuous supply of oxygen. But a more serious problem is pressure increases during descent. Pressure is created by the weight of the water pressing down from above.

FIG. 15–7. A diver using scuba equipment guides a platform carrying cameras on its way to the sea floor. (Carl Roessler/Alpha)

Naturally, this pressure increases as the diver goes deeper and there is more water above. Pressure nearly doubles at 10 meters of depth and increases the same amount for each additional 10 meters. At a depth of 1000 meters, the pressure is 100 times that of the surface.

The human body can withstand fairly high pressures if breathing air is provided at close to normal pressure. Breathing air at above normal pressures can dissolve gases in the blood and in other body fluids. When returned to normal pressure, these dissolved gases form bubbles in the body. These bubbles then produce the often fatal condition known as "bends." To prevent this from happening, air must be supplied to a person in such a way that its pressure is nearly normal, no matter how great the water pressure. An apparatus usually called *scuba* (self-contained underwater breathing apparatus) is designed to be carried by an individual diver. See Figure 15–7. It supplies air at the correct pressure for periods up to about an hour. But it must be used only by trained divers and is safe only for relatively shallow water. At depths greater than about 50 meters (about 164 ft), it is necessary to enclose the diver completely in a protective suit. See Figure 15–8. With this outfit, a diver can work efficiently in depths up to about 300 m (984 ft).

Divers are able to dive to greater depths through the use of more complicated methods. Gases such as helium can be mixed with oxygen and breathed by divers under high pressure. Such a mixture of gases will not dissolve in the body as readily as ordinary air. This allows divers to remain under the higher pressures at greater depths for longer periods. Other techniques have divers slowly exposed to high pressures in special chambers on the mother ship. Since their bodies are already adjusted to the high pressures, they can quickly descend to working depth. In these chambers, they can slowly be brought back to normal pressure or be kept at the higher pressure for later dives. With the use of these systems, divers have been able to reach depths of over 300 meters without protective suits.

Divers can also live for long periods of time in undersea chambers that are kept at high pressures. They can come and go from these chambers without any problems caused by pressure changes.

To reach very great depths in the sea, a human must be enclosed in some kind of submarine vehicle. It is possible to reach bottom in the deepest parts of the sea in a vehicle

FIG. 15–8. A diver in an armored suit performs work thousands of feet down on the ocean floor. (Commercial Diving Center)

that acts as a kind of underwater balloon. Gasoline, which is lighter than water, is used along with ballast to control the craft under water in the same way a balloon rises or sinks in air. Such a vehicle has reached the record depth of about 10,900 meters (about 35,700 ft). But this type of undersea vessel is not able to move around very much once it has reached the bottom. Other research submarines have been designed that allow deep diving along with the ability to move over considerable distances while underwater. See Figure 15–9.

FIG. 15–9. This research submarine, called "Alvin," is able to carry people and instruments to very great depths. (John Porteous/ Woods Hole Oceanographic Institute)

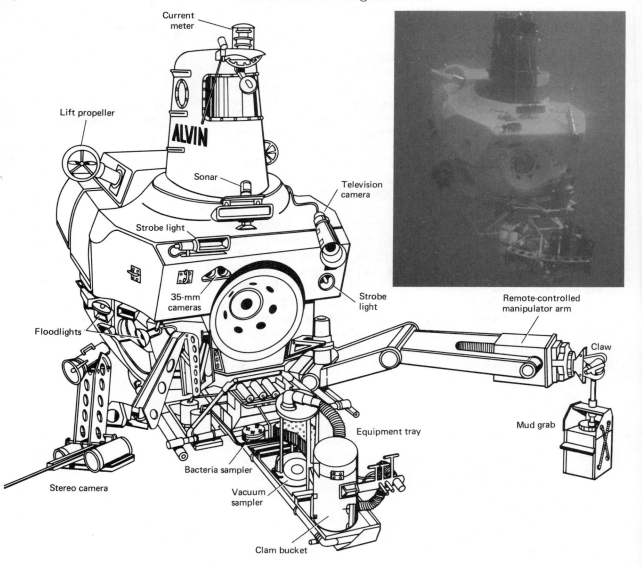

ALVIN

Current meter

Lift propeller

Sonar

Strobe light

Television camera

35-mm cameras

Strobe light

Floodlights

Remote-controlled manipulator arm

Claw

Mud grab

Equipment tray

Bacteria sampler

Stereo camera

Vacuum sampler

Clam bucket

THE SHAPE OF THE SEA FLOOR

Ocean basins. If the earth's oceans dried up and the sea floor could be seen as part of the solid crust, the earth's landscape would be very different. Mountains taller than any on the continents would become visible. The deepest canyons and longest range of mountains on the earth would be seen for the first time. See Figure 15–10. But the mountains on the sea floor are very different from those on land. Continental mountains are made mostly of layers of sediments that have been pushed up and folded. The world-circling mountain range of the sea floor, the mid-ocean ridge, is built of igneous volcanic rock coming directly from the earth's interior. It is evidence that the ocean basins were created as the continents were broken apart by the action of the moving crustal plates.

The study of the oceans has slowly revealed the true shape of the crust beneath the sea. Thus it is possible to divide the ocean basins into three major areas. These are

FIG. 15–10. Main features of the Atlantic sea floor are shown in this diagram.

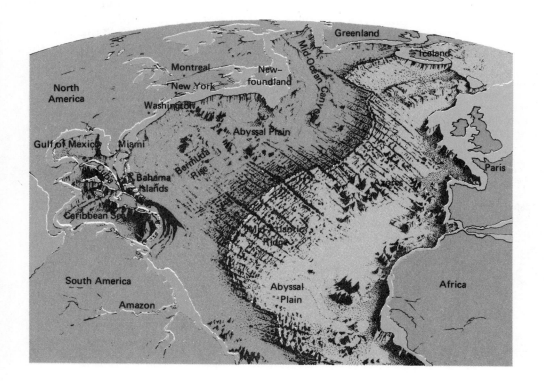

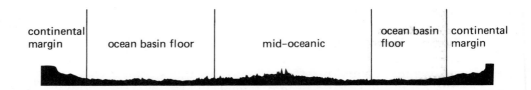

| continental margin | ocean basin floor | mid-oceanic | ocean basin floor | continental margin |

FIG. 15–11. A profile of the floor of the North Atlantic Ocean from North America to Africa shows the major divisions. The features have been exaggerated to make them more visible. The actual sea floor is not nearly as rugged as this diagram shows.

Explore
Recent deep-sea expeditions of the ocean floor have made many interesting discoveries. Look up some recent magazine and newspaper articles that give accounts of deep-sea findings.

FIG. 15–12. This diagram shows the features found along continental margins. Notice how sediments have accumulated on the continental shelf and moved down to form the continental rise.

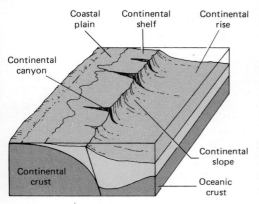

the *continental margins*, the *ocean basin floor*, and the *mid-ocean ridges*. Figure 15–11 outlines these three regions as they are found in a typical ocean basin like that of the North Atlantic.

The edges of the continents. The line that divides the land masses from the sea is not as sharp as it seems. Shorelines, where the water washes against the continents, are not always the true boundaries between sea and land. The real boundary usually lies some distance off the shoreline, beneath the water. Every continent is bounded in most places by a zone of shallow water where the sea has flooded the edges of the continental land masses. This zone of shallow water is called the *continental shelf*. It is present beneath the water around the margins of all the continents.

Continental shelves may have come into existence as a result of changes in sea level. During an ice age, when great amounts of water are locked up in glaciers, sea level falls. Later, with passage of the ice age, glaciers melt and sea level rises. Apparently, the earth is now in a warm period when the sea level is relatively high. Thus the sea covers a part of the continental edges.

The continental shelf is part of the continent itself, rather than part of the ocean basin. At one time or another, parts of the shelf's surface were exposed to the forces that carved the land surface into hills and valleys. This means that the surface of the continental shelf is not likely to be perfectly flat. In general, the surface of the continental shelf is a smoother version of the land surface at the shoreline. The shelf usually slopes gradually away from the shoreline until it reaches a steeper slope. This marks the edge of the continent and is called the *conti-*

nental slope. See Figure 15–12.

The width of the continental shelf is not the same at all places. In some places, like off the west coast of South America, it is only a few kilometers wide. Off the west coast of Florida, in the Gulf of Mexico, the shelf reaches out 760 km. The widest continental shelves in the world extend about 1280 km off Siberia and Alaska into the eastern Bering Sea. But a more typical continental shelf is that found along the east coast of the United States. It averages 170 km in width.

Because the continental shelf generally slopes very gently as it extends out to sea, the water is shallow. Average water depth over the world's continental shelves is about 60 meters. However, at the continental slope, there is usually a very sharp increase in depth. Within a distance of a few kilometers, the depth increases to several thousand meters. At the base of the continental slope, there may be an accumulation of sediments that have moved down the slope. This bulge at the base of the slope is known as a *continental rise.* The continental slope itself may be cut by valleys running along its surface. Occasionally, there can be an unusually deep valley called a *submarine canyon.* See Figure 15–12. The origin of these deep canyons is not completely known. They may be caused by turbidity currents that move down the continental slopes. See page 354.

In some places, the continental slopes plunge downward into the deepest parts of the sea floor. These long, narrow depressions at the base of some continental slopes are called *trenches.* Their depth is usually greater than 6000 meters. Most of the trenches of the world are found in the Pacific Ocean. See Table 15–1.

Ocean basin floor. Between the margins of the continents and the mid-ocean ridges, lies the ocean basin floor. Most of this part of the sea floor is made up of *abyssal plains.* These are extremely level regions that cover about half of all the deep ocean floor. The abyssal plains are the flattest regions on the earth. Some parts have changes in elevation of less than 3 m over distances greater than 1300 km. Sound signals used to study the sea floor reveal that the abyssal plains consist of thick layers of sediment deposited on the rougher features of the ocean bottom. The sediments appear to have been deposited by currents that carried them out into the deep sea from the con-

Table 15–1 Major Ocean Trenches

Trench	Depth	
	Meters	Feet
Pacific Ocean		
(1) Aleutian	7672	25,165
(2) Kurile	10,543	34,580
(3) Japan	9800	32,153
(4) Mariana	10,915	35,800
(5) Philippine	10,033	32,907
(6) Tonga	10,853	35,597
(7) Kermadec	10,003	32,809
(8) Peru-Chile	8057	26,427
Atlantic Ocean		
(9) South sandwich	8262	27,100
(10) Puerto Rico	9392	30,184
Indian Ocean		
(11) Sunda (Java) Trench	7252	24,442

FIG. 15–13. The Hawaiian Islands
are the most modern part of a chain
of volcanic peaks that have grown
over a permanent hot spot in the
crust and upper mantle.

tinental margins. Trenches found along the edges of some continents act as traps for the sediments carried down the continental slopes. Thus, abyssal plains are most widespread in parts of the sea floor where there are no trenches along the continental margins. The Atlantic Ocean, with few trenches, has much of its floor covered with abyssal plains. The Pacific, with more trenches, has less extensive abyssal plains.

Scattered across the ocean basin floor are a number of separate volcanic mountains called *seamounts*. Most of them have a height of at least 1000 m. Hundreds of seamounts dot the Pacific Ocean floor but are less common in the other oceans. Many of these underwater volcanic peaks are formed near the spreading plate boundaries marked by mid-ocean ridges. Volcanic activity is very common in this part of the sea floor. Seamounts often grow high enough to raise their tops above the water surface, making an island. Some of these volcanic islands have their tops eroded away by waves. Then they slowly sink back beneath the sea as they are carried farther from the mid-ocean ridge area. These flat-topped seamounts are called *guyots*.

In a few places along the sea floor, a chain of volcanic islands and seamounts were produced by a *hot spot*. A hot spot exists where a continuous stream of magma erupts through the crust. Volcanoes are periodically built where hot spots are located. The crustal plate making up the sea floor moves, while the hot spot remains fixed. As a result, a chain of seamounts or volcanic islands form as the moving plate carries the volcanic peaks away from the hot spot. The Hawaiian Islands are part of a long chain of volcanic peaks. They were built over millions

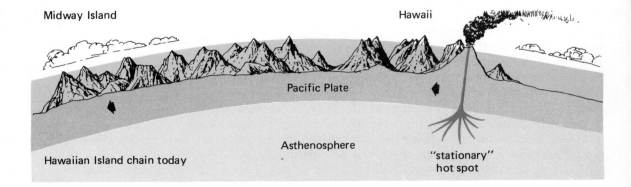

Midway Island   Hawaii

Pacific Plate

Asthenosphere

Hawaiian Island chain today   "stationary"
                                                 hot spot

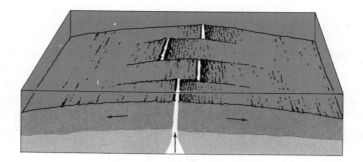

of years as the Pacific plate slowly moved over the hot spot now marked by the island of Hawaii. See Figure 15–13.

Mid-ocean ridges. The most prominent features of the sea floor are the *mid-ocean ridges*. See Figure 15–10. This continuous series of underwater mountain ranges runs through all of the ocean floors. See page 323, mid-ocean ridges exist where new sea floor is formed. This happens when molten rock is forced up from below the crust and spreads out. Because the molten rock is forced up, the central region of the sea floor occupied by the ridges is elevated. See Figure 15–11. Mid-ocean ridges usually have a narrow depression, or *rift,* running along their centers. The rift is evidence that the crust pulled apart as the plates moved away on each side of the ridge. Volcanic activity is very common along the rift zone, since this is where magma reaches the surface. Each side of the mid-ocean ridge system is covered by a series of parallel hills and valleys. A rugged landscape has been produced because the plate movement breaks the surrounding crust into tilted blocks. In many places, the mid-ocean ridges are cut across by large faults. See Figure 15–14. The faults are probably produced by the molten rock that rises in neighboring parts of the ridges. This causes the rate of spreading out to be greater in some parts of the ridges than in others. Due to the unequal rate of spreading, the ridge breaks into sections.

SEA FLOOR SEDIMENTS

Origin of marine sediments. The kinds of sea bottom

activity

Fill a large test tube about one fourth full of sandy soil. Add water to the test tube until it is filled to within 1 cm of the top. Cover the top of the test tube and shake the mixture until the soil is mixed throughout the water.

Set the test tube upright and observe the soil particles as they settle to the bottom. Use a magnifier. When the soil has nearly settled to the bottom, answer these questions:

1. How does the particle size at the bottom of the test tube compare with that found farther up?

2. What size particles settled to the bottom first?

3. What size particles settled to the bottom last?

4. Do any pieces of material float to the surface?

5. How does the size of the particles that remain suspended in the water, but not floating on the surface, compare to those that settle out?

Rivers carry soil particles, and even rocks, downstream and to the ocean. When the river water reaches the calm of the ocean, the soil particles settle out.

6. Where, in the ocean, would you expect to find the larger soil particles carried by rivers? Explain.

7. What kind of river-carried particles would you expect to find on the ocean floor, far from shore? Explain.

Investigate

The first Atlantic telephone cable was mysteriously ripped apart by some unknown powerful force on the sea floor. Further investigation led to the discovery of deep-sea turbidity currents. See what you can find out about how these powerful currents could have torn the cable apart.

sediments differ a great deal from one part of the sea to another. Over the continental shelves and slopes, the bottom is covered mainly with fragments of rock. These come mostly from rivers emptying into the sea and from wave action against the shoreline. Some of the sediments also seem to have been produced before the area was covered by the sea. Usually, the sediments are fairly well separated according to size. Since it takes more force to carry larger rock fragments, coarse gravel and sand are usually found closer to the shore. Fine particles (mud) usually cover the bottom farther from shore.

In the deep sea, beyond the continental slopes, the bottom sediments are different from those found in shallower water. Most of the sediments in the deep sea are formed from materials that settle from above. These materials come from a variety of sources. A brief description of the origin of the main types of deep sea bottom sediment follows.

1. Particles carried from the land. The total amount of sediment carried into the oceans by streams can be accurately estimated. When this amount is compared to the amount already known to be deposited along the shore and on the continental shelf, a large quantity of sediment is found to be missing. What has happened to it? Careful study of the problem revealed that great quantities of sediments occasionally slide down continental slopes to the deep sea floor. These movements are like great underwater landslides and may be set loose by earthquake shocks. In a single movement, millions of tons of material can be involved. The force of the material sliding down creates powerful currents. These currents probably cause the moving sediments to be spread over large areas of the deep sea bottom.

Wind is another way that land sediments reach the deep sea. Fine particles of rock blown great distances out to sea by the wind finally settle on the water's surface. Eventually, they sink to the bottom and become part of the sediment accumulation. Volcanic dust thrown into the air often becomes part of sea floor sediment after the wind has carried it some distance. Volcanic activity that occurs on the sea floor also forms part of the bottom sediments.

Icebergs also carry material from the land into the sea. When the iceberg melts, any glacial material that it happens to be carrying is dropped and sinks to the bottom.

2. **Organic sediments.** Many of the plants and animals that live in the sea have hard parts, including their skeletons. When these organisms die, their skeletal remains become part of the bottom sediment. In many places on the sea floor, almost all the sediment comes from this source. The two most common substances from organic sources are silica (SiO_2) and calcium carbonate ($CaCO_3$). Silica comes mainly from microscopic

FIG. 15–15. Common microscopic organisms that contribute to ocean floor sediments. (A) radiolarians, (B) diatoms, and (C) foraminifera specimens.

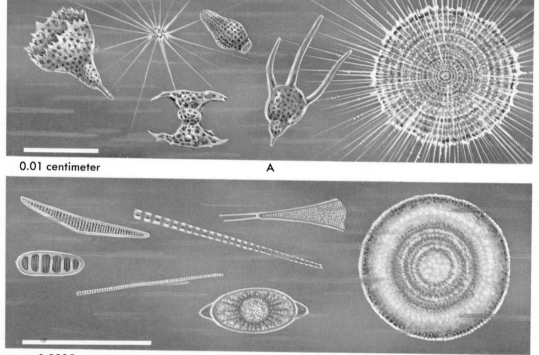

0.01 centimeter   A

0.0005 centimeter   B

0.1 centimeter   C

FIG. 15–16. A concentration of manganese nodules on the floor of the Atlantic Ocean at a depth of 4100 meters. (Lamont Geological Observatory)

FIG. 15–17. A core sample of sea floor sediments is being removed from the hollow coring device at the right. (ESSA)

animals called *radiolaria*. Microscopic plants called *diatoms* (di-a-toms) may also have silica skeletons that form sediments. Calcium carbonate comes mostly from the skeletons of tiny animals called *foraminifera*. See Figure 15–15. Other animals, like corals and clams, can also contribute calcium carbonate to sea floor sediments.

3. **Chemical deposits.** Many chemical changes take place in the sea. Some of these chemical reactions form solid materials that settle to the bottom. An example is the presence of lumps or nodules over large areas in certain regions of the sea floor. These nodules are composed mainly of oxides of manganese, nickel and iron. See Figure 15–16. They seem to have been produced from these substances dissolved in sea water. Other minerals found in sea bottom sediments were formed directly from materials in the sea water. However, these minerals are not abundant and the way they were formed is not completely known.

Results of sediment accumulation. Many devices are available, such as the one shown in Figure 15–17, to obtain samples of sea bottom sediments. Examination of such samples shows that two general types of sediments are common. One kind of sediment is *red clay*. This is composed of at least 40 percent clay particles.

The remainder is made up of silt, sand, and organic material. Some red clays also contain material from meteorites. Actually, red clay is usually brown in color and is sometimes called "brown clay." The first samples obtained happened to have a red color and the name "red clay" became established. Sediments similar to red clay but containing more silt are called *muds*. Various types of muds are common near the continental shelves and slopes.

About 40 percent of the ocean floor seems to be covered with a soft organic sediment called *ooze*. Again, the name is misleading since most ooze looks and feels like sand. However, ooze is not made up of sand. At least 30 percent of it is made up of the remains of radiolarians, foraminifera, and diatoms.

VOCABULARY REVIEW

Match the word or words in the column on the right with the correct phrase in the column on the left. *Do not write in this book.*

1. The scientific study of the oceans.
2. The name given to the water covering most of the earth's surface.
3. Used to determine the depth of the water.
4. The horizontal movements of water.
5. A self-contained underwater breathing apparatus.
6. Samples that show sediment layers.
7. The zone of shallow water around a continent.
8. The true edge of the continent.
9. Name of the bulge caused by an accumulation of sediments at the base of the continental slopes.
10. An unusually deep valley on continental slopes.
11. Long narrow depression at the base of some continental slopes.
12. Extremely level region of half the sea floor.
13. A series of great underwater mountain chains.
14. Sharp volcanic peaks on the sea floor.
15. A very common sea floor sediment.

a. submarine canyon
b. temperature
c. continental slope
d. core samples
e. mid-ocean ridges
f. oceanography
g. continental rise
h. sound waves
i. abyssal plains
j. scuba
k. red clay
l. the sea
m. seamounts
n. currents
o. depth
p. continental shelf
q. trench

QUESTIONS

Group A

Select the best term to complete the following statements. *Do not write in this book.*

1. About how many years ago was the first voyage made for a scientific study of the oceans? (a) 25 (b) 50 (c) 75 (d) 100.

2. Which oceans were included in the first scientific study of the oceans? (a) Arctic, Antarctic, and South Pacific (b) Arctic, Atlantic, and Indian (c) Atlantic, Antarctic, and South Pacific (d) Atlantic, Arctic, and South Pacific.

3. Oceanography is a (a) blend of several sciences (b) blend of only two sciences (c) science independent of other sciences (d) brand new science.

4. All the water in the oceans weighs about (a) 1560 million tons (b) 1560 million million tons (c) 1560 million billion tons (d) 1560 billion billion tons.

5. What part of the earth's mass is water? (a) about 10% (b) about 5% (c) about 1% (d) much less than 1%.

6. What part of the earth's surface is covered with water? (a) 70% (b) 50% (c) 30% (d) 10%.

7. Which of the following is *not* one of the seven seas? (a) Arctic (b) Antarctic (c) Baltic (d) Indian.

8. One of the characteristics of the polar oceans is that they (a) both surround continents (b) are both surrounded by land (c) are more salty than the others (d) are less salty than the others.

9. The temperature tends to remain constant for a certain mass of water because (a) the temperature of water changes slowly (b) the oceans are quite deep (c) currents tend to mix the water (d) thermometers react slowly.

10. Sea water is a solution that (a) is very simple (b) has a few dissolved mineral substances (c) is very complex with many dissolved substances (d) contains mostly gases.

11. An electronic sensing device is used to record (a) temperature (b) dissolved gases (c) depth (d) water motions.

12. Beneath the sea surface, the pressure doubles every (a) 10 meters (b) 32.8 meters (c) 100 meters (d) 1000 meters.

13. A diver enclosed in a protective suit can work in depths up to (a) 36,000 m (b) 11,000 m (c) 200 m (d) 50 m.

14. Bubbles that produce the condition called ''bends'' can be prevented ,by (a) breathing air at high pressure (b) breathing air at nearly normal pressure (c) breathing only nitrogen (d) breathing deeply.

15. If the earth's oceans dried up, the sea floor would look (a) similar to earth's landscape (b) exactly the same as earth's landscape (c) different from earth's landscape.

16. Which of the following features is *not* found as part of the ocean floor? (a) barrier islands (b) abyssal plains (c) trenches (d) submarine canyons.

17. The continental shelf is a part of (a) the ocean basin (b) a mountain ridge (c) the continental slope (d) the continent.

18. The continental slope is (a) part of the sea floor (b) the edge of the continent (c) a shallow area near shore (d) the end of a glacier.

19. How wide are the widest continental shelves in the world? (a) 760 km (b) 800 km (c) 1280 km (d) 170 km.

20. The average water depth over the continental shelves is (a) 60 m (b) 105 m (c) 200 m (d) 10,000 m.

21. The depth of trenches is usually greater than (a) 6000 m (b) 60,000 m (c) 600,000 m (d) 6,000,000 m.

22. The level region of the sea floor is called (a) a continental shelf (b) a mid-ocean ridge (c) an abyssal plain (d) a seamount.

23. Separate volcanic mountains on the ocean basin floor are called (a) continental shelves (b) mid-ocean ridges (c) abyssal plains (d) seamounts.

24. Which of the following is *not* a major way that sediment is carried from land to the sea? (a) streams (b) rain (c) wind (d) icebergs.

25. Sea floor sediments containing silica and calcium carbonate come from (a) rocks (b) rain water falling into the sea (c) both plants and animals (d) plant skeletons only.

26. The largest particles of sediment on the sea floor are found (a) near shore (b) far from shore (c) on abyssal plains (d) on top of seamounts.

27. Compared to the total amount of sediment carried by streams, the amount of sediment found on continental shelves is (a) too small (b) about the same (c) slightly greater (d) much too great.

28. Nodules on the sea floor are *not* composed of (a) manganese (b) nickel (c) silver (d) iron.

29. Red clay is usually colored (a) red (b) brown (c) blue (d) green.

30. About 40% of the ocean floor is covered with a soft, organic sediment called (a) sand (b) clay (c) ooze (d) nodules.

Group B

1. Why is the earth known as the "water planet"?
2. Name the "seven seas."
3. Why is the British Navy ship, *H.M.S. Challenger,* famous?
4. What characteristics of the sea are most commonly measured?
5. Why is the science of oceanography said to be a blend of other basic sciences?
6. Describe several instruments used to study the sea.
7. What basic difference is there between the Arctic and Antarctic oceans, other than that they are at opposite poles of the earth?
8. Why is gasoline used in the tanks of some underwater research vehicles?
9. How do the mountains and valleys in the ocean compare to those on land?
10. How are the continental shelf, the continental slope, and the continental rise related?
11. How are abyssal plains formed?
12. Describe a seamount and how it is formed.

13. What is the relationship between seamounts and guyots?

14. Explain how a chain of seamounts can be formed from a single source.

15. Why is a mid-ocean ridge usually broken into sections?

16. Name three types of sediment deposits found on the sea floor.

17. How are most sediments that are found on the continental shelf formed?

18. Describe the nature and origin of the common sea floor sediments.

19. What is meant by the statement, "The total volume of sediments on the sea floor seems to be too small"? How do scientists account for this?

20. What is red clay?

As far as we know, the earth is the only planet that has liquid water on its surface. All other planets in the solar system are either too hot or too cold. Surface temperatures on the earth are temperate and allow water to remain in a liquid state. If the earth were too cold, its water would be another of the solid materials in its crust. If it were too hot, water would be found mostly as a gas. Only in liquid water do we find the properties that make the oceans one of earth's most distinctive features.

All the properties of liquid water in the sea can be divided into two main groups. The characteristics of sea water which permit it to dissolve other substances are generally listed as *chemical properties*. Other characteristics, not connected with dissolved materials, are called *physical properties* of sea water. For example, one of the most important physical properties of sea water is its temperature. Together, the chemical and physical properties of sea water help to give us some idea of what a complicated liquid sea water actually is. A brief study of these properties can also reveal some of the many processes which continuously take place in the sea. A study of the liquid water of the oceans will help us understand the complex relationships which exist between the land, the air and sea.

objectives

☐ Describe the physical properties of sea water.

☐ Describe several conditions that change these properties.

☐ Describe the chemical properties of sea water.

☐ Explain how the salt content and dissolved gases in sea water affect sea life.

☐ Explain how the sea can be a valuable resource.

PHYSICAL PROPERTIES OF SEA WATER

The sea and energy from the sun. The sun is one of the most important single influences on the sea. Energy from the sun reaches the earth in the form of light. This light travels in various wavelengths. Since the sea covers a major part of the earth's surface, most of the sun's energy falls upon the oceans. Nearly all of this solar energy penetrates the sea surfaces and is absorbed into the water. Although water appears to be transparent in small amounts, it is actually able to quickly absorb visible light and most other forms of radiant energy as well.

Of the various wavelengths in visible light, only the blue wavelengths are able to travel very far into the water before they are absorbed. At depths greater than 10 meters. (about 33 feet), only a blue-green light from the sun can be seen. All other wavelengths or colors of light usually present in sunlight have already been absorbed. It is the ability of blue light to penetrate water that gives the sea and other large bodies of water a blue color. If the water is clear and you look almost straight down into the sea, the water appears blue because it is the last color to be absorbed. The blue color is not so apparent to a viewer looking horizontally across the water surface. Light rays which strike the water at much of an angle are reflected rather than absorbed. See Figure 16–1. Many times the natural blue color of the water is clouded by small particles suspended in the water. No light of any kind can penetrate the sea to depths below a few hundred meters. All but the upper layers of the sea are in total darkness.

Experiment

Obtain a sample of sea water in a small jar. In a second jar place an equal volume of tap water. Place them both an equal distance from an uncovered 100 watt bulb. After intervals of 10 minutes for a period of 1 hour, record the temperature in both jars. Report your findings to the class.

FIG. 16–1. Some of the light falling on the sea surface is reflected causing the sparkling effect shown in this photograph. The angle of the sun's rays does much to determine whether the light is reflected or penetrates the water. (Ramsey)

Table 16–1 Average Surface Temperature (°C) of the Oceans Between Parallels of Latitude

North latitude	Atlantic Ocean	Indian Ocean	Pacific Ocean	South latitude	Atlantic Ocean	Indian Ocean	Pacific Ocean
70°–60°......	5.60	.....	.....	70°–60°......	− 1.30	− 1.50	− 1.30
60°–50°......	8.66	.....	5.74	60°–50°......	1.76	1.63	5.00
50°–40°......	13.16	.....	9.99	50°–40°......	8.68	8.67	11.16
40°–30°......	20.40	.....	18.62	40°–30°......	16.90	17.00	16.98
30°–20°......	24.16	26.14	23.38	30°–20°......	21.20	22.53	21.53
20°–10°......	25.81	27.23	26.42	20°–10°......	23.16	25.85	25.11
10°– 0°......	26.66	27.88	27.20	10°– 0°......	25.18	27.41	26.01

Surface temperature of the sea. In addition to its ability to absorb the visible wavelengths of solar energy, sea water is capable of also absorbing the longer infrared wavelengths. Energy from these infrared rays reach the earth in the form of heat. Thus they play an important role in determining the surface temperature of the sea. Like visible light, infrared rays are completely absorbed within the upper layers of the sea water. This means that the sun can only heat the upper part of the oceans directly. In the deeper parts of the sea the temperature of the water is always close to freezing (0°C).

The total amount of solar heat falling upon the surface of the sea is much greater at the equator than at the poles. At high latitudes, near the poles, the sun's most slanted rays strike the earth. See page 66. These are weakest in ability to heat the water. Surface temperatures in polar seas usually drop below 0°C. Table 16–1 lists some average surface temperatures for the three major oceans. For sea water to freeze, it must be chilled to about −2°C (28.5°F). The dissolved salts in sea water lower its freezing point below that of water. Vast areas of sea ice exist in both arctic and antarctic waters. If the floating layer of ice completely covers the sea surface it is called *pack ice*. The layer of ice is usually not more than 5 meters (about 17 feet) thick because the bulk of the ice acts as an insulating cover to prevent the water below from freezing. The Arctic Ocean is covered by pack ice during most of the year. The ice pack is brittle and will crack and buckle under pressure when the forces of wind and currents become too great. Broken-off pieces of the ice pack called

FIG. 16–2A. Pressure ridges cause buckling of the pack ice in the sea off the coast of Antarctica. (Black Star-Schulthess)

ice floes are sometimes strewn about stretches of open water. These passages of open water quickly freeze over in winter but usually remain open during the summer. Buckling of the ice pack disturbs its smooth surface and builds up *pressure ridges,* as shown in Figure 16–2A.

In tropical waters near the equator, surface temperatures around 30°C (86°F) are not unusual. One of the most important effects of the high surface temperatures of tropical water is rapid evaporation.

When water absorbs the longer infrared rays from the sun, the heat energy increases the movement of water molecules. It is this increased amount of molecular motion which is recorded by thermometers as an increase in temperature. When the temperature of sea water rises, many water molecules move fast enough to enter the atmosphere. The process of water molecules leaving liquid water and becoming water vapor in the air is called *evaporation.*

Notice that evaporation produces two results. First, the liquid water that is lost by the sea is taken into the atmosphere as water vapor. Second, heat energy is transferred. Water molecules moving from the surface of the sea into the atmosphere removes heat energy represented by their motion. Thus evaporation creates a loss of heat in the sea and a gain for the atmosphere. The energy that is transferred in this process creates a very close relationship between activities in the sea and the atmosphere.

The large amount of water evaporated near the surface also influences the amount of dissolved salts present in sea water. During evaporation only water molecules are removed. Dissolved salts remain behind. This results in an increase in the relative amount of dissolved salts in surface water where evaporation is high. For this reason, tropical waters will have higher concentrations of dissolved salts at the surface than will polar waters.

Temperature and depth. Mainly because the sun does not directly heat sea water below the surface layers, the temperature dips sharply with increasing depth. To understand the reason for this sudden drop in temperature, it is necessary to consider the factors which affect the *density* of sea water. Density expresses how much mass a given volume of sea water will have. It is generally described in grams per cubic centimeter (g/cm³).

Two factors have a major effect on the density of water in the oceans. First, there are the dissolved salts. These add mass to the water in direct proportion to the amount present. The large amount of dissolved substances in sea water make it more dense than fresh water under the same conditions. A second factor affecting sea water density is temperature. When liquid water is heated, it expands slightly; when it is cooled, it shrinks a small degree. This means that if liquid water is warmed, its density generally will be reduced. Cooling will have the opposite effect. Its density will increase. Actually, pure water has a maximum density at about 4°C. Below this temperature liquid water expands as it reaches the freezing point. In addition to dissolved salts and temperature, pressure is considered a minor factor in determining the density of water.

In most places in the sea, measurements indicate a sudden temperature drop not far below the surface. This zone of rapid temperature change, called the *thermocline,* marks the distinct separation between a warm surface layer and colder deep water. See Figure 16–2B. A thermocline exists because heat which enters the water lowers its density as warming takes place. This warm water cannot mix easily with the cold, dense water below. A thermocline is established at the boundary zone between the upper and lower layers of water. The differing densities help to keep them separate. Changing conditions of heat or currents may alter the depth of the thermocline or cause it to disappear completely. Regardless of these possibilities, a thermocline is usually present beneath much of the sea surface. Thermoclines that can definitely be measured are also observed in many lakes.

Describe
In a room with the windows and doors shut, record the temperature near the floor. Take a few readings. Now repeat, taking the temperature near the ceiling. How is this distribution of temperature the same or different from that found in the ocean?

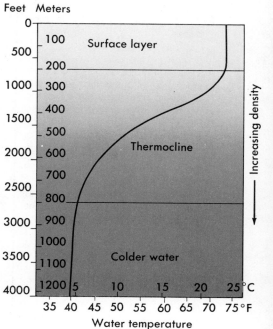

FIG. 16–2B. An idealized graph of changes in temperature and density with increasing depth in the open ocean.

CHEMICAL PROPERTIES OF SEA WATER

activity

Part I

Make a millimeter scale on a piece of masking tape. Make every fifth line extra long. Stick the tape to one end of a plastic straw and run it lengthwise along the straw. Clog up the other end of the straw with clay. Keep adding clay to this end until the straw floats upright in water with about 5 cm of the marked tape still above water. This instrument is a hydrometer.

Make some salt water similar in density to sea water. (Add 3.6 grams of table salt to each 100 ml of distilled water.)

Use a large test tube to hold the water being tested. Record the readings when you float the hydrometer in distilled water, tap water, and sea water.

1. How does the hydrometer level differ in the three liquids? (The hydrometer floats highest in the liquid that is denser.)
2. Which of the liquids is the most dense?

Part II

Add several drops of dark food color to about 100 ml of imitation sea water. Fill a plastic box about half full of tap water. Raise one end and place a book under it. Carefully pour the colored sea water into the raised end of the box and observe what happens.

1. Do the tap water and colored water mix immediately?
2. Where a river enters the ocean, would the water be saltier at the surface or near the bottom?

Dissolved salts in sea water. During the millions of years the oceans have been in existence, sea water has been continuously evaporating. After condensing, this very same water falls as rain. Some of this rain forms the streams and rivers that wash over the land on their way back to the oceans. Large amounts of mineral matter are dissolved in the water as it runs over the land surfaces. Each year the world's rivers carry about 400 million tons of dissolved minerals, including salts, into the sea. These dissolved salts remain trapped in the sea as the water molecules evaporate again in their endless cycle.

Not all of the dissolved matter in the sea originated in minerals found in the rocks. Some of the substances now in sea water probably came from gases produced by volcanoes. Gases that were present in the earth's early atmosphere may also have dissolved in the sea. Other unknown processes have played a role in making the sea a huge reservoir of dissolved substances. If all of these materials could be recovered, they would form a layer 136 meters (about 450 feet) thick over the surfaces of all the continents.

During the great span of time that the sea has been receiving dissolved materials, it has probably reached a state of balance. The processes which are constantly adding new salts to the oceans seem to be balanced by the processes which remove dissolved materials. The formation of sediments on the sea floor removes millions of tons of material from the sea each year. Living things also use up the salts as part of their life processes.

The dissolved salts found in sea water are evidence of the many complicated chemical processes which are at work. At the present time, we have only a very limited knowledge of the chemical reactions which control the composition of sea water. When a sample of sea water is analyzed, a few substances are always found to make up more than 99.9% of the dissolved salts. These substances are listed in Table 16–2. Notice that these materials are represented in the form of ions, a result of the dissolving action of the water molecules. This was described in Chapter 7. Almost all of the known chemical elements have been found in sea water. Most of these are present in very small amounts as ions. All of the known elements are probably present in sea water, but in many cases have not been detected because they exist in such small amounts.

The total amount of dissolved solids present in a sample of sea water is described as its *salinity*. For example, suppose that one kilogram of sea water (1000 grams or about 2.2 pounds) is dried and the total weight of salts which remains weighs 35.0 grams. The salinity of this sample of sea water would then be very near to 35 parts per thousand. This is usually written as: salinity = 35 %o. Sea water with a salinity of 35 %o would have almost 3.5 per cent of its total weight made up of dissolved salts. See Figure 16–3.

The process of evaporation, which removes water, increases the salinity of sea water. On the other hand, heavy rainfall or large amounts of fresh water from rivers bring about a decrease in salinity. Figure 16–4 gives the average surface salinity at different latitudes between the equator and poles. You will notice that because of heavy rainfall near the equator, salinity will be lower in this area than in dry regions. Another factor which produces low salinity near the poles is that the great masses of melting ice at the poles are almost entirely fresh water. Over most of the sea surface, salinity usually ranges between 33 and 36 %o, with an average value for all the oceans of 34.7 %o.

Although the salinity or total dissolved salts of sea water may differ from one location to another, the relative amounts of the dissolved salts do not change significantly. This means that the amount of any of the principal ions compared to any other is a nearly constant ratio. This state of equilibrium is brought about by the continuous mixing of all parts of the sea with all other parts. Any dissolved substance emptied into the sea is eventually

Table 16–2 Principal Dissolved Substances in Sea Water.

Ion	Percent of total dissolved solids (By weight)
Chloride (Cl^-)	55.04
Sodium (Na^+)	30.61
Sulfate (SO_4^{-2})	7.68
Magnesium (Mg^{+2})	3.69
Calcium (Ca^{+2})	1.16
Potassium (K^+)	1.10
Bicarbonate (HCO_3^-)	0.41
Bromide (Br^-)	0.19
Borate ($H_2BO_3^-$)	0.07
Strontium (Sr^{+2})	0.04

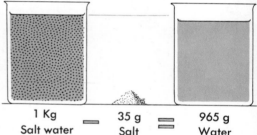

1 Kg   =   35 g   =   965 g
Salt water   Salt   Water

FIG. 16–3. A kilogram of sea water which contains 35 grams of dissolved salts has a salinity of 35 %o.

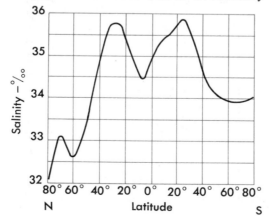

FIG. 16–4. Average salinity for surface water at various latitudes.

FIG. 16–5. A sample of sea water for analysis is being taken from the sampling bottle. (Ramsey)

FIG. 16–6. Potash salt deposits remain after sea water evaporates. (Magnum-Brian Brake)

spread around uniformly through all the water. Thus each dissolved salt will usually be found in the same proportion everywhere in the sea.

In addition to substances found naturally in sea water, some materials are present as a result of human activities. Until recently, people were able to safely use the oceans as a dump. Wastes would be made harmless by being spread through the sea water and become very dilute or be destroyed. But the growth of population and increase in industry has seriously changed the situation. The ability of the oceans to absorb wastes is far less than the ever increasing amounts that are discharged all over the world. Coastal waters are particularly in danger but pollutants can now be found everywhere in the sea.

Traces of the insecticide DDT have been detected in various places in the open ocean. Measurable amounts of lead material in the Pacific have increased by ten times since it first began to be used in gasoline about fifty years ago. Radioactivity from nuclear fallout can be detected in water samples taken from any ocean. While none of these materials has built up to harmful levels in sea water, their presence is a warning that the ocean can not forever be used as a garbage dump.

Dissolved gases in sea water. In addition to the ions which form the solid salts left when sea water evaporates completely, the oceans contain large amounts of dissolved gases, almost all of which enter the sea from the atmosphere. The principal gases of the atmosphere are nitrogen (N_2), oxygen (O_2), argon (Ar), and carbon dioxide (CO_2). All four of these gases are found in the sea, but unlike the dissolved solids, generally not as ions. The dissolved gases usually appear in molecular form just as they do in the atmosphere.

An exception is carbon dioxide which chemically combines with the water to form carbonic acid (H_2CO_3). The reaction itself dissolves large amounts of carbon dioxide. Carbon dioxide is found in sea water in a variety of forms. The total amount of carbon dioxide dissolved in the sea is much greater than the amount which exists in the atmosphere. The other atmospheric gases cannot dissolve as easily in water and are present in the sea in smaller amounts.

Temperature has a strong effect on the amount of gas that dissolves in water. For instance, cold water can dissolve gases better than warm water. In colder regions, water at the surface of the sea will usually dissolve larger amounts of gases from the air than warm tropical waters.

Certain conditions allow dissolved gases to leave the sea and return to the atmosphere. Any time the amount of a dissolved gas becomes greater than the amount which can be dissolved at a particular temperature, excess gas will leave the water. The sea and the atmosphere are continuously exchanging gases as conditions change with time from place to place.

Life and the chemistry of the sea. Once a substance is dissolved in sea water, it becomes part of a very complex, always changing system. Many factors are at work in the sea to change the composition of its water. Evaporation has already been mentioned as a process which concentrates the dissolved substance. Chemical deposits which become part of the bottom sediments remove certain substances and thus also change the composition of sea water. Volcanic action both adds and removes materials from sea water. Many other natural but unknown processes which occur in or near the sea probably act to change the makeup of sea water.

Among the most outstanding of these processes are the changes caused by living things in the sea. These animals and plants must receive from the water, directly or indirectly, all of the materials required for their life processes. And finally, the substances produced in their bodies return a tremendous variety of by-products to the water. Each step in this continuous series involves changes in the chemical composition of the surrounding water.

All life in the sea is regulated by its plant life. Plants remove certain dissolved substances from the water needed for growth and turn them into organic or living matter. To meet these needs, plants absorb large amounts of various substances containing the elements carbon, hydrogen, oxygen, and sulfur. These materials are present in such large amounts in sea water that biological processes do not have any significant effect on their concentrations. More critical to the growth of plants are compounds containing the elements nitrogen, phosphorus, and silicon. Since these substances are not as abundant, heavy plant growth reduces their concentration to nearly zero. In this case the growth of plants in the depleted water will be greatly slowed down or stopped completely. This situation can be compared to the problems of plant growth in poor soil on the land.

Investigate
Put some water in a dish and then cover the dish with clear plastic food wrap. Place the bowl in direct sunlight for at least two hours. How could a method similar to this be used for getting fresh water from the oceans?

In addition to certain dissolved substances from the water, almost all plants in the sea require sunlight. This means that plant growth in the sea is restricted to the upper few meters into which light penetrates. Below about 80 meters (265 feet), there is never enough light to meet all the needs of plants. However, within the depth zone where there is sufficient light, most regions of the sea contain large amounts of free-floating, microscopic living forms called *plankton*. See Figure 16–7. There are two main types. The microscopic plants in plankton are known as *phytoplankton*. These tiny plants remove dissolved materials from the water and use energy from light to carry on photosynthesis. In the cycle of sea life, the phytoplankton then serve as the source of food for microscopic animals or *zooplankton*.

Both forms of plankton are eaten by larger life forms such as small fishes and squid. These, in turn, become the food of adult fish and other large marine animals. However, some large animals, certain whales for example, feed directly on plankton. In any case, phytoplankton are always the first link in the complicated food chain which supports life in the sea. See Figure 16–8. In addition, because phytoplankton require light to carry on photosynthesis, most of the chemical changes associated with life in the sea are confined to the upper layers of the water.

All of the elements needed to support life in the sea are removed from the sea water. In time these elements are returned to the water when the plant or animal remains decay. Bacteria in the water attack the organic remains and again release the trapped elements. The decay processes occur at all depths, but gravity will pull the organic

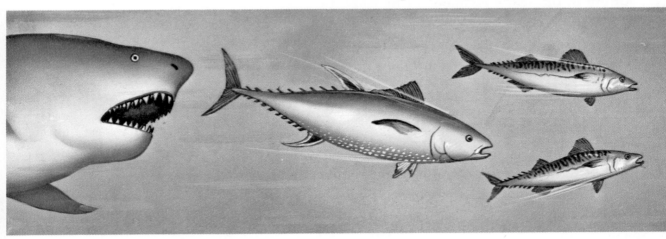

remains slowly downward from the near-surface layers where biological activity is greatest. A tendency exists for the necessary elements in water to be consumed near the surface but released at greater depths. Thus deeper water becomes a storage region for the vital materials needed to support life. Some deep-dwelling animals use the material coming from above as a source of food. Yet most living forms in the sea depend upon the needed substances being returned again to the surface.

Several processes seem to be at work in the oceans causing deep water to move upward. Among the processes which appear to play an important role in returning the needed materials in deep water to the surface are:

1. *Upwelling.* When wind blows steadily away from the shore along a coastline, surface water is moved out to sea. Deep water then moves upward to replace the surface water. See Figure 16–9. This situation is commonly found off the west coast of South America, the California coast, and the northern coast of Florida.

2. *Overturn.* When surface water is chilled it becomes more dense and will sink. Warmer water at a greater depth then moves upward.

3. *Mixing.* In shallow water wave action on the shore may be powerful enough to cause deep water to mix with surface water. Tides can also serve this function.

Other processes are probably also at work to make deep water move upward. The search for these processes is of great interest to oceanographers. The distribution of life in the sea depends, to a large extent, on the way these life-supporting materials return from the depths.

FIG. 16–7. Microscopic photograph of typical plankton organism. (Dr. Roman Vishniac)

FIG. 16–8. A typical food chain in the sea is shown beginning with microscopic plankton and ending with a shark.

THE SEA AS A SOURCE OF WEALTH

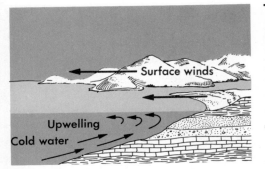

FIG. 16–9. Upwelling may cause cold water currents present in coastal areas. It tends to cool the land nearby and at the same time force to the surface, microscopic marine animals which help supply the food chain.

FIG. 16–10. A desalting plant located at Freeport, Texas. This plant can produce one million gallons of fresh water per day. (U.S. Department of the Interior—Office of Saline Waters)

Fresh water from the sea. The sea has always been a source of food and a means of transportation. Recently, however, we have begun to realize its importance as a source of vital resources needed to support our civilization. Increased knowledge of the oceans, combined with the earth's expanding population has caused new interest in the sea as a source of water, minerals, and food.

One of the most important resources of the sea is its water. The world's need for water is increasing at a rapid rate. Developing countries need huge new supplies of water for industry and irrigation. Nations like the United States, that formerly had abundant water supplies are facing shortages. The demand for water in the future can be met in two ways: Most important, the water now available must be used carefully to avoid waste. The long term solution to the problem is to increase the amount of water available. This can be done, if we find a way to convert sea water to fresh water, at reasonable cost.

Several methods can be used to extract fresh water from salt water. The most common method now used is *distillation,* which involves heating sea water. This causes the water to be changed into vapor that is carried off to be condensed into pure fresh water. See Figure 16–10. However, changing liquid water into vapor requires a great deal of costly heat energy. Making fresh water by distillation will be expensive unless a way to use cheap solar or nuclear energy can also be developed.

Another method for desalting water involves freezing it. When sea water freezes, the first ice crystals that form are free of salt. The salt remains in pockets of liquid water in the ice. The ice can then be melted to obtain fresh water, using only about one-sixth as much energy as is needed for distillation. Other methods for desalting sea water use special membranes that allow water to pass through, but block the dissolved salts. Chemicals can also be used to combine temporarily with either the water or salt to separate them.

Minerals from the sea. Since the first oceanographic research voyage made by the *Challenger,* strange black lumps of material called *nodules* have been discovered on the sea floor. Investigation has shown that huge areas of the ocean basin floor, particularly in the Pacific, are covered with these nodules. See Figure 16–11. They contain valuable amounts of manganese that is used in making certain kinds of steel. They also contain a large amount of iron, and small amounts of copper, nickel, and cobalt. Others contain phosphates that are useful as fertilizers.

Recovery of nodules from the sea floor is difficult. They are found mostly in very deep water. Several methods of mining these minerals are shown in Figure 16–12. The way the nodules are formed is not completely understood. However, measurement of their rate of growth suggests they grow fast enough to supply the world's present needs for some of the metals they contain. In the future, the mineral-rich nodules from the sea floor may largely replace the disappearing mines on land.

Some materials of value are now extracted easily from the oceans. Common salt has been obtained by evaporation of sea water for many centuries. The sea is the main source of magnesium metal and bromine. These two materials are manufactured from the dissolved salts in sea water. However, most of the useful minerals dissolved in the oceans are in such small concentrations that it is not practical to extract them. For example, sea water contains about 6 kg of gold in each cubic km. It would require processing about 4 million liters of sea water to obtain a few cents worth of gold.

The most valuable mineral resource taken from the ocean is the petroleum found beneath the sea floor. Huge deposits of oil and natural gas are found along the continental margins in many parts of the world. As the need for energy grows, these fuel resources have become very important. New techniques of drilling have allowed oil and gas to be produced from water depths as great as 100 meters as far as 100 km offshore. In the future, it will be possible to drill wells at depths of several thousand meters along the continental shelves and on the continental rise.

Food from the sea. Of all the resources that the sea is capable of supplying, the one in greatest immediate demand is protein food. At present, a large part of the world's population must rely on a diet of starchy food. While such a diet can maintain life, the lack of protein to build strong tissue allows disease to become a more serious problem. Perhaps half a billion people in the entire world suffer from some form of disease due to a lack of protein in their diets. The only source of high protein food able to meet this mounting need is fish from the sea.

But the present methods for fishing in the sea are not

Fig. 16–11. Nodules found on the floor of the Pacific Ocean. (Kennecott Copper Company)

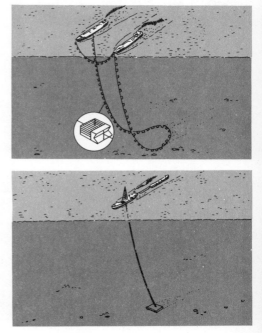

FIG. 16–12. Top. Two ships carry a cable with a continuous line of dredge buckets that scoop up nodules. Bottom. Air pumped into a pipe running from a ship to a dredge creates a pressure that brings nodules to the surface.

FIG. 16–13. An aquatic farm where trout are raised for food. (Thousand Springs Trout Farm)

satisfactory. While humans have learned to manage the resources of the land to grow animals for food, they still hunt the animals of the sea. The full food resources of the oceans cannot be utilized until humans learn to cultivate the sea the way farmers do the land.

In the future, the practice of *aquaculture* or farming of the sea may become as important as agriculture to provide food for the world's population. Aquaculture involves developing and raising special breeds of water dwelling animals and plants that yield large amounts of food. The principles of aquaculture have already been used to grow fish such as trout in large aquatic farms. See Figure 16–13. Similar methods might be used to breed fish, shellfish, and plants especially suited to grow in ocean farms. These farms would be closed off areas of the sea that provide a suitable environment for aquaculture.

Under the best conditions, water farming could produce more valuable protein food than an equal amount of land. Fish and shellfish are more efficient than most land animals in changing their food supply into protein. Water dwelling animals do not waste energy keeping an even body temperature or in having to support their weight against the force of gravity. Also, agriculture on the land can only use the top layers of soil, while ocean farms will be able to use the entire depth of water to produce crops. The sea may even be "plowed" by pumping the bottom water that is rich in nutrients to the surface. This kind of artificial upwelling would greatly speed up the growth of the "crop." With additional research and experience, it will be possible to develop the food resources of the sea far beyond the limited methods now used.

VOCABULARY REVIEW

Match the word or words in the column on the right with the correct phrase in the column on the left. *Do not write in this book.*

1. Properties of the sea connected with its ability to dissolve other substances.
2. Properties of the sea which include temperature and depth.
3. A floating layer of ice completely covering the sea surface.
4. Broken off pieces of an ice pack.
5. Disturbances on the smooth surface of an ice pack due to buckling.
6. Is measured in units of grams per cubic centimeter.
7. Zone of rapid temperature change which separates a warm surface layer of water from colder deep water.
8. Describes the total amount of dissolved solids present in a sample of sea water.
9. The general name of free-floating microscopic plants and animals in the sea.
10. The food for most zooplankton.
11. Free-floating microscopic animals in the sea.
12. Deep water moving upward to replace the surface water moved away from shore.
13. A method for removing dissolved salts from sea water.

a. density
b. pack ice
c. sea water
d. pressure ridges
e. salinity
f. chemical properties
g. upwelling
h. ice floes
i. plankton
j. physical properties
k. distillation
l. thermocline
m. overturn
n. phytoplankton
o. glacier
p. zooplankton
q. mixing

QUESTIONS

Group A

Select the best term to complete the following statements. *Do not write in this book.*

1. The temperature of sea water is an (a) important chemical property (b) unimportant chemical property (c) important physical property (d) unimportant physical property.
2. The most important single influence on the sea is (a) the sun (b) the wind (c) mixing (d) upwelling.
3. Which color of visible light penetrates farthest into the ocean? (a) red (b) orange (c) yellow (d) blue.
4. The sun heats the upper layers of the sea (a) less than the deeper layers (b) about the same as the deeper layers (c) more than the deeper layers (d) by deep penetration.
5. For sea water to freeze it must be chilled to about (a) 0°C (b) −2°C (c) −28°C (d) +2°C.
6. The process of water molecules escaping into the atmosphere after receiving sufficient energy from the sun is called (a) evaporation (b) density variation (c) condensation (d) rain.

7. Evaporation in the sea causes heat to be (a) gained by the sea and gained by the atmosphere (b) lost by the sea and gained by the atmosphere (c) gained by the sea and lost by the atmosphere (d) lost by both the sea and atmosphere.

8. Water reaches its maximum density at (a) 0°C (b) 0°F (c) 4°C (d) 4°F.

9. The amount of dissolved materials in the sea is probably (a) increasing (b) decreasing (c) remaining about the same (d) very small.

10. Analysis of sea water shows that the substances that are found in greatest amounts are ions of (a) sodium and sulfate (b) chloride and sulfate (c) magnesium and sodium (d) sodium and chloride.

11. The total weight of salts in one kilogram of sea water is 33 grams. The salinity of this water is expressed as (a) .33 ‰ (b) 3.3 ‰ (c) 33 ‰ (d) 330 ‰.

12. The gas found dissolved in the sea in an amount greater than its amount in the atmosphere is (a) nitrogen (b) oxygen (c) argon (d) carbon dioxide.

13. Which of the following usually has the greatest effect on the chemical properties of sea water? (a) plants and animals (b) currents (c) temperature (d) depth.

14. Which of the following groups of elements may have their concentration in the sea reduced as a result of heavy plant growth? (a) carbon, hydrogen and oxygen (b) carbon, nitrogen and sulfur (c) sulfur, phosphorus and silicon (d) nitrogen, phosphorus and silicon.

15. Plant life in the sea is restricted to the upper few meters because most plants require (a) low temperatures (b) sunlight (c) rain (d) high pressure.

16. The primary reason that zooplankton are found mostly in the upper layers of the sea is because (a) they need sunlight (b) the temperatures are higher (c) the pressure is less (d) phytoplankton are restricted to this region.

17. Which of the following terms is *not* closely related to the others? (a) upwelling (b) overturn (c) mixing (d) freezing.

18. Desalting of sea water is a solution for the problem of (a) fresh water not always being located where it is needed (b) salt water fish being relocated (c) rain reducing the salt concentration in sea water (d) river water increasing the salt concentration in sea water.

19. Which of the following is *not* a practical method of desalting? (a) evaporation (b) filtering (c) freezing (d) none of these.

20. Desalting of sea water has not been done on a large scale because (a) it is too dangerous (b) no one knows how (c) it costs too much (d) energy cannot be produced to do it.

21. Which of the following is *not* presently obtained from the sea in large amounts? (a) salt (b) bromine (c) gold (d) magnesium.

Group B

1. Explain why the earth is likely to be the only planet in the solar system with oceans of liquid water.

2. Explain what is meant by physical and chemical properties of sea water.

3. Why is desalting of sea water important?

4. Explain the process of evaporation.

5. Why are fish found primarily in the upper 200 meters of the ocean?

6. Why is the ocean usually blue?

7. A sample of sea water is found to contain 34 grams of dissolved materials in one kilogram of the sample. What is the salinity of the sample?

8. A rock has a mass of 350 grams and a volume of 70 cubic centimeters. What is its density?

9. Why do we say that living things have a great influence on sea water?

10. Explain why heating of the ocean by the sun is greater near the equator than near the poles.

11. Give two reasons why the upper part of the sea is warmer than the deeper part.

12. Four hundred million tons of dissolved materials are carried to the oceans each year by rivers. Why is there probably no great increase in the salinity of the sea?

13. A sample of sea water whose mass is 2000 grams is heated until only solids remain. These solids have a mass of 72 grams. What was the salinity of the sample?

14. When water in a pond freezes, the water at the bottom is warmer than at the surface. Therefore the surface freezes first. Explain why this happens.

15. An object will float if its density is less than the density of the fluid it is to float in. The density of sea water is about 1 gram per cubic centimeter while that of steel is about 7.5 grams per cubic centimeter. Under these conditions how can a steel ship float?

16. The volume of a sphere is given by the formula $V = 4/3\ \pi\ r^3$, where r is the radius. A spherical rock has a diameter of 4.0 cm and a mass of 200 grams. Find its density.

17. Vast amounts of minerals are found dissolved in the oceans, including gold, bromine and magnesium. Why are bromine and magnesium obtained from ocean water while gold is not?

18. Why is the ocean considered to be a valuable source of food?

19. Discuss the proportion, types, uses, and origin of mineral matter in sea water.

20. Compare and contrast ice floes and icebergs.

21. Discuss the cause and properties of the thermocline.

22. What kind of chemical change does life in the sea cause?

23. Distinguish between plankton, phytoplankton, and zooplankton.

24. Describe the food chain found in the sea, ending with the large fish.

17
motions of the sea

objectives

☐ Identify the causes of sea water movements.

☐ Describe the patterns of circulation near the sea surface.

☐ Describe the characteristics and effects of the Gulf Stream.

☐ Compare deep currents with surface currents.

☐ Describe the characteristics of ocean waves.

☐ Explain how ocean waves change near shore.

☐ Identify the causes and effects of a tsunami and tides.

Forces acting upon the surface waters of the ocean stir the seas into constant motion. Energy from the sun powers the circulation of air in the atmosphere. Winds are created in the lower atmosphere by the sun's energy and are affected by the earth's rotation. These winds push the surface waters of the ocean into an ever-changing pattern of near-surface currents called "circulation cells" or *gyres* (*ji*-ers). Gravitational forces of the moon and sun add to the swirl of the waters through the rhythm of the tides. Even the rotation of the earth on its axis contributes to changes in the direction of current movement in the sea.

Many of the motions of the sea are quite complex and difficult to trace. It is hard to follow the movements of sea water to find patterns that might lead to the discovery of all the causes. Moving masses of water can be identified by their physical and chemical characteristics. Then their movements must be tracked until it is clear where the water has come from, where it is going, and what causes it to move. Frequently there is not enough information available. Many of the movements of water in the sea are poorly understood, but as oceanographers slowly collect accurate and detailed observations, more pieces begin to fit into the puzzle. Although the picture is far from complete, the general motions of the sea and their possible causes can be outlined.

GENERAL CIRCULATION IN THE SEA

Causes of sea water movements. The liquid water of the sea can be set into motion only if it receives energy. For the oceans, the most important source of energy is the

352

sun. Almost all water movements in the sea can be traced back to the sun as the original source of energy. But most of its energy of motion is not received "directly" from the sun. It comes to the sea by way of the atmosphere.

The atmosphere, like the oceans, is in continuous motion. These atmospheric movements are caused by absorption of solar energy by the gases of the air. This atmospheric motion creates a pattern of winds over the surface of the earth. Global winds will be studied in detail in Chapter 20. Some of the energy possessed by moving air is transferred to the water of the sea. Winds blowing across a body of water will move the water in the direction of the wind. Steady winds that blow constantly in the same direction can move large masses of water near the surface of the sea. Almost all of the surface currents of the oceans are the result of established wind patterns over the entire globe. See Figure 17-1.

Some solar energy is directly absorbed by sea water and is eventually returned to the atmosphere. This heating and cooling of the water as a direct result of the sun's radiant energy also results in sea water movement. This movement may occur in one of two ways. First, when water is heated or cooled its density changes. Cool water becomes denser and sinks. Warm water is lighter and rises. Thus colder regions of the sea have a downward movement of cold water. Warmer water then moves in near the surface to replace the colder sinking water. Temperature differences near the upper layers create a steady cyclical movement of cold water downward close to the ocean floor, as the warmer water moves upward close to the sea surface.

Second, water motion may result from evaporation of water by solar energy. When this happens the relative concentration of dissolved salts increases. Then the higher relative density of the water makes it sink. In regions of the sea where evaporation is high, salty water at the surface will sink and be replaced by less dense water from below.

Winds and differences in density, already mentioned as causes of sea water movement, operate at all times to produce ocean currents. Other factors also produce water motion but only occasionally. For example, a very strong temporary current may be caused by an underwater landslide. These occur when large masses of sediment accumulated on a sloping part of the sea bottom suddenly break loose and slide downward. This seems to be a fairly

FIG. 17-1. The earth from 22,300 miles in space. The swirling cloud formations (white areas) are the result of the continuous motion of the atmosphere, which also affects surface movements of the sea. (NASA)

Demonstrate

Fill a small pyrex beaker with water. Place the beaker on a stand. Use an alcohol or bunsen burner to heat one side of the beaker. Pour a few drops of food coloring on the cold side of the beaker. Describe what you see. How is this similar to convection currents?

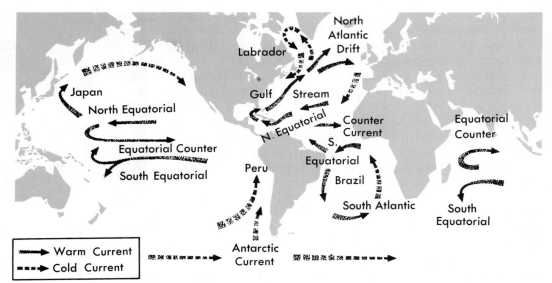

FIG. 17-2. Ocean currents of the world.

activity

When light shines on water, it heats the water. In this activity you will determine if the heating is the same at all depths of water.

Fill a beaker nearly full of water. Use rubber bands to hold three thermometers together. The bulb of one should be at the bottom of the beaker, one in the water near the surface, and one midway between the other two.

Make a data table to record the temperature readings of each thermometer every two minutes.

With the thermometers in place, move the apparatus into the sun (or use a 250 watt infrared lamp). Cover the side of the beaker so that light enters only at the top.

Describe the effects of sunlight on the heating of the ocean at various depths.

common event on the continental slopes. Because a large amount of sediment carried along by the water turns it cloudy or turbid, these movements are called *turbidity currents.* Since these currents are difficult to observe directly, little is known of their effects. They may, however, play a major role in the movement of sediment away from the continental shelves to the deep sea floor.

Earthquakes and volcanic activity on the sea floor may also be a temporary factor in the motions of sea water. Of special interest because of their destructiveness are the poorly named "tidal" waves, or large-scale water movements caused by earthquakes on the sea bottom. These waves will be discussed later in this chapter.

Patterns of circulation near the sea surface. Water within the upper layers of the sea moves mainly in response to winds. But it is not free to move in the direction influenced by the global wind patterns. Land masses of the continents acts as barriers to the motion of wind-driven currents. The rotation of the earth also has an effect on the system of ocean currents. Because of the earth's rotation, the path of a moving object is deflected from a straight line. This shift in direction is to the right in the Northern Hemisphere and to the left in the Southern Hemisphere. This is called the *Coriolis effect,* after the nineteenth-century French mathematician who first described it. Ocean currents are subject to the Coriolis effect and move in a direction that is partly determined by it. However, the effect is more noticeable in the atmosphere. The Coriolis effect in relation to winds will be discussed in Chapter 20.

In all the oceans there is a powerful surface current near the equator moving toward the west. It is driven by the steady winds characteristic of the warm equatorial regions. If there were no continents to deflect this current, it would continuously circle the earth like a great river in the sea. The continents, however, form barriers which turn the current to the north or south. Equatorial currents are found in the Atlantic, Pacific, and Indian Oceans. In each ocean there are two westward flowing parts, the *North Equatorial Current* and the *South Equatorial Current*. Separating these is the east-flowing Equatorial Counter Current. See Figure 17-2.

In the Atlantic and Pacific Oceans, the North and South Equatorial Currents are turned to the north and south along the shores of the continents. At higher latitudes, steady winds push the water from west to east, opposite the direction of current flow near the equator. This action, along with the Coriolis effect, creates great circles of moving water in each ocean. These circular movements, plus many smaller movements form the basis for the principal currents. See Figure 17-3.

In the North Atlantic, the North Equatorial Current piles water against the east coast of North America around the Gulf of Mexico. This water then moves north along the east coast of the United States as a warm, swift current called the *Gulf Stream*. The Coriolis effect then forces the Gulf Stream to move to the right, toward the

FIG. 17–3. Major wind patterns of the earth shown in (A) combine with the effects of the earth's rotation to produce a generally clockwise movement in northern oceans and a counterclockwise one in the Southern Hemisphere (B).

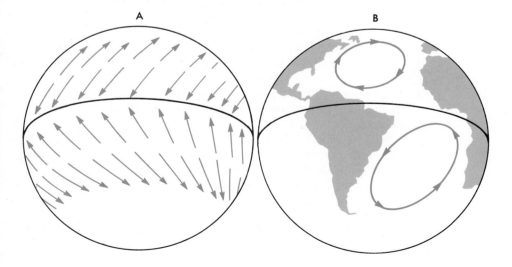

northeast. This takes it into the North Atlantic. There it branches into three weaker currents. One branch, the *Labrador Current,* doubles back to the south and carries colder water back along the northeast coast of the United States. Another branch turns southward in the direction of the mid-Atlantic, where it eventually disappears. A third branch is the *North Atlantic Drift.* A drift is a weak current. It crosses the North Atlantic and then turns south along the west coast of Europe, finally rejoining the North Equatorial Current.

These currents completely circle the North Atlantic, leaving in the center a vast still area which is relatively free of current action. This is called the *Sargasso Sea.* Great quantities of seaweed are found floating in this quiet region of the Atlantic, named for the *Sargassum* seaweed typical of these waters.

In the Pacific, the pattern of currents is similar to those in the Atlantic. The North Equatorial Current of the Pacific sweeps all the way across the mid-Pacific from Panama to the Philippines. When it meets the island barriers in the east Pacific, the greater part of the current turns to the north and becomes the *Japan Current.* The warm waters of the Japan Current, the Pacific equivalent of the Gulf Stream, pass along the Asian coast. They then swing to the northeast as the *North Pacific Current.* This current carries the cool water from the North Pacific down the west coast of North America to join the North Equatorial Current off the coast of Mexico.

In the southern parts of the Atlantic and Pacific, the current patterns also appear in the form of huge circles. Whereas the currents circle in a clockwise direction in the northern oceans, they move in a counterclockwise direction in the southern oceans. See Figure 17–2.

In the most southerly regions of the Atlantic and Pacific, constant west winds produce the *Antarctic Current.* This powerful current completely circles the Antarctic continent. There are no land masses to interfere with its movement.

Within the Indian Ocean, surface currents follow a two-part pattern. In the southern part of the ocean, the South Equatorial Current is deflected to the south along the coast of Africa. It then flows east toward Australia, where it mingles with the many smaller currents among the islands of the South Pacific. Currents in the northern region of the Indian Ocean are governed by winds that change with the seasons.

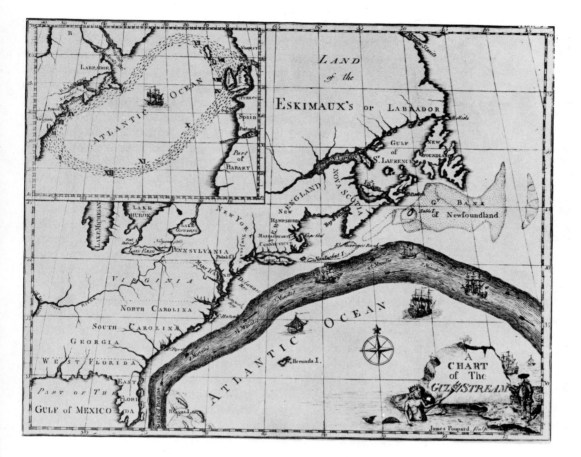

FIG. 17-4. Franklin's chart of the Gulf Stream. (Black Star—Werner Wolff)

The Gulf Stream. Sometime before 1770, Benjamin Franklin became aware that many of the New England whaling ships crossed the Atlantic from England about two weeks faster than the regular mail ships. The whaling captains told Franklin of a swift eastward flowing current in the North Atlantic. They avoided this current by sailing south rather than on a straight line from England. From what he could learn from sailors, Franklin was able to draw and publish in 1779 the first chart of this current. See Figure 17-4. It is still known as the Gulf Stream, the name Franklin gave it, and illustrates the characteristics of many of the sea's surface currents.

The Gulf Stream begins where the North Equatorial Current is deflected by the Panama ridge connecting North and South America. The Equatorial current piles up water in the Gulf of Mexico with such force that sea level there is significantly higher than in the Atlantic.

activity

Density currents like those in the ocean can be produced in the laboratory.

Place an ice cube in a cup of cold water. Then, add some dark food color.

Fill a clear rectangular box (shoe box) about half full of warm (not hot) tap water. Raise one end of the box by placing it on a book. Carefully pour the colored ice water into the raised end of the box, allowing it to flow down the end of the box, so it does not splash. Observe the results.

1. Did the two water masses mix immediately?

2. Where does the colder water tend to go?

3. In the ocean or a deep lake, where would you find the coldest water?

FIG. 17–5. Relative speed of movement of water in the Gulf Stream. Darker colors indicate higher speeds.

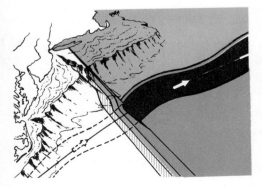

The mass of trapped water discharges in a swift flow past the tip of Florida entering the Atlantic as the Gulf Stream.

Once in the open sea, the Gulf Stream flows swiftly in a wavy path along the southeast coast of the United States. Its speed of movement does change, but it usually moves along at the rate of about 10 to 15 kilometers per day. The moving stream of water that makes up the Gulf Stream averages about 100 kilometers in width and at least one kilometer in depth. Because of its size, it carries an enormous amount of water. The Gulf Stream is estimated to transport between 75 to 90 million cubic meters of water per second. By comparison, the Mississippi River averages 20,000 cubic meters per second.

Because it originates in the equatorial regions, the water carried by the Gulf Stream is warmed at its source. The diagram in Figure 17–5 shows how the warm water flows along just off the continental shelf at the latitude of North Carolina. Strangely, this warm water flowing off the Atlantic coastline is often the cause of a drop in temperature in the eastern United States. This climatic change occurs because cooler air moves toward the Gulf Stream to replace the rising air warmed by the current. Thus cold air from inland may move seaward bringing cool weather to the coast. Some of the severest winters along the Atlantic coast have occurred when the water of the Gulf Stream was at its warmest.

Moving north, the stream is deflected to its right by the Coriolis effect. The force that twists the stream off to the right piles water up along the right side of the current. Because of this, water along the Cuban coast is about 50 centimeters (around 20 inches) higher than on the side of the current toward the mainland.

Near Greenland the warm water meets the southward moving cold water of the Labrador Current. Great banks of fog are created when the warm and cold water meet. The sea off Labrador is one of the foggiest regions of all the oceans. Collision with the Labrador Current helps to deflect the Gulf Stream and slow it down. It splits into three parts and sends one branch north to the western shore of Greenland. Another branch passes along the southwest coast of Iceland. But the main branch proceeds eastward as the North Atlantic Drift. This slow movement of warm water across the North Atlantic greatly influences the climate of northern Europe and the British Isles. After warming Europe, the water moves slowly south and eventually becomes part of the westward-flowing Equatorial Current. In this way the water is carried back to its beginning in the Gulf of Mexico.

Deep Currents. In addition to wind driven surface currents, there are slower but equally powerful currents flowing deep beneath the surface. An increase in density due to cooling or an increase in salt content will make water sink. These slower currents seem to be produced as surface water is pulled downward. Very little is known about the way water moves at great depths in the various oceans. However, the deep circulation of the Atlantic has been worked out in sufficient detail for us to understand how water moves at great depths.

In a small region of the North Atlantic, just south of Greenland, the water is very cold and salty. Its salt content is increased by the formation of ice. This cool water with high salinity sinks and moves to the south as a deep current. It flows southward beneath the Gulf Stream. Near the equator, part of this deep water begins a return flow back north again. The above facts tell us that the North Atlantic has a circling deep current which runs in a direction opposite that of the surface currents.

Part of the deep water from the North Atlantic moves into the South Atlantic. As it nears Antarctica, the deep Atlantic water meets similar cold, salty water formed off the coast of Antarctica. The sinking Antarctic water seems to move outward in all directions from the polar region. Deep water from the Atlantic mixes with the Antarctic water and is probably carried into other oceans. Some of the Atlantic water eventually finds its way back to the surface. The complete pattern of deep circulation in the Atlantic is shown in Figure 17–6.

Deep currents are generally more slow moving than those near the surface. An amount of cold, salty water sinking in the polar regions seems to take about 30 years to find its way to the equator. At the surface, the same amount of water would take only a few years to complete the entire circle of an ocean such as the North Atlantic.

WAVES

Wind and waves. Waves on the surface of the water are created when energy is added to a body of water. Energy

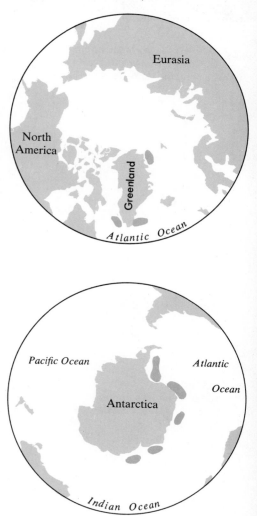

FIG. 17–6. Dark blue areas show where cold water is sinking to cause deep currents in the oceans of the two hemispheres.

FIG. 17–7. Waves breaking on this beach are the result of swells approaching the shore. (Greene-FPG)

Observe

Tie one end of a spring to a stationary object. Shake the other end. Watch the way the spring moves. Identify the different parts of the wave. Does tension affect the speed and size of the wave. How would the spring move if the stationary end could hang free and then shaken?

to create waves is added to the sea in many ways. Earthquakes and undersea landslides, gravity of the moon and sun, changes in atmospheric pressure, and the movement of ships are examples of ways that waves are raised on the sea. But one source of energy far exceeds all the others as a force creating ocean waves. That force is the wind.

If the air movements generated by solar heat did not exist, the surface of the sea would be almost as smooth as glass. Even the smallest breeze creates ripples as a result of friction between moving air and the water. The wind is able to push directly against the side of a small wave. As the wave grows larger, more energy can be transferred to it from the moving air. Thus as long as the wind blows the wave increases in size. Actually the wind seldom blows constantly from the same direction. Usually the sea surface is covered with a confused pattern of small waves moving in all directions in response to the shifting winds. However, the larger waves are better able to capture energy from the wind than the small waves. This means that although the wind usually produces waves of all sizes, the larger ones continue to grow while the smaller waves quickly die out.

The usual pattern of waves on the sea surface becomes one of small waves and ripples continually being formed and disappearing on the slopes of larger waves. Groups of large waves, which are alike in size, are called *swells*. See Figure 17–7. They are able to move great distances over the sea surface with very little change. Swells move in groups, one following the other away from the area of their origin. The swells which finally arrive at the shore may have been produced a thousand miles out at sea. From careful analysis of the waves, oceanographers can usually determine where the swells originated. They can also predict the height of waves likely to occur at any point on the water's surface or along the ocean shore.

Characteristics of an ocean wave. A wave on the surface of the water has two basic parts. A ridge which is elevated above the surrounding water surface is called a *crest*. On either side of the crest is a depression or *trough*. Any particular wave can be described by three characteristics. One is its *height*: the distance from the bottom of the trough to the top of the next crest. The second characteristic is its *wavelength;* the distance between crests. The third is its *period:* the time it takes for two consecutive crests to pass a given point. See Figure 17–8. The speed at which a wave moves is determined by a simple relationship between its wavelength and period. This relationship is given by the following formula:

$$\text{wave speed} = \frac{\text{wavelength}}{\text{period}}$$

For example, if a wave has a wavelength of 225 meters and a period of 12 seconds, its speed would be 225/12 = 18.8 meters/second (about 42 miles per hour).

A wave moving through water does not carry water along with it. To illustrate, observe the behavior of a floating object as a wave passes under it. A cork floating in water moves forward slightly as the crest of a wave approaches. Then if falls back an almost equal distance as the wave passes into a trough, ending up almost in its starting position. Analysis of this movement will lead us to conclude that waves are produced when the individual particles of water move in circles.

As the wave passes a given point, the water particles at the surface trace a circle whose diameter is equal to the height of the wave. Each water particle makes one complete circuit as the wave passes by, ending up almost exactly where it started. See Figure 17–8. Only the wave itself moves any distance over the surface of the water. Thus, the surface water remains in place, transmitting only the motion of the wave by the circular movements of individual water particles.

Waves near the shore. When a wave approaches the shore it undergoes change. The reason is found in the circular motion of the water particles which produce the wave. When a wave reaches shallow water, the movement of the water particles is slowed as they rub against the sea floor. If the wave approaches the shore at an angle, the slow-down causes *refraction*. This means the path of the wave

FIG. 17–8. The mechanics of motion in a water wave. As a wave moves forward, the water particles (represented by black dots) move in circles.

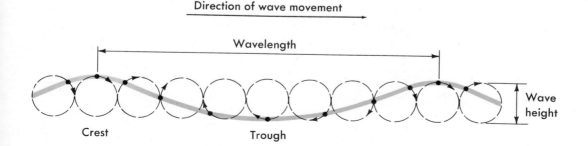

Direction of wave movement

Wavelength

Crest

Trough

Wave height

FIG. 17–9. When waves approach the shore they are refracted, as shown in the diagram above. Refraction causes the waves approaching the curved shoreline in the photo to hit all parts directly. (Ramsey)

FIG. 17–10A. Wave stages a to d show the development of a breaker. The wave advances on the shore in the direction indicated by the black arrows above. The white arrows indicate the undertow, and the lower black arrows show the movement of bottom sediments below the surf.

is bent. Bending is produced when one part of the angled wave strikes the shallow bottom first. A part of the wave slowing down before the rest causes a bending or refraction in the entire wave as it moves toward the shore. See Figure 17–9.

Wave refraction causes the incoming waves to line up parallel to the shore. Thus waves tend to approach the shoreline head-on, no matter what direction they had originally. This is an important fact in the development of shorelines.

Directional changes in waves are due to refraction; in addition, changes in height occur as waves meet the shore. The series of height changes in an incoming wave are shown in Figure 17–10A. When a wave reaches water where depth is about twice the height of the wave, the circular paths of the water particles are squeezed upward by the resistance they meet at the bottom. The circular motion is changed to an ellipse, raising the crest of the wave to a greater height (as shown at position a). As the wave moves into more shallow water, the crest rises higher and higher. Finally it tumbles forward into the trough and the wave becomes a *breaker* (at position b and c).

After the wave breaks, the water is usually thrown up on the shore as a foamy sheet. This action uses up the wave's energy and the water runs back into the breakers (as in position d). The size and violence of the breakers is determined by the original wave height and the steepness of the sea floor close to shore. If the bottom is steep, the height of the wave increases rapidly and the wave breaks violently. On the other hand, if the shore slopes gently, the wave rises slowly, spilling forward with a rolling motion that continues for some distance as the wave advances.

Water carried up on the shore by the breaking waves escapes back to deeper water in the form of a usually weak and irregular current that is called the *undertow*. Where the undertow exists, it is a current running seaward along the bottom far below the breakers. An undertow current is seldom very strong and would almost never

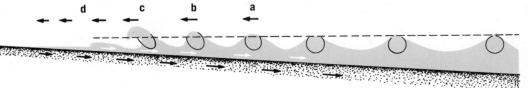

be a problem to swimmers. Some people confuse the relatively weak undertow with a more dangerous but lesser known type. Swift *rip currents* are caused by water returning to the ocean through breaks in underwater sandbars close to the beach. See Figure 17–11. Rip currents are a hazard to swimmers because they carry the person out to deeper water in the midst of the breakers. Generally, rip currents occur in spurts, lasting only a few minutes at any particular spot along the shore. Their presence can be detected by a gap in the line of breakers and by the yellowish color of the water where sand is being stirred up by the current. Anyone caught in a rip current should swim to one side or the other to escape the current and get to a region where the action of the breakers will carry him back to shore.

Giant waves. When the wind creates waves on the sea, three factors determine their size. There is (1) the speed at which the wind blows, (2) the length of time the wind blows and (3) the distance the wind blows across the open water (often called the *fetch*). Very large waves are produced by strong, steady winds with a long fetch. Such conditions are most likely to occur during a storm when a few waves are able to gather enough energy from the wind to reach very great size. Waves more than 33.5 meters in height (about 100 feet) have been observed during severe storms. Even such large waves as these are seldom dangerous to ships. A floating object tends to move in the same circular pattern as the water itself. A ship is not usually damaged unless it moves in a different direction from the water. This happens only if the crest of a wave is blown off by the wind. Then the waves break in the open sea, forming what is usually called *whitecaps*. Breaking waves that collide violently with ships can create severe damage.

Tsunamis. The most destructive waves in the sea are not created by wind action. By far the most dangerous are the giant seismic sea waves or *tsunamis* ((t) su-*na*-mees) that are produced by earthquake or seismic disturbances on the sea floor. In the past, these waves have been called "tidal" waves. The name is misleading because they have no connection with the tides. In an effort to remove this confusion, the Japanese word "tsunami" has been substituted.

FIG. 17–10B. The spilling breakers shown in the photo begin rather far from shore and continue landward as the crest gradually collapses. (Photo-Researchers-Winch)

FIG. 17–11. Rip currents (darker areas in the surf) are clearly visible in this aerial view of the beach near Monterey, California. (Hydraulic Research Laboratory, University of California)

Tsunamis seem to originate close to the deep sea trenches around the edges of the Pacific Ocean. They are probably caused by two kinds of events that accompany an earthquake on the sea floor. Faulting may bring about a sudden drop or rise in a part of the sea bottom. Then a large amount of water drops or rises at the same time. This mass of water circulates up and down as it tries to adjust again to sea level. A series of low waves is sent out by this water movement. Tsunamis may also be triggered by a severe underwater landslide set off by an earthquake. When this happens, a large mass of water above the landslide is thrown into an up and down motion creating a series of tsunamis. Volcanic eruptions may also disturb great amounts of water and produce tsunamis.

Tsunami waves are not very tall but they have a long wavelength. Their height is usually less than a meter in deep water but the wavelength may be as much as 240 kilometers (about 150 miles). A period of about 1000 seconds is common. Because their heights are so small in the open sea, tsunamis cannot be felt aboard ships or seen from the air. But the amount of energy they possess is tremendous. The movement of a tsunami represents motion of the entire mass of water from the surface to the bottom. As the tsunami approaches a coastline, it delivers its destructive energy against the shore, causing great damage to populated regions.

Near the shore the height of the tsunami greatly increases as its speed diminishes. The arrival of a tsunami may be announced by a sudden pull-back of the water along the shore. This occurs when a trough arrives before a crest. If a crest arrives first, there will be a sudden, rapid rise in the water level. The coastline itself seems to affect the height of the tsunami. It may be small enough to cause little harm at one point but large enough to cause great destruction at nearby locations.

More than two hundred tsunamis are known to have caused damage within the span of recorded history. Their effects have been observed mostly along the coasts of Japan, southeast Asia, the Caribbean Sea, Mexico, South America and Alaska. The most destructive tsunami in modern times took place in 1883 when the volcanic island of Krakatoa blew up. More recently, an earthquake on the sea floor near Alaska brought on a series of tsunamis. The waves struck the Hawaiian Islands on April 1, 1946, killing 159 people. See Figure 17–12. Following these disastrous Hawaiian tsunamis, a warning network was set up to observe unusual water movements and give advance notice of any onrushing tsunamis.

However in May of 1960, a tsunami arising from an earthquake off the coast of Chile caused extensive damage and loss of life in Chile, Hawaii, the Philippines, Okinawa, and Japan. In the Hawaiian city of Hilo, 61 people who ignored the warning given six hours before were killed. In Japan and the eastern Pacific, no general alert was issued since it was thought that tsunami waves could not cross the entire Pacific. The waves did strike, resulting in great loss of life and damage to property. These experiences led to expansion and improvement of the Seismic Sea Wave Warning System to function over the entire Pacific area. In the United States the warning network is operated by the National Oceanic and Atmospheric Administration. Tsunami warnings for threatened regions are issued to the general public through the Weather Bureau's links with radio and television stations.

Tides. The basic reason that tides occur is due to the gravity of the moon and sun. However, the tides that come in at any particular place along the shore are determined mainly by conditions here on the surface of the earth. If the moon and sun were the only influences producing tides, there would still always be two high and two low tides each day. Both are linked to the passage of the two tidal bulges as the earth rotates. Along the shores of the Atlantic, tides generally follow the basic pattern of two high and two low tides every 24 hours and 50 minutes. At some locations along the shore of the Gulf of Mexico, there is only one high and one low tide each day. On the Caribbean side of the Panama Canal, there is only one tide change a day and it is very small. On the Pacific side, the average height is fourteen times greater and there are tides twice a day.

There are, then, great differences in the timing and range of tides at different places on the earth. Apparently, varying parts of the sea respond in their own way to the rhythmic gravitational pull of the moon and sun. Remember that the sea is all one body of water. But the unevenness of the sea floor and the position of the continents divide it into several parts. For each part, there seems to be a particular tidal pattern.

Another important factor in determining the tides in any particular part of the sea is the size and depth of its basin. Both of these factors have great influence on a

FIG. 17–12. Tsunami approaching the shore near Hilo, Hawaii. Photos show oncoming waves destroying a pier. Arrow points to a man succumbing to the torrent of water. (World Wide Photo)

Observe

All bodies of water undergo the affect of tides. If there are any small lakes in the area see if you can pick a point at the water level, and after a few hours see if the water level is still at the same point.

FIG. 17–13. These two photographs, taken at different hours on the same day, show the difference between high and low tides in the Bay of Fundy. (Photo Researchers-Russ Kinne)

certain kind of motion in the water. *Tidal oscillations* (*os*-uh-*lay*-shuns) are very slow rocking motions which occur in various parts of the sea in response to the movement of tidal bulges caused by the sun and moon. Similar oscillations can be seen in any container filled with water which is stirred with just the right rhythm. The water in a bathtub, for example, can be kept rocking back and forth easily if it is stirred up with a rhythmic motion suited to its particular size. In parts of the sea, the rhythm of the passing tidal bulges also creates such oscillations. These may add to or cancel out the up and down motion created by the tidal bulges themselves.

A good example of oscillations which greatly add to the flow of tides occurs in the Bay of Fundy. This bay is located on the shore of New Brunswick, Canada, at the end of the Gulf of Maine. The water in the Bay of Fundy is set into rapid oscillations by the twice-daily passage of the tidal bulges. As the water rocks back and forth in the Gulf, it first floods the Bay of Fundy, raising tides 50 feet high. Then, as the water rocks slowly back to the other end of the Gulf, the resulting low tide almost completely drains the bay. See Figure 17–13. Nantucket Island, not far from the Bay of Fundy, does not receive the effect of the tidal oscillations in the Gulf of Maine. Its tidal range is only about 1½ feet. Along straight coastlines and in the open sea, tidal oscillations are not as apparent as they are in smaller bodies of water.

The rise and fall of tides often produce very strong currents. These *tidal currents* are most powerful when there are two regions close together which have large differences in tidal height. A narrow connection between the areas, such as the entrance to a bay, will increase the effect of the tidal current. Ships may have great difficulty in navigating some narrow passages when the tidal current is running.

When a river enters the ocean through a long bay, a *tidal bore* may be produced. This is a wave of water which passes up the river from the sea as the tide rises. See Figure 17–14. In a few rivers the tidal bore moves rapidly upstream in the form of a large wave which eventually exhausts itself. The Amazon River has a tidal bore that is said to resemble a small waterfall, moving rapidly upstream for a great distance.

The power of tides in moving large amounts of water represents a source of energy that could be useful to man. To make use of tidal power, dams must be constructed to trap the water at high tide. A location must be chosen where such a dam can be constructed without too much difficulty. Then water can be trapped behind the dam at high tide and released at low tide. As the water flows through the dam it could be used to generate electricity.

An experimental tidal power plant has been constructed at the mouth of the Rance River in France. This is the first in a series of tidal power projects which the French hope will help supply their increasing need for electricity. A similar project has been considered as a joint Canadian-American venture in the region of the Bay of Fundy. Perhaps the tides will be called upon to supply a part of the increasingly great world need for electricity in the years to come.

FIG. 17–14. The tidal bore of the River Severn near Gloucester, England. The force of the high tide is strong enough to push the wave of water far upstream. (Weston Kemp)

VOCABULARY REVIEW

Match the word or words in the column on the right with the correct phrase in the column on the left. *Do not write in this book.*

1. Ocean current caused by underwater landslides.	**a.**	Gulf stream
2. A change in direction of current flow due to the earth's rotation.	**b.**	Sargasso Sea
3. An eastward flowing ocean current found near the equator.	**c.**	swells
4. A swift warm ocean current found off the Eastern coast of the U.S.A.	**d.**	wavelength
5. A weak ocean current flowing eastward across the North Atlantic.	**e.**	crest
6. A vast, almost motionless area in the center of the Atlantic.	**f.**	turbidity current
7. Groups of large waves which are alike in size.	**g.**	period
8. The elevated portion of a wave.	**h.**	trough
9. The depressed portion of a wave.	**i.**	Coriolis effect
10. The distance between crests of adjacent waves.	**j.**	Japan Current
11. The change in direction of motion of waves on reaching shallow water.	**k.**	North Atlantic Drift
12. Swift currents of water returning to the sea from the beach area.	**l.**	Equatorial Counter Current
13. How far the wind blows across open water.	**m.**	tidal bore
14. Seismic sea waves incorrectly called "tidal waves."	**n.**	tsunami
15. A wave of water passing up a river as the tide rises.	**o.**	fetch
	p.	refraction
	q.	rip current

QUESTIONS

Group A

Select the best term to complete the following statements. *Do not write in this book.*

1. Most of the energy which sets the seas in motion comes from the (a) sun (b) moon (c) rotation of the earth (d) many earthquakes.

2. Almost all of the ocean's surface currents are caused directly by the (a) rotation of the earth (b) differences in water temperature (c) global wind patterns (d) gravity effects from the moon.

3. As the ocean water cools it will (a) migrate toward the poles (b) move downward (c) move upward (d) spread out equally in all directions.

4. Where evaporation is occurring rapidly, the saltiest sea water is found (a) near the ocean floor (b) near the ocean surface (c) at moderate depths (d) at all depths.

5. Which one of the following causes of ocean currents is *not* considered an occasional or temporary cause? (a) density differences (b) underwater landslides (c) earthquakes (d) volcanic activity.

6. The rotation of the earth on its axis causes ocean currents in the northern hemisphere to be deflected (a) northward (b) southward (c) to the left (d) to the right.

7. The eastward moving Equatorial Current is bounded by the North Equatorial Current and the South Equatorial Current which flow (a) north and south respectively (b) eastward (c) south and north respectively (d) westward.

8. The Gulf Stream is caused by the piling up of water in the Gulf of Mexico from the (a) Labrador Current (b) Mid-Atlantic Current (c) North Equatorial Current (d) South Equatorial Current.

9. A current which is considered a branch of the Gulf Stream is the (a) Japan Current (b) Labrador Current (c) North Atlantic Drift (d) North Pacific Current.

10. The current in the Pacific Ocean which is equivalent to the Gulf Stream is the (a) Pacific North Equatorial Current (b) Japan Current (c) North Pacific Current (d) Sargasso Current.

11. The large current circles found north of the equator move (a) clockwise (b) counter-clockwise.

12. There is no Arctic Current similar to the Antarctic Current due to the (a) warmer climate in the Arctic (b) land barriers found in the Arctic (c) slower speed of winds in the Arctic (d) fact that there is no land near the north pole.

13. Deep currents are powerful ocean currents thought to be caused by (a) typhoons (b) volcanoes (c) sinking of heavier water (d) the earth's rotation.

14. Although the Gulf Stream moves much slower than the Mississippi River, it transports (a) twice as much water (b) one hundred times as much water (c) one thousand times as much water (d) four thousand times as much water.

15. The Eastern Seaboard frequently has colder weather than expected due directly to the (a) North Atlantic Drift (b) Gulf Stream (c) Sargasso Sea (d) North Equatorial Current.

16. Great fog banks form off Labrador as a result of (a) the meeting of the Gulf Stream and the Labrador Current (b) the unusually warm climate of Labrador (c) the unusually cold climate of Labrador (d) the vast amount of ocean surrounding Labrador.

17. The North Atlantic Drift influences the climate of Europe and the British Isles causing (a) much cooler weather than usual (b) very stormy weather (c) unusually clear skies (d) much warmer weather than usual.

18. The source of almost all waves on the seas is (a) gravity of the moon (b) wind (c) earthquakes (d) undersea landslides.

19. Large waves naturally grow larger due to (a) more surface area being exposed to the push of the wind (b) the elastic nature of water (c) the absorption of the smaller waves (d) the formation of whitecaps.

20. The period of a wave can be defined as (a) the distance between two adjacent wave crests (b) the distance between the crest and trough (c) the time it takes for two consecutive wave crests to pass a given point (d) the distance from the bottom of the trough to the top of the next crest.

21. A wave whose wavelength is 100 meters and whose period is 5 seconds would have a speed of (a) 100 meters/sec (b) 500 meters/sec (c) .05 meters/sec (d) 20 meters/sec.

22. Waves cause the individual particles near the surface of deep water to move primarily (a) in the direction of the wave (b) back and forth (c) up and down (d) in circles.

23. Waves approaching the shore are refracted (a) when they come straight in (b) when they approach at an angle (c) only when the shoreline curves inward (d) only when the shoreline curves outward.

24. Waves usually approach the shore head-on no matter what direction they originally had. This is due to (a) refraction (b) reflection (c) rip currents (d) the action of breakers.

25. Breakers form when waves (a) approach the beach from shallower water than that near the beach (b) tend to speed up in the more shallow water (c) enter shallow water, causing crests to build so high they tumble into the trough ahead (d) cause rip currents.

Group B

1. Sea water is set in motion only if it receives energy. (a) What is the main source of this energy? (b) Describe the ways water receives this energy.

2. Describe the ways temporary ocean currents are produced.

3. How does the rotation of the earth influence the current patterns of the oceans in (a) the Northern Hemisphere (b) the Southern Hemisphere?

4. It is said that water normally spirals clockwise as it goes down a sink drain. Do you think this might be due to the Coriolis effect? Explain your answer.

5. Using drawings, show the ocean current pattern in the Atlantic Ocean for both hemispheres.

6. How is the ocean current pattern of the Pacific Ocean similar to that of the Atlantic Ocean? How do they differ?

7. What causes deep currents? How does their direction of flow compare to the surface currents?

8. How does the Gulf Stream form from the Atlantic North Equatorial Current?

9. Describe the results of the Coriolis effect on the Gulf Stream.

10. Large surface waves on the ocean grow larger while the small waves usually die out rather quickly. Explain why this is so.

11. What are the three characteristics which describe any ocean wave?

12. You notice a group of swells passing by a stationary object. Measurements which you make indicate their wavelength is 200 meters. You notice that six pass the stationary object in one minute. What is the speed of the swells.

13. Analysis of the movement of a free floating object shows it to have a very egg-shaped motion. Describe the condition which would cause the wave to have this motion.

14. Why do waves approaching a beach at an angle tend to line up parallel to the beach?

15. What is necessary for an undertow to become a rip current?

16. During World War II many concrete cargo vessels were built due to the shortage of steel. These vessels were soon found to be unsatisfactory because they were not flexible enough to bend in the waves. It was also found that the engineers chose the wrong length to make them. Why would their length be a critical factor?

17. Why is the name "tsunami" being used more often today in place of the term "tidal wave"?

18. In the Bay of Fundy on the East Coast of North America, 50-foot-high tides are common. Explain why this is so.

19. Some scientists in Russia have recently suggested that the Bering Straights should be partially dammed with an underwater rock ridge from the Alaskan coast to coast of Siberia. The idea is to shut off the deep current of cold water that enters the Pacific Ocean from the Arctic Ocean but allow the warm water to enter the Arctic. What would be the purpose behind this project?

earth, the water planet

A Super Flood

Approximately 100,000 years ago, a tongue of ice that reached out from the continental ice sheet made a dam across a large river. The water collected behind the ice dam and formed a giant lake. Eventually the water reached the top of the ice dam. As the water flowed over the ice, rapid melting caused the dam to suddenly collapse. During a short time, a few days at most, all of the water was released from the lake. This sudden flood as the water made its way to the sea had a flow ten times greater than all the rivers of the world. It was probably the greatest flood the earth has ever experienced.

Today the evidence of that flood can be found in a region of eastern Washington called the "scablands." It is an area of about 40,000 km² made up mostly of bare rock cut with large channels, great potholes, dry waterfalls, giant ripple marks, and other erosion features created by running water. At top is a picture of the eastern Washington scablands. Compare its appearance to features seen on the surface of Mars (bottom). There is no water on that part of the Martian surface at the present time. However, the presence of scabland features on the surface of Mars indicates that sudden great floods have occurred during its past.

Sending Icebergs to Solve the Water Supply Problem

The water supply for coastal regions of the world could in part be supplemented by icy resources from Antarctica. These regions could be supplied with fresh water from one of the numerous icebergs that constantly breaks off the Antarctic continental glacier. Although approximately 70% of the earth's surface is covered with water, most of it is found in the sea. No inexpensive method for obtaining fresh water from sea water is available. The iceberg method would compare favorably with other methods, including distillation (Chapter 16).

Icebergs as large as 16 km high and 0.8 km wide could be towed from Antarctica to almost any coastal location in the world. A route would be chosen that would take advantage of ocean currents. The trip would take about one year. Even if the iceberg melted to half its original size during the trip, it would still contain one billion cubic meters (about 980 billion liters) of fresh water. When it reached its destination, the iceberg would be surrounded by a floating dam. The less dense melted water from the iceberg would float on the surrounding sea water behind the dam and could be drawn off as needed.

Explorers of the Sea

Below, a research team out of Woods Hole Oceanographic Institution looks over dredgings from the ocean floor. From the first voyages of H.M.S. Challenger to modern undersea vessels, oceanographers such as these have been exploring the ocean with instruments specially equipped for the watery realm of the ocean.

When more direct contact with the ocean is required, a diver such as the one at the right performs the task. Diving with SCUBA gear has also become a popular sport in areas such as the tropics where colorful fish and unusual life forms populate the sea.

18
the rock record

objectives

- [ ] Explain the principle of superposition.

- [ ] Describe how this principle relates to conformity, unconformity, and disconformity.

- [ ] Explain how the half-life of a radioactive element is used to learn the age of a rock.

- [ ] Describe three means other than radioactivity used to measure geologic time.

- [ ] Describe several ways fossils are formed.

- [ ] Explain how fossils can be used to relate rock layers to one another.

Within every piece of rock there is a story. Part of that story involves minerals and the way in which these minerals combined to form the rock. Actually, rocks begin to change the moment they come into existence. Through careful study, geologists can often reveal the history of change that a particular rock has undergone. To unravel this mystery very patient detective work is needed. In addition, a knowledge of the way various processes act upon the rocks of the earth's crust is needed. The researcher finds that his reward is a better understanding of the history of the earth.

However, the clues that lead to a complete history of the rocks are not easily found. The search is only successful when certain principles which control the patterns of structures are understood. When these principles are understood, the arrangement of rocks in the crust becomes a record of their history.

THE PATTERN OF THE PAST

Rock layers. Throughout most of the earth's history there has been a steady weathering and erosion of rocks at the earth's surface. These powerful erosive forces have steadily built up an accumulation of rock fragments deposited in layers or strata of sedimentary rocks. It is in these rock strata that clues to much of the earth's past can be found. An estimate of the total thickness of sedimentary rocks deposited since the earth was formed is more than 100 kilometers. This does not mean that the total thickness of sedimentary strata was laid down in any one place. Conditions in some locations and at certain

376

times have not been favorable for the deposition of sediments. At other times however, thick deposits have been laid down. For example, a series of strata up to 30 kilometers in depth were built up in geosynclines. These geosynclines were formed as the floors of ancient shallow seas slowly sank. If the greatest known thickness of each layer in all these different places were added together, we could get some idea of the total record available in these rocks.

The most basic principle used in understanding the rock record is a simple one. It is generally believed that when the layers are horizontal, or nearly so, each overlying bed is younger than the one beneath it. This is called the *principle of superposition*. When strata of sedimentary rock are known to have been deposited by a body of water or as wind-carried material, there can be no doubt that the build-up was according to the principle of superposition.

In addition, the earth's crust is twisted and deformed by internal movements. The evidence that is found in these rocks indicates that periods of violent disturbances brought changed conditions over a wide region. These periods of disturbance, called *revolutions*, have caused large scale folding and faulting of the rock layers. We must keep this in mind when applying the principle of superposition to any rock layers being studied. Violent movements, such as those which take place during a revolution, might push older layers up over younger ones. If overturning does occur, then the upper strata would actually be older than the layers below.

Interpret

An exposed section of sedimentary rock layers are visible along a road cut. On identification of the rock types, conglomerate and sandstone were the first two layers. These were followed by limestone and shale. What may have occurred in the early history of the rock layers?

FIG. 18–1. Sedimentary rock layers are clearly visible in this photograph of Badlands National Monument. (Ramsey)

FIG. 18–2A. Cross-bedding in sandstone near Zion National Park. (Ramsey)

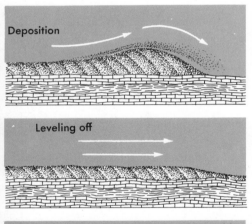

Deposition

Leveling off

Deposition

FIG. 18–2B. Cross-bedding generally originates as tilted layers of sand on a migrating dune. As the wind changes its direction and speed, the layers are tilted in different directions.

FIG. 18–2C. Ripple marks in an exposed layer of shale. (Ramsey)

To prevent errors in the use of the principle of superposition, it will be necessary to observe very carefully certain details of the rock structure. In some sedimentary layers, for example, *cross bedding* has occurred, as shown in Figure 18–2A. The angled layers in cross-bedded strata were produced by the movement of wind or water in a certain direction. See Figure 18–2B. When we understand fully the nature of this process, the appearance of crossbedding will be very useful in determining whether or not a rock layer has been overturned. Ripple marks and the position of fossils can also be used to determine the original position of many sedimentary layers. See Figure 18–2C.

The principle of superposition does not give us any clue to the actual age of the various rock layers. However, if the rock layers were deposited in an uninterrupted sequence, it would accurately show the order of deposition. Then the relative age of the rocks could be determined. The boundary between any two of these layers is called a *conformity*. On the other hand, if the rock layers were folded, faulted and tilted, they would eventually have been eroded. This would leave the disturbed layers exposed. If the area underwent a second period of deposition, new horizontal rock layers would form on the eroded surface. When the beds on either side of the eroded surface are no longer parallel, the boundary between two such layers is called an *unconformity*. See Figure 18–3A.

Sometimes after a period of erosion the rock layers are left in a relatively horizontal position. Then any new sediments will be deposited on top of the almost parallel but

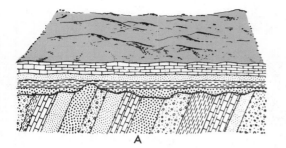

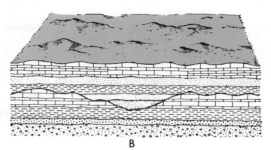

eroded surface. Where the boundary between an eroded surface and the younger overlying layers is nearly horizontal, it is called a *disconformity*. See Figure 18–3B.

Thus, when either an unconformity or disconformity is found in the rock layers, it indicates that a gap exists in the rock record between two adjacent layers.

Evidence from the rocks. In addition to their layered arrangement, the make-up of the rocks themselves reveals some of their early history. Again, it is the sedimentary rocks that provide the most clues to their origins. By examing the texture and composition of sedimentary rocks, we can learn about the conditions which produced them. Interpretation of ancient sediments is made possible by a knowledge of the conditions under which similar sediments are being deposited today.

The last traces of ancient shallow seas, extinct lakes, desert regions, glaciers and other features of the distant past lie buried in these sediments. Rocks that are sedimentary in origin contain information about the original rocks, the relief of the land, the climate of the region, and the kind of life that once existed. These layers can be compared to the pages in a book of geologic history. However, reading this book requires skill and knowledge.

Igneous rocks are also a useful tool in understanding earth history. The rocks which form dikes and sills in sedimentary rocks for example, are younger than the rocks they intrude. Likewise, extrusive formations such as lava flows are obviously younger than the rocks they cover. It is not always easy to tell which rocks are older. However, it is fairly simple to determine whether a formation of igneous rock found horizontally between layers of sedimentary rock, is intrusive or extrusive. If it is intrusive, the igneous formation must have invaded the existing structure, and should of course be younger than the rock layers above and below. In the case of an extrusive

FIG. 18–3. A at the left is an unconformity where tilted, folded, or faulted layers have been eroded, then covered again. B at the right is a diagram of a disconformity where parallel layers are separated by an old erosional surface.

Identify

Refer to Figure 18–3. Identify at least two geologic changes that took place before the sedimentary layers were deposited on the unconformity. How many changes have occurred below the disconformity?

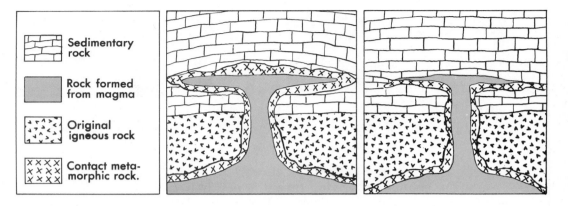

Sedimentary rock	
Rock formed from magma	
Original igneous rock	
Contact metamorphic rock.	

FIG. 18-4. Which diagram illustrates an igneous extrusion; which illustrates an igneous intrusion? Upon what evidence do you base your conclusions?

Construct

Refer to the diagram on page 395. Construct your own cross-section and have someone try to interpret the proper order in which the processes you indicated took place.

FIG. 18-5. From the evidence shown here it is easy to see why the rock in intrusion B is younger than that in intrusion A.

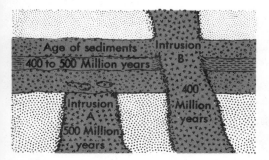

Age of sediments 400 to 500 Million years

Intrusion B

400 Million years

Intrusion A 500 Million years

lava flow, the underlying rock structure is older and the younger rock layers are found above. The situation becomes more complicated when the hot magma changes the existing rocks with which it comes in contact. This process is called *contact metamorphism*. See Figure 18–4.

In some situations, it is difficult to determine the order in which these processes have taken place. But in other cases, certain time relationships can usually be worked out if the types of rocks present are carefully examined. For example, a conglomerate may contain granite pebbles identical to those in a granite batholith located close by. It is almost certain that the granite in the batholith would be older than the granite in the conglomerate. Suppose a second mass of igneous rock, such as a dike, cut through the layers of this conglomerate and the granite. This line of thought is illustrated in Figure 18–5.

The above examples are intended as simple and very brief illustrations of how it is possible to judge the age relationships of various rock formations. With actual rocks, it is often very difficult to observe these relationships. Many times, even trained scientists disagree on the relative age of rock formations. For the most part, however, geologists have obtained a fairly clear picture of the rock relationships over the earth's surface. Their experience has led them to make some conclusions about the age and history of the rocks.

Gaps in the record. The general history of the earth as revealed in the rock record is clear up to a point. Beyond that, scientists must search painstakingly for clues to the earth's early history. For this reason, the sedimentary rocks that were laid down in the earliest geological periods are more difficult to identify and study. Our knowledge of the earlier geologic history of the earth is an incomplete one. Nearly all of the rocks formed during the first one and a half billion years cannot be studied directly. These rocks may have been deeply buried, lost through erosion, or else greatly metamorphosed. The early geologists made no attempt to work out the history of the most ancient rocks. Instead, they merely classified the oldest rocks as the *basement complex.*

Modern geologists, however, are gradually developing a more complete picture. In recent years they have gathered additional information from all parts of the earth. This has enabled them to compare individual types of rocks with related strata in other regions. Recently developed methods for drilling deep into the crust on both land and on the sea floor have helped close these gaps in the rock record. However, a part of the earth's most ancient history will probably remain permanently sealed as a record left in the rocks.

MEASURING GEOLOGIC TIME

Methods based on radioactivity. Our knowledge of the ways that rocks are laid down makes it possible to compare the relative age of one group of rocks to another. The actual age of the rocks in terms of the number of years before the present must be found by other means. One of the most accurate ways used to determine the absolute age of rock material is based on the rate of radioactive decay of certain kinds of atoms.

Atoms of the various chemical elements differ in the stability of their nuclei. Some atoms have an unstable nucleus, which will, of its own accord, undergo change by emitting some electrically charged particles along with electromagnetic energy. For example, the nuclei of the uranium isotope with a mass of 238 are radioactive (symbol, U, atomic number = 92). One of these nuclei may throw off an *alpha particle.* The make up of this electrically charged particle includes two protons and two neutrons. When the alpha particle is ejected the loss of protons and neutrons in the uranium nucleus brings about a decrease in both atomic number and mass. A new nucleus is then formed with atomic number 90 and mass of 234. It is called an atom of the element thorium.

activity

Determining whether a given radioactive atom will or will not decay is similar to predicting whether a coin, thrown into the air, will come to rest "heads up" or "tails up." Therefore, you can simulate the rate of radioactive decay by using coins or discs marked so that one side can be identified from the other.

Obtain at least 100 identical coins or discs. Place them in a covered box, like a shoebox. Shake to mix them well. Then remove all coins that come to rest in the box "heads up." Coins that come up "heads" will represent atoms that have decayed radioactively to become stable atoms. Record this as shake 1, together with the number of coins remaining. See the table below. Continue shaking and removing coins until only 2 or 3 coins remain. Graph your results showing the shake number on the horizontal axis and the coins remaining on the vertical axis. Compare the shape of your curve with that in Figure 18–7, page 383.

Number of Shake	Coins Remaining
0. 1. 2. etc.	

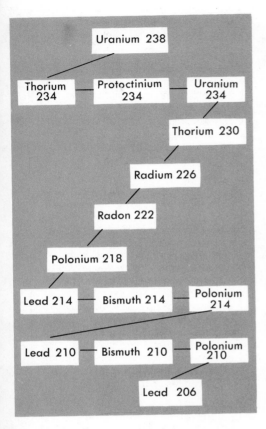

FIG. 18–6. Series of elements formed when uranium-238 undergoes radioactive decay forming lead-206.

The thorium nucleus is also radioactive and it gives off a *beta particle* consisting of one electron. When the beta particle is lost, a neutron in the thorium (Th) nucleus becomes a proton. The atomic number is increased by one and thorium is changed to protactinium (symbol, Pa, atomic number = 91). Protactinium is also radioactive, and the processes of radioactive change continue until a stable nucleus of the element, lead (Pb), is produced. This isotope of lead has a mass of 206 and is the final product of this radioactive chain. Thus an atom of U-238 eventually will become an atom of Pb-206 by radioactive decay. See Figure 18–6.

The conversion of uranium into lead goes on at a very slow but constant rate. Measurements show that for half of a given amount of uranium it takes 4.5 billion years for it to change into lead. This constant rate of atomic decay is not affected by changes in temperature, pressure, or other environmental conditions. At the end of 4.5 billion years half of the original uranium remains. By giving off electrified particles and radiant energy, the other half is eventually changed to lead. The remaining uranium continues to decay at the same rate as before. At the end of another 4.5 billion years just half of that (one-fourth of the original uranium) is left. We can say the *half-life* of uranium is 4.5 billion years. Its half-life, regardless of the amount, is the length of time it takes for half of its atoms to decay. A graph showing the rate of radioactive decay of uranium is given in Figure 18–7.

Knowing that the half-life for uranium is 4.5 billion years makes it possible to determine the age of any rock which contains uranium. Each year that passes $1/7,700,000,000$ of the amount of uranium is changed into lead. If the amount of uranium and lead in a mineral is measured, comparison of their amounts will show how many years the mineral crystal has existed. To determine the uranium-lead ratio in rocks requires complicated equipment and very accurate measurement.

The U/Pb ratio can only be used with rocks where there is reason to believe that all lead in the rocks has come from decay of uranium. Since uranium is most common in igneous rocks, the method would be less useful in finding the age of sedimentary rocks. Other radioactive atoms with long-lives which are more abundant than uranium can also be used in this way. Two such elements are rubidium-87, which changes into strontium-87 with half-life of 50 billion years, and potassium-40/argon-40 with a half-life of 12.5 billion years.

For solving age problems involving shorter periods of time (1000 to about 60,000 years), a form of radioactive carbon is often used in place of uranium. The value of this method lies in the fact that radioactive carbon-14 is constantly being produced in the upper atmosphere. Carbon-14 is formed when cosmic rays from outer space occasionally strike nitrogen atoms and change them into this radioactive form of carbon. Living things absorb small amounts of carbon-14 along with ordinary carbon atoms all during their lives. After death, the carbon-14 in the organic matter decreases as the radioactive carbon atoms decay and are not replaced. To establish the age of a small amount of organic material, the amount of carbon-14 must be measured and compared with the normal amount in the atmosphere. The half-life of carbon-14 is known to be about 6,000 years. We are able to calculate the age of the organism because it stopped taking in carbon-14 at the time of its death. Carbon-14 dating methods are a valuable tool in establishing the age of such things as wood, bones, shells, and the remains left by early people.

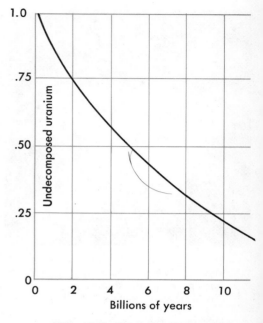

FIG. 18–7. The rate of decay of uranium is shown on this graph.

Other methods for measuring geologic time. Radioactivity measurements for dating of rocks and other materials have become a vital means for finding the age of the earth. They do not, however, accurately measure the age of the rock itself. Amounts of both the original radioactive substance and its final produce may have undergone change after the rock was formed. In that case, the radioactivity measurements would show only when the rock was changed. For example, measuring the age of a mineral crystal in a sedimentary rock does not tell us when the rock itself was created. Nevertheless, radioactive methods are still the most accurate available for establishing the age of the earth's older rocks.

Before the discovery of "radioactive clocks" several decades ago, other approaches were used to estimate geologic time. Some of these methods include the following:

1. *Rate of erosion.* If we can determine the rate at which a stream cuts its channel or erodes other features in its bed, we can determine its age. For example, the rate at which Niagara Falls is retreating by erosion of its rocky

Discuss

Explain why radioactive dating is not as useful for determining the age of sedimentary or metamorphic rocks as it is for igneous rocks.

FIG. 18–8. Niagara Falls was formed by weak rocks that rapidly wore back, cutting a long gorge. (Photo Researchers-Gianni Tortoli)

FIG. 18–9. The age of some rock strata can be estimated by counting the varves found in some sedimentary rocks. (U.S. Geological Survey)

ledge has been studied for about two hundred years. Geologists have found that the falls move back at an average rate of four feet a year. See Figure 18-8. A gorge about seven and a half miles long runs just below Niagara Falls. A simple calculation shows that it has been 9900 years since the Niagara River started cutting its gorge back from Lake Ontario. The cut-back probably began when ice from the glaciers of the last ice age started to melt.

Estimating geologic age by rates of erosion is most practical with relatively recent geologic features such as Niagara Falls. For older features, like the Grand Canyon, the method is much less dependable. The Grand Canyon developed over millions of years and conditions change during such long intervals of time. There is no way to be sure that the rate of erosion remains constant over many millions of years.

2. *Rate of deposition of sediments.* Before the use of radioactive methods, the most accurate way to date rocks was to study the time it took for them to form sedimentary layers. Attempts were made to determine the average rates of deposition for the different sediments which form the common sedimentary rocks, such as limestone, shale, and sandstone. On the average, it takes between 4000 and 10,000 years for a one-foot layer of sedimentary rock to be deposited. With these figures, it would be possible to discover how long it took to form a layer of a particular thickness. However, there is no way to be certain that any given sedimentary layer was deposited at an average rate. A flood can deposit many feet of mud in just one day. On the other hand, the same kind of mud could accumulate slowly on a lake bottom.

Some sedimentary deposits show definite annual layers, or *varves*, as seen in Figure 18–9. Since each layer represents one year's deposition, a series of varves indicate the number of years it took for the deposits to form. This is much the same way that the annual growth rings in the trunks of trees show the age of the tree when it was cut.

Varves are often found in lake deposits. Some were formed from sediments that were deposited on the bottom of a lake at the edge of a melting ice sheet during the last ice age. During the summer the ice probably melted rapidly, resulting in a rush of water that carried large amounts of sediment into the lake. Most of the coarser particles settled quickly to form a thick layer on the bottom. With the coming of winter, the melting stopped and the lake froze over. The finer clay particles, still suspended in the water, settled slowly to form a thin layer above the coarser sediments. A summer layer and the overlying winter layer make up one varve. Thus, each varve represents one year's deposition.

3. *Amount of salt in the sea.* The third method for estimating the age of the earth also includes finding the age of the oceans. Here the geologist calculates the amount of salt that is now present in all the oceans. Next the yearly average amount of salt carried into the oceans by all of the rivers on earth is figured. From these calculations it is then possible to estimate how many years the oceans have been receiving salt. However, this method contains two serious errors. First it assumes that the rate at which salt has been added to the oceans has remained the same. This is probably not true. In the past salt was very likely added at a different rate than at the present. A second problem is the rate at which salt is removed from the oceans by formation of bottom sediments and various life processes of sea creatures. It is not known how much allowance should be made for the salt lost from the oceans over millions of years. Thus we cannot use the saltiness of the sea as an accurate guide in dealing with the problem of geologic time.

Investigate
The rate of salt accumulation in the oceans throughout geologic time has not remained constant. Discuss at least two factors that may be responsible for varying the rate of salt accumulation.

THE FOSSIL RECORD

What can be learned from fossils? As we have seen, piecing together the record of the rocks is largely a matter of discovering their relation to each other. The study of fossils of ancient plants and animals, called *paleontology* (*pay*-lee-on-*tol*-uh-jee), is an important source of evidence for the age of events in the geologic past. Fossils are the remains of ancient plants and animals or indications of their presence, preserved in the rocks. The paleontologist compares the appearance, composition, and fossil content of rocks in the various layers.

One reason that fossils are so useful is that they can be used to establish the age of similar rocks from different places. For instance, we could assume that a sandstone of similar composition and appearance, in a particular region was formed at the same time. But similar sandstone coming from another continent may have an entirely different age. This is because sandstones were formed in nearly all periods of earth history. Thus we cannot assume that similar sandstones from different places have the same age.

But suppose it is discovered that rocks from many different locations contain the same general kinds of fossils. This would prove that the rocks were formed under similar conditions and would therefore be nearly the same age. Rocks of the same age need not necessarily contain identical fossils. Allowance must be made for differences between related plants and animals which lived in different parts of the world at the same time. In general, there is a similarity but no exact resemblance between fossils formed on different continents during the same geologic period.

However, certain fossils are found in rock layers of only one geologic age. These fossils are called *index* or *guide fossils*. They serve to identify specific rock layers no matter where they are found. These index fossils must meet certain requirements. First, they must be found in rocks scattered over a wide area of the earth's surface. Second, they must have features which clearly distinguish them from other forms. Third, the organisms that formed the index fossils must have lived during only a relatively short span of geologic time. Fourth, they must occur in fairly large numbers within the rock layers. A fossil that meets most of these requirements gives important clues to the history of a particular series of rocks. The presence of index fossils is especially important because they quickly establish the relative ages of rock layers. You will find this illustrated in Figure 18–10.

An economically valuable use of index fossils is in the location of oil and gas formations. Petroleum and natural

FIG. 18–10. Isolated outcrops of rocks containing index fossils are very useful to geologists. Even though the rock layers are in widely separated areas, as shown here, the relative ages of the rocks can be established. (Adapted from the Fossil Book, copyright 1958 by Fenton and Fenton. Used by permission of Doubleday and Co., Inc.)

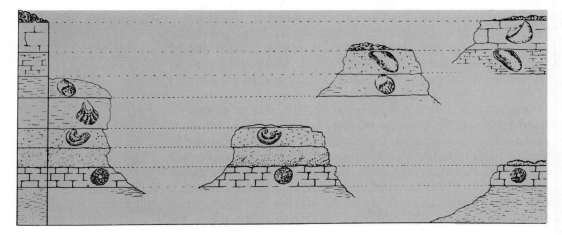

gas are apparently produced from ancient plants and animals which have been changed by complex chemical processes over millions of years. Index fossils of certain types are commonly found in the layers of sedimentary rock above the gas and oil formations. They often aid geologists in locating the oil-bearing rocks. Scientifically trained field and laboratory workers are needed to locate and identify these fossils.

Fossils also furnish us with important clues to the environments that have existed in the past. For example, fossils of tropical plants have been found in the polar regions. This indicates that the climates over the earth's surface have changed greatly from one geologic period to another. The boundaries of ancient seas have been traced through the remains of marine animals and plants. Analysis of oxygen isotopes in certain fossil shells can even tell the relative temperature of the water the animals lived in.

Formation of fossils. Many different conditions that exist when certain animals or plants die allow fossils to be formed. In rare cases, an entire animal may be completely preserved. The frozen bodies of an extinct elephant and several woolly mammoths have been found in Siberia and Alaska. Frequently however, only the harder parts of organisms are preserved in a relatively unchanged condition. Some fossils are so altered that little of the original material remains. Only shapes and textures can be recognized. Other fossils are not remains at all, but evidence that life has existed in some form. Even so, such evidence is often remarkably clear and detailed. In general, fossils can be classified in one of the following groups, according to the way they were formed:

1. *Unaltered remains of organisms.* A dead plant or animal must be protected for it to remain intact so long. Usually it must be buried to prevent other organisms from eating it. It must also be protected from decay caused by bacteria. Low temperatures found in frozen soil and solid ice will effectively protect and preserve organisms. Unaltered remains of animals have also been found in soil saturated with oil, deposits of volcanic ash, and in caves. Some remains, such as those often found in caves, are preserved by drying out and are said to be mummified. See Figure 18–11. Tar beds, formed by petroleum coming to the surface, are also effective in trapping animals and preserving their bones. See Figure 18–12.

Discuss
Explain how fossils of different ages have helped provide evidence for a theory of evolution. Find out about fossils found in the layers of the Grand Canyon. How do they change from bottom to top?

FIG. 18–11. Above, the mummified remains of Trachodon, the duck-billed dinosaur; left, closer detail of the skin. (The American Museum of Natural History)

2. *Altered remains of organisms.* Under certain conditions, fossil specimens may be completely altered. In many cases the original parts have been replaced by mineral substances. This process is said to *petrify* the organism. This does not mean that the original organic matter has been changed to stone. Actually, what happened was that minerals replaced the original cell materials which had long since disappeared. Some of the common petrifying minerals are silica, calcite, and pyrite. Usually the replacement of the original materials is a very slow process. It probably takes place molecule by molecule. The substitution of mineral for organic material is often nearly

FIG. 18–12. An artist's conception of a scene at the Rancho La Brea tar pits in prehistoric times. (The American Museum of Natural History)

perfect, such as the fossil shown in Figure 18–13. Seen under a microscope, the detailed cell structure of the original tissues is clearly visible in thin sections of some petrified fossils. In other fossils, only the general outline of the organism remains.

The vast coal deposits of today are fossil remains of plants which have been changed by the process of *carbonization*. This takes place when the remains of trees and other plants decompose in swamp water and then are buried. Some of the plant material changes to marsh gas or methane (CH_4), water (H_2O), and carbon dioxide (CO_2). In the formation of these coal deposits much carbon remained because such products carried away more oxygen and hydrogen than carbon. The original, complex chemical compounds present in the plants were gradually changed by a concentration of this carbon. After a long period under pressure, the final product we know as coal was produced.

3. *Indirect fossil evidence of life.* Some fossils have left definite imprints, including such things as the footprints of an animal or the outline of a leaf. Let us trace the history of such an imprint. Suppose a giant dinosaur left deep footprints in the soft mud. What happened then? Later sand or silt may have blown or washed into the footprints so gently that they were not destroyed. Still later, more sediment may have been deposited above the prints. Then, as the ages passed, the mud containing the footprints hardened into solid rock. Thus the footprints were preserved. An imprint like this one is shown in Figure 18–14. The imprints of fossil leaves, stems, flowers, and fish were often laid down in soft mud or clay and then preserved in a similar way. Footprints of amphibians, birds, and mammals, and even early human-like creatures have been found.

FIG. 18–13. Fossil crinoids, perfectly preserved by the process of petrifaction. Crinoids are marine animals often resembling plants. (Allan Roberts)

FIG. 18–14. Fossilized dinosaur tracks made in mud more than a hundred million years ago. (The American Museum of Natural History)

FIG. 18–15. An ant, perfectly preserved in amber. Amber is a fossil gum derived from the sap of prehistoric plants. (The American Museum of Natural History)

FIG. 18–16. Prehistoric worm trails, believed to be the fossil burrow of worms. (E. R. Degginger)

Shells of snails, parts of trees and plants, and similar organic remains were often buried in sediments which later formed rock. Eventually these remains decayed or dissolved. In some of them an empty cavity, called a *mold*, remained. The mold reveals many characteristics of the original organism. Natural molds may be studied just as they are. Or they may be filled with plaster or some other substance to make duplicates of the original organism. These replicas are called artificial casts. Sometimes natural casts were formed when sand or mud hardened in a natural mold. An unusually perfect type of mold was formed when insects millions of years ago were trapped in sticky resin flowing from trees. Later the resin was changed to hard amber, with the forms of the insects perfectly preserved as cavities in the transparent amber. See Figure 18–15.

Trails and burrows may also be preserved as fossils. See Figure 18–16. Some ancient snails, sponges, and crabs are known as much by their borings or burrows as they are by their remains. For instance, ancient sea worms may have swallowed sand or mud. Then they digested the small organisms contained in the sand. After the food was extracted, the sand or mud was left in the form of castings. These are found as fossils in some marine sediments. *Coprolites* are fossilized masses of waste materials from animals. They can be cut into thin sections. Analysis of these sections reveals the feeding habits of animals long since extinct.

Most people are familiar with the gizzard of a bird such as a chicken. The gizzard contains small stones which help grind the food. Some dinosaurs also had such gizzard stones or *gastroliths* to grind their food. They can often be recognized by their smooth, rounded, and polished surfaces. However, their identification is certain only if they are found within the remains of a dinosaur.

Finding fossils. Anyone with sufficient interest can successfully hunt fossils. The first requirement is to look in places where fossils are likely to be found. Maps of good fossil-collecting areas are available for most regions. Information about the best collecting areas can be obtained from museums, colleges, and other collectors. Many students have found fossils in the exposed rocks of gorges and canyons, quarries, and rock cuts along railroads and highways. However, permission must be obtained from owners before collecting from private property such as railroad rights-of-way.

Fossils may be found in almost any kind of sedimentary rock. Shales and limestones are especially likely sources. Dolomite and sandstone are also often good. Other sources of unusual or well-preserved fossils include chalk, diatomaceous earth (remains of microscopic water plants) and coal. Fossils are not found in igneous rock. Those originally present in metamorphic rocks were almost always destroyed or changed into unrecognizable forms. Among sedimentary rocks, conglomerates and chert generally give poor results. Any fossils that exist in these rocks are usually difficult to remove.

The collector should note the locality and rock layer from which the fossils come. A label with this information should be made out for each specimen as soon as possible. Collecting tools might include a hammer with one pointed end, such as a geologist's hammer. This is used to loosen rock-containing fossils from their surroundings. An ordinary hammer with a cold chisel can also be used for this purpose. A stout knife blade is useful for separating small rock layers such as those found in soft shale. A large iron bar is also useful for moving large pieces of rock. Other valuable equipment includes a notebook, knapsack and a magnifier. Small boxes and sacks are convenient.

Specimens should be cleaned at home so that they can be handled carefully. The cleaning can best be done with plain water. The rock surrounding them can be trimmed with a knife, chisel or hack saw. Needles are often used to clean small specimens.

Some rock structures resemble fossils in appearance, but they were not made by animals or plants. These are called *pseudofossils* (false fossils). One type of pseudofossil is a rounded accumulation of mineral matter known as a *concretion*. They often look like petrified eggs or potatoes. Others have cracks filled with minerals and may be mistaken for fossilized turtle shells. Although concretions are not fossils, some do contain fossils.

If you decide to make a collection of fossils, you will want to classify them into their proper groups. Each group is composed of organisms that have some common characteristics. The scheme of classification can be found in most paleontology reference books.

VOCABULARY REVIEW

Match the word or words in the column on the right with the correct phrase in the column on the left. *Do not write in this book.*

1. Period of violent earth crust disturbance.
2. The boundary between layers of rock deposited at greatly differing times.
3. An early classification of the oldest rocks.
4. An electrically charged particle consisting of two protons and two neutrons.
5. An electron given off by a nucleus.
6. Alternating dark and light layers found in sedimentary rock.
7. The study of fossils.
8. Fossils used to give the age of a rock layer.
9. Changing of an organism to stone.
10. The method by which coal is produced.
11. An empty cavity left in rock by an organism.
12. Undigested sand or mud fossils left by small organisms.
13. Gizzard stones left by dinosaurs.
14. Rock structures resembling fossils.

a. petrification
b. coprolites
c. pseudofossils
d. carbonization
e. revolution
f. basement complex
g. gastroliths
h. unconformity
i. concretions
j. alpha particle
k. varves
l. index fossils
m. paleontology
n. beta particle
o. castings
p. mold

QUESTIONS

Group A

Select the best term to complete the following statements. *Do not write in this book.*

1. The type of rock which gives the most distinct clues to the earth's past is (a) igneous (b) sedimentary (c) metamorphic (d) contact metamorphic.
2. A basic principle used in interpreting the earth's history through rock layers is the principle of (a) superposition (b) sediment deposition (c) contact metamorphosis (d) unconformity.
3. Violent disturbances evidenced by the twisted and deformed layers of the earth's crust are called (a) castings (b) concretions (c) revolutions (d) varves.
4. An unconformity in rock layers results when (a) a revolution occurs (b) layers of different thicknesses are formed (c) magma intrudes between the layers (d) a long period of time passes before the next layer of rock is formed.
5. What else besides a layered arrangement tells us most about the history of the rock? (a) the composition of the rocks (b) the thickness of the rock layers (c) the location of the rocks on the earth's surface (d) the present condition of the rock's surroundings.
6. When hot magma changes the make-up of nearby rocks, the process is known as (a) revolution (b) unconformity (c) contact metamorphosis (d) radioactivity.

7. Of the following, which type of igneous rock formation is most useful in interpreting the rock record? (a) extrusive formations (b) intrusive formations (c) lava flows (d) laccolith.

8. The greatest gap in the earth's rock history occurs (a) in the first 1.5 billion years (b) about the time of the great flood (c) during the time of the first ice age (d) about the time of the last ice age.

9. An atom whose nucleus is unstable undergoes change by throwing off electromagnetic energy and (a) heat (b) electrical particles (c) magnetism (d) light.

10. The particle given off by an atom, which changes both the atomic mass and atomic number is the (a) alpha particle (b) beta particle (c) gamma particle (d) electromagnetic particle.

11. The particle given off by an atom, which increases the atomic number by one is the (a) alpha particle (b) beta particle (c) gamma particle (d) electromagnetic particle.

12. When an atom of U-238 undergoes radioactive decay, eventual stability is reached when it becomes an atom of (a) thorium (b) protactinium (c) lead (d) strontium.

13. The half-life of uranium-238 is (a) 50 billion years (b) 12.5 billion years (c) 4.5 billion years (d) 100,000 years.

14. Another element having a long half-life which can be used for radioactive dating of rocks is (a) thorium-234 (b) strontium-87 (c) carbon-14 (d) protactinium-234.

15. Carbon-14 is most useful in radioactive dating to determine the age of (a) igneous rocks (b) sedimentary rocks (c) mineral crystals (d) remains of plants and animals.

16. Carbon-14 is produced by (a) lightning striking carbon dioxide (b) cosmic rays striking oxygen (c) cosmic rays striking nitrogen (d) lightning striking nitrogen.

17. The age of Niagara Falls has been estimated by a method using (a) rate of erosion (b) rate of deposition of sediments (c) U-238 half-life (d) C-14 half-life.

18. Determining geologic age by counting varves is an example of the method using (a) rate of erosion (b) seasonal change in the deposition of sediments (c) strontium-87 (d) carbon-14.

19. The dating method which is believed to be most accurate is (a) rate of erosion (b) rate of deposition of sediments (c) amount of salt in the sea (d) radioactivity.

20. A paleontologist is concerned mostly with the (a) layers of the earth (b) fossil content of rocks (c) relationship of earthquakes to the structure of the earth (d) age of igneous rocks.

21. Index fossils serve to identify (a) specific rock layers no matter where they are found (b) the composition of different rock layers (c) the location of ancient oceans (d) the origin of rock layers.

22. A practical use of index fossils is (a) to use them as fuel (b) to help locate oil and gas deposits (c) to determine the composition of a given rock layer (d) in understanding the origin of the continents.

23. One reason paleontologists believe the polar regions once had a tropical climate is that (a) fossils of tropical plants have been found there (b) evidence shows that the polar continents were once at the equator (c) the temperatures of polar regions sometimes are as high as in the tropics (d) evidence indicates that snow was absent at one time in the polar regions.

24. Analysis of oxygen isotopes in certain fossil shells has enabled scientists to determine (a) the age of the fossil (b) the relative temperature of the water where the shells grew (c) the kind of water where the shells grew (d) the growth rate of the shells.

25. In rare cases entire plant or animal bodies preserved as fossils have been found, such as (a) trees in the Arizona petrified forest (b) wooly mammoths in Alaska (c) dinosaur imprints in mud flats (d) shell molds in limestone.

26. When an organism's body has been completely replaced by minerals, it is (a) petrified (b) carbonized (c) a casting (d) a coprolite.

27. Carbonization of plants results in the formation of (a) oil and gas (b) petrified forests (c) coal (d) castings.

28. An example of an imprint fossil would be (a) the wooly mammoths found in Alaska (b) the castings left by worms (c) coprolites left by large numbers of animals (d) the outline of a fern leaf.

29. Preserved waste materials left by ancient animals are called (a) gastroliths (b) imprints (c) coprolites (d) molds.

30. Concretions are a good example of (a) pseudofossils (b) castings (c) coprolites (d) gastroliths.

Group B

1. Why is it that sedimentary rock layers offer the most clues to the earth's history?

2. What is the principle of superposition?

3. What is the principal reason that errors occur in the use of the principle of superposition?

4. Why is an unconformity a problem in determining the age of two neighboring layers of rock?

5. List two general ways cited in the text of how igneous rock formations can be useful in understanding earth history.

6. Why is the rock record for the earth's first 1.5 billion years not available?

7. Describe how an unstable atomic nucleus becomes stable.

8. What is meant by the term "half-life?"

9. How can the half-life of a radioactive substance be used to determine the age of rock in which it is found?

10. Explain why carbon-14 can be used in dating a piece of wood but is not generally used for dating rock.

11. Briefly describe three methods not based on radioactivity which have been used for measuring geologic time.

12. In what ways does the work of the paleontologist differ from that of the geologist?

13. What are the four requirements a fossil must meet to be used as an index fossil?

14. What explanation can you give for the fact that gas, oil, and coal are called "fossil fuels?"

15. In what ways besides determining the relative ages of rock layers are fossils useful in understanding the earth's history?

16. Why are fossils consisting of an entire animal or plant rarely found?

17. How do organisms become petrified?

18. Give several examples showing how indirect fossil evidence proves the existence of life.

19-20. The following questions refer to the cross section below. Formations are indicated by letters A-H.

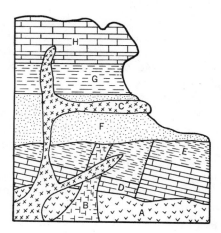

19. Between which two lettered formations does an unconformity exist? Explain.

20. Using the letters for each formation, list the correct order in which the formations were deposited. (From oldest to youngest)

19

a view of the earth's past

objectives

☐ Give an example of the way a rock layer may contain a record of part of the earth's history.

☐ List the kinds of information that may be obtained from exposed rock layers.

☐ Explain how the Geologic Column is used.

☐ Name the geologic eras in order of their age, beginning with the oldest.

☐ Explain why maps of the earth's surface would not look the same throughout its history.

☐ Give a brief description of the conditions and life that existed on North America during each geologic era.

Early one morning, a few years ago, a snow-covered mountain in the Cascades of southern Washington exploded. Its top was blasted away in a huge volcanic eruption. Surrounding forestland was changed almost instantly into a lifeless and barren landscape. A sudden and dramatic event such as a volcanic eruption, an earthquake, or a flood vividly reminds us that the earth's surface is in a constant state of change. Such occurrences take place in minutes, hours, or days. It is relatively easy to understand events that alter the earth's surface in such short periods of time.

Other processes that shape the major features of the earth's crust operate very slowly. For example, changes in the continents and oceans caused by the movement of crustal plates take place over millions of years. A million years is a period of time that may be difficult for our minds to grasp. Perhaps this is because our lifetimes seem so brief when compared to a scale of a million years. It strains our imaginations to think of a change that occurs over millions or billions of years. However, almost all scientists believe that there is evidence in the earth's crust to support the idea that the earth's history requires a time scale of about four billion years. In this chapter, we will study how earth scientists have used a record in the rocks of the crust to develop a plan that describes the earth's long history. The history of geologic changes and development of life forms described in this chapter is based on scientific evidence that has been used as clues in recognizing events that happened very long ago. At times there are pieces of the picture that are missing, or parts that are difficult to interpret and open to question. Just as the history of human civilization is told in written records, the history of the earth is recorded in rocks. However, unlike the history of civilization, which is based on observations made by people that lived in that time, no person ever looked upon the scenes that make up the history of the earth.

FIG. **19–1.** Millions of years ago a dinosaur left this footprint in soft sand. The sand has now become sandstone, with the footprint preserved as a fossil. (Ramsey)

THE GEOLOGIC TIME SCALE

History in rock layers. Our only scientific record of the earth's past is in the rocks found at or near its surface. Materials from which the rock is made hold part of the record. For example, limestone is almost always formed in bodies of water. Its presence in a particular location usually means that water once covered that area. A certain type of limestone is made from the remains of reef-building corals. Its presence among crustal rocks can indicate that warm, shallow water once covered that part of the land. When fossils are found in rocks, they often tell of the kind of environment that existed when the rock was formed.

Of the three basic kinds of rocks, sedimentary rocks contain the most complete record. They make up about 75 percent of the rocks found close to the earth's surface. Sedimentary rocks have also formed under the conditions normally found at the surface. As a result, their materials have not been greatly changed by heat or pressure. Generally, only sedimentary rocks can be expected to contain fossils. For these reasons, the record of the most recent parts of earth history comes mostly from study of sedimentary rocks.

Sedimentary rock layers are exposed at many places. These exposures, called *outcrops*, may be only a small area on a weathered hillside or a large cliff along the wall of a valley. Where highways or railroads are cut through a hill, otherwise hidden rock layers are often exposed. One of the greatest outcrops has been made naturally. It is the Grand Canyon in Arizona. Here the Colorado River has opened a huge gorge 6.5 to 29 km (4.0 to 18 mi) wide and more than 1.6 km (1.0 mi) deep in places. The sedimentary layers exposed in the Grand Canyon provide an opportunity to study a record of millions of years of earth history.

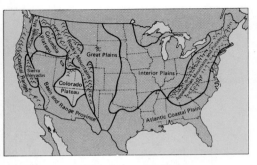

FIG. 19–2. The Colorado Plateau makes up one of the natural regions of the United States.

FIG. 19–3. Many layers of rock are exposed in the Grand Canyon. (Ramsey)

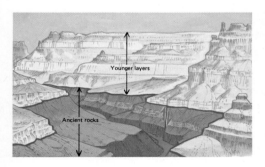

FIG. 19–4. Younger rock layers rest on ancient rocks at the bottom of the Grand Canyon.

History recorded in the Grand Canyon. The Grand Canyon must have begun as a small valley cut by the Colorado River. Slowly, the land was lifted to form a high plateau. This uplifted region, now called the Colorado Plateau, occupies a large area in the southwestern United States. See Figure 19–2. As the Colorado Plateau was uplifted, the river cut through the existing rock layers, and weathering processes acted on the slopes to help widen the great gorge we know as the Grand Canyon. It probably took the river at least 10 million years to erode its way through the slowly rising plateau. This erosion has produced a spectacular natural feature that attracts many visitors. See Figure 19–3.

An earth scientist studying the colorful scenery of the Grand Canyon is able to see in the rocks a record of part of the earth's history. This information comes from three main sources. First, it is possible to assume that the rock layers were deposited at different times, with the youngest on top. This is an application of the principle of superposition. There is no evidence that the great sheets of sedimentary rocks, whose edges are exposed in the walls of the canyon, have been overturned. Thus the upper layers can be judged to be younger than any layers found below. The actual age of the layers must be determined from other evidence. The superposition principle describes only the relative ages of the beds of rocks. See Figure 19–4.

A second source of information is the composition of the rocks found in the separate layers. For example, the topmost, light-colored layer seen in many photographs of the Grand Canyon is made of sandstone. This sandstone bed shows the type of cross-bedding that is characteristic of sand dunes. There can be little doubt that this layer was laid down when the region was an arid desert similar to the Sahara or Arabian deserts of today. Several of the lower layers are made of limestone and shale that must have been deposited when the region was covered by the sea. Thus the composition of the various rocks of the Grand Canyon reveals that the history of the earth's surface we now know as the Colorado Plateau includes changes of climate and the advance and retreat of the sea.

A third source of information comes from the fossils found in many of the rock layers of the Grand Canyon. These plant and animal remains are evidence of the kind of life that existed during various periods of the region's history. Limestones, for example, often contain fossils of marine-life forms, such as clams, fishes, and corals. The fossils found in each rock layer are different from those found in beds above and below. See Figure 19–5. This indicates that there was a characteristic group of plants and animals that lived during each of the time periods represented by different rock layers. Paleontologists, specialists in the study of fossils, are able to trace the changes in many of the life forms as their remains are found in successive rock layers. Analysis of all the fossils found in the older, lower rocks through the younger, upper rocks yields a sequence of events in which early, simple life forms gradually disappear or seem to develop into higher forms.

Altogether, the position, composition, and fossil content of the rocks of the Grand Canyon can tell a detailed and lengthy story of that part of the earth. Scientists believe that application of the general principles that reveal the history of the Grand Canyon can also reveal much of the history of the entire earth's surface.

A Geologic Column. Suppose that you are able to examine a particular outcrop of a limestone layer on one side of the Grand Canyon. You find that the rock contains certain kinds of fossils. Wherever you find that limestone layer along the canyon wall, there are always the same fossils present. Now you travel across to the opposite canyon wall. There you find a limestone layer with exactly the same kind of fossils. How do you think that the two limestone layers on opposite sides of the canyon are related to each other? You would probably decide that the limestone layer on each side of the canyon had been formed at the same time. It must have been laid down as a continuous sheet that was eroded through as the river cut the canyon.

Scientists use the same kind of reasoning whenever they find rock beds that contain identical fossils at different locations on the earth. It can be judged that the age of that particular rock layer is the same everywhere. See Figure 19–6. This conclusion holds true even if the particular kind of rock is found in widely separated parts of the world. For example, the arrangement of many of the sedimentary layers, with their characteristic fossils, found in the Grand Canyon is almost identical to the arrangement of layers found in outcrops in England.

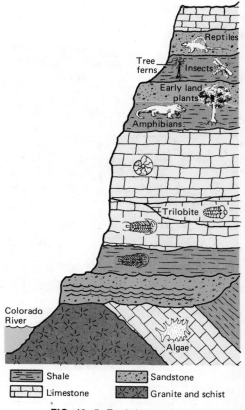

Shale   Sandstone

Limestone   Granite and schist

FIG. 19–5. Each layer of the Grand Canyon carries its characteristic fossils.

FIG. 19–6. The dark bed in these exposed layers can be traced in both the near and distant outcrops shown in this photograph. (Ramsey)

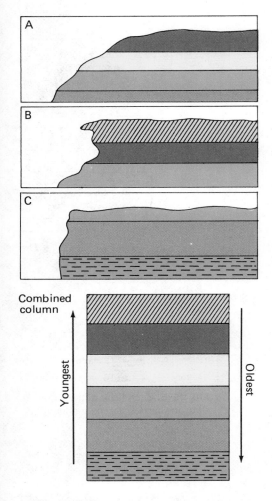

FIG. 19–7. Suppose that you investigated three different outcrops as shown in A, B, and C above. Evidence shows that layers with the same color are the same age. You could then make an imaginary combined column like the one above with the layers combined in order of their relative ages.

The work of thousands of scientists during the past 150 years has shown that rock layers of the same age, as determined by their fossils, are arranged in the same pattern, regardless of their locations. There is no known place that contains all the rock layers. However, scientists believe that the observations from many different places can be combined to provide a picture of the complete arrangement of layers in order of their relative age. See Figure 19–7. It is like an imaginary Grand Canyon, where all the different rock layers laid down at the same time everywhere on the earth could be exposed. This imaginary sequence of layers is called the *Geologic Column.* The Geologic Column consists of layer upon layer of rocks in the same order in which they were formed on the earth's surface. Scientists compare the complete Geologic Column with the actual arrangement of rock layers that they observe in their work. The rocks can then be classified as old or young, according to where they are found in the Geologic Column.

A geologic time scale. The Geologic Column is a description of actual rock layers that have been studied at many different locations. These rock layers are distinguished from each other mainly by the kind of fossils preserved in each layer. The kind of fossils changes when comparing one layer with another. These changes follow a general pattern. Fossils in the upper, therefore the younger, layers are most like modern plants and animals. Lower, older layers contain fossils that are very different from modern types of plant and animal life. Many species observed in older layers have become extinct. The fossils found in different rock layers of nearly the same age are most similar. Thus the various layers that make up the Geologic Column have been divided into groups according to similarities in their characteristic fossils. Each group is made up of a sequence of layers, all produced during a period of time when certain groups of plants and animals were abundant. As conditions on the earth's surface changed, and the life forms also changed or became extinct, a new and younger sequence of rock layers was left as a record.

The largest section of the Geologic Column is called an *era.* The rock layers that make up an era represent a very long period of time. Because the changes that affect living things on earth have not occurred at definite times, the different eras are not all the same length.

For example, the oldest and longest era is the *Precambrian*. It began when the earth came into existence and ended about six hundred million years ago when the fossil record shows that life became abundant. It is the largest of the geologic eras, and it represents the greatest amount of geologic time. The Precambrian rocks make up the bottom layers of the Geologic Column. It is believed that many events must have occurred during the Precambrian era. However, because it was so long ago, evidence of these changes is very difficult to find in Precambrian rocks. The Precambrian era can be divided into two major parts. They are the *Early Precambrian era,* which has no record of life, and the *Late Precambrian era,* during which the first, faint record of life appears.

The Geologic Column from the end of the Precambrian era to the present can be divided into three more eras. Following the Precambrian was the *Paleozoic era.* The word "Paleozoic" comes from Greek word roots that mean "ancient life." Rocks formed during the Paleozoic era show fossil evidence of a wide variety of primitive plants and animals. After the Paleozoic era came the *Mesozoic era.* Its name means "middle life." The Mesozoic fossils tell of the development of higher life forms such as reptiles and birds. Last in the Geologic Column is the *Cenozoic era.* Cenozoic means "recent life." One of the most characteristic kinds of fossils in Cenozoic rocks is the remains of mammals.

Enough is known about these three most recent eras to divide them into smaller segments called *periods*. Within an era, one period is separated from another by the occurrence of characteristic fossils. A period is usually named for the location in which the rocks containing the identifying fossils were best observed. For example, there is a Devonian period (from Devon, England) in the Paleozoic era. When the term "Devonian" is applied to certain rock layers in the Geologic Column, it means that these layers were deposited in the same time span as the rocks in England containing the fossils that first identified the period.

Since the Cenozoic era is the most recent, there is enough evidence to divide its two periods into even shorter time spans called *epochs*.

When scientists first developed the Geologic Column more than a century ago, they had no known method for finding the absolute ages of the rocks. The evidence indicated that a long time was necessary for the earth pro-

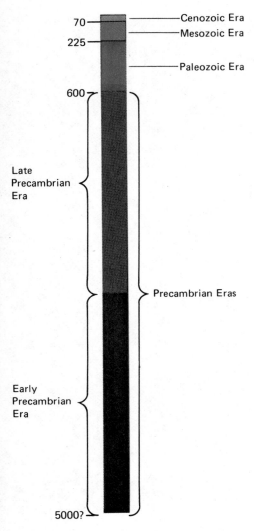

70 —
225 —

——— Cenozoic Era
——— Mesozoic Era

——— Paleozoic Era

600 —

Late
Precambrian
Era

— Precambrian Eras

Early
Precambrian
Era

5000? —

FIG. 19–8. This graph compares the lengths of the geologic eras. The number of years ago that each era ended is given on the left.

Investigate
Refer to Figure 19–8. Using the bar graph that depicts the relative time period for each era, set up a similar scale based on portions of a 24-hour day. Approximately what part of a day would be represented by the Precambrian era?

cesses to cause the sedimentary layers to accumulate. Scientists guessed that the various groups of rock layers each represented a very long time. But a guess it was, for there was no scientific way to actually measure the time intervals represented by the eras, periods, and epochs. Eventually, however, research uncovered the existence of radioactive atoms whose presence in rocks can be used to measure the age of the rock. Scientists now believe that they can accurately determine the time intervals represented by the various sequences of rocks in the Geologic Column. Measurements indicate that the Paleozoic era lasted about 375 million years, the Mesozoic about 155 million years, and the Cenozoic from about 70 million years ago to the present. The remainder of the earth's history is hidden in rocks of the Precambrian era, which lasted from the earth's beginning until about 600 million years ago. Figure 19–8 illustrates the length of the different eras.

When the actual ages of its rocks were determined, the Geologic Column became a kind of geologic calendar, which may be used to describe when events in the earth's history occurred. A complete summary of this geologic calendar is given in the Appendix, pages 530–531. It shows the eras, periods, and epochs with their approximate time durations and characteristic plant and animal life. You should refer to this chart frequently while studying the following sections.

THE HISTORY OF A CONTINENT

Ancient geographies. The scientific principle of uniformitarianism, which says that the present is the key to the past, has a special meaning for the history of the earth's continents and oceans. Today there is convincing scientific evidence for the movement of crustal plates on the earth's surface. Because plate tectonics is a continuous process, it must have constantly altered the face of the earth throughout much of the planet's history. From some very early time, the earth's surface has been made of shifting crustal plates. There has been no permanent or even typical arrangement of continents and oceans.

The positions of the continents during the last 225 million years, which includes the Mesozoic and Cenozoic eras, are well understood. The pattern of magnetic stripes in the sea floor permits the movement of continents during the most recent eras to be traced and dated. This evidence indicates that at the beginning of the Mesozoic era all of the present continents were grouped together into the supercontinent called Pangaea. During the Mesozoic and Cenozoic eras, Pangaea broke up. New ocean basins were created, and the fragments of the ancient supercontinent moved toward their present positions.

Tracing the continental positions during the 375 million years of the Paleozoic era is more difficult. However, by combining evidence from many scientific fields, a general picture of the Paleozoic continental movements can be constructed. A world map made at any time during the Paleozoic era would look nothing like a map of the world today. The evidence shows that six major continents have existed at various times during the Paleozoic era. At the beginning of the Paleozoic, there was one very large continent and five smaller ones, all separated from each other. These continents slowly moved in different directions, colliding with each other until, by the end of the Paleozoic era, they were all joined into the supercontinent Pangaea. The continents that collided to make Pangaea are not the same continents that were formed when Pangaea broke apart. Thus the modern continents are made of pieces of the Paleozoic continental blocks. See Figure 19–9.

Moving farther back in time into the Precambrian era, the evidence for continental movements becomes very difficult to interpret. There can be little doubt that continents and oceans existed during the most ancient periods of the earth's history. Modern continents contain large

FIG. 19–9. Each part of a modern continent that comes from one of the separate Paleozoic continents is shown in a different color. Names of the Paleozoic continents are matched with their colors in the key at the lower left.

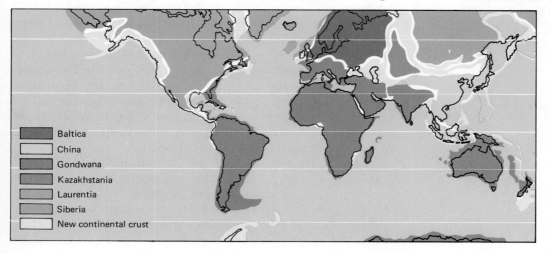

Baltica
China
Gondwana
Kazakhstania
Laurentia
Siberia
New continental crust

Construct

Make a flip-card movie to show the movements of earth's continents as they are thought to have occurred from the beginning of the Paleozoic era to the present. Use Figure 19–9, p. 403, as a guide. On 3 × 5 index cards, make drawings of the known positions of the continents for each time period described in the text. Then make many drawings to show a gradual change from one position to the next. Fasten the cards into one deck in proper sequence. When you flip the card deck, you will see a movie of the continents as they collide and move around.

areas of Precambrian rocks that may represent very old continents around which the modern continents have grown. These regions of very old rocks found within the present continents are called *cratons*. The rocks found in the cratons show evidence of mountain building, volcanic activity, and sediment formation. But the outlines of continents and oceans that may have existed during the Precambrian era remain unknown.

All available scientific evidence indicates that the world as we know it is a single stage in a long series of changing surface patterns. Continents have collided and moved apart. Mountains have grown and been eroded away, leaving thick layers of sediments. Life has spread over the land surfaces and left its record. Each continent holds a record of the movements of crust and living things that populated its surface. Careful study of the rock structure of modern continents can reveal the general outlines of their history. In the following sections, the development of the modern continent of North America will be followed as a part of the history of the entire earth.

GROWTH OF A CONTINENT

The Precambrian era. Since Precambrian rocks are always the oldest in the Geologic Column, they are usually covered by younger layers. For example, Precambrian rocks in the Grand Canyon are exposed only in the very bottom of the gorge where the river has cut through younger sedimentary layers that are more than 1 km (0.6 mi) thick. In North America there is a large area of eastern Canada and parts of the northeastern United States where Precambrian rocks are exposed. This region is called the *Canadian Shield*. See Figure 19–10. The Canadian Shield is the nucleus of ancient rocks, or craton, around which the modern continent has grown. For 600 million years, since Paleozoic times, the Canadian Shield has been the only part of the North American continent that has remained largely unchanged.

To the south and west of the Canadian Shield is a central, stable region where the undisturbed Precambrian rocks are buried under layers of sediments up to about 2.5 km (about 1.5 mi) thick. These sediments were deposited during Paleozoic, Mesozoic, and Cenozoic eras. Surrounding the stable, continental nucleus formed by

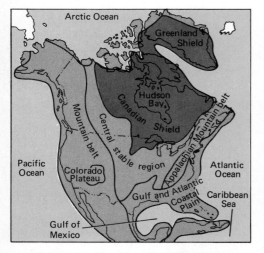

FIG. 19–10. The Canadian Shield is a region of very old rocks that is the foundation for the North American continent.

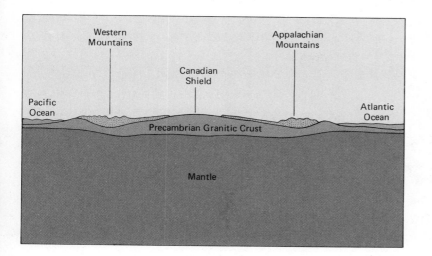

FIG. 19–11. This diagram shows the relationship of the eastern and western mountains, or mobile belts, to the Canadian Shield and central stable regions of North America.

the Canadian Shield and the platform around it are wide, mobile belts that have been created where the crust has been folded and faulted by plate movements and the invasion of magma from below. Figure 19–11 shows a cross section of the continent that illustrates the general relationship between its Precambrian foundation and surrounding younger structures. The heart of the modern continent of North America is the Canadian Shield, where the exposed Precambrian rocks show evidence of being part of still older continents whose outlines are unknown. The remainder of North America as we find it today is made mostly of sediment layers that have been folded, faulted, uplifted, and pushed down by the tectonic forces that deform the earth's crust.

Like similar rocks found all over the world, the Precambrian rocks of North America show fossil evidence of only the simplest forms of life. Rocks of the Canadian Shield contain traces of microscopic plant cells called bacteria. They are believed to be at least 2 billion years old. There is also evidence that toward the end of the Precambrian era the primitive water-dwelling plants called algae had become common. No fossil remains of Precambrian animals have been found in North America. However, some impressions that suggest the presence of a kind of worm and jellyfish have been found in late Precambrian rocks in Australia. Animals may have been abundant during the last part of the Precambrian era, but they lacked hard body parts that would be preserved as fossils. Judging from the Precambrian fossil record, simple plants

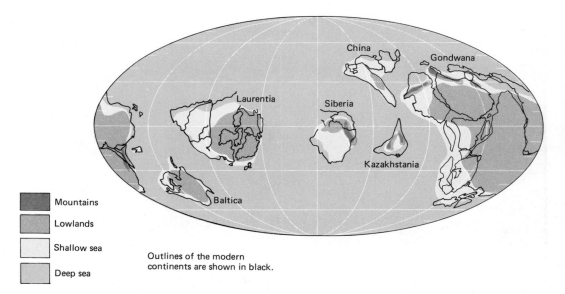

Mountains

Lowlands

Shallow sea

Deep sea

Outlines of the modern
continents are shown in black.

FIG. 19–12. This map shows how the Paleozoic continents were spread over the earth's surface about 550 million years ago. North America was part of Laurentia. Can you find its general outline and those of the other modern continents in the ancient continents?

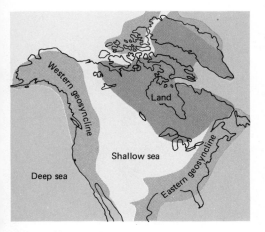

FIG. 19–13. Positions of the Paleozoic geosynclines and the region not covered by shallow seas are shown on this map of modern North America.

like bacteria and algae have constituted the only living environment of the earth during all but the last 600 million years, which represents less than 20 percent of the earth's history.

The Paleozoic era. During the nearly 400 million years of the Paleozoic era, the geography of the earth's surface was completely different from what it was to be at later times. Most of the land area that was to become North America was included in an ancient continent that no longer exists. See Figure 19–12. All of the Paleozoic continents were isolated from each other. They were scattered around the globe in the tropical regions close to the equator. Sea level was high, and much of the low-lying parts of the continents were covered by shallow sea. Such a shallow sea covered much of what is now the central part of the United States. Huge amounts of sediments were carried into the shallow water as a result of erosion of the exposed land areas. The weight of these sediments caused the sea floor to gradually sink. More sediments were deposited and continued to accumulate on the sinking, shallow sea floor until a thickness of over 12 km (7.5 mi) was reached. These huge, Paleozoic, sediment-filled troughs, or geosynclines, now make up much of the older rock structure found along the east and west coasts of North America. See Figure 19–13. Thinner layers of Paleozoic sediments were also laid down over the stable region that is now the central United States.

FIG. 19–14. The fossil record indicates that undersea life might have looked like this during the Cambrian period. (American Museum of Natural History)

As the Paleozoic era unfolded, the separate continents shifted their positions. By about the middle of this era, the separate continents had begun a series of collisions from which the great, single continent of Pangaea was to emerge. This tectonic activity caused periods of mountain building and uplift of the land. The shallow seas advanced and retreated. During parts of the Paleozoic era, such as the Devonian period, most of the future North America was covered by the shallow seas. However, by the Permian period, at the end of the Paleozoic, much of the land had been lifted above sea level by widespread mountain building.

The fossil record in Paleozoic rocks shows that there was an "explosion" in the number of different kinds of plants and animals at the beginning of this era. The Cambrian period, which initiated the Paleozoic era, marked the sudden appearance of a variety of more advanced, sea-dwelling forms of life that replaced the simple life of the Precambrian. One part of the explanation for this Cambrian explosion of life may be the presence of the warm, shallow seas that covered much of the continents. Conditions were favorable for the forms of life that were able to live in the marine environment. There is no evidence of land-dwelling plants and animals during the Cambrian period.

The number of fossils found in Cambrian rocks indicates that the sea flourished with life. See Figure 19–14. There was a variety of marine plants similar to many modern seaweeds. All Cambrian animals were *invertebrates*. An invertebrate animal does not have a bony internal skel-

FIG. 19–15. Fossil remains of trilobites. (Breck P. Kent/Earth Scenes)

FIG. 19–16. Late Ordovician life as it may have appeared on the floor of an ancient sea. (Allan Roberts)

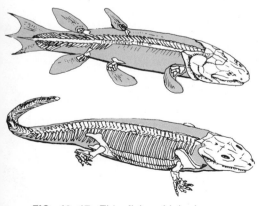

FIG. 19–17. This fish, which developed in the late Paleozoic, had boned fins that may have eventually become the limbs of the land-dwelling amphibian shown below.

eton, although it may have different types of hard, outer shells. The most common invertebrate fossil found in Cambrian rocks is the *trilobite* (*try*-loh-byt). See Figure 19–15. There were thousands of species of this small animal, which is so named because its body form is made of three parts. Trilobites became extinct by the end of the Paleozoic era. The large number of trilobites in Cambrian rocks makes them an excellent index fossil for identifying rock layers formed during that period.

The second most common Cambrian animal was a clam-like group called brachiopods or "lamp shells." Some brachiopod shells resemble ancient Roman lamps, even including a hole for the wick. A few kinds of brachiopods exist today, although they are not common. About 90 percent of all Cambrian fossils are trilobites and brachiopods. The remaining fossils include species of snails, jellyfish, sponges, and worms.

Moving ahead by 100 million years from the first of the six Paleozoic periods, the Cambrian, to the second, the Ordovician, there is still no evidence of any life form existing on land. However, sea life continued to develop during the Ordovician period. Among invertebrates, the trilobites were replaced by the mollusk group as the chief form of life. Mollusks include clams and similar hard-shelled animals, as well as octopuses and squids. See Figure 19–16. One important development during the Ordovician period was the appearance of the first vertebrate. It was a kind of fish covered with bony plates. It seems to be the most ancient ancestor of all the many kinds of vertebrate animals found in later times.

Marine life, including the oldest vertebrates, continued to thrive during the next period, the Silurian, which began 440 million years ago. The earliest life form known to exist on land appeared during the Silurian period. Simple plants without roots or leaves spread over moist soils. Small ancestors of insects, air-breathing animals very much like scorpions, lived at least a part of their lives on land.

The following period, the Devonian, is often called the Age of Fishes. Many fossil remains of bony fish are found in rocks of the Devonian period, which began 400 million years ago. Some of the Devonian fish show signs that their fins were beginning to develop into the legs needed to move on land. See Figure 19–17. Scientists think that these fish also developed the ability to breathe air. They were apparently the earliest of the amphibian animals,

who spend part of their life in water and part on land.

Invasion of the land by animals and plants was completed during the Mississippian and Pennsylvanian periods, which began 350 million years ago and lasted for 80 million years. Great swampy forests grew over much of the land surface. See Figure 19–18. Fossil remains of these forests of ferns, horsetails, and distant ancestors of modern trees are the coal deposits now found in North America, as well as in many other continents. Because of the vast coal deposits formed during these two periods, they are together often called the *Carboniferous period.* Carboniferous means "carbon (or coal) bearing."

Amphibious animals continued to develop during the Carboniferous period. Many kinds of insects, such as cockroaches and giant dragon flies, also appeared. A new group of animals descended from amphibians, the reptiles, became the first vertebrates to dwell entirely on land. Reptiles represented the first group of backboned animals to be exclusively air breathers with no need to return to the water to lay their eggs. Sea life also continued to develop, as sharks made their first appearance.

Two hundred seventy million years ago, as the Paleozoic era ended with the Permian period, a great change occurred. The marine invertebrates that had been so abundant for more than 400 million years almost entirely vanished. This great extinction of Paleozoic life forms could have been related to the joining of separate continents into the supercontinent Pangaea. Most of the warm, shallow seas that provided such a favorable environment for the sea dwellers disappeared. For example, the large inland sea that covered much of what was to become North America dried up. Left behind were the huge deposits of salt now found buried deep in Permian layers from Kansas to New Mexico. It is possible that the earth's climate also changed. However, it seems certain that the forma-

activity

You will be given an imaginary cross section of rock that contains layers and fossils. The cross section shows the history of a particular part of the North American continent. You may be able to discover what part of the continent the cross section represents by doing the following:

1. Use the Geologic Calendar in the Appendix, pages 530–531, to fill in the blanks in the column labeled Geologic Period for each layer based on the fossils present.

2. For each layer fill in the blanks in the column labeled Land/Water to indicate whether the fossils are from a land environment or a water environment.

3. Use the information from GROWTH OF A CONTINENT, pages 404–413, describing the land and sea environments found during the periods of the rock layers shown in the cross section. Which one of the following parts of the North American continent does the history of the cross section of rock most closely match? Canadian Shield area, the eastern coastal area, northeastern United States, southern coastal states, the midwestern United States, western Canada and Alaska, northwestern United States, western coastal area, the Rocky Mountain area, southwestern United States.

A

end of Paleozoic
200 million years ago

B

Mesozoic—65 million years ago

C

Cenozoic—present

FIG. 19–19. (A) Pangaea separated into two smaller fragments at the beginning of the Mesozoic. (B) and (C) During the Mesozoic era, the continents took on their modern shapes.

tion of Pangaea was the main factor that ended the Paleozoic era and set the stage for the next era of the earth's history.

The Mesozoic era. Beginning about 225 million years ago, the supercontinent of Pangaea began to break up. First it separated into a northern part called Laurasia and a southern part named Gondwanaland. See Figure 19–19A. Pangaea broke up as a result of the opening of crustal zones from which magma poured out. As the parts of Pangaea separated, new sea floor was created between the fragments. The Atlantic Ocean was created as North America came into existence by the splitting off of the western part of Laurasia. See Figures 19–19B and C. At the same time, Gondwanaland spread apart to form the African, South American, Antarctican-Australian, and Indian Plates. Further separation created Australia, with India taking off on a northward trip that finally caused it to collide with Asia. By the end of the Mesozoic era, about 70 million years ago, the earth's continents had taken the general position we find now.

The burst of tectonic activity accompanying the breakup of Pangaea played a large role in giving North America its modern shape. The basic form of the continent was set by the position of the rift that split it off Laurasia and then spread apart to produce the Atlantic. The North America that separated from Laurasia during the Mesozoic era does not have exactly the same shape as its Paleozoic ancestor that became part of Pangaea. A part of the present state of Georgia was left attached to the African Plate, and some of the European Plate became welded onto parts of present-day New England and Canada. In addition, some new crust was added to both the east and west sides of the continent.

Early in the Mesozoic, a large mountain range, the ancient Appalachians, existed along the eastern United States. On the western side, a sea covered the geosynclinal trough created during Paleozoic times. Toward the middle of the continent, the land was sometimes covered by a shallow sea and at other times exposed. See Figure 19–20. As time went on, tectonic activity within the continent caused the situation to be almost reversed. By the late Mesozoic, the eastern mountains had been worn down but were still evident. The Rocky Mountains had begun to form where the western geosynclinal sea had been. This western uplift divided the continent into eastern and western land areas separated by an inland sea. See Figure 19–21. For the first time, the general landscape of the United States, which is made up of an eastern and a western highlands separated by a central region of low elevation, began to take shape.

The sudden display of tectonic activity during the Mesozoic had a great influence on plant and animal life. The formation of Pangaea and its later separation created new environments. Among the animals, it was the reptiles that proved to be most successful in the Mesozoic world. They became the dominant life form for more than one hundred million years. The Mesozoic era is often called the Age of Reptiles.

More than any other kinds of animals that have ever lived, the giant reptiles of the Mesozoic era are thought of with fear and wonder. They are all usually called dinosaurs, a word that comes from the Greek for "terrible lizard." Indeed, some were huge, fierce hunters with long jaws, sharp teeth, and two or four legs. However, many were only large plant-eaters, and some were no bigger than a small dog. See Figure 19–22. All were descendants of a single family of ancient reptiles that lived during the Triassic period, which began the Mesozoic era. In the following periods, the Jurassic and Cretaceous, the fossil evidence shows that a tremendous variety of dinosaurs developed. At the same time, tropical forests that provided a plentiful food supply covered the land. The gigantic reptiles also invaded the sea and were particularly common in the shallow inland waters that cut through North America. One branch of the reptile group developed wings. These flying reptiles, the ancestors of birds, first appeared during the Jurassic period. By the Cretaceous period, birds had become common.

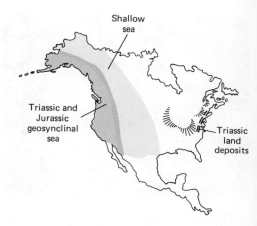

FIG. 19–20. During the Triassic and Jurassic periods of the Mesozoic era, most of western North America was covered by the sea.

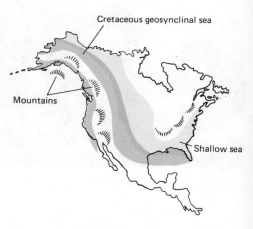

FIG. 19–21. During the Cretaceous period in the late Mesozoic era, North America had a central inland sea.

FIG. 19–22. Examples of the dinosaurs of the Mesozoic era. From left to right: Deinoychus, a vicious hunter; Brachiosaurus, a plant eater that weighed more than 100 tons; Pteranodon, a flying meat-eater; Tyrannosaurus Rex, a six-ton predator; Hypselosaurus, a swamp dweller; and Tylosaurus, a marine lizard.

1,5 meters
(about five feet)

FIG. 19–23. Animals of the Tertiary period. (American Museum of Natural History)

FIG. 19–24. A kind of fur-covered elephant called the wooly mammoth existed during the Pleistocene Ice Age. It became extinct about 10,000 years ago. (American Museum of Natural History)

Suddenly it all ended. At the end of the Cretaceous period, dinosaurs, and many of their fellow creatures, abruptly disappeared from the fossil record. The only reptiles that survived were the smaller ones, familiar to us today as lizards, snakes, turtles, and crocodiles. No one knows what caused this extinction of a life form that dominated the earth for such a long time. Various theories have been proposed by scientists. Some theories are based on changes in the earth's climate caused by plate motions that affected the temperature of the sea. Other scientists think that there is evidence that the earth collided with a large object such as an asteroid. Such an event might have temporarily changed the temperature over the entire earth by throwing up a global dust cloud. It is possible that the dinosaurs were killed off by a combination of factors that may never be completely understood. Whatever its cause, the disappearance of the dinosaurs marked the end of the Mesozoic era.

The Cenozoic era. By the beginning of the Cenozoic era, about 70 million years ago, North America had reached the final stages in becoming the familiar continent we know today. The inland sea that covered much of the central regions during Mesozoic times had drained away. Only the present coastal plain along the Atlantic and Gulf coasts remained submerged. Tectonic disturbances that continued during the Cenozoic era were responsible for most of the major land forms of modern North America. Violent movements along plate boundaries in the western part of the continent created the mountains we know as the Sierra Nevadas in California. These mountains, which first appeared in the Jurassic period, were lifted closer to their present height during the Cenozoic. East of the

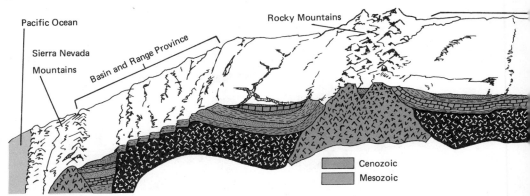

Pacific Ocean

Sierra Nevada Mountains

Basin and Range Province

Rocky Mountains

Cenozoic
Mesozoic

Sierra Nevadas, a series of tilted blocks created a large region of low mountains and wide valleys we now call the Basin and Range Province. Uplift of a broad area around the Colorado River produced the Colorado Plateau. The Grand Canyon was created as the river cut into the rising land. Farther to the east, the Rocky Mountains continued to be pushed up. On the level platform to the east, sediments washed off the Rockies were deposited as the Great Plains. Near the east coast, the ancient Appalachian Mountains continued to be worn down.

Much volcanic activity took place during the Cenozoic. Large areas of the states of Washington, Oregon, Idaho, and northern California are covered with lava plains and volcanic mountains created by volcanic eruptions.

One of the most important events of the Cenozoic era occurred during the past million years. This was the Ice Age. At least four times during the last million years, great sheets of ice have advanced southward over North America and then retreated. The ice was a powerful force in giving final shape to the landscape of much of the continent. When compared to the total time span of the earth's history, only a moment has passed since the glaciers last retreated.

The Ice Age also had a great influence on the plant and animal life that had developed since the beginning of the Cenozoic era. Just as the Mesozoic era is the Age of Reptiles, the Cenozoic era is the Age of Mammals. Ancestors of modern mammals existed during the late Mesozoic, but they became the dominant form of animal life during the *Tertiary* (*tersh*-i-ary) *period.* The five epochs of the Tertiary period, covering about 60 million years, saw the gradual appearance of the families of dogs, cats, horses, and almost all other modern animals. See Figure

Explore
In 1938 a type of fish believed to have been extinct for over 70 million years was caught alive in the Indian Ocean. The fish was called coelacanth. Find out more about this species and its special characteristics.

FIG. 19–25. A cross section of the United States shows the ages of the rock structures that control the surface features.

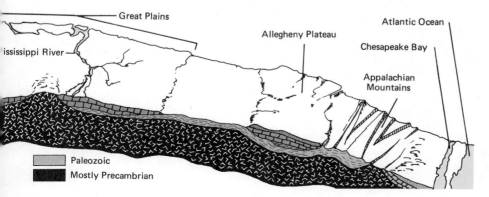

Great Plains

Mississippi River

Allegheny Plateau

Atlantic Ocean

Chesapeake Bay

Appalachian Mountains

Paleozoic

Mostly Precambrian

Construct
Prepare a diorama of the
Cenozoic era, depicting the
major epochs and the
corresponding mammal types
that dominated this era.

19–23. Mammals even moved into the sea, becoming whales and similar marine species. The tropical forests of the Mesozoic were replaced with the characteristic plants of the Cenozoic, which became the common grasses, flowering plants, and trees of today. A colder climate marked the beginning of the *Quaternary* (*kwah-ter-na-ree*) period. The one complete epoch of this period, the *Pleistocene,* includes the time in which we now live. Harsh, Pleistocene climatic conditions forced mammals and other animals in most of the world to develop protection against the cold or move to warmer climates. Many animals developed special characteristics that allowed them to survive. See Figure 19–24. Others moved to new regions. But many species became extinct. Those that survived are the animal populations of the modern world. Evidence indicates that humans also first appeared during the Pleistocene epoch.

Thus the continent of North America as it now exists has had a long history. Most of that history is recorded in the rock structure that gives the surface its modern shape. Figure 19–25 shows this basic rock structure and its general age along an east-west cross section of the United States at about the location of Washington, D.C.

VOCABULARY REVIEW

Match the word or words in the column on the right with the correct phrase in the column on the left. *Do not write in this book.*

1. Places where layers of sedimentary rock are exposed to view.
2. The complete arrangement of rock layers in order of their relative age.
3. The largest sections of the Geologic Column.
4. Oldest and longest era.
5. Means "ancient life."
6. The divisions of an era.
7. Means "recent life."
8. The divisions of a period.
9. Precambrian rocks around which modern continents may have formed.
10. Ancient Precambrian rocks around which the North American continent is thought to have formed.

a. Paleozoic
b. periods
c. epochs
d. Geologic Column
e. Canadian Shield
f. outcrops
g. Cenozoic
h. eras
i. Mesozoic
j. Precambrian
k. Quaternary
l. cratons

QUESTIONS

Group A

Select the best term to complete the following statements. *Do not write in this book.*

1. Almost all of the scientific record of the earth's past is found in (a) written records (b) rocks found at or near its surface (c) ancient tree rings (d) its atmosphere.

2. A limestone rock layer almost always indicates that (a) corals lived in the area (b) warm, shallow seas once covered the area (c) water once covered that location (d) mountain building occurred at that location.

3. In addition to the materials that make up rock, the kind of environment that once existed can often be told when (a) fossils are found in the rock (b) the radioactive level of the rock is determined (c) an outcrop is made (d) a geologic column is discovered.

4. Of the three basic kinds of rocks, the most complete record of earth's past is found in (a) metamorphic rocks (b) igneous rocks (c) sedimentary rocks.

5. The Grand Canyon of Arizona was produced by the erosion of the Colorado Plateau by the Colorado River in about (a) one million years (b) two million years (c) five million years (d) 10 million years.

6. The Grand Canyon represents an excellent record of earth's past because the information it gives (a) allows the principle of superposition to be applied (b) shows a large range of the composition of its rock layers (c) comes from fossils found in many of its rock layers (d) applies to all three of these sources and makes the Grand Canyon an excellent record of earth's past.

7. The work of thousands of scientists during the past 150 years has shown that rock layers of the same age are usually arranged (a) in different patterns in different locations (b) in the same pattern when found in many different places (c) in a haphazard manner at any location (d) in a definite, but different pattern at different places.

8. Observed rock layers can be classified by comparing them to (a) the Geologic Column (b) radioactive scales (c) the Geologic Calendar (d) tree rings.

9. For the Geologic Calendar to be formed, it was necessary to (a) know when life began on earth (b) find out the earth's age (c) determine the actual age of the Geologic Column's rocks (d) find out how many Geologic Columns existed.

10. The eras, periods, and epochs were first determined from the (a) Geologic Calendar (b) Geologic Column (c) Grand Canyon rock layers (d) dating of rock layers.

11. Evidence shows that during the Paleozoic era there were (a) six continents, one very large one and five small ones (b) two continents, one very large one and one very small one (c) many continents about equal in size (d) many equally sized continents that soon collided and stayed together as one supercontinent.

12. After the supercontinent Pangaea separated into two parts, Laurasia and Gondwanaland, the North American continent resulted from the (a) splitting off of the western part of Gondwanaland (b) splitting off of the northern part of Gondwanaland (c) splitting off of the western part of Laurasia (d) splitting off of the southern part of Laurasia.

13. The oldest and longest era in the earth's history is the (a) Precambrian (b) Cenozoic (c) Mesozoic (d) Paleozoic.

14. Which of the eras listed shows sufficient evidence for its periods to be divided into epochs? (a) Precambrian (b) Mesozoic (c) Cenozoic (d) Paleozoic.

15. The record left by rocks of the early Precambrian era is (a) very difficult to read (b) moderately difficult to read (c) easy to read (d) impossible to read.

16. During the late Precambrian era, (a) no life forms existed (b) vertebrates probably existed (c) the simplest forms of life were probably present (d) mollusks were present, as shown in the fossils found.

17. Sedimentary rocks show that animal life of the Cambrian period included many (a) snakes (b) land animals (c) fish (d) invertebrates.

18. The earliest known land animals appeared during the (a) Cambrian period (b) Ordovician period (c) Silurian period (d) Devonian period.

19. Reptiles developed during the (a) Devonian period (b) Permian period (c) Silurian period (d) Carboniferous period.

20. The Mesozoic era was a time of great changes that are attributed mostly to (a) great tectonic activity (b) the ice ages (c) meteor bombardment (d) the earth changing its orbit.

21. The Rocky Mountains were created at the end of the (a) Paleozoic era (b) Mesozoic era (c) Cenozoic era (d) Precambrian era.

22. The North American continent first began to take on its present familiar shape during the (a) Paleozoic era (b) early Mesozoic era (c) late Mesozoic era (d) early Cenozoic era.

23. The period in which the great dinosaurs dominated the earth was the (a) Jurassic (b) Triassic (c) Permian (d) Cretaceous.

24. A fossil commonly found in Paleozoic rock formation is (a) eohippus (b) trilobite (c) mastodon (d) dinosaur.

25. The era called the "Age of Mammals" is the (a) Cambrian (b) Paleozoic (c) Mesozoic (d) Cenozoic.

26. The Ice Age was one of the most important events of the (a) Precambrian era (b) Paleozoic era (c) Mesozoic era (d) Cenozoic era.

27. The familiar modern animals such as dogs, cats, and horses appeared during the (a) Teritiary period (b) Jurassic period (c) Cretaceous period (d) Quaternary period.

28. The Quaternary period marked the beginning of (a) the Age of Mammals (b) a warmer climate (c) the Age of Reptiles (d) a colder climate.

29. The most important development of the Pleistocene epoch was probably the appearance of (a) trilobites (b) human beings (c) reptiles (d) mammals.

30. The way the land surface is changed by the many forces at work is controlled mainly by (a) humans (b) the underlying basic rock structure (c) the climate (d) action of plants and animals.

Group B

1. Explain why the scientific view of the earth's complete history cannot be thought of as made up of proven facts.

2. Which one of the three basic kinds of rocks contains the most complete record of earth's history? Why?

3. Describe at least three ways in which outcrops may occur.

4. Describe how the Grand Canyon of Arizona is thought to have been formed.

5. List the three main sources of information about earth's history as read from the rocks of the Grand Canyon, and tell what information each source gives.

6. Explain how scientists developed the idea of a Geologic Column.

7. How is the Geologic Column related to the geologic calendar?

8. List the divisions in the Geologic Column in order from largest to smallest with respect to period of time represented.

9. Name the two major parts of the Precambrian era and tell why they are convenient to use.

10. List the eras of the earth's history in the order they occurred and give the approximate length of time each lasted.

11. Describe the positions of the continents during the Paleozoic era.

12. What relationship exists between cratons and continents?

13. Why is it difficult to find Precambrian fossils?

14. Describe the major event that marked the end of two different geologic eras. Name these eras.

15. What is a trilobite? Why are their fossils important?

16. Humans were one of the last organisms to appear on the earth. How do you account for the fact that they now dominate the earth and, to some degree, control their environment?

17. Why do you suppose the earliest forms of life developed in the sea rather than on land?

18. Briefly describe the tectonic activity that occurred from the beginning of the Mesozoic era to its end.

19. What period is known as the "Age of Reptiles"? Describe the conditions that existed during this time.

20. Why are the Pennsylvanian and Mississippian periods commonly combined under the title "Carboniferous"?

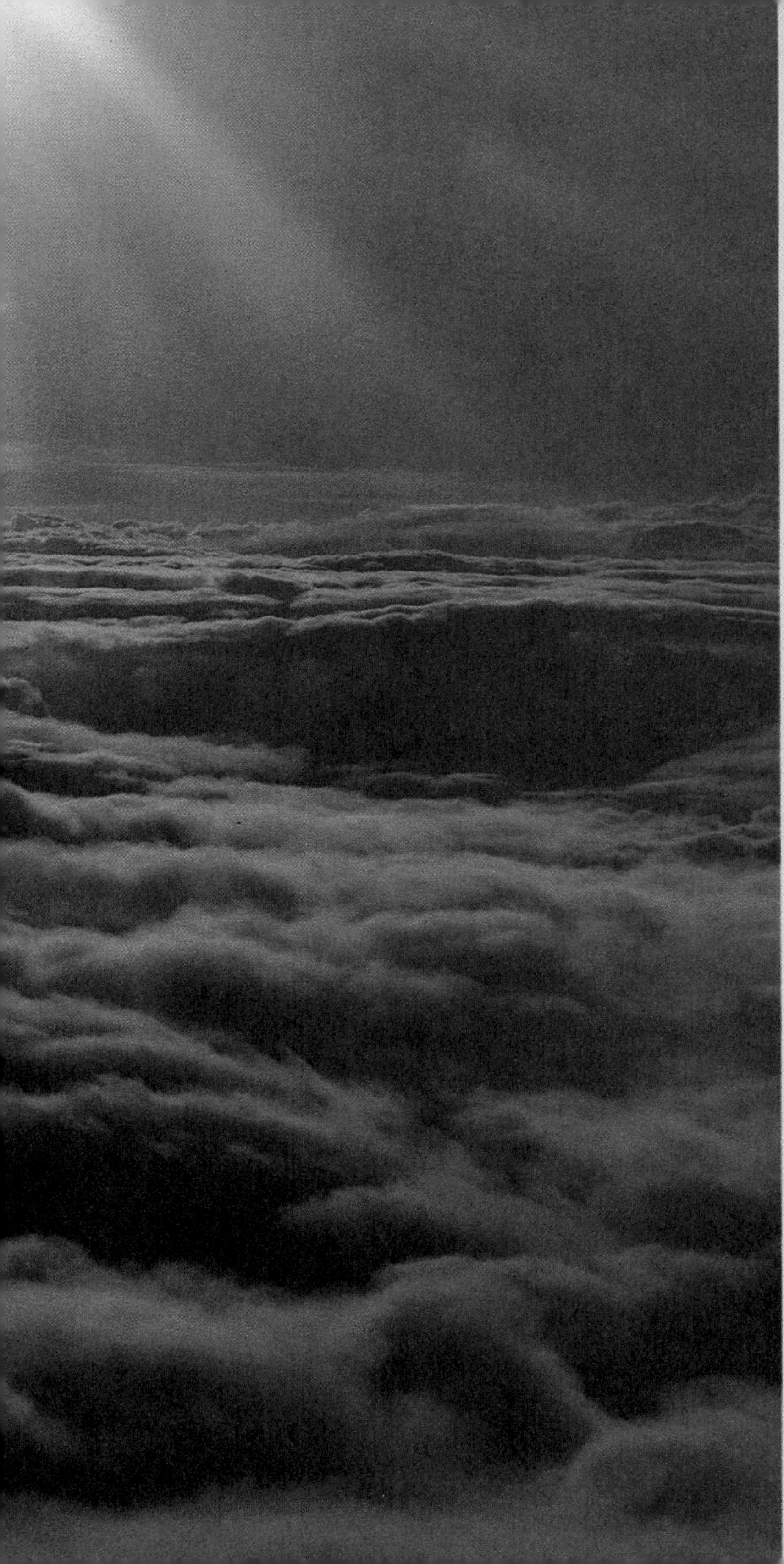

unit 6 the earth's atmosphere

20
air and its movements

objectives

- [ ] List the materials that make up the earth's atmosphere.

- [ ] Describe air pressure and the instruments used to measure it.

- [ ] Describe the chemical balance of the atmosphere.

- [ ] List several types of air pollution and their effects on earth.

- [ ] Describe the effect of the atmosphere on the solar energy received by the earth.

- [ ] List the layers of the atmosphere.

- [ ] Explain how movement of the air is caused by heating the atmosphere.

- [ ] Describe how the rotation of the earth affects the movement of air.

We are accustomed to believing that we live on the surface of the earth. Seldom do we think of ourselves as creatures who must live beneath a blanket of gases, as a fish must live in water. But man is designed to live only within the narrow region where two of the earth's layers meet. Above this area is a layer of gases, the *atmosphere,* which surrounds the entire planet. Below are the earth's oceans or *hydrosphere,* covering $7/10$ of the earth's solid crust or *lithosphere.* At the boundary between the atmosphere and the lithosphere is the place where we spend most of our lives.

Since we are always submerged in the atmosphere, it is important that we know which gases and other materials are present. It is equally important that we understand the processes which affect their composition. The study of the gaseous region above the solid earth is called *meteorology.* One of the greatest challenges in meteorology is to understand the effects of heat energy in the atmosphere. The atmosphere is actually a giant heat engine. An engine is a device arranged for changing one form of energy into another. The atmosphere is an engine because it changes radiant energy from the sun into heat energy. The workings of the atmospheric heat engine fuel the winds and cause the ever changing patterns of weather.

THE AIR

Composition of the atmosphere. If a sample of air were dried and all foreign matter removed, chemical analysis would uncover two main gases. The gas nitrogen (N_2)

would account for 78 percent of the total volume of the sample; oxygen (O_2) would make up 21 percent. The remaining 1 percent would be mostly argon (Ar) and carbon dioxide (CO_2). Table 20–1 shows the normal composition of air near sea level.

Some of the atmospheric gases, including those listed in Table 20–1, are always present in approximately the same amounts in air near the earth's surface. Other gases are also likely to be present, but not in definite amounts. Water vapor is by far the most important of these variable gases. In very moist air, water vapor makes up as much as three percent of the total volume. Another important variable gas in the atmosphere is ozone (O_3), which is a form of oxygen. The ordinary oxygen that we breath (O_2) has only two atoms per molecule while poisonous ozone has three. There is very little ozone found in the lowest parts of the atmosphere. It is found in large amounts at altitudes of 20 to 50 km.

In addition to the true gases which are composed of molecules, air usually contains larger solid particles which include dust, smoke, and salt crystals. A large number of dust particles are small enough to mix with the gaseous molecules and remain suspended almost indefinitely. Much of the dust in the atmosphere is composed of mineral particles carried by the force of the wind. Smoke is a common source of dust. Volcanic eruptions and meteors which have vaporized in the air also contribute to the amount of dust in the atmosphere. Tiny drops of sea water thrown into the air evaporate, leaving behind minute particles of salt crystals which are important in the formation of clouds.

Air pressure. The gases which make up the atmosphere are held in place by the earth's gravity. Altogether, the atmosphere presses down with a weight of about 6×10^{15} tons. For each square centimeter of surface near sea level, air pressure exerts a force of 1.03 kg. (14.7 pounds/sq. in). An average-size person at sea level must support a force of 10 to 20 tons due to air pressure. Usually you can't feel any pressure because it is the same both inside and outside your body. At higher altitudes the air is thinner and the pressure is reduced. You would become conscious of the change because of the effect of reduced pressure on the outside of your ear drum. The "popping" sensation in your ears as you rise quickly in an elevator or in an automobile is also due to a sudden drop in pressure.

Table 20–1 Gases in Pure Dry Air

Gas	Symbol or Formula	Percent by Volume
Nitrogen	N_2	78.084
Oxygen	O_2	20.946
Argon	Ar	0.934
Carbon dioxide	CO_2	0.033
Neon	Ne	0.00182
Helium	He	0.00053
Krypton	Kr	0.00012
Xenon	Xe	0.00009
Hydrogen	H_2	0.00005
Nitrous oxide	N_2O	0.00005
Methane	CH_4	0.00002

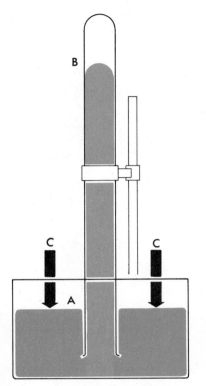

FIG. 20-1. This diagram shows the principle of the Torricellian barometer. The tube containing the mercury has been shortened for convenience.

An instrument which measures atmospheric pressure is called a *barometer*. The type of barometer most commonly used was invented in 1643 by an Italian named Torricelli (tohr-ree-*chel*-ee). The principle of the Torricellian barometer is illustrated in Figure 20–1. Air pressure is indicated by arrow C, pressing down on the surface of mercury in the dish. The height of the mercury column in the tube, A to B, is determined by the atmospheric pressure. The space in the tube above the level of the mercury is nearly a perfect vacuum. Mercury is used because it is very heavy, about 14 times more so than water. Thus a relatively small amount of mercury in the tube is sufficient to balance the pressure exerted by a column of air extending from sea level to the top of the atmosphere. Torricelli found that air pressure at sea level would balance a column of mercury about 760 mm (30 in) high.

The modern version of Torricelli's barometer is shown in Figure 20–2. There is a scale to measure the height of the mercury and a thermometer to indicate corrections for temperature changes. This type of barometer is called a *mercurial barometer*.

Air pressure measured by a mercurial barometer is expressed by how high the mercury rises. For example, the atmospheric pressure at a certain time might be given as "758 mm of mercury" which is the same as "29.87 inches of mercury." A pressure equal to 1 mm of mercury may be called 1 Torr after Torricelli. Thus the pressure mentioned above might also be given as "758 Torr."

Official weather maps may use still another measurement of air pressure called millibars (mb). This is a way of expressing air pressure which is understood all over the world. Millibars are complicated to define and it is usually only necessary to know that normal air pressure at sea level is equal to 1013.25 millibars. Average sea level pressure may also be expressed as 760 mm of mercury or 29.92 inches of mercury.

Another type of barometer is also commonly used. The *aneroid* (*an*-uh-royd) *barometer*, also shown in Figure 20–2, is more portable and rugged than the mercurial barometer. Unlike the mercurial type, an aneroid barometer contains no liquid. The heart of this instrument is a sealed metal container from which most of the air has been removed. When the air pressure increases, the sides of the container bend inward. When the pressure decreases, the sides bulge out again. These changes are indicated by a moving pointer on a dial. The movement of the pointer is regulated by a system of gears and levers. The dial can be marked off to show the pressure in inches of mercury or any other units desired. Aneroid barometers can be constructed so that a continuous record of air pressure is made on a revolving drum. Such an instrument, shown in Figure 20-2, is called a *barograph*.

Aneroid barometers can also be used to measure approximate altitude above sea level. When it is used for this purpose it is called an *altimeter*. At higher altitudes the atmosphere is thinner and exerts less pressure. Thus a lowered pressure reading can be interpreted as an increased altitude reading. To accurately measure altitude,

FIG. 20-2. Types of barometers. Left, a *modern mercurial barometer*. Top, an *aneroid barometer*. Compression and expanison of the sealed metal box provide a measurement of air pressure. Below, a *barograph*. The readings of the aneroid barometer are recorded on the drum. (Bendix Aviation Corporation; U.S. Weather Bureau)

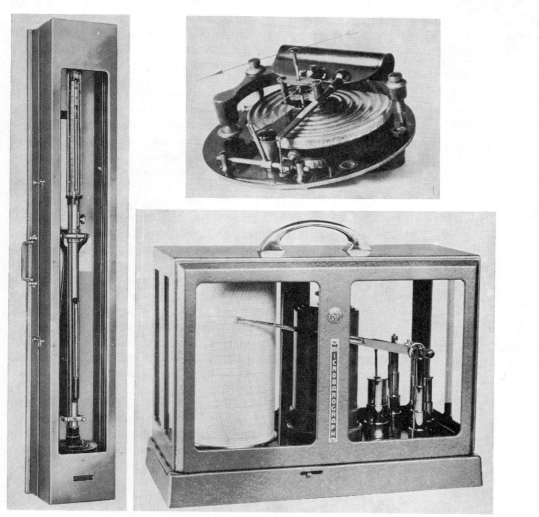

FIG. 20-3: Animals, such as these cattle, play a part in the nitrogen cycle. First nitrogen is removed from the air by nitrogen-fixing bacteria. Secondly nitrogen is removed from the soil by plants, which are in turn eaten by animals. Decomposing plant and animal remains return the nitrogen and its compounds to the atmosphere. (Ramsey)

an altimeter based on air pressure must be corrected to local conditions. Changes in pressure due to weather changes require continual correction adjustments. Aircraft receive information for these corrections from ground stations.

The chemical balance of the atmosphere. The gases of the atmosphere seem to be in a state of chemical balance. Although the amount of each main gas in the air remains almost constant, these gases are always being added and removed from the atmosphere. Nevertheless, processes which remove and produce the principal gases, nitrogen, oxygen and carbon dioxide, seem to be in balance with each other.

The balance of oxygen in the atmosphere is an example of the chemical balance that keeps the composition of air the same. Many of the processes involved in rock weathering and mineral formation remove oxygen from the air. Animals also remove oxygen from the air as part of their life processes. If there were no means of replacement, the supply of oxygen in the atmosphere would probably be exhausted in a few thousand years. Green plants create a counterbalance by giving off oxygen in the manufacture of their food. In the process of photosynthesis, plants use sunlight, water, and carbon dioxide to produce their food. Oxygen is released as a by-product of photosynthesis.

The balance of nitrogen in the atmosphere is maintained by the *nitrogen cycle*. Nitrogen is removed from the air mainly by the action of nitrogen-fixing bacteria. These microscopic plants live in the soil and in the roots of plants of the legume family (beans, peas, clover, alfalfa, and others). They take nitrogen from the air and chemically change it into nitrogen compounds. These nitrogen compounds are vital to the growth of all plants. Through plants, nitrogen compounds enter the bodies of animals. These compounds are then returned to the soil through animal excretions or by decay of their bodies after death.

In the soil, decay processes release nitrogen and return it to the atmosphere as a gas. See Figure 20-3. A similar cycle takes place among water-dwelling plants and animals. To a lesser extent, lightning also removes nitrogen from the air. The weathering of certain rocks results in the formation of nitrogen compounds which are carried out of the atmosphere by rain and snow. Volcanic activity is yet another way nitrogen is released into the air.

The carbon dioxide cycle in the atmosphere is regulated by three great reservoirs: the sea, rocks, and living things. Of these, the sea is the largest storage area for carbon dioxide. The oceans contain about 50 times more carbon dioxide than the entire atmosphere. Most carbon dioxide is in the form of carbonate compounds. However, quantities of this gas are only dissolved in water. Each year the sea and the atmosphere exchange 200 billion tons of carbon dioxide. This exchange seems to keep the amount of carbon dioxide in the atmosphere fairly stable. If the concentration of carbon dioxide in the air is increased, the sea will probably absorb more of this gas. If the concentration of carbon dioxide in the atmosphere falls, the sea will probably release more gas. Figure 20–4 shows all of the processes which maintain the carbon dioxide concentration in the atmosphere.

FIG. 20–4. The process which maintains the carbon dioxide concentration in the atmosphere is shown in this diagram. The numbers indicate how many tons of carbon dioxide are involved in the process.

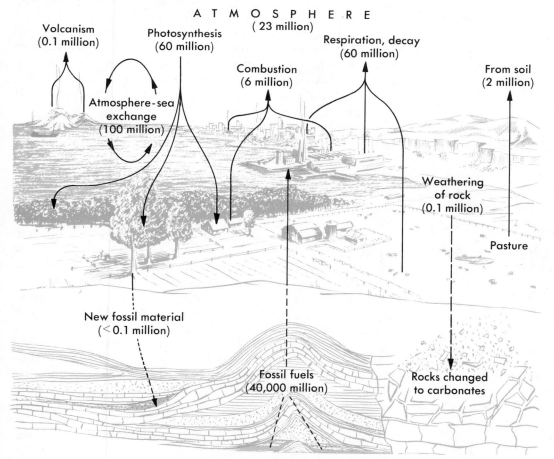

A T M O S P H E R E
(23 million)

Volcanism
(0.1 million)

Photosynthesis
(60 million)

Respiration, decay
(60 million)

Combustion
(6 million)

From soil
(2 million)

Atmosphere-sea exchange
(100 million)

Weathering of rock
(0.1 million)

Pasture

New fossil material
(<0.1 million)

Fossil fuels
(40,000 million)

Rocks changed to carbonates

Air pollution. It has taken billions of years for the earth's atmosphere to reach its present form. Sensitive natural cycles now tend to maintain its normal composition. However, like many of the earth's delicate natural cycles, this balance can be upset by man's activities. Consider the case of the atmosphere. Thousands of tons of gases and solids are dumped into the atmosphere each day. These troublesome and often dangerous foreign substances in the atmosphere are called *air pollutants*.

There are two main types of air pollutants. First are the gases which irritate eyes, throats, and lungs, harm plants, and even attack stone and metal. The chief air pollutant gases are produced mostly by burning of fuels. The sulfur which is present in a fuel such as coal or oil will produce irritating and destructive sulfur dioxide gas when either fuel is burned. The operation of automobile engines produces two kinds of pollutant gases. Unburned gasoline containing hydrocarbon-type compounds is usually given off in automobile exhausts. In the air, hydrocarbons mix with other gases to produce many additional substances known to be irritating when breathed. Besides the non-lethal hydrocarbons, automobile exhausts also contain quantities of poisonous nitrogen oxides. Also present is the very deadly gas carbon monoxide.

Another source of unwanted gases in the air are some manufacturing processes. Compounds of fluorine and chlorine may be discharged into the air around certain kinds of industrial plants. Generally speaking, however, this is a very limited problem. The burning of fuels, particularly in automobiles, remains the most serious problem.

A second type of air pollutant is composed of solid particles. Here too, the automobile is a serious problem. The gasoline engine changes about 0.08 percent of the fuel it burns into solid material. About half of this solid material given off in automobile exhausts is in the form of lead and lead compounds. The friction of automobile tires wearing against pavement also throws tiny particles of rubber into the air. Other than motor vehicles, the chief source of solid air pollutants is smoke. If all the solid pollutants in the air over a large city during one month's time settled out at once, it would amount to an average of more than 10 tons per square mile.

Air pollution becomes more serious when weather conditions and the topography of a region prevent the dispersal of pollutants. Under normal daytime conditions,

Investigate

Call your town government and find out if they are keeping a check on local air pollution and how it is done. Report your findings to the class.

the temperature of the air decreases as the altitude above the earth's surface increases. However, there are occasions when, instead of decreasing, the temperature increases with altitude. This is often what occurs when a layer of warm air exists above a layer of cool air next to the ground. Meteorologists refer to this condition as a *temperature inversion*. The inverted layer of warm air traps the pollutants and holds them close to the ground because cool air has a tendency to sink. Temperature inversions or other weather conditions which produce stagnant air commonly occur in a valley topography. The hills forming the walls of a valley trap the inversion layer and prevent the polluted air from drifting away. Some cities, such as Los Angeles, have frequent temperature inversions resulting from the cooler Pacific air being held near the ground by the arrangement of mountains. In locations where these factors are present air pollution is an ever increasing problem. See Figure 20–5.

The total effects of air pollution are difficult to determine. It is obvious that polluted air is much less clear than clean air. Solid particles produce a hazy effect which has led to the term "smog" to describe the appearance of polluted air around many cities. Of more importance is the effect of polluted air on the people who must breathe it. This problem receives much attention on the rare occasions when unusually heavy air pollution causes deaths. In London, England, in 1952, at least 4000 deaths occurred because of exposure to five days of a heavily polluted fog. In Donora, Pennsylvania, in 1948, twenty deaths were apparently caused by an attack of severe air pollution. Less dramatic but perhaps more serious are the still unknown effects on the health of city-dwellers who breathe polluted air for their entire lives. But the problem of air pollution will not be solved quickly or easily. It is deeply rooted in industrial processes and the increasing numbers of automobiles and their exhausts. Only the combined efforts of government and interested citizens can help solve the problem of increasing pollution of the atmosphere.

FIG. 20–5. This view of Los Angeles shows the definite layer of polluted air, or smog, which is caused by a pronounced temperature inversion. (Los Angeles Air Pollution District)

Explore
Find out exactly what caused the air to become deadly in Donora, Pennsylvania in 1948.

THE ATMOSPHERE AND THE SUN

Solar energy and the atmosphere. Remember that almost all of the energy reaching the earth from the sun is

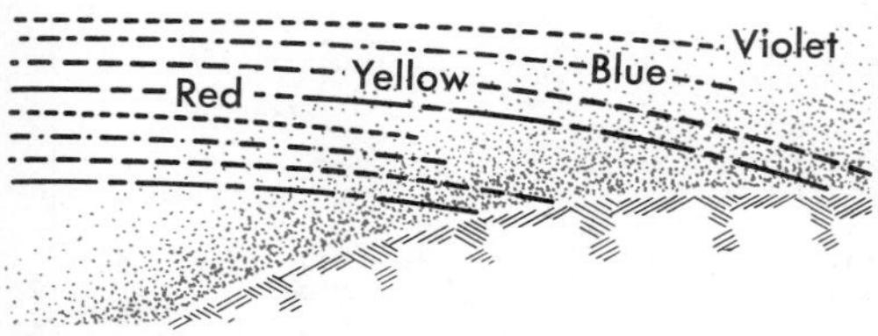

Sunset

FIG. 20-6. This diagram shows how sunlight which contains all colors can appear red or yellow when the sun is near the horizon.

Table 20-2 Reflection of Solar Energy

Surface	Percent of incoming solar radiant energy which is reflected
Water	
Incoming rays straight down	2
Incoming rays 40° from overhead	2½
Incoming rays at low angle	Very large
Clouds	40–80
Snow, Ice	46–86
Grass	14–37
Fields	3–25
Bare ground	7–20
Forest	3–10

in the form of electromagnetic waves. The earth's atmosphere is the first part of the earth to receive this radiant energy from the sun. As the electromagnetic waves from the sun pass downward they are affected by the atmosphere. First, the very short wavelength X-rays, gamma and ultraviolet rays are absorbed by molecules of nitrogen and oxygen. This happens in the upper parts of the atmosphere above 50 km (30 miles). As solar energy is absorbed, molecules and atoms of nitrogen and oxygen lose electrons and become positively charged ions. At a lower level, from about 20 km up to 35 km (12 to 21 miles) above the earth's surface, ultraviolet rays act upon oxygen molecules (O_2) to form the ozone layer, (O_3).

In their passage through the upper atmosphere, almost all of the shorter wavelengths of solar radiation are lost through absorption. The small number of ultraviolet rays which penetrate to the earth's surface are responsible for sunburn of skin exposed to sunlight for too long a period. Only the longer wavelengths of visible light and the infrared rays reach the lower atmosphere. In this region the air is denser and the gas molecules crowd closer together. These gas molecules are responsible for scattering light. Scattering means that the light rays are bent in all directions. Some of the rays are turned back into space and are lost. Since blue light is most affected by scattering of molecules, a clear sky always has a blue color. When the sun or moon is seen near the horizon, the light from its surface appears yellow or red. This is because dust and water droplets in the air near the earth's surface allow yellow and red light to come through while the blue light is scattered. See Figure 20-6.

The longer infrared rays are not scattered by the atmosphere as is visible light. However, infrared rays may be partly absorbed by water vapor in the air. Clouds also absorb and reflect back into space a part of the solar energy which reaches the lower atmosphere.

Of the total amount of solar energy which enters the atmosphere, an average of about 19 percent is absorbed at some level. Another 35 percent is turned back into space either by reflection or by scattering. See Table 20-2. This leaves 46 percent which actually reaches the land and water surfaces of the earth. The percents given are average figures for the entire earth's surface. The amount of solar energy reaching the ground is affected by clouds, dust in the air and, of course, by the latitude and the season of the year.

Trapping the sun's energy. When the rays carrying solar energy finally reach the earth's surface, two effects are possible. The radiations are either absorbed or reflected. The kind of surface on which the radiation falls determines to a large extent which of the above will take place. The amount of incoming solar radiation that is reflected by various surfaces is shown in Table 20–2. Of the total solar energy arriving from the sun, 35 percent is reflected back into space. Thus the earth is said to have an average reflectivity or *albedo* of 0.35. By comparison, the albedo of the moon is only 0.07, indicating that it reflects 7 percent of the total solar energy it receives.

The solar radiation that is not reflected by the earth's surface is absorbed. Part of the radiation absorbed by the earth's surface is composed of the infrared rays that have penetrated the atmosphere. The rocks, soil, water, and other earth materials absorbing these infrared rays are heated. Then, the heated materials produce their own infrared rays from the heat energy. But the infrared rays produced by the materials of the earth's surface have a much longer wavelength than the infrared rays in sunlight. The gases in the atmosphere allow the shorter infrared rays from the sun to pass through. The longer wavelength infrared rays sent out from the warmed materials of the earth's surface are mostly absorbed by the atmosphere. Water vapor and carbon dioxide in the atmosphere are chiefly responsible for absorbing the longer infrared rays.

The final effect of this action is to trap heat energy from the sun and prevent it from escaping back through the atmosphere into space. This process acts very much like a greenhouse. The principle is the same. In a greenhouse the glass allows the short wavelengths of infrared rays coming from the sun to pass through. But the glass prevents the longer infrared rays leaving the warmed surfaces in the greenhouse from escaping. Since the two effects are so similar, the process of the atmosphere trapping the sun's heat over the earth's surfaces is called the *greenhouse effect*. This is illustrated in Figure 20–7.

By the greenhouse effect, the energy from the sun warms the air after having first been absorbed at the earth's surface. Thus the atmosphere is heated mostly in its lower parts.

Layers of the atmosphere. Gravity pulls the gases of the atmosphere toward the earth's surface. The molecules of

Investigate
Ask your teacher how you can find out how the albedo is effected by the construction of cities, highways, airports, and large housing developments.

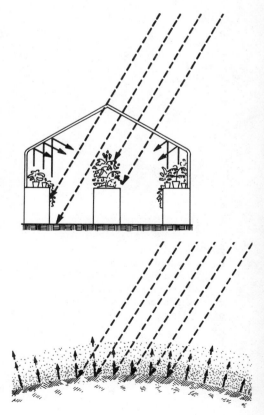

FIG. 20–7. The greenhouse effect. The atmosphere, like the glass in a greenhouse, allows infrared rays to enter, but prevents the escape of longer wavelength infrared rays.

gases move apart and take up as much space as possible. If gravity did not attract the gaseous molecules in the atmosphere, they would eventually move out into space and be lost. Half of the total weight of atmospheric gases is found within 5.6 km (3.5 miles) of the earth's surface. The remaining half of the atmosphere extends upward for hundreds of miles but gets increasingly thinner. Thus air pressure falls rapidly but smoothly with increasing altitude. There are no sharp pressure changes which separate the atmosphere into layers.

However, the atmosphere does show distinct changes in temperature and composition with increasing altitude. Basically, this is a result of the greenhouse effect. The air closest to the earth's surface becomes heated, which in turn warms the rest of the atmospheric zone or layer closest to the earth's surface, called the *troposphere*. Temperature within this layer decreases with height at the average rate of 6.5° C/km. The word "troposphere" comes from a Greek root meaning "change" because almost all weather changes occur in the troposphere. The upper boundary of the troposphere, called the *tropopause,* extends to an average height of 10 km. However, this layer changes with latitude and the season of the year. The tropopause is located higher above the equator than it is above the poles.

Above the tropopause the temperature remains almost constant, even as height increases. This steady temperature zone marks the beginning of the second layer called the *stratosphere*. The only important warming effect upon the air in the stratosphere is the direct absorption of solar rays by the ozone layer. Here the temperature rises slowly with increasing height. The temperature continues to rise steadily above the ozone layer to a height of about 48 km (30 miles). The elevation marking the upper boundary of the stratosphere, is the *stratopause.*

Above the stratopause is the atmospheric layer called the *mesosphere*. In this layer temperature decreases rapidly with height until its upper boundary, the *mesopause,* is reached. Above the mesopause, the temperature increases steadily. Information gathered about temperature changes in this layer, known as the *thermosphere,* has been too incomplete to determine its upper boundary. In this region of very thin air the molecules are so far apart that a thermometer could not accurately indicate the energy of the moving molecules. The air is much too thin to warm any object that might pass through. At high

Construct
Make a chart showing the relationship between the height of our atmosphere and (1) the density of air and (2) the amount of air.

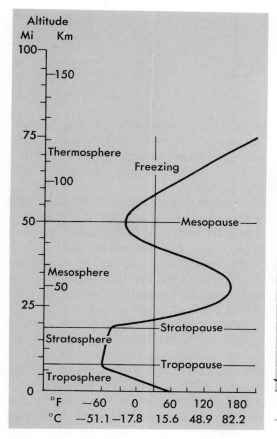

Altitude
Mi   Km

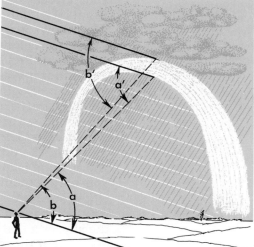

FIG. 20–8. Models illustrating the layers of the atmosphere have been made based on temperature differences at various altitudes.

FIG. 20–9. For a rainbow to form, only those raindrops that refract and reflect the sun's rays at a certain angle (a=a', b=b') can produce a spectrum that will reach the eye of the observer in the form of an arch. The angle of the reflected rays determines the band of color produced.

altitudes, objects are warmed almost entirely by the solar rays they absorb directly. A thermometer is placed in this layer and shaded from the sun, would register a temperature far below zero in spite of the high speeds at which the few surrounding gas molecules move.

The region of the thermosphere and upper mesosphere is often described as the *ionosphere*. In this zone the effects of absorption of solar rays by the atmospheric gases produce the ions responsible for the auroras (northern and southern lights). Above the thermosphere is the region where the earth's atmosphere blends into the almost complete vacuum of deep space. This indefinite zone is called the *exosphere* because individual atoms and molecules escape the earth's gravity and drift away into space. A summary of the characteristics of the various layers of the atmosphere is given in Figure 20–8.

Visible light from the sun passing through the atmosphere is affected by the gases in the air itself and other

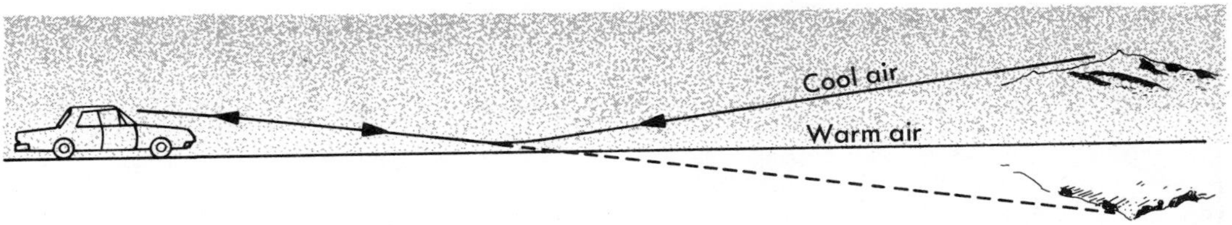

FIG. 20–10. In an inferior mirage light rays are bent in such a way that the ground seems to be a mirror. Distant objects may appear upside down below the horizon, as if reflected from the surface of water. When this type of mirage reflects the sky, the familiar illusion of shimmering water is seen.

FIG. 20–11. The effect of a superior mirage is that of a mirror suspended in the sky, reflecting the image of objects that are out of sight over the horizon.

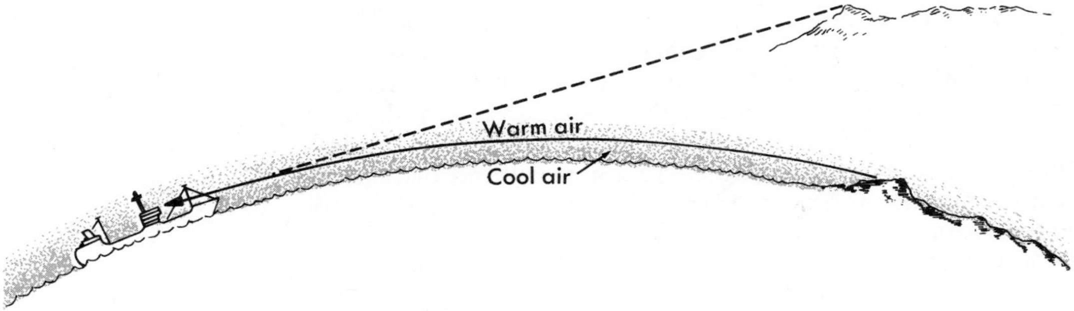

particles suspended in the atmosphere. It has already been mentioned that the sky is colored by the scattering of visible light by molecules of gas or other particles. Varying color effects are produced by separation of visible light into its individual wavelengths.

Rainbows are caused by the separation of sunlight into a range of colors. In this case, raindrops are the means of separation. Bending and reflection of light rays in the falling raindrops separate each ray of white light into a spectrum of colors. This spectrum produced by raindrops is visible only from certain angles. We can see a rainbow only when we are facing the falling raindrops and the sun is behind us. The correct position is shown in Figure 20–9.

Among the effects of light passing through the atmosphere, none is more impressive than a *mirage* (mi-*rahzh*). You may remember seeing a mirage on a summer day as water seemed to appear on the dry pavement of a highway. Perhaps a distant building appeared suspended in the sky, or a distant landscape seemed to be reflected on a lake. All of these are forms of mirages. An optical illusion is created by the bending or refraction of light rays passing through layers of air at different temperatures. The density of air varies according to temperature. The path of a light ray is refracted when it passes through several such layers.

The most common type of mirage is an *inferior mirage*. The mirage appears to the observer to be inferior or below his own eye level. Such a mirage is caused by the bending upward of light rays as they enter warmer air near the ground. See Figure 20–10.

In another type of mirage, the image seems to be above the observer. This is called the superior mirage, or *looming mirage*. It appears most commonly over the sea. Ships, icebergs, shorelines, cities, and other objects which would normally be out of sight over the horizon appear suspended in the sky. Sometimes double mirages are seen, one upright and the other upside down. Superior mirages occur when light rays entering cool surface air are bent downward. See Figure 20–11.

THE WINDS

Movement of air caused by heating. There are three ways that heat can be transferred to the atmosphere. Heating by radiation has already been described earlier in this chapter. The atmosphere is warmed when radiant energy is directly absorbed by its gases. Much less important is heating of the atmosphere by *conduction*. In this process heat flows directly to the atmosphere when the air comes in contact with anything that contains more heat. A third way that heat is transferred is by *convection*. This term is used to describe the movement of gases or liquids when they are heated unevenly. It is a very important atmospheric process and is the basic factor controlling the movement of air over the entire earth. It is the main cause of the *planetary circulation system*.

Movement of air due to convection takes place when air is heated by radiation or conduction. Air which has been heated, being less dense and lighter, presses down on the earth with less force. Thus the atmospheric pressure is generally lower beneath a body of warm air than under cooler air. As denser, cooler air, moves into a low pressure region, the lighter, warmer air is forced to rise. The general movement of air is always toward regions of lower pressure. These pressure differences, caused by unequal heating and resulting convection, create winds.

At the equator the earth receives more solar energy than at the poles. Consequently, atmospheric pressure at the equator is lower. The heated air in the region of the equator is constantly rising. At the poles the colder air is heavier and tends to settle, thus creating regions of high atmospheric pressure.

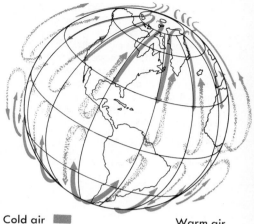

Cold air ▬ Warm air

FIG. 20–12. If the earth did not rotate, the circulation of the atmosphere would be as shown here.

activity

In this activity, you will feel heat transferred by three methods: conduction, convection, and radiation. Be careful, you should just feel the heat, not get burned.

Light a candle and drip some wax on the center of a index card. Blow out the candle and stand it upright in the soft wax. Light the candle.

Hold a finger about 40 cm above the flame. Bring it closer until you can feel the heat from the rising hot air.

1. At what distance did you feel the heat?

2. Was the heat transferred by conduction, convection, or radiation?

Hold your fingers 40 cm to the side of the lighted candle. Bring your fingers closer until you can feel heat.

3. At what distance did you feel the heat?

4. How was the heat transferred?

Hold a nail (16d finishing nail) near its head. You are going to determine the time it takes for heat to travel the length of the nail. Put the point in the flame for 15 seconds. Then move the nail and your hand until they are more than 40 cm from the candle. Determine how long it takes heat to travel along the nail until you can feel it near your fingers. Be careful!

5. How long did it take?

6. How was the heat transferred?

Pressure differences between air at the equator and the poles create a general movement of the atmosphere. Since air moves from high toward low pressure, there is a general flow of air near the earth's surface from the poles toward the equator. At higher levels there is a general return flow of warm air from the equator toward the poles. See Figure 20-12.

The atmosphere of the earth would probably move as just described if the earth did not rotate and if it were always heated equally on all sides. But the earth does rotate and its shape causes it to be heated unequally. These factors have a strong influence on the circulation of the atmosphere. The general movement of air from the poles toward the equator and its return because of convection results in the planetary circulation of winds.

Effects of the earth's rotation on winds. Everything that moves over the surface of the earth is affected by the rotation of the earth on its axis. To illustrate this, consider what would happen to a rocket fired from the North Pole and aimed toward New York. As the rocket moved south, the earth beneath it would be turning to the east. Since the earth's rotation carries points nearer the equator around faster than points near the poles, the earth would turn faster under the rocket as it moved south. If the rocket took about one hour for its flight, the earth's rotation would cause it to finally land near Chicago. The rocket would appear to have been pushed off to its right during the journey. See Figure 20-13.

All objects moving over the earth's surface will veer off to the right in the Northern Hemisphere and to the left in the Southern Hemisphere. This motion is called the *Coriolis effect,* after the nineteenth-century French mathematician who first described it. A baseball when thrown will curve only slightly due to the Coriolis effect. The ball is usually in the air much too short a time for it to be affected. The world's ocean currents and winds, however, are strongly affected. Winds, instead of blowing only in a direction from high toward lower pressure are turned by the Coriolis effect. This tends to divide the general north-south movement of air over the entire earth into several smaller parts. Within each division the winds generally move in a set direction, determined mainly by pressure differences and the Coriolis effect.

FIG. 20-13. A rocket fired from the North Pole and aimed directly at New York would land near Chicago because of deflection due to the Coriolis effect.

The earth's wind patterns. Near the equator, winds are controlled mainly by the lifting action of heated air. This is a region of calms, and weak undependable winds near the surface. The motion of the air in this region of low pressure is mostly upward. Because of the general lack of surface winds, the equatorial regions are known as the *doldrums*. The name originated in the days when low pressure and little wind often becalmed sailing ships for long periods.

As warm equatorial air rises and begins to move toward the poles, the Coriolis effect deflects it (to the east in the Northern Hemisphere). Around 30° latitude, the air begins to sink again toward the earth's surface. Part of it descends at these latitudes, forming a belt of high pressure. At the surface the settling air turns and flows both to the north and south. The portion moving back toward the equator creates surface winds called *trade winds*. In the Northern Hemisphere, the trade winds are deflected to the right and become the *northeast trades* because they flow from the northeast. In the Southern Hemisphere they are turned to the left and become the *southeast trades*.

The belt of high pressure mentioned above, which is created by descending air in the vicinity of 30° latitude, is called the *subtropical high*. The surface winds here are weak and changeable. The name "horse latitudes" was given to this region in the days when sailing ships carried horses from Europe to the New World. Horses often had to be thrown overboard to save food and water.

Some of the descending air in the subtropical highs that moves toward the poles is also deflected. It forms a belt of surface winds known as the *westerlies*. In the Northern Hemisphere the westerlies are southwest winds; in the Southern Hemisphere they are northwest winds. The westerlies are located in a belt between 40° and 60° latitude. By comparison they are much less steady than the trade winds.

North and south of the belt of westerlies around latitude 65°, there is another belt of low pressure. The *subpolar lows* are caused by the lifting of warmer air by cold polar air moving toward the equator. Over the polar regions themselves cold sinking air creates areas of high pressure. Within the cold polar air masses the general surface movement is toward the equator. Winds created by this movement are deflected by the Coriolis effect, becoming the *polar easterlies*. Because of the uniformly low temperatures in the polar regions and the resulting lack of

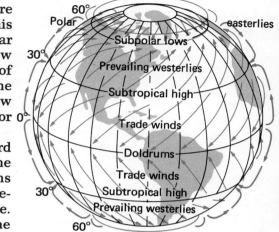

FIG. 20–14. The wind and pressure belts of the earth. Note how wind direction is deflected from straight north and south by the Coriolis effect.

pressure differences, these are usually very weak winds. Wind and pressure belts for the entire earth are shown in Figure 20–14.

As the sun's vertical rays shift north and south of the equator during the year, the positions of the pressure belts and wind belts also shift. The amount of change in their latitude is much less than the 23½° movement of the sun's vertical rays. The average shift for the pressure and wind belts during the year is about 6° of latitude. However, even this relatively small change is sufficient to cause some places to be in different wind belts during the year. This change may greatly affect the climate of an area, as is the case in southern California, where westerly winds prevail in the winter, and trade winds are dominant in the summer.

VOCABULARY REVIEW

Match the word or words in the column on the right with the correct phrase in the column on the left. *Do not write in this book.*

1. The study of the atmosphere.
2. An instrument which measures atmospheric pressure.
3. The process by which the atmosphere traps the sun's heat.
4. The atmospheric zone nearest the earth's surface.
5. The atmospheric zone in which the temperature increases slowly with height.
6. The atmospheric layer just above the stratosphere.
7. The atmospheric layer named for its high temperatures.
8. The region of the earth's atmosphere which blends into the vacuum of space.
9. The transfer of heat by movement due to uneven heating.
10. The apparent push to the right as objects move in the Northern Hemisphere.
11. A name often given to the equatorial regions because of their general lack of surface winds.
12. Winds moving toward the equator from the subtropical high.
13. Winds moving toward the poles from the subtropical high.
14. A belt in which warmer air is lifted by cold polar air, around latitude 65°.
15. The cold polar air masses moving toward the subpolar low.

a. greenhouse effect
b. stratosphere
c. thermosphere
d. barometer
e. exosphere
f. meteorology
g. mesosphere
h. troposphere
i. albedo
j. doldrums
k. convection
l. polar easterlies
m. trade winds
n. conduction
o. Coriolis effect
p. westerlies
q. subpolar lows

QUESTIONS

Group A

Select the best term to complete the following statements. *Do not write in this book.*

1. Which of the following gases occurs in the atmosphere in the greatest amount? (a) oxygen (b) nitrogen (c) argon (d) carbon dioxide.

2. The most important and usually most abundant of the variable gases in the lower atmosphere is (a) ozone (b) oxygen (c) water vapor (d) argon.

3. Which of the following is closest to the number of pounds of air which press down upon an average person at sea level? (a) 10 (b) 10,000 (c) 20,000 (d) 100,000.

4. The height of the mercury column in a mercury barometer is determined by the (a) atmospheric pressure (b) surface area of the mercury (c) density of water (d) amount of mercury used.

5. The height of a column of mercury which will be balanced by average air pressure at sea level is (a) 76 mm (b) 30 cm (c) 760 mm (d) 760 cm.

6. Aneroid barometers can be made to read any of the following except (a) millimeters of mercury (b) inches of mercury (c) feet of altitude (d) water vapor pressure.

7. An aneroid barometer can be used as an altimeter because at higher altitudes (a) air pressure is greater (b) there is less water vapor (c) the temperature is less (d) air pressure is less.

8. The amount of oxygen in the atmosphere remains relatively constant because oxygen is given off by (a) rocks (b) animals (c) plants (d) the ocean.

9. Plants take nitrogen from the air. It is returned by (a) decaying plants and dead animals (b) breath of animals (c) plants giving it off (d) water evaporating.

10. Compared to the amount of carbon dioxide in the atmosphere, the amount in the oceans is (a) greater (b) less (c) about the same (d) unknown.

11. The danger of air pollution becomes more serious when there is (a) a storm (b) low pressure (c) a clear sky (d) a temperature inversion.

12. In passing through the upper atmosphere, the electromagnetic waves from the sun lose most of their (a) long wavelengths (b) visible wavelengths (c) short wavelengths (d) infrared waves.

13. A clear sky is blue because (a) red light is scattered more than blue (b) blue light is scattered more than red (c) ultra-violet light has been absorbed (d) infrared light has been absorbed.

14. Of the total solar energy which reaches the earth's atmosphere about what percent actually reaches the earth's surface? (a) 19 (b) 35 (c) 54 (d) 46

15. The albedo of the earth is about how many times that of the moon? (a) .07 (b) .2 (c) .35 (d) 5

16. After being heated by solar infrared rays, the materials at the earth's surface produce infrared waves whose wavelengths are (a) longer (b) shorter (c) about the same (d) near ultraviolet.

17. The infrared waves produced by the warmed materials of the earth's surface are absorbed in the atmosphere chiefly by (a) water vapor and oxygen (b) carbon dioxide and oxygen (c) water vapor and carbon dioxide (d) oxygen and nitrogen.

18. Atmospheric pressure changes with altitude. This change in pressure is (a) uniform without sudden change (b) uniform but with several sudden changes (c) uniform but with one sudden change (d) not uniform and not predictable.

19. About half of the total weight of the atmosphere is within what distance of the earth's surface? (a) 5600 mm (b) 5600 mi (c) 5600 m (d) 5600 km.

20. The boundary between the troposphere and the stratosphere is called the (a) ionosphere (b) mesopause (c) stratopause (d) tropopause.

21. The upper mesosphere and the thermosphere are often called the (a) ionosphere (b) mesosphere (c) thermopause (d) exosphere.

Group B

1. List the layers of the atmosphere and their boundaries in proper order.

2. Why is the sky blue?

3. Explain the "greenhouse effect."

4. One inch equals 2.54 centimeters. A barometer reading of 30 inches would equal how many centimeters?

5. Describe the nitrogen cycle.

6. Describe the carbon dioxide cycle.

7. Describe how a mercury barometer measures air pressure.

8. How does a temperature inversion increase the problem of air pollution?

9. What are the principal gases in the atmosphere?

10. Explain the three ways that heat is transferred to the atmosphere.

11. Of what value is a barograph over a barometer?

12. Name and give the approximate locations for the earth's wind belts.

13. Describe the way solar energy is distributed after it enters the earth's atmosphere.

14. Why do surface winds generally blow toward the equator and away from the polar regions?

15. Explain the Coriolis effect.

16. How does the tilt of the earth's axis affect the planetary winds?

If all the water vapor contained in the earth's atmosphere were to fall suddenly as a worldwide rain, it would add a layer of water only about one inch deep to the earth's surface. It may be somewhat surprising, but water vapor is not one of the most abundant materials in the atmosphere. It is never present in large amounts, yet it is one of the most important components in the atmosphere.

If the earth's atmosphere did not contain water vapor, the greenhouse effect would be much less efficient and the earth would then be a much colder planet. Weather changes that carry life-giving moisture over the land depend upon water vapor in the air. With an atmosphere void of water, this planet's surface would probably be a barren desert swept by clouds of dust and exposed each night to bitter cold. Thus even the small amount of water vapor present is vital to our environment.

Driven by the energy of the sun, billions of tons of water each year pass into the atmosphere as vapor. Tracing the action of this water vapor in the air is the key to understanding the ever-changing weather.

ATMOSPHERIC MOISTURE

How water vapor enters the air. Water can exist as a solid, liquid, or gas (vapor). The difference between water in these three phases is in the amount of energy contained by the water molecules. For example, to change water from liquid to gas, enough energy must be added to overcome the forces holding them together. See Figure 21–1. Then the molecules are able to separate and move around independently as a gas. For 1 gram of liquid water (at

objectives

- ☐ Explain how water vapor enters the atmosphere and how it is measured.

- ☐ Explain how water leaves the atmosphere.

- ☐ Describe cloud formation.

- ☐ List and describe the various types of clouds.

- ☐ Describe several causes of precipitation.

- ☐ Describe the conditions needed for the various types of precipitation to occur.

- ☐ Describe several methods used to measure precipitation.

439

FIG. 21-1. Solar energy which causes water to evaporate from the sea surface supplies most of the moisture found in the air. (Photo Researchers—Granitas)

Investigate

Find out more about the process of transpiration and design an experiment that will show that a potted plant carries on transpiration. Report your findings.

40°C) to evaporate, about 600 calories of heat energy must be added. A *calorie* (or gram-calorie) is the amount of heat required to raise the temperature of 1 gram of water through 1°C at near freezing temperatures. It is important to keep in mind that when water vapor enters the atmosphere heat energy is carried along with it. This is called *latent energy* and can be released to the air if the water vapor changes back to liquid water.

Millions of tons of water evaporate each day from the surface of the sea. Because of energy absorbed from the sun, separate molecules of pure water pass into the air. Most of this evaporation takes place in the regions around the equator where the largest amounts of solar energy are received. While the sea is the principal source of atmospheric moisture, evaporation from lakes, ponds, streams and soil also supplies some water vapor to the atmosphere. Plants also give off water vapor as they carry on a part of their life process called *transpiration*. A small amount of moisture enters the air from volcanoes and from burning of fuels.

Measurement of atmospheric moisture. The amount of water vapor in the air is referred to as its *humidity*. If a sample of air is passed over a certain chemical, its moisture will be absorbed and can then be weighed. This kind of investigation would show that the total weight of water vapor in a certain volume of air is relatively small. For example, a cubic meter of saturated air (35.3 cubic feet) at 10°C will contain only about 9 grams of water vapor. This direct measurement of the amount of water vapor in a certain volume of air is said to express its *absolute humidity.*

Further investigation would show that the capacity of air to hold water vapor changes with its temperature. A given volume of warm air holds more moisture than the same volume of cool air. This means that the maximum absolute humidity of a certain volume of air increases with temperature. The graph in Figure 21–2 shows the maximum absolute humidity of air in relation to temperature changes. The amount of water vapor an amount of air can hold at any particular temperature is called its *saturation value.*

The humidity of air is often expressed by comparing its absolute humidity to its saturation value. Suppose a cubic meter of air at a temperature of 21°C is found to have an absolute humidity of 13.9 grams per cubic meter (13.9 g/m³). The saturation value for air at 21°C is about 18.5 g/m³. Then,

$$\frac{13.9 \text{ g/m}^3}{18.5 \text{ g/m}^3} = 0.75 \text{ or } 75\%$$

The formula states that air must contain 75% of the moisture it can possibly hold at that temperature. This method of stating the moisture content of the air is called the *relative humidity.* Relative humidity is always given as a percent. It shows the amount of moisture in the air (absolute humidity) compared to the amount which could be contained at the given temperature (saturation value). Notice that relative humidity must change if the air temperature changes and no additional water vapor enters the air. As the temperature increases, relative humidity decreases, if the moisture content remains the same. This means that on most days relative humidity is highest in the cool hours of night and morning and lowest in the afternoon.

The actual amount of water vapor in the air is difficult to measure accurately. The only way absolute humidity can be measured directly is by trapping the water with a chemical drying agent, then weighing the water collected from a known volume of air. Generally this is not practical because it takes too much time and complicated equipment. A less accurate but more convenient method for

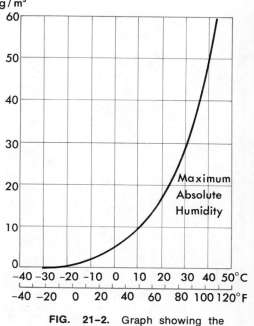

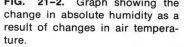

FIG. 21–2. Graph showing the change in absolute humidity as a result of changes in air temperature.

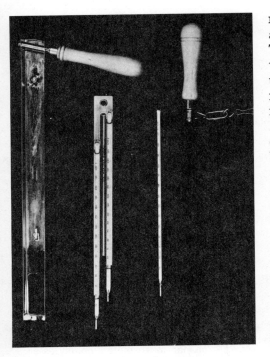

FIG. 21-3. A psychrometer is an instrument used to measure humidity. The wick of the thermometer shown at the left is wet. Evaporation of moisture from the wick cools the thermometer bulb. (U.S. Weather Bureau)

activity

Water vapor condenses on the outside of a glass of ice water because the temperature of the air that comes in contact with the glass is lowered below its dew point. The following technique can be used to measure dew point:

a. Fill a shiny tin can or plastic glass half-full of water.

b. Place an ice cube in the water and stir carefully with a thermometer.

c. When dew first appears, record the temperature. Immediately remove all the ice from the water. (See the photo, top of page 443.)

measuring humidity makes use of two thermometers in an instrument called a *psychrometer*. See Figure 21-3. The bulb of one thermometer is covered with a damp wick while the other remains dry. The wet-bulb thermometer usually gives a lower temperature reading because evaporating water has a cooling effect. When the air is dry, rapid evaporation cools the wet-bulb thermometer faster. Its temperature reading, therefore, differs most from the dry-bulb thermometer when the relative humidity is low. When the air is very moist, there is little evaporation and the two thermometers indicate about the same temperature. Using tables or a special slide rule, the temperature difference between the two thermometers can be translated into approximate relative humidity.

Another type of humidity measuring instrument is based on the fact that human hair stretches when the moisture of air increases. A piece of human hair will stretch about 2½ percent when the relative humidity increases from 0 to 100 percent. The *hair hygrometer* is an instrument which records the changing length of a bundle of hairs when humidity changes. The variation in length moves a pointer on a scale or a pen on a graph to show changes in relative humidity.

To measure humidity at high altitudes an electrical instrument connected to a balloon is used. This measuring device is triggered by pressing an electrical current through a moisture-collecting chemical substance. The amount of moisture collected determines the electrical conductivity. This can then be measured and expressed as the humidity of the surrounding air.

CONDENSATION OF WATER VAPOR

How water leaves the air. A water molecule will remain a part of the water vapor in the air as long as it maintains its supply of latent energy. However, if the air is cooled, its water vapor molecules may release this energy, forcing them to join together. This means that when the temperature of the air is lowered, the capacity of the air for holding water is reduced. If the temperature continues to drop, a point is reached at which the amount of water vapor the air holds is equal to its capacity. At this point the air is said to be saturated and the relative humidity is 100 percent. The temperature to which air must be cooled to reach this level of relative humidity is called its *dew point*. At

any temperature lower than the dew point water vapor may begin to change to a liquid or solid in the process called *condensation.*

Dew and frost. On almost any clear autumn night, radiation of heat from grass and leaves close to the ground is relatively rapid, and the temperature of their surfaces soon reaches the dew point of the surrounding air. The resulting form of condensation called *dew* is seen as tiny water droplets on many surfaces which have become cooled by radiation. See Figure 21–4.

If the dew point falls below the freezing temperature of water (0°C), water vapor will condense directly into solid ice crystals to make *frost.* Since frost forms directly from water vapor without first becoming liquid, it is not frozen dew. Frozen dew, which is relatively uncommon, forms as clear beads of ice. Orchards and valuable crops must be protected from frost, since it will kill growing plants. One method of frost protection uses heaters or blowers that circulate the cold air.

Adiabatic cooling. Air may be cooled by two other processes besides radiation. If a quantity of air mixes with a mass of colder air, a drop in temperature will result. A second cooling process is caused by the upward movement of air. When air rises, it expands because pressure is decreased. A gas such as air is cooled when it expands causing collisions between its molecules to take place less often. Less frequent collisions mean less energy and lower temperatures for the individual molecules. Thus, a force that causes air to rise will also cause it to expand and become cooler. Downward movement of air has the opposite effect.

This cooling and heating effect of air can be seen with the kind of hand pump often used to fill bicycle tires. Pushing the pump handle down compresses the air which releases enough heat to make the barrel of the pump quite hot. Yet the escaping compressed air feels cool. This is because it expands rapidly as it escapes. Temperature changes that take place in air with no addition or withdrawal of heat from the outside are called *adiabatic changes.*

Dry air registers an adiabatic temperature change of about 1°C for every 100 meters (5.5°F for each 1000 ft) of altitude change. This means that *rising* air will show a decrease in temperature of about 1°C for every 100 meters

d. Continue to stir until the dew disappears. Record the temperature of the water at this time.

The two temperatures you have recorded should not differ by more than a few degrees. If they do, it would be advisable to repeat the above procedure until they differ by only a few degrees. The average of these two temperatures is the dew point.

FIG. 21–4. Dew forms on objects that are cold enough to lower the temperature of the surrounding air below the dew point. (Allan Roberts)

FIG. 21–5. Cumulus clouds form when rising air is cooled. (Ramsey)

FIG. 21–6. Banner clouds often form around mountain tops. (Ramsey)

it rises. The temperature of *sinking* air will increase by an equal amount for every 100 meters it descends. But the adiabatic rate of 1°C/100 meters applies only to dry air. If there is water vapor in the air, there will also be absorption or release of heat due to evaporation or condensation. For air in which water is changing its phase, the temperature change for either rising or sinking air is about 0.6°C per 100 meters.

Cloud formation. Adiabatic changes are an important factor in cloud formation. These changes provide a simple method for the necessary cooling of air to its dew point. Air rises mainly because of temperature differences. A body of air that becomes warmer than the surrounding air also becomes lighter and rises. For example, on sunny days a small region of air near the ground may receive more heat than the surrounding air. This will happen if the heated air is spread over a part of the ground that reflects more of the sun's energy. The heated parcel of air will rise and undergo adiabatic cooling until the dew point is reached. Then its moisture will condense to form a cloud. See Figure 21–5.

Air is also cooled as it is lifted in its passage over mountains. Entire mountaintops are often covered with clouds formed in this way. See Figure 21–6. Large areas of clouds connected with storms are formed when a mass of warm air is pushed up over a heavier body of cooler air. This type of action will be taken up in Chapter 22.

A cloud can only be formed when a body of air is cooled below its dew point. However, cooling of moisture-laden air below its dew point is not the only requirement a forming cloud must meet. The individual water molecules which make up the water vapor in the air must clump together. Only in this way can the gaseous water in the air form the droplets of liquid water that usually make up a cloud. This means that the rapidly moving gaseous water molecules must collide with other water molecules and collect until a droplet of liquid water has accumulated. This process cannot take place until the temperature has dropped low enough so that the water molecules will stay clumped together.

After the air is swept clean of all foreign particles such as dust, it can be cooled until the relative humidity is much greater than 100 percent. Only then do cloud droplets begin to appear. Air with a relative humidity above 100 percent is said to be *supersaturated*. Only clean air

Cirrus (UNESCO)

Altocumulus (UNESCO)

Cirrocumulus (UNESCO)

Stratus (UNESCO)

Cirrostratus

Nimbostratus

can be supersaturated with water vapor and still not contain cloud droplets.

Usually the air is filled with many small particles which prevent supersaturation. These particles include dust, salt particles from the sea, smoke, and perhaps particles showered on the earth by disturbances on the sun. Each particle can serve as a nucleus for condensation of water vapor. These particles are called *condensation nuclei.* As soon as the dew point is reached, water molecules begin to collect on the various condensation nuclei present in air and cloud droplets are produced. If the dew point is below the freezing temperature of water, an ice crystal will form. Each separate cloud droplet or ice crystal usually contains the particular condensation nucleus around which it grew. Because air normally contains large numbers of condensation nuclei, clouds are usually formed when the air is cooled to its dew point. For this reason supersaturation of air with water vapor does not often occur naturally.

Types of clouds. The most convenient method of classifying clouds is based on altitude. The major cloud types and their principal subdividions are as follows:

1. *High clouds.* The base of these clouds is at an altitude well above 7 km. They are generally thin and composed of ice crystals. Through them the outline of the sun or moon can be seen. Their principal forms are:
   a. *Cirrus.* Thin, featherlike with a delicate appearance, frequently arranged in bands across the sky; sometimes called "mare's tails."
   b. *Cirrocumulus.* Like patches of fluffy cotton or a mass of small white flakes, frequently in groups or lines; sometimes called "mackerel sky."
   c. *Cirrostratus.* Whitish layers, like a sheet or veil, giving the sky a milky appearance. They often produce a halo or ring around the sun or moon. This is a result of the bending of light by ice crystals in the cloud.

2. *Middle clouds.* These clouds range from 2 to 7 km. Their principal forms are:
   a. *Altocumulus.* White or gray patches, or layers of clouds having a rounded appearance.
   b. *Altostratus.* Gray to bluish colored layers, often with a streaked appearance.

3. *Low clouds.* The bases of these clouds range from near the surface to about 2 km.

Investigate
Call your local weather station every day at the same time for a week, to find out the cloud types and heights. Try to relate this information to the kind of precipitation that may be occurring.

Cumulus (UNESCO)

Cumulonimbus (UNESCO)

High clouds 　Cirrus (Ci) 　Cirrocumulus (Cc) 　Cirrostratus (Cs)	12 km. 10 km. 8 km.
Middle clouds 　Altocumulus (Ac) 　Altostratus (As)	6 km. 5 km. 3 km.
Low clouds 　Stratocumulus (Sc) 　Stratus (St) 　Nimbostratus (Ns)	1.5 km. Altitude

Clouds with vertical development
　Cumulus (Cu)
　Cumulonimbus (Cb)

FIG. 21-7A. Classification of cloud types and their elevations.

Altostratus (Keith Gunnar/Bruce Coleman)

Stratocumulus (UNESCO)

FIG. 21–7B. A typical upslope fog caused by adiabatic cooling in a glacial valley. (Photo Researchers—Mappes)

a. *Stratus.* Low, uniform, sheetlike clouds similar to fog but not resting on the ground.

b. *Stratocumulus.* Large rounded clouds with a soft appearance, usually arranged in some pattern with spaces in between.

c. *Nimbostratus.* Low, shapeless thick layers, dark gray in color. They usually bring rain or snow.

4. ***Clouds with vertical development.*** These clouds extend from a lower level of about 2 km to a maximum of over 7 km.

a. *Cumulus.* Thick, dome-shaped clouds, usually with flat bases and many rounded projections reaching out from the upper parts. Cumulus clouds are often widely separated from one another.

b. *Cumulonimbus.* Large, thick, towering clouds with cauliflower-like tops; often crowned with veils of thick cirrus giving the entire cloud a flat top. These are the thunderhead clouds, frequently associated with violent weather.

The various cloud types and forms are summarized in Figure 21–7A.

Fog. Like clouds, fog is the result of condensation of water vapor in the air. The chief difference is that fog forms when air is cooled by some means other than lifting. This generally occurs by contact with a cool surface. For example, one type of fog results from the nightly cooling of the earth. The layer of air in contact with the earth becomes chilled below its dew point and condensation of water droplets occurs. This type of fog is called a *radiation fog* or ground fog because it is caused by the radiation heat lost by the earth. Radiation fogs form most often on calm, clear nights. The fog is thickest in valleys and low places because the dense, cold air in which the fog forms drains to the lower elevations. Fogs are often unusually thick around cities because of the greater amount of smoke and dust particles which act as condensation nuclei.

Another common condition which produces fog is the movement of warm, moist air over cold surfaces. A fog produced in this way is called an *advection fog,* referring to the horizontal air movements. Advection fog is very common along seacoasts as the warm moist air from the water moves in over the cooler land surface. Dense fogs may form on the sea when warm, moist air is carried over cold ocean currents.

A third type of fog, called an *upslope fog,* is formed by the adiabatic cooling of air as it sweeps up rising land slopes. This is really a kind of cloud formation at ground level. See Figure 21–7B.

PRECIPITATION OF MOISTURE

Causes of precipitation. The process of condensation cannot go beyond a certain point without causing precipitation. A cloud produces precipitation when its droplets or ice crystals become large enough to fall as rain or snow. Most cloud droplets have a radius of about 10 microns (1 micron = 0.00004 in). Droplets of this size easily remain suspended in the air. Even slight air movements prevent them from settling out. Before it will fall as a raindrop, a cloud droplet must grow about 100 times larger in radius. See Figure 21–8.

Cloud droplets seem to reach the precipitation stage in two ways. The first involves differences in size between separate cloud droplets. To a large extent the original size of a cloud droplet depends on the kind and size of its condensation nucleus. Large nuclei tend to form large cloud droplets. The size of various cloud droplets is generally matched to those of the condensation nuclei. The larger droplets do not remain suspended in the cloud as well as the smaller ones. Instead they drift downward, forcing the larger droplets to collide and combine with the smaller ones. This process is called *coalescence* (koh-uh-*les*-ens). See Figure 21–9. The larger droplets continue to grow by coalescence until they contain several million times as

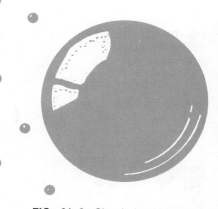

FIG. 21–8. Cloud droplets, compared with a growing raindrop.

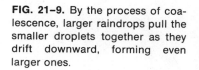

FIG. 21–9. By the process of coalescence, larger raindrops pull the smaller droplets together as they drift downward, forming even larger ones.

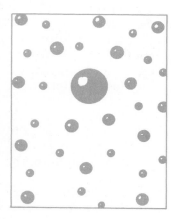
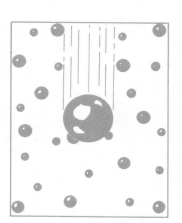

FIG. 21–10. The breaks in these clouds were produced by seeding with pulverized dry ice. (U.S. Weather Bureau)

FIG. 21–11. Types of ice crystals. (a) flat plate, (b) six-sided crystal, (c) needle-like. (Courtesy of Carl Zeiss, Inc., N.Y.)

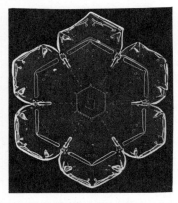

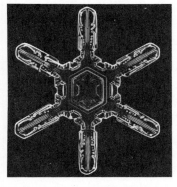

much water as a single cloud droplet. By this time, their weight is such that they fall as raindrops.

A second way that precipitation develops is related to the presence of ice crystals in a cloud. Each microscopic ice crystal grows by capturing and freezing water which evaporates from its neighboring water droplets. In this way the ice crystal grows large enough to fall. The falling ice crystal may melt in passage and reach the earth as a raindrop, or it may fall as a snowflake.

Knowledge of the way in which ice crystals grow in clouds has made it possible to take the first steps toward controlling the weather. This is done by scattering tiny crystals in a cloud with the correct temperature and moisture conditions. The artificial crystals most used are the compound silver iodide. Crystals of this compound resemble ice crystals and are able to act as "seeds" for the formation of ice. Silver iodide can be released into clouds by special rockets fired into the cloud. Flares dropped from aircraft which release silver iodide can also be used. Powdered dry ice has also been used successfully as seeds. These techniques for causing clouds to form precipitation are usually called "cloud seeding." See Figure 21–10.

The seeding of clouds under certain temperature conditions may cause them to release more water than they would naturally without being seeded. Other temperature conditions in seeded clouds can produce less precipitation. Thus cloud seeding seems to be a way of getting some control over the amount of precipitation coming from clouds under the right conditions. Cloud seeding may be a way to make droughts less of a problem as well as to help control floods by reducing snowfall when heavy runoff might cause flooding.

Seeding of clouds may also offer the possibility of reducing lightning which starts many forest fires. How clouds produce lightning is not understood very well but one theory requires the presence of ice crystals in the correct amounts. Cloud seeding might be able to change the production of these crystals and thus reduce lightning. It also may be possible to change hurricanes by cloud seeding. This could be done by causing heavier precipitation in the storm's early stages. Loss of this moisture would remove some of the cause of the strong winds and reduce the strength of the storm.

Increasing ability to control the weather will bring new problems as well as the solutions to some old ones. It is difficult to get everyone at a particular place to agree about the amount of precipitation most helpful to all. Generous amounts of rain or snow, or the lack of it, might please farmers while causing trouble for people with other interests. Weather control also raises serious questions about the future effects on the natural environment. It will be necessary to move very carefully in changing weather patterns in order to avoid upsetting the natural plant growth and stream flow of the land.

FIG. 21–12A. Glaze ice resulting from ice storms may do tremendous damage in woods and forests. (E. R. Degginger)

FIG. 21–12B. Hailstones covering a cornfield. (J. C. Allen)

Other types of precipitation. The only other type of liquid precipitation is *drizzle*. This occurs when cloud or fog droplets fall to earth because the air is very still. All other forms of precipitation are in solid form.

The most common solid precipitation is, of course, snow. Snow will fall if the ice crystals fail to melt before they reach the ground. These ice crystals come in many interesting shapes. There are three main types: needles, plates, and branching crystals. See Figure 21–11. Temperature and humidity conditions control the type of ice crystal which forms. Lowest temperatures favor the growth of needles and plates, while branching crystals grow at higher temperatures. Naturally, temperatures must be below freezing in the cloud for snow to be produced.

However, if temperatures near the ground are above freezing, some of the falling ice crystals are likely to melt. Under the resulting film of water, clumps of them will stick together, forming large snowflakes. Colder temperatures near the ground produce a hard, fine snow, since the individual ice crystals remain separate.

Extreme temperatures near the ground may sometimes produce solid precipitation from raindrops. The ice pellets which are formed when rain falls through a layer of freezing air are called *sleet*. Occasionally rain does not

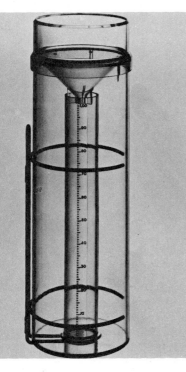

FIG. 21-13. A simple but accurate rain gauge. (Taylor Instrument Co.)

FIG. 21-14. A tipping bucket rain gauge with the tipping buckets visible through the open door. (Bendix Aviation Corporation)

freeze until it actually strikes the ground. In this event it forms a thick and very destructive layer of *glaze ice* over everything. Conditions which produce glaze ice occur in what is generally called an ice storm. See Figure 21–12.

Hail is also produced when raindrops are frozen into ice pellets. Unlike sleet however, hail is formed by layers of ice. Rain freezes, then falls through the warm air again, accumulating a layer of water. Then the hail is carried up again by strong updrafts into freezing air, or falls back through another layer of cold air. The water coating freezes and another layer of ice is formed. If the process is repeated a number of times, the hailstones may accumulate many layers of ice and grow quite large. These large hailstones cause great damage.

Measurement of precipitation. A *rain gauge* consists of a container which measures the depth of the rain water it collects. In one type of rain gauge a wide-mouthed funnel catches the rain and empties it into a cylindrical container below, as shown in Figure 21–13. The mouth of the funnel is much larger than the measuring container beneath it. This magnification of the actual amount of rain makes it possible to accurately measure the rainfall.

In another type of rain gauge, a small divided bucket catches the water from the funnel, fills on one side, then tips, dumping the water and allowing the other half of the small bucket to fill. Each time one side of the tipping bucket fills with exactly $1/100$ inch of rain, it tips and sets off an electrical device that records the amount. The rainwater dumped from the tipping bucket is collected and weighed as a means of checking the accuracy of the record. Another type of rain gauge catches the water in a large bucket which is weighed continuously. The weight is recorded directly on a graph as inches of rain. See Figure 21–14.

Rain gauges can also be used to measure snowfall. The collected snow is melted and the water is weighed to determine its rain equivalent. A light fluffy snow may yield one inch of rain for every 15 inches of snow; a dense well-packed snowfall may produce one inch of rain for every six inches of snow.

Because rain gauges measure only the precipitation falling in one spot, it is difficult to establish the total amount of rain or snow which falls over a large area. The amount of precipitation may differ by large amounts for even short distances. This is especially true in mountainous regions.

VOCABULARY REVIEW

Match the word or words in the column on the right with the correct phrase in the column on the left. *Do not write in this book.*

1. The conversion of a vapor into a liquid or a solid.
2. Ice pellets formed when rain falls through a layer of freezing air.
3. A device used to measure relative humidity.
4. The way large water droplets become larger by adding smaller droplets.
5. Air with a relative humidity above 100%.
6. The measurement of the weight of water vapor in a certain amount of air.
7. Comparison of the absolute humidity of air to its saturation value and expressed as a percentage.
8. Temperature to which air must be cooled to raise its relative humidity to 100%.
9. The only form of liquid precipitation other than rain.
10. Ice pellets having a number of layers of ice.
11. A very high cloud composed of ice crystals.
12. Low clouds which usually bring rain or snow.
13. A cloud belonging to the "middle" group.
14. Clouds which have great vertical development.
15. The weight of water vapor an amount of air can hold at any given temperature.

a. psychrometer
b. supersaturated
c. drizzle
d. cirrostratus
e. cumulus
f. sleet
g. hail
h. altostratus
i. nimbostratus
j. calorie
k. saturation value
l. absolute humidity
m. ground fog
n. condensation
o. dew point
p. coalescence
q. relative humidity

QUESTIONS

Group A
Select the best term to complete the following statements. *Do not write in this book.*

1. To change liquid water to solid water (a) energy must be added (b) the temperature must increase (c) energy must be taken away (d) the dew point must be reached.

2. When water vapor enters the air (a) energy is released (b) energy is carried along (c) the relative humidity decreases (d) the saturation value changes.

3. The principal source of moisture in the air is (a) evaporation from lakes, ponds, streams, and soil (b) living plants and animals (c) volcanoes and burning fuel (d) evaporation from oceans.

4. The major source of energy causing moisture to enter the air is (a) the sun (b) the earth (c) volcanoes (d) living substances.

5. A humidity reading of 75% was obtained from an instrument. This is a measurement of (a) relative humidity (b) absolute humidity (c) saturation value (d) dew point.

6. The graph in Figure 21–2 shows the relationship to be (a) an increase in temperature decreases the amount of moisture a given volume of air will hold (b) an in-

crease in volume increases the amount of moisture a given volume of air will hold (c) a decrease in volume increases the amount of moisture a given volume of air will hold (d) an increase in temperature increases the amount of moisture a given volume of air will hold.

7. The most accurate way to measure humidity is to (a) use a psychrometer (b) trap the water in a given volume of air with a chemical drying agent (c) use a human hair hygrometer (d) measure the electrical current which flows through a chemical substance which collects moisture.

8. The temperature to which air must be cooled to give a relative humidity of 100% is called its (a) saturation value (b) condensation point (c) dew point (d) absolute humidity.

9. If the temperature of the air is below the freezing temperature of water (a) the dew point cannot be reached (b) when the dew point is reached, the moisture will condense into a liquid (c) the moisture in the air will condense into ice (d) frost will form when the dew point is reached.

10. Frost (a) does not actually cause plants to die (b) kills growing plants (c) forms only above the dew point (d) is the result of dew freezing.

11. Which one of the following would be *least* likely to cause moisture to condense from the air? (a) radiation of heat at night (b) mixing of warm air with cold (c) downward motion of air (d) upward motion of air.

12. An adiabatic change is one in which air temperature changes due to (a) radiation (b) condensation (c) mixing of warm and cool air (d) rising or sinking air.

13. A cloud may form when a body of air is (a) cooled below its freezing point (b) cooled below its dew point (c) mixed with warmer air (d) sinks 100 meters.

14. For each 100 meters a parcel of dry air rises, it (a) warms 1°C (b) cools 1°C (c) warms 5½°F (d) cools 0.6°C.

15. For clouds to form out of moist air, the air must (a) cool, with condensation nuclei present (b) rise and cool adiabatically until the dew point is reached (c) be cooled below its dew point (d) sink adiabatically until its dew point is reached.

16. The reason water droplets usually form in air when the relative humidity reaches 100% is that (a) the air contains dust, smoke, salt particles, etc. (b) the air can hold more moisture (c) the dew point has not been reached (d) heating is taking place.

17. The type of cloud which does not belong with the rest is the (a) cirrocumulus (b) cirrostratus (c) cirrus (d) stratocumulus.

18. The type of cloud which would be found at the lowest altitude is the (a) altocumulus (b) altostratus (c) cirrus (d) stratus.

19. Clouds composed of ice crystals share a common part to their name, which is (a) alto (b) cirro (c) strato (d) nimbo.

20. Low, shapeless, thick-layered, dark gray clouds usually bringing rain or snow are called (a) stratus (b) altostratus (c) nimbostratus (d) stratocumulus.

21. The "thunderhead" cloud belongs to which of the following cloud types? (a) cumulus (b) altocumulus (c) stratocumulus (d) cumulonimbus.

22. The kind of fog produced from the nightly cooling of the earth is called (a) sea fog (b) advection fog (c) radiation fog (d) night fog.

23. The kind of fog produced by horizontal motion of air is called (a) sea fog (b) advection fog (c) radiation fog (d) wind fog.

24. Before water droplets in clouds can fall as rain or snow, they must grow to about (a) 100 times their normal size (b) 10 microns (c) 0.000004 inches (d) 100 microns.

25. An example of coalescence in precipitation is (a) the formation of large droplets from one large condensation nucleus (b) ice crystal capture of moisture from air (c) seeding with silver iodide (d) the collision of two droplets to form one.

26. Methods of artificially producing rain by "cloud seeding" are based on our knowledge of how (a) clouds form (b) ice crystals grow (c) cloud droplets coalesce (d) fog is formed.

27. Which one of the following forms of precipitation is not like the others? (a) drizzle (b) snow (c) hail (d) sleet.

28. A type of precipitation which does not start in raindrop form is (a) sleet (b) glaze ice (c) hail (d) snow.

29. A tipping bucket measures liquid precipitation to the nearest (a) inch (b) centimeter (c) 0.1 centimeter (d) 0.01 inch.

30. The amount of rainfall which would be equal to 2.5 feet of fluffy snow would most likely be (a) 15 inches (b) 2 inches (c) 6 inches (d) 30 inches.

Group B

1. Water vapor is not very abundant in the atmosphere. If it were completely missing, what effect would this have on the earth's surface?

2. If you spilled 454 grams (approximately 1 pint) of water and it all evaporated into the air, how many calories of energy would it carry with it due to the evaporation?

3. Compare the role of plants, lakes and rivers, burning fuels, and seas as sources of the moisture in air.

4. One liter (one thousandth of a cubic meter) of air is passed over the chemical calcium chloride which absorbs all the moisture. It is found to be .01 gram of water. What is the absolute humidity of the sample of air?

5. The saturation value of a given sample of air is known to be 10 grams at room temperature. The absolute humidity is found to be 8.0 grams. What is the relative humidity of the air sample?

6. What effect does increasing the temperature of a sample of air have on each of the following? (a) absolute humidity (b) saturation value (c) relative humidity.

7. Describe three different methods for measuring relative humidity.

8. What are three ways that air is generally cooled?

9. A parcel of air decreased in temperature by 5°C (9°F). If this change was adiabatic, how high did the parcel rise?

10. What two conditions must be met before clouds will form?

11. Name the form(s) of clouds which would go with each of the following: (a) thunderhead (b) rain clouds (c) high fog (d) ring around the moon (d) mare's tails (f) mackerel sky.

12. Very early in the morning, a coastal city is enveloped in a dense fog. At the same time 100 miles inland, a farm valley is also packed with a dense fog. Explain how these two fogs most likely formed.

13. Describe two ways cloud particles may grow large enough to precipitate as rain or snow.

14. Describe a method of artificially producing precipitation from clouds and discuss the possible benefits and problems which might be encountered.

15. The tipping bucket rain gauge cannot usually be used to measure snowfall. Why?

16. How would you explain the fact that source regions for many violent storms originate over the oceans near the tropics?

17. The normal flow of air is sometimes reversed over certain areas of the continents, causing an unusual hot wind to blow. Such a wind is the "Santa Ana" in the southwestern region of the United States. The Santa Ana winds originate over the high desert areas and blow toward the coast. Using your knowledge of the geography of southern California and the characteristics of sinking air, explain why these winds sometimes result in temperatures soaring as high as 110°–115°F in such cities as Los Angeles and San Diego.

18. At present many cities are having problems with a condition in which smoke and a very light fog combine to cause "smog." Smog is most likely to occur when there is a layer of warm air at about 1,000 to 2,000 feet, with cooler air extending down to the ground. (a) Why would this contribute to the production of smog? (b) Why is this layer called an "inversion" layer?

Weather is the sum of all the properties of the atmosphere — temperature, pressure, humidity, and winds — at any particular time. Of all the forces of nature, weather influences people's environment most. Its properties may have far reaching effects. Thunderstorms, blizzards, rain or snow, all affect our daily lives in one way or another.

Almost all weather changes are the result of the unequal heating of the earth's atmosphere. These temperature differences in the atmosphere play a key role in the earth's ever-changing weather patterns.

Many of the earth's changes in weather originate at high altitudes, about halfway between the equator and the poles. In this zone cold air moving down from the poles collides with warm air from the equatorial regions. Cold air clings close to the earth. It pushes in under the warm air, thrusting cold fingers toward the warm tropics. At the same time, warm air slides up over the cold and moves toward the poles. The collision of these bodies of warm and cold air produces storm centers. These are responsible for most of the weather experienced in the middle latitudes.

Scientists are only beginning to understand the origins of weather changes. Much remains a mystery. The shifting patterns of the weather are often hard to analyze. However, one important principle is clear as a result of careful study of weather patterns. This principle is the basis for understanding most of the weather of North America. It states that weather changes are largely brought about by movement of giant bodies or masses of air moving in response to the earth's general wind patterns.

objectives

- [ ] Describe the origin of air masses.
- [ ] Name the air masses that influence North American weather.
- [ ] Explain how weather fronts form.
- [ ] Describe the methods and instruments used to observe weather.
- [ ] Describe how to make a weather map and forecast weather.
- [ ] Explain how local weather phenomena develop.

AIR MASSES

Origin of air masses. A body of air covering millions of square miles of the earth's surface, with nearly the same temperature and humidity throughout at any particular altitude, is called an *air mass*. See Figure 22–1. An air mass is created when a large body of air remains relatively stationary over a body of land or water for some time. It then takes on the temperature and humidity characteristic of that region. For example, air remaining over the cold arctic plains of North America becomes very cold and dry. Over a tropical ocean a quiet body of air becomes warm and moist. A region that serves as a source of air masses must be fairly uniform, such as the arctic plains or the sea. Mixed land and water areas are not suitable. The source region must be free of continuous strong winds.

Air masses are classified according to their source region. A system of letters is used to designate the source and characteristics of the various air masses. The principal source regions are the cold polar (P) areas and the warm tropical (T) areas. Air masses are described as coming from the sea, maritime (m), and continental (c), if the source is over land. Naturally, maritime air masses tend to be moist, and continental air masses generally dry. It follows that there are four main types of air masses, described by the following letter combinations: *mP, mT, cP* and *cT*.

Once formed, an air mass moves away from its source region because of the general movements of the atmosphere. An air mass moving away from its source region is said to be *cold* if it is cooler than the surface over which it is moving. The temperature of a *warm* air mass is higher than the surface over which it passes. Several days are usually required for an air mass to move past an area. During this time, the region will have a period of *air mass weather*, meaning the weather usually remains the same for several days. When the air mass moves on, there is a change in the weather. The winds, clouds, and precipitation accompanying these weather changes are the result of the collision of the different air masses.

North American air masses. Air masses that strongly influence North American weather come from seven main source regions. The origins of these air masses and their general direction of movement are shown in Figure 22–2.

FIG. 22–1. A polar air mass is shown moving from Canada into the United States. A cross-section along the line AB shown above indicates the uniform conditions that prevail within the air mass.

Polar Canadian (cP) air masses are formed in northern Canada. They generally move in a southeasterly direction across Canada and into the northern United States. In the winter they create intense cold waves. Occasionally these cold air masses penetrate as far south as the coast of the Gulf of Mexico. In the summer, they usually bring cool, dry weather.

Polar Pacific (mP) air masses originate in the northern Pacific Ocean region. They are very moist and not extremely cold. To the Pacific Coast, the area most affected, they bring rain and snow. In summer, the polar Pacific air brings cool, often foggy, weather.

Polar Atlantic (mP) air masses are formed over the northern Atlantic Ocean. The general direction of their movement is eastward toward Europe. But they may pass over the northeastern portion of North America. In winter they usually bring cold, cloudy weather and light precipitation. In summer these air masses often produce cool weather with low clouds and fog.

Tropical Continental (cT) air masses affect North America only in the summer. They form over Mexico and the southwest United States and generally move northeast. They usually bring clear, dry, and very hot weather.

Tropical Gulf and Tropical Atlantic (mT) air masses form over the warm water of the Gulf of Mexico and the South Atlantic. They move north across the eastern United States. In the winter they bring mild and often cloudy weather. In the summer, they produce hot, very humid weather and frequent thunderstorms.

Tropical Pacific (mT) air masses collect over the warm parts of the Pacific Ocean and reach the Pacific Coast only rarely in the winter. On these infrequent visits they bring very heavy precipitation.

WEATHER FRONTS

Formation of a front. When two air masses meet, no general mixing of the air contained within the two masses takes place. Temperature differences keep the two air masses separate. The cooler air of one mass is denser and does not mix with the warmer air. Thus a definite boundary is usually formed between air masses. It is called a *front*. The colder of the two air masses is denser, and thus it will push under the warmer air and lift it. Therefore the front always slopes up over the cooler air. See Figure 22–3.

FIG. 22–2. The types of air masses that influence weather in North America. Arrows indicate the general direction in which these air masses usually move.

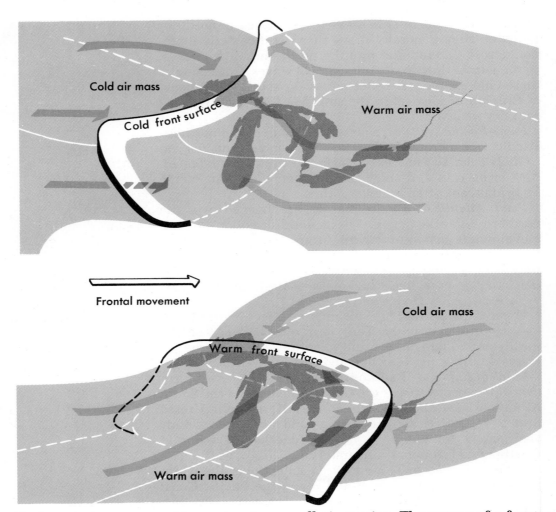

FIG. 22-3. A side-view of air masses along a cold and a warm front. Note that the frontal surface separating the two air masses exists not only on the ground but continues upward along the boundary of the air masses.

Air masses are usually in motion. The nature of a front depends upon how the air masses are moving. When a cold air mass overtakes warmer air, a *cold front* is formed. The advancing cold air is held back by friction against the ground so that cold air tends to pile up in a steep slope. The moving cold air lifts the warmer air and will create a heavy cloud formation if the warm air is moist. Large cumulus and cumulonimbus clouds are typical of a rapidly moving cold front. See Figure 22-4. Storms created along a cold front are usually short and violent. A long line of heavy thunderstorms, called a *squall line,* may advance just ahead of a fast moving cold front. A slowly moving cold front lifts the warm air ahead less rapidly, producing a less concentrated area of cloudiness and precipitation.

When warm air advances over the edge of a mass of colder air, a *warm front* is produced. The slope of a warm front is very gradual, as shown in Figure 22–5. Because of this gentle slope, the clouds may extend far beyond the base of the front. Stratus clouds and heavy, though not violent precipitation over a large area, are conditions generally associated with a warm front. Occasionally a warm front will produce violent storms if the warm air advancing over the cooler air is very moist.

Both a cold and a warm front may come to a halt for several days. When this happens, they form a *stationary front*. The weather around a stationary front is not very different from that produced by a warm front.

Storm centers formed by fronts. Bulges often develop along the edges of slowly moving cold or stationary fronts. A bulge is created when the air on one side moves slightly ahead of the front itself. These bulges or waves in the frontal boundary can signal the beginning of a storm center called a *wave cyclone*. A frontal wave cyclone is not the same kind of storm as a tropical cyclone or hurricane. Nor should it be confused with a tornado that is a small but very violent local storm. A wave cyclone consists of a very large body of air. Its winds blow in circular paths toward a low pressure region at the center. Such storms may cover a large part of an entire continent.

The stages in the development of a typical wave cyclone are illustrated in Figure 22–6. At the beginning (A), there

FIG. 22–4. Weather conditions along a cold front. Clouds shown along a cross-section of the frontal surface are typical of those that accompany an approaching cold front.

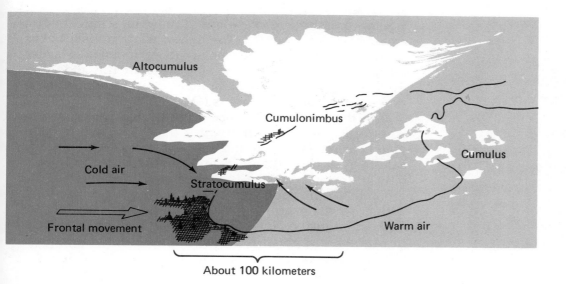

Altocumulus

Cumulonimbus

Cumulus

Cold air

Stratocumulus

Warm air

Frontal movement

About 100 kilometers

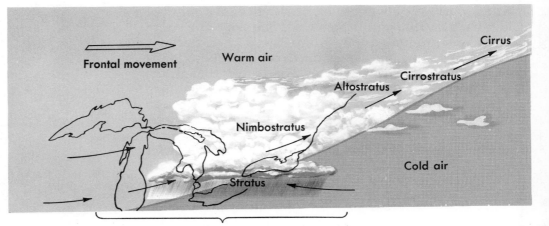

Frontal movement

Warm air

Cirrus

Cirrostratus

Altostratus

Nimbostratus

Cold air

Stratus

About 500 miles

FIG. 22–5. Weather conditions along a warm front. To an observer on the ground the approach of a warm front can be detected by the appearance of cirrus clouds.

is only the frontal boundary between the two air masses. At this stage, their winds usually move parallel to the front. This is due to the fact that air in the two masses does not mix. The winds on each side of the front are blowing opposite each other at different speeds. Thus they tend to develop a spinning motion. Try the following activity and you will see why this is so. Place a pencil across your two hands, then slide them in opposite directions.

This tendency toward a spinning movement causes slight waves or bulges to form along the front. This action also causes cold air to push into the warm. At the same time, warm air is pushed forward into the colder air. See Part (B) of Figure 22–6. The result is a warm front moving slowly ahead of a rapidly moving cold front.

A low pressure area develops at the point where the two fronts come together (C). The lighter warm air is lifted as it presses into the cold air along the warm front. It is also pushed up by the advancing cold air along the cold front. As a result of these lifting actions, clouds and precipitation spread along both fronts (D).

Soon the swiftly moving cold front overtakes the warm front and the warm air trapped between them is completely lifted off the ground (E). The front at that place is said to be *occluded*. This means that the front then exists only at upper levels and is completely closed off from the ground. At this point, the effects from the strong lifting action on the warm air usually bring the storm to its highest intensity. Now, the winds are moving in a circle toward the low pressure region at the center of the storm (F). In this area pressure is low because air is rising.

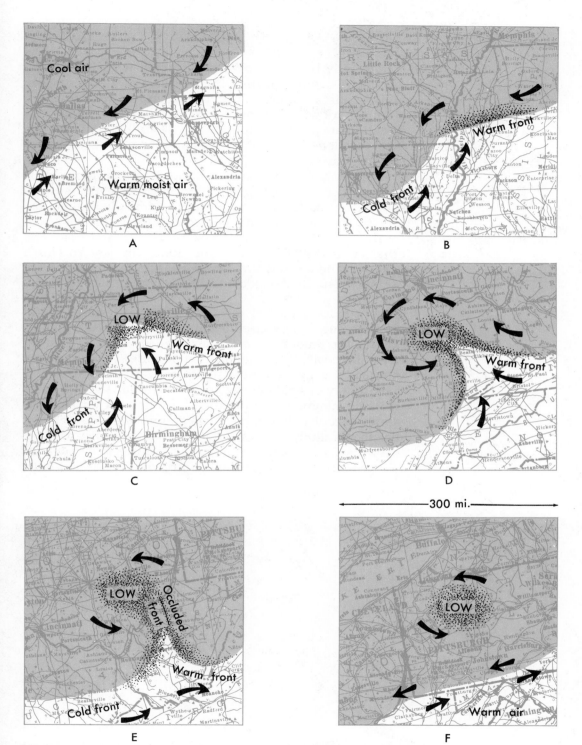

FIG. 22–6. The development of a typical wave cyclone or low pressure area as seen from above. The underlying maps indicate that this storm area moved northeastward.

It usually takes 12 to 24 hours for a wave cyclone to develop. During this time the air masses are in motion, and the disturbance moves with them. In the Northern Hemisphere wave cyclones generally move in an easterly direction at speeds of around 32 to 64 km/hr (20-40 mi/hr). Most cyclones follow well-known paths established through careful observation.

Figure 22–7 shows some of the most common paths that wave cyclones take. Notice that there are two well established paths in the Northern Hemisphere. One begins in the North Pacific and moves toward Alaska and south along the western coast of North America. The other common cyclone tract originates in eastern North America and crosses the North Atlantic to Europe. Wave cyclones usually develop one after another in a common path along a front.

Without its supply of moist air a cyclonic storm would soon die out. Winds circling into the low pressure center provide the amount of moist air needed. As this moist air rises and cools, the condensation of water vapor releases the heat energy which originally converted it into vapor. This supply of heat energy keeps the cyclone going. Later in its development the cyclone becomes cut off from its supply of moist air and begins to gradually die out. Then it disappears altogether as a separate weather disturbance.

As a separate storm center, the cyclone is a region of low pressure. The general wind direction near the ground

Investigate

Find out how you can tell the directions to high and low pressure regions by just knowing the wind direction. Then report your discovery to the class.

FIG. 22–7. Arrows indicate the most common paths for cyclonic storms.

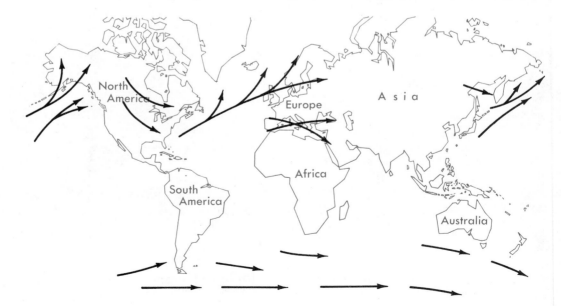

within such a storm is inward toward the low pressure. The Coriolis effect and friction of the winds against the earth's surface create a whirling motion. This twisting motion of winds inward toward a low pressure center moves *counter-clockwise* in the Northern Hemisphere.

In high pressure regions the direction of wind flow is outward from the high pressure center. The air spreads out away from the center. The Coriolis effect and surface friction again deflect the winds. In the Northern Hemisphere the outward moving winds of a high pressure region move in a *clockwise* whirl. The high pressure centers are usually called *highs* or *anticyclones*. See Figure 22–8. Since the air in an anticyclone is sinking, there is little tendency for precipitation or for clouds to form. Thus anticyclones or highs usually bring clear weather.

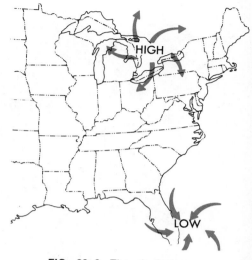

FIG. 22–8. The clockwise movement of winds from a high, contrasted with the counter-clockwise flow of winds into a low.

The polar front. Over the entire earth there is a general movement of warm air rising from the equator and moving toward the poles. Due to the Coriolis effect, air moving in the direction of the poles is turned to the right. In the Northern Hemisphere this means that it is deflected to the east. Then, about halfway between the equator and the North Pole, the eastward moving air meets cold polar air flowing south toward the equator. This zone where warm tropical air meets cold polar air is a more or less continuous weather front circling the entire earth. It is often called the polar front. The eastward motion of the air in the region of the polar front results in a wide band of high altitude winds. This is known as the *circumpolar whirl.* See Figure 22–9. These winds are generally strongest in the region of latitude 30° and weaken toward the equator and poles.

In the Northern Hemisphere, near the southern boundary of the circumpolar whirl, a band of very strong winds often exist. These winds are called the *jet streams.* They are found at an altitude of 10 to 15 kilometers (about 6 to 9 mi), and measure about 100 kilometers wide and two to three kilometers deep. When temperature differences between the two air masses separated by the polar front are greatest, the jet streams may have speeds up to 480 km/hr (300 mi/hr). The jet streams do not blow steadily, but change in speed, direction, and position. In summer when the circumpolar whirl contracts, the jet streams are found closer to the North Pole at 35° to 45° N latitude. In

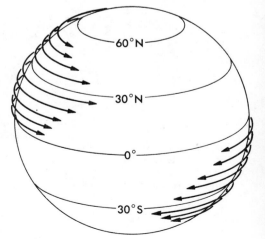

FIG. 22–9. The circumpolar whirl. The length of the arrows indicates the relative speed of the winds.

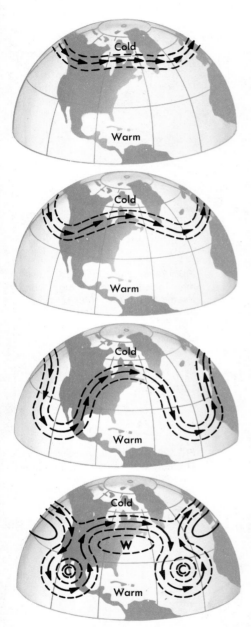

FIG. 22–10. A wave of jet streams. The waves tend to isolate pockets of warm and cold air far north and south of their average locations.

winter as the circumpolar whirl expands, they shift to the south and are found then at 20° to 25° N latitude.

The mixing of polar air and tropical air does not apparently take place continuously at the polar front. Mixing occurs only when waves develop in the circumpolar whirl. These waves carry cold polar air to the south and warm equatorial air to the north. The development of a wave in the circumpolar whirl can be traced by shifts in the position of the jet streams. A wave cycle of the jet streams is shown in Figure 22–10.

The behavior of the jet streams is closely connected with the development of wave cyclones. The cyclonic storms exist at any time in the air layers nearest the earth's surface. They seem to be formed and move under the general control of the jet stream cycles. But the details of this relationship are not completely understood.

WEATHER INSTRUMENTS

Observing weather conditions. One of the most practical applications of meteorology, the science of the atmosphere and the study of changing weather conditions, is the ability to make fairly accurate weather observations. To do this, meteorologists rely on various weather instruments and on techniques that have been developed largely within the last hundred years. The use of new instruments and improved techniques has increased the accuracy of observing the following weather conditions:

1. *Temperature.* Air temperature at ground level is measured on either the *Fahrenheit* (F) or *Celsius* (C) (also called Centigrade) scale. In the United States both scales are now commonly being used. See the appendix for a description of these temperature scales. Special thermometers are constructed to record the highest or lowest reading since the previous observation. See Figure 22–11.

2. *Pressure.* The instruments used to measure atmospheric pressure were described in Chapter 20. In weather reporting and forecasting, pressure readings are usually expressed in *millibars* (mb). A pressure of 2.54 cm of mercury equals 33.86 millibars.

3. *Precipitation.* The amount of precipitation since the last observation is recorded in hundredths of inches or millimeters. Solid forms of precipitation are melted and the depth of the liquid is recorded.

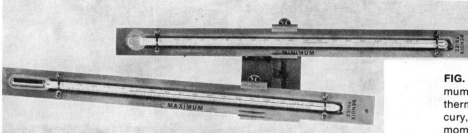

FIG. 22–11. Maximum and minimum thermometers. The maximum thermometer is filled with mercury, whereas the minimum thermometer uses alcohol. The special holder allows the maximum thermometer to be whirled and the minimum thermometer to be tilted in resetting. (Bendix Aviation Corporation)

4. *Wind speed and direction.* To measure wind speed, an instrument called an *anemometer* (an-uh-*mom*-uh-ter) is used. A typical anemometer consists of several small cups attached by spokes to a shaft that is free to rotate. See Figure 22–12. This rotation is then converted into an electrical signal which registers the wind speed on a conveniently located dial. An arrangement which produces a permanent record of wind speed on a graph is especially useful.

To determine wind speed and direction at higher altitudes, meteorologists send up a hydrogen or helium filled balloon. The balloon rises at a constant speed. Its movement in response to prevailing winds is followed with a small telescope, which also measures the angle of the line to the balloon. With this information the speed of the balloon can be calculated as it is carried along by the wind.

Wind speed is expressed in meters per second, miles per hour, or in knots (1.15 mi per hour). The direction of a wind is described according to the direction from which it comes. Thus a wind from the west, blowing toward the east, is called a west wind. In some weather reports, exact directions for winds may be given as one of 32 directions (points) on the compass. Wind direction is also recorded in degrees. They start with 0° at north and move clockwise around to 360° at north again. These two commonly used systems for describing wind direction are shown in Figure 22–13.

Electronic weather instruments. Conditions of the atmosphere near ground level are only a part of the complete picture of weather changes. Conditions at upper levels must also be known. An instrument commonly used to investigate weather conditions in the upper atmosphere is the *radiosonde*. See Figure 22–14. Measurements of temperature, pressure, and humidity are sent out by a radio

FIG. 22–12. An anemometer. (Bendix Aviation Corporation)

FIG. 22–13. The two systems by which wind directions are given.

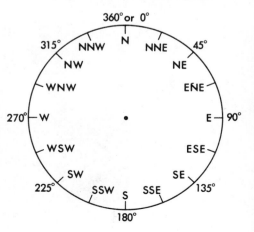

FIG. 22–14. The radiosonde being lifted by a balloon with a parachute which will return the instrument to earth. (Environmental Sciences Services Administration)

FIG. 22–15. Spacecraft launching of a TIROS weather satellite. Close-up of the satellite shows some of its principal parts. (ESSA)

signal as a helium-filled balloon lifts the instrument. A special radio is set up to receive and record the information. The path of the balloon is also followed by radar so that direction and speed of high altitude winds may be determined. When it reaches a very high altitude, the balloon finally bursts. The radiosonde instrument is then parachuted back to earth.

Another valuable electronic weather instrument is *radar*. Particles of water in the form of cloud droplets or precipitation reflect radar waves. Thus weather disturbances are visible on a radar screen. Radar information can also be used to give the precise location and extent of a storm. On a radarscope, one can watch the origin and growth of a storm system and track its movements across the land.

Recent advances in space science have given meteorologists an important instrument for future study of the weather. The weather satellite carries equipment for making the kind of detailed observations that are needed. The first weather satellite was known as TIROS (Television and Infrared Observation Satellite). A number of these satellites were launched in 1960 and orbited the earth from west to east. Later models were launched into a polar orbit allowing television pictures to be made of the entire earth. See Figure 22–15.

NIMBUS, the first of a more advanced series of weather satellites, was launched in 1964. From the results of the TIROS and NIMBUS experiments, a plan for continuous satellite observation of the earth's weather was developed. As time goes on, information obtained from satellites will greatly increase our knowledge of the earth's atmosphere and result in improved weather forecasting.

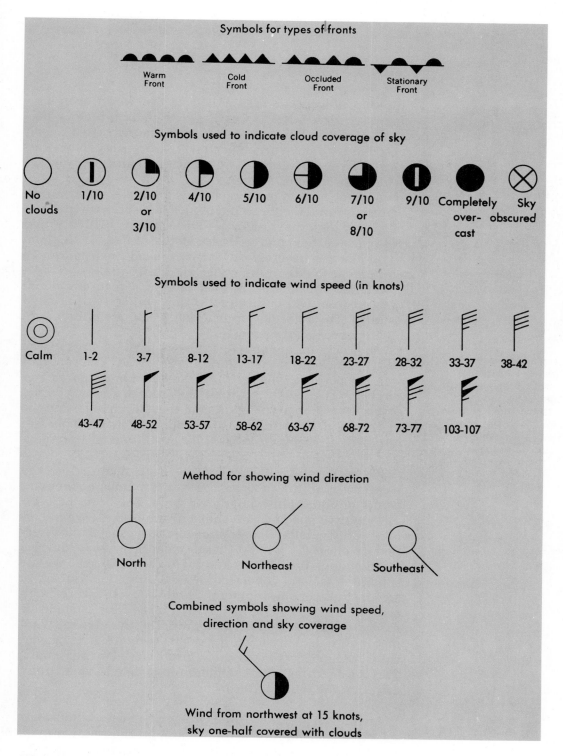

Symbols for types of fronts

Warm Front Cold Front Occluded Front Stationary Front

Symbols used to indicate cloud coverage of sky

No clouds 1/10 2/10 or 3/10 4/10 5/10 6/10 7/10 or 8/10 9/10 Completely over-cast Sky obscured

Symbols used to indicate wind speed (in knots)

Calm 1-2 3-7 8-12 13-17 18-22 23-27 28-32 33-37 38-42

43-47 48-52 53-57 58-62 63-67 68-72 73-77 103-107

Method for showing wind direction

North Northeast Southeast

Combined symbols showing wind speed, direction and sky coverage

Wind from northwest at 15 knots, sky one-half covered with clouds

FIG. 22–16. Symbols used in the construction of station models.

THE WEATHER MAP

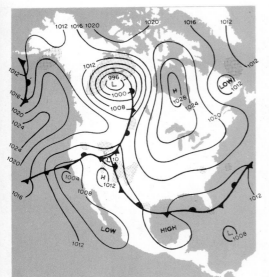

FIG. 22–17. A simplified weather map usually shows isobars, highs and lows, and fronts. It may also indicate other conditions, such as wind direction, temperature, and precipitation. Front symbols: triangle—cold front; half-circle—warm front; both on opposite sides—stationary front; both on same side—occluded front.

Construct
Make a form that could be used to keep a record of local weather observations.

Making a weather map. All over the world, every six hours, observers report weather conditions at their respective locations. Barometers are read and corrected to correspond with pressure at sea level. All reports show pressure to be on the same basis regardless of altitude. Surface winds speed and direction are noted. Precipitation is measured, temperature read, and humidity determined. The information collected includes a description of cloud covering and the height to cloud bases. Visibility and general weather conditions are also recorded. In addition, many larger observation stations send up radiosonde equipment to determine upper air conditions.

All this information is then put into an international code and sent to central collection centers within each country. The information is then exchanged internationally by government agencies.

Weather maps are usually prepared at the centers where the coded weather information is received. The reported weather observations are first translated into figures and symbols. Then they are grouped around a small circle, drawn on a map at the position of the station reporting the information. The circle on the map, with its symbols and numbers describing the weather conditions at that location, is called a *station model*. See Figure 22–16.

When the station models have been recorded on the weather map, the next step is to draw lines connecting points of equal atmospheric pressure. See Figure 22–17. The relative spacing and shape of the isobars, when correctly drawn, tells us something about the speed and direction of observed winds. Closely spaced isobars indicate rapid change in pressure and higher wind speeds. Widely separated isobars generally mean light winds. Isobars in rough circles enclose centers of high or low pressure. Such centers are usually marked with a large H or L. Since air moves toward regions of low pressure, the general wind direction is usually toward low pressure areas. However, the Coriolis effect makes winds flow parallel to the isobars. Surface features, such as mountains, also have a strong effect on the wind pattern.

Principles of weather forecasting. The basic tool used by meteorologists in forecasting weather is the weather map. Weather forecasters try to predict the intensity and path of weather systems plotted on the map. They also try to forecast the formation of new weather disturbances. The forecaster relies on the fact that in the middle latitudes the upper air moves in a general easterly direction. The low level weather is usually carried along with this upper air movement. Most cyclonic storms that affect North America enter from the west. They move eastward across the middle of the continent and pass out to sea off the North Atlantic coast.

In determining the speed and direction of a storm center, forecasters make use of charts showing pressures and winds, particularly at upper levels. For their estimates of the future path of the disturbance, they note upper movements, pressures, previous rates, and direction of movement in the system. To learn about other features of the coming weather, such as temperature, humidity, and cloudiness, they must have more information. From their knowledge of how steep the surfaces of other weather fronts have been, they can draw some conclusions about future conditions. The steepness of a front's slope is a very important factor. If the slope is steep enough the lifting of air may be sufficient to cause condensation of moisture producing cloudiness and precipitation. Information obtained from satellites provide pictures of general cloud patterns. These are very useful in helping discover and follow weather systems.

At the present time it is possible to accurately predict general weather conditions up to about five days in advance. Detailed forecasts cannot be made beyond 48 hours. Extended-range forecasts up to 30 days are made by computer analysis of the average movements of the atmosphere. In this procedure slowly changing large-scale movements of the air, such as the jet stream cycles, are revealed. These changes help to predict the general weather pattern, but they cannot forecast the exact weather at a particular place. Accurate and detailed long-range forecasting will not be possible until all of the factors which control the weather are better understood.

LOCAL WEATHER

The scale of weather changes. A weather map shows only large-scale weather systems that affect large areas of the earth's surface. The reason for this is the way that weather observations are made. Weather instruments usually

activity

On a weather map, a line that connects points with the same temperature is called an **isotherm**. Isotherms can be used to help identify air masses. You are to draw isotherms in this activity.

Get a map on which the temperatures at stations in the United States are written. The number above the circle is the temperature. The number below is the number of the station.

The 15° isotherm starts at station 14 in the northeast United States. It goes southwest into Texas, across to Arizona, then north to Washington, and ends between stations 1 and 2.

Notice that the 15° isotherm goes through station 13 because the temperature is 15°C there. It also goes between stations 20 and 21. Since the temperature must be 15°C, the line must pass somewhere between 13.9°C and 15.6°C, the temperatures at stations 20 and 21. The line also goes between stations 40 and 22 as well as between 36 and 37. These are examples of places you can use to draw the 15° isotherm.

Draw the 15° isotherm on the map lightly, in pencil, then go over it and smooth out any bumps.

The 15° isotherm is a single line. Another isotherm can be in two parts, or it may come back to where it started, forming a closed loop.

On your map, draw isotherms for 9°C, 11°C, 13°C, 15°C, 17°C, 19°C, 21°C, 23°C, and 25°C. Label each isotherm with the correct temperature.

1. What is the lowest temperature for which you have drawn an isotherm? what is the highest?

2. Which forms a closed loop, the low-temperature isotherm or the high-temperature isotherm?

3. Is the air mass identified by the closed isotherm a cold air mass or a warm air mass?

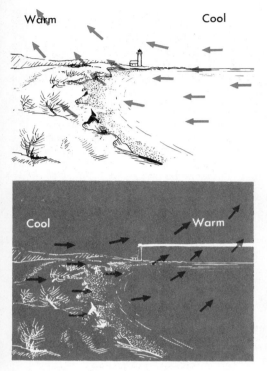

FIG. 22–18. Unequal heating of land and water results in land and sea breezes.

register only the slower changes in atmospheric conditions. For example, most anemometers are designed to record only general wind speed. Short gusts of wind are not registered. Weather observations are also made at widely spaced locations and must be averaged to give a picture of the general weather pattern. This averaging process eliminates small-weather changes from a common weather map. Only the weather disturbances that last for days are described by weather maps.

At any particular place, the movements of air are influenced by local conditions. A breeze that suddenly springs up is not usually part of a large-scale weather pattern. Small-scale winds that extend over distances of less than 100 kilometers are more often caused by some local feature that produces temperature differences. Some examples of this type of air movement are:

1. *Land and sea breezes.* Equal areas of land and water may receive the same total amount of energy from the sun. But the land surface is soon heated to a higher temperature than the water. During daylight hours this causes a sharp temperature difference to develop between the water and land along a shore. This temperature difference also appears in the air above the land and water. The warmer air above the land rises and the cool air from above the water moves in to replace it. A cool sea breeze will begin generally in the late morning hours. In late afternoon or early evening the sea breeze is replaced by a land breeze as the temperature situation is reversed. See Figure 22–18.

2. *Mountain and valley breezes.* In the daytime, mountains heat faster than surrounding valleys. This is mainly because the exposed slopes of mountains easily absorb the sun's energy. The valleys, because they are generally covered with forests and other vegetation, absorb energy more slowly. In the daylight hours, a gentle valley breeze blows up the slopes. It is caused by cooler air from the valleys moving up to replace the warmer mountain air. At night the mountains cool faster than the valleys. Then cooler air descends on the mountain slopes creating a mountain breeze.

Hurricanes. For reasons that are not yet clearly understood, hurricanes frequently develop over warm, tropical sea water near the equator. These storms have some resemblance to the wave cyclones of the middle latitudes and may develop in a similar way. However, hurricanes are seldom more than 700 kilometers (450 mi) in diameter, compared with the diameter of a wave cyclone which is about 2,500 kilometers. A full-scale hurricane carries an unusually large amount of energy. This energy comes almost entirely from the heat that is released as moisture condenses from warm tropical air.

A hurricane begins when very warm moist air over the sea rises rapidly. When moisture in the rising warm air condenses, large amounts of heat are released. This forces the air to rise even faster. More moist tropical air is drawn into the column of rising air at lower levels. A continuous supply of water vapor is needed to keep the process going. The entire system develops a spin which is thought to be caused by the Coriolis effect. See Figure 22–19A.

A fully developed hurricane consists of a series of thick cloud bands spiraling upward into the storm's center. See Figure 22–19B. Precipitation is very heavy in these clouds. Winds increase in velocity toward the center or eye of the storm, reaching maximum speeds of over 100 miles/hour. Their lifetime is around nine to twelve days. During this time the hurricane moves in the direction of the existing large-scale wind pattern.

The principal paths of hurricanes over the entire earth are shown in Figure 22–20.

FIG. 22–19A. A cyclonic storm developing over the Pacific as seen from space. (Shostal)

FIG. 22–19B. Diagram of a hurricane.

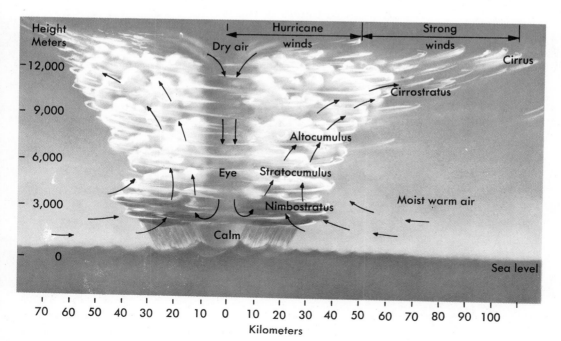

FIG. 22–20. The principal hurricane and typhoon areas of the world are shown in color. Arrows indicate the normal paths of the storms. Note that these storms all originate over water in tropical regions.

FIG. 22–21. The first stage (A) in the development of a thunderstorm is the cumulus stage. A swiftly moving current of warm air rises to high altitudes. The second, (B) or mature, stage of a thunderstorm begins as precipitation falls from the upper levels of the cloud. The precipitation causes violent up-and-down movements of air within the cloud. Final stage of a thunderstorm (C). Upward movement air has stopped and precipitation has slowed or stopped. Ice crystals at the top of the cloud spread out into the anvil-shaped top typical of a well developed thunderhead cloud.

Thunderstorms. When a small-scale upward movement of warm moist air takes place, a thunderstorm may result. Such an event is likely when the air in a moist, warm (mT) air mass is heated in one location. The air may also be lifted by mountains or by contact with an occluded front.

A local thunderstorm develops in three distinct stages. These stages are illustrated in Figure 22–21. How some clouds develop the electrical features necessary to produce lightning is not completely understood. The electrical

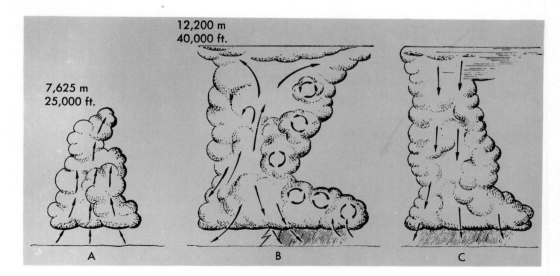

12,200 m
40,000 ft.

7,625 m
25,000 ft.

A　　　　　　　　B　　　　　　　　C

charges separate, causing the upper part of the cloud to carry positive electricity while the lower part carries both positive and negative charges. What seems to cause this separation is that large water droplets carrying negative charges fall to the lower parts of the cloud. Small droplets carrying positive charges are lifted up by strong updrafts to the top of the cloud. See Figure 22–22.

Each year lightning kills a large number of people. A few simple precautions would lessen the risk of being struck by lightning. The first rule is to avoid any prominent feature or high location in open land. Lightning always follows the shortest path between cloud and ground. Therefore, it is most likely to strike the highest part of any particular location. Trees, tall metal objects, and bodies of water are all likely targets and should be avoided during a thunderstorm. The interior of buildings, particularly those with a metal frame, are relatively safe, as are the interiors of automobiles.

FIG. 22–22. Lightning results from the accumulation of electric charges within a cloud.

Tornadoes. The smallest, most violent and short-lived of all storms is the tornado. A tornado is most likely to occur on a warm, humid day when the sky is filled with heavy thunderclouds. One of the cloud bases may suddenly develop a funnel-shaped, rapidly turning extension that reaches down toward the ground. The tip of the funnel may reach the ground or it may not. If the tip of the funnel does touch the earth, it generally moves in a wandering path at a speed faster than a man can run. Frequently the funnel rises, then touches down again a short distance away. The tornado generally sweeps a path 100 meters or less in width. But within that path there is usually complete destruction of everything on the ground surface. See Figure 22–23.

The destructive power of a tornado is a result of the speed of the winds whirling within the funnel. These winds have never been measured, but their great power indicates speeds as high as several hundred miles per hour. Buildings are also damaged because the pressure within the funnel is very low. The sudden pressure drop often causes buildings to burst open as the air at normal pressure trapped inside expands. Most of the injuries caused by tornadoes occur when people are struck by objects flung around by the winds. It is usually safest to lie down as low as possible with the head covered. However, only a hole in the ground can offer much real protection.

FIG. 22–23. Tornado photographed as it approached a small Canadian town. (Environmental Science Services Administration)

How tornadoes develop their tremendous energy is not understood. To form, they seem to require moist warm air at low levels and cool dry air at upper levels. Some sudden lifting action, such as that caused by an advancing cold front, is frequently associated with tornadoes. Electrical charges formed within the funnel may also be a source of energy for these storms.

Tornadoes are most common in the Midwest region of the United States during the late spring or early summer. They have been known to occur in many locations. Tornadoes over the sea are called *waterspouts*.

FIG. 22–24. A waterspout at sea. (Pix—Hoflinger)

VOCABULARY REVIEW

Match the word or words in the column on the right with the correct phrase in the column on the left. *Do not write in this book.*

1. A large body of air of about the same temperature and humidity throughout.
2. A definite boundary between different air masses.
3. A long line of heavy thunderstorms which may advance just ahead of a fast moving cold front.
4. A large body of air with winds tending to blow in circular paths toward a low pressure region at the center.
5. A front which exists only at upper levels.
6. High pressure center.
7. A band of very strong winds near the southern boundary of the circumpolar whirl.
8. Centigrade scale.
9. An instrument used to measure wind speeds.
10. Measures temperature, pressure, and humidity at upper levels.
11. Symbols and numbers grouped around a circle on a map to describe weather conditions.
12. Some resemblance to the wave cyclone of the middle latitudes but seldom more than 700 km in diameter.
13. Electrically charged clouds resulting from a small-scale upward movement of warm moist air.
14. The smallest, most violent, and short lived of all storms.
15. Tornadoes over the sea.

a. squall line
b. anticyclone
c. front
d. occluded front
e. air mass
f. Celsius
g. wave cyclone
h. radar
i. jet stream
j. Fahrenheit
k. radiosonde
l. waterspouts
m. anemometer
n. tornadoes
o. station model
p. thunderstorm
q. hurricane

QUESTIONS

Group A

Select the best term to complete the following statements. *Do not write in this book.*

1. The sum of all the properties of the atmosphere at any particular time is called (a) the weather (b) a storm (c) the temperature (d) energy.

2. The main cause of the earth's changing weather patterns is (a) the oceans (b) temperature differences in the atmosphere (c) pressure differences (d) humidity.

3. An air mass is created when a large body of air (a) moves across land (b) moves across the ocean (c) remains stationary over a uniform region for a time (d) remains stationary over land only.

4. Which of the following air masses would most likely be dry and warm? (a) mP (b) mT (c) cP (d) cT.

5. Which of the following air masses would most likely be moist and cool? (a) mP (b) mT (c) cP (d) cT.

6. Which of the following is most likely a polar Pacific air mass? (a) mP (b) mT (c) cP (d) cT.

7. Which of the following is most likely a tropical gulf air mass? (a) mP (b) mT (c) cP (d) cT.

8. The weather at a stationary front is generally the same as that produced by a (a) squall line (b) warm front (c) cold front (d) anticyclone.

9. Wave cyclones result in an area of (a) low pressure (b) anticyclones (c) high pressure (d) clear weather.

10. The twisting motion of the winds toward a low pressure center in the Southern Hemisphere is (a) southward (b) northward (c) clockwise (d) counterclockwise.

11. Which of the following is *not* related to the others? (a) temperature (b) Celsius (c) Fahrenheit (d) millibar.

12. An anemometer would most likely be read in (a) degrees Celsius (b) knots (c) millibars (d) centimeters.

13. Which of the following does *not* belong with the others? (a) temperature (b) pressure (c) precipitation (d) knots.

14. A northwest wind would be from a compass direction of (a) 315° (b) 225° (c) 135° (d) 45°.

15. On a weather map, lines which connect points of equal atmospheric pressure are called (a) isotherms (b) thermoclines (c) isobars (d) pressure barriers.

16. On a weather map, closely spaced isobars mean (a) rapid temperature change (b) higher wind speeds (c) lower wind speeds (d) slow change in pressure.

17. The Coriolis effect tends to make winds blow (a) perpendicular to isobars (b) clockwise around a low pressure region in the Northern Hemisphere (c) clockwise around a high pressure region in the Northern Hemisphere (d) toward low pressure regions.

18. Most cyclonic storms that affect North America move (a) eastward (b) westward (c) northward (d) southward.

19. International weather maps are drawn once every (a) hour (b) 2 hours (c) 6 hours (d) 24 hours.

20. Detailed weather forecasts can be made accurately up to (a) 5 days (b) 2 days (c) 30 days (d) 60 days.

21. How many weather details are shown in a complete station model? (a) exactly 5 (b) less than 5 (c) more than 5 but less than 10 (d) more than 15.

22. Sea breezes are most likely to begin in (a) late morning (b) mid-afternoon (c) late afternoon (d) late evening.

23. Which of the following is likely to occur at about the same time of day and for the same reason as a mountain breeze? (a) valley breeze (b) sea breeze (c) land breeze (d) early morning breeze.

Group B

1. What role do satellites have in weather forecasting?

2. Why is a polar orbit an advantage for a weather satellite?

3. Name the seven source regions for air masses which strongly influence North American weather. Name the type of air mass from each region.

4. What effect does the earth's rotation have on winds?

5. What are the steps in constructing a weather map?

6. Explain the formation of a warm front.

7. Explain the formation of a cold front.

8. Change a wind speed of 16 knots into miles per hour.

9. Change a pressure of 29.75 inches of mercury into millibars.

10. Draw a station model with the following information: (a) wind: NW at 16 knots (b) temperature: 64°F (c) sky: overcast (d) clouds: stratus (e) pressure: 1002.4 millibars.

11. Temperature in degrees Fahrenheit (F) can be changed to degrees Celsius (C) by using the formula $C = \frac{5}{9} (F - 32)$. How many degrees Celsius is normal body temperature (98.6°F)?

12. How many degrees Fahrenheit is −40° Celsius?

13. Change 1015.8 millibars pressure into inches of mercury.

14. What advantage does the Celsius temperature scale have over the Fahrenheit scale?

The day-to-day changes in the weather are part of a greater pattern of changes in solar energy, wind, and moisture. This overall pattern is called *climate*. The word climate comes from a Greek word meaning "slope." The ancient Greeks believed that the earth sloped toward the North Pole and that differences in climate depended upon this factor.

Climate is the average weather in a locality over a number of years. It is the result of many influences. Places on the same latitude around the globe may have widely different climates. Local conditions alter the temperature or modify moisture-bearing winds. Thus the surface of the earth is subject to a countless variety of climates. These blend and form every possible environment between the extremes of hot, humid tropics and cold, dry polar regions.

Basic to all climates are the forces that govern the weather. The nature of climate lies in the actions of heat, moisture, and air movements. These all make up the daily conditions of weather. In this chapter the influences of the sun, the winds, and the moisture of the air will be studied in relation to their effect on climate as well as their role in the classification of climates.

CONTROLS OF CLIMATE

Temperature. The total solar energy received by any location on the earth's surface is the major influence affecting its average temperature. At any particular place the amount of solar energy received is determined by two factors: (1) the angle at which the sun's rays strike, and (2) the length of time the sun shines during a day. Both of

objectives

☐ Define the term climate.

☐ Explain how temperature, moisture, and air movement control climate.

☐ Name and briefly describe the earth's three main climatic zones.

☐ Describe several different climates found in each of the three main climatic zones.

☐ List the three major climate controls.

☐ Describe how these climate controls influence North American weather.

Table 23–1

Latitude	Longest Night
0°	12 hours
17°	13 hours
41°	15 hours
49°	16 hours
63°	20 hours
66°30′	24 hours
67° 21′	1 month
69° 51′	2 months
78° 11′	4 months
90°	6 months

activity

Using an almanac, list the dates of the year on which your area has 14 or more hours of daylight. Also list the dates of the year on which your area has 10 or less hours of daylight. Obtain from your local weather station or from newspaper files at the library the high and low temperatures recorded for each day. Make a separate graph for each set of data. Use the vertical axis to show temperature and the horizontal axis for the date. Plot the high temperatures in red and the low temperatures in blue. Join the red points with a red line; the blue points with a blue line. Write an explanation of the results shown on your two graphs.

these factors depend on the latitude of the place in question.

Because of the inclination or tilt of the earth's axis, the sun's rays are always nearly perpendicular at the equator. For the same reason the length of day and night near the equator does not change very much throughout the year. Thus the equatorial regions have steady high temperatures all year long. In the higher latitudes, closer to the poles, the sun's rays are more slanted and less effective in heating. Thus temperatures are lower. Winter nights are long and the days short. Table 23–1 shows how the length of the longest night increases as the latitude increases. Very long nights in regions near the poles allow very low temperatures to be reached. This reduces the yearly average temperature in the polar regions to a low level.

It should be remembered that the length of the longest night in winter is the same as the length of the longest daylight period in summer. During the summer the high latitudes have very long daylight periods. The sun's rays then are more direct, giving these regions summer temperature maximums which are a great deal higher than the winter readings.

Latitude is not the only factor which determines the average temperature at a particular place. If it were, all locations at the same latitude would have the same average temperature conditions. When the average temperatures over the earth's surface are examined by means of *isotherms* drawn on a map, we find that this is not the case. An isotherm is a line which connects all locations having the same average temperature. See Figure 23–1. Notice that the isotherms do not follow the parallels of latitude. This indicates that factors other than position relative to the equator and poles influence the average temperature.

The following are some of the other principal influences on the average yearly temperature at a specific place:

1. *Land and water differences.* A quantity of solar energy heats land faster and to a higher temperature than it does a water surface. One reason for this difference is that water is mixed by waves, currents, and other motions. The warm surface water is thus continuously replaced by cooler water from below. This prevents the surface temperature from increasing rapidly.

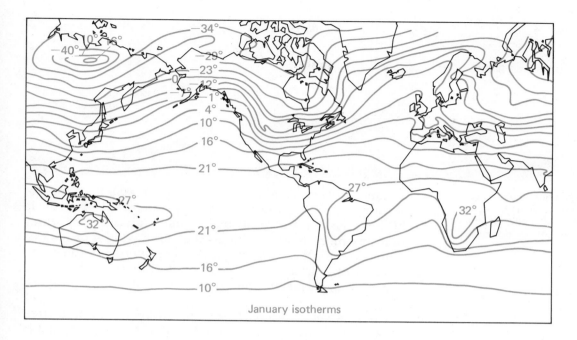

January isotherms

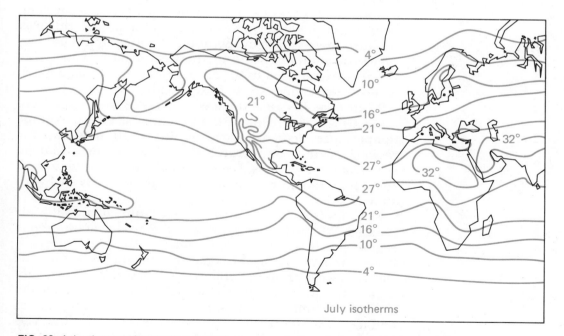

July isotherms

FIG. 23–1. Isotherms of average temperature: above in January: and below, in July. Why are differences much more pronounced in the Northern Hemisphere than in the Southern Hemisphere? Temperatures are in degrees Celsius.

Descending air warms at the dry adiabatic rate: 1°C per 100 m

FIG. 23–2. Foehn or chinook winds are warm and drying because of the difference between the moist adiabatic and dry adiabatic changes.

Land and water also have different capacities for absorbing and releasing heat. The *specific heat* of water is higher than that of land materials. A quantity of water requires more heat than does the same weight of rock or other land materials to increase its temperature the same number of degrees. Even if it is not in motion, water will warm up more slowly than land. Water also cools off more slowly than land.

There is yet another reason for differing temperatures of land and water at the same latitude. Some of the solar energy received by water is released in the process of evaporation. This heat cannot be used to raise the temperature of the water. The land is not affected in this way because there is less evaporation of water from land surfaces. In addition, some of the heat used in vaporizing water is transferred to the air masses over the land when water vapor condenses to form clouds.

2. *Altitude.* Because the lower layers of the troposphere are warmed as a result of the Greenhouse effect, average temperature decreases with altitude. In mountainous regions average temperatures are always lower than in surrounding lowlands. Even along the equator the tops of some high mountains are cool enough to be covered with snow.

Mountains also affect the temperature of passing air masses. When a moving air mass enters a mountainous region, it is lifted and cooled adiabatically. As the air rises, it loses most of its water vapor by condensation. Then as the air descends on the other side, it is warmed at the dry adiabatic rate of about 1°C per 100 meters. This means that air flowing down mountain slopes is usually warm and dry. See Figure 23–2. Such a wind occurs in the Alps Mountains and is known as a *foehn* (fain). On the eastern slopes of the Rocky Mountains, these warm winds are called *chinook winds*.

3. *Ocean currents.* Nearness to the sea has a strong influence on temperature. This is particularly true if ocean currents passing close to land masses have a different temperature than the land. Such currents have a stronger effect on the coastal air masses as winds consistently blow toward the shore. For example, the average temperature of northwestern Europe is unusually high for its latitude. Here we see the combined effects of a warm current, the North Atlantic Drift, and the steady westerly winds. On the other hand, the warm Gulf Stream has less of an effect along the east coast of the United States. Here winds come more often from the west, blowing away from the coast.

Wind and moisture. Other than average temperature, the climate of a region is determined by its usual winds and precipitation. For the most part, the general direction of the wind in a particular place is determined by its location in one of the global wind belts described in Chapter 20. Storms and local weather disturb and mask this effect. But in most places winds are able to overcome these factors and blow in the direction set by the global wind pattern.

Some locations experience two distinct and directly opposite wind directions during a year. They are caused by differences in heating between large land masses and oceans. During the summer, the land is heated more than the sea. If a low pressure center develops over the land the air will move away from the water toward the land. During the winter, the land receives less heat than the water. Then the wind direction is reversed, as air flows out to the oceans. Such seasonal winds are called *monsoons*. They are strongest over the larger continental land regions near the equator. For example, monsoons occur in southern Asia as a result of the heating and cooling of a large land area in the northern Indian peninsula. See Figure 23–3. Southerly winds in the summer usually bring heavy rainfall as they carry moisture from the sea. In winter the northeast winds bring dry weather. Monsoon conditions also occur in eastern Asia and in other places near the equator where a long stretch of coastline faces toward the equator.

While precipitation may be controlled by factors such as a monsoon, it usually remains under the main influence of the global pattern of air movements. Within the world's belts of average low pressure, the doldrums and the subpolar lows, there is a continuous lifting of air. As a result,

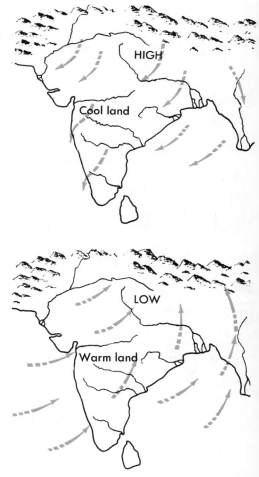

FIG. 23–3. Monsoons are seasonal winds. Why are they more pronounced over large land masses near the equator?

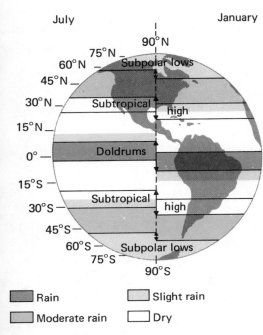

July ... January

90°N
75°N
60°N — Subpolar lows
45°N
30°N — Subtropical
15°N   high
0° — Doldrums
15°S
30°S — Subtropical
                        high
45°S
60°S — Subpolar lows
75°S
90°S

■ Rain   □ Slight rain
■ Moderate rain □ Dry

FIG. 23–4. The general rainfall pattern over the earth at two different seasons. Note that this pattern shifts with seasonal shifts in the planetary wind belts and pressure areas.

the regions of the earth covered by these areas of low pressure generally experience heavy precipitation. The most abundant rainfall of the earth occurs in a belt around the equator. Moving from the equator toward the poles, the amount of rainfall steadily decreases. It reaches a minimum at around 20° to 30° latitude, in the region of the sinking air of the subtropical high. Here there is hardly any rainfall since the air is mostly sinking and there is little condensation.

Closer to the poles, at the subpolar lows around 45° to 55° latitude, there is another belt of high average precipitation. Here warm air meets cold polar air and wave cyclones frequently develop. Above latitudes 50° to 55°, average precipitation decreases in the cold, dry polar air masses.

With the changing seasons, the global wind pattern shifts in a north-south direction. As the wind and pressure belts shift, the belts of precipitation associated with them also change position. The positions of the precipitation belts and their seasonal shifts are shown in Figure 23–4.

CLIMATIC REGIONS

Classification of climates. The warm zone immediately around the equator is the zone of *tropical climates*. To be classed as tropical, a region must have an average temperature of at least 18°C (64°F) during the coldest month of the year. Tropical climates fall under the influence of the continental tropical (cT) and maritime tropical (mT) air masses formed in the source regions close to the equator.

At the other extreme are the *polar climates*. In these regions average temperature never climbs higher than 10°C (50°F) during the warmest month. Continental polar (cP) and maritime polar (mP) air masses are produced in the source regions of these areas.

Between the tropical and polar climate zones is the belt of *middle latitude climates*. In these climates the average temperature of the coldest month is below 18°C (64°F) and the temperature of the warmest month is above 10°C (50°F). Weather is changeable in the middle latitude climates since both the tropical and polar air masses invade this zone. They are also exposed to frequent cyclonic storms which are produced along the polar front. The general boundaries of the three major climate zones are shown in Figure 23–5.

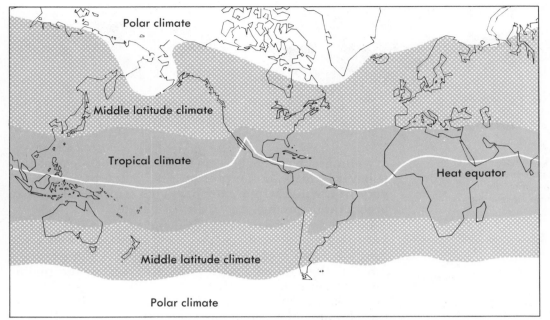

Polar climate

Middle latitude climate

Tropical climate

Heat equator

Middle latitude climate

Polar climate

Within each of the principal climate zones, there are many types of climates. These various types of climates result mainly from differences in precipitation. For example, one type of tropical climate can be found in regions with abundant rain. Such warm, humid climates occur near the equator, where the generally rising air produces an annual rainfall that is usually greater than 250 cm (100 in). The constant, warm, humid weather in these equatorial regions produces the heavy plant growth called the "rain forest," common in central Africa, the Amazon River Basin of South America, parts of Central America, and southeast Asia. See Figure 23–6. A second type of tropical climate is very dry. It is found in regions farther north or south of the equator. These parts of the tropical climate zone lie in the sinking air masses of the subtropical highs. Annual rainfall is less than 25 cm (10 in). Land areas within these regions contain some of the world's driest deserts. See Figure 23–7. A belt of tropical deserts extends across North Africa, the Near East, and southwest Asia. Smaller tropical deserts are also found in southwest Africa, the interior of Australia, northern Mexico, the west coast of Peru and northern Chile, and the southwestern United States.

A third type of tropical climate occurs between the rain forests and tropical deserts. These areas have a wet-and-

FIG. 23–5. General boundaries of the major climate zones are shown on this world map. The "heat equator" connects those places with the highest yearly average temperature.

FIG. 23–6. A tropical rain forest in Puerto Rico. (Shostal)

FIG. 23–7. This sand-covered region of the Sahara Desert in Africa is typical of the very dry tropical deserts. (Photo Researchers—Thomas Hollyman)

dry climate cycle each year. It is a result of the seasonal poleward shifts of the precipitation belts. Summers are very wet, while winters are dry. This seasonal change in precipitation produces a *savanna* climate.

In the same way as the tropical zone is divided into various climate regions, the middle latitude and polar zones are also made up of different regions. For example, most of the United States is located within the middle latitude climate zone. However, residents of the United States know very well that there are great differences in the climates of various parts of the country. In the next section, North American climates will be discussed. You will learn how various factors produce differences within a general climate zone.

North American climates. The most important influence on the climate of North America is the position of the continent. It stretches from Central America, in the tropical zone, through the middle latitudes to northern Canada and Alaska, in the polar zone. But the United States and most of Canada lie in the middle latitudes. These regions are influenced by air masses that come from both tropical and polar regions. Meetings of the warm and cool air masses produce the cyclonic storms that furnish most of the precipitation falling on the United States and Canada. This precipitation, however, is not evenly distributed over the middle latitude regions of North America. Thus several types of middle latitude climates are found in the United States and Canada.

In the United States, air masses and weather disturbances usually move from west to east because most of the land area is in the belt of westerly winds. As moist air moves toward the east from the Pacific, the west coast receives precipitation. Along the northwest coasts of Washington, Oregon, and northern California, where mountains block the eastward movement of moist air, abundant

precipitation falls. These northern parts of the coastline have a *marine west coast* climate. Average yearly precipitation is 50 to 75 cm (20 to 30 in). Most regions having marine west coast climates are covered with heavy forests of cone-bearing trees. See Figure 23–8.

The southwest coast, including Southern California, is affected by the drier air in the subtropical high precipitation belt during the summer. Winter brings the southward shift of the westerlies and their cyclonic storms to this part of the coast. The result is a *Mediterranean* climate with a yearly average rainfall of about 25 cm (10 in) falling almost entirely during the usually mild winters.

The air becomes drier as it moves inland. Thus much of the land area in the far western United States, other than the coasts and mountains, is desert. Precipitation is very low in these deserts, usually less than 25 cm (10 in) per year. Unlike the tropical deserts, the middle latitude deserts have a definite winter season, which may be cold. Summers are warm to very hot.

When the dry, eastward-moving air reaches the central part of the United States, it begins to pick up moisture from the tropical air moving northward. The almost barren western deserts gradually merge into *steppes* regions where rainfall is sufficient to support heavy growth of grasses. Much of the midwestern United States is made up of steppes.

Farther to the east is a large region whose northern portion has a climate controlled mainly by both cold, dry

FIG. 23–8. A marine west coast forest in Oregon. Coastal fog is approaching through the valley. (Ramsey)

polar air masses and moist, tropical air masses. Summers are usually warm and humid as tropical air masses move north. But winters are commonly very cold as polar air masses move south. This is said to be a *humid continental* climate. Most of the northeastern part of North America has a humid continental climate. Yearly average precipitation is at least 75 cm (about 30 in) from cyclonic storms and summer thunderstorms. Heavy forests of both hardwood and softwood trees are common in these regions.

Southeastern United States has a *humid subtropical* climate. Here the climate is controlled in summer by moist, tropical air masses moving north. They usually bring warm, humid weather, often with heavy rains. In winter, continental polar air masses moving from inland may bring brief but sharp cold waves. Annual precipitation is between 75 to 165 cm (about 30 to 65 in). The land is usually covered with heavy plant growth and forests.

Figure 23–9 shows all of the climate regions of the United States except Alaska and Hawaii. Alaska is in the polar climate zone. Most of this state has a *subarctic* climate. Winters are very cold and summers are short.

FIG. 23–9. The types of middle latitude climates in the continental United States.

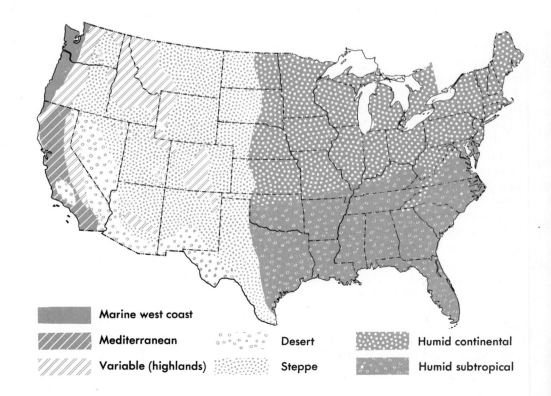

Marine west coast

Mediterranean

Variable (highlands)

Desert

Steppe

Humid continental

Humid subtropical

However, the summer may be surprisingly warm due to the very long periods of daylight at these high latitudes. Annual precipitation is between 25 and 50 cm (10 to 20 in). Plants found in subarctic regions make up an unusually thin forest of pines, spruce, fir, and smaller cone-bearing trees. The most northern part of Alaska has a *tundra* climate. Here the deeper soil is permanently frozen. During the summer, the surface thaws to form mud. Plant life in the tundra consists of mosses and related plants, with a scattering of small shrubs.

The Hawaiian Islands have both tropical rain forests and savanna climates. Because of their location in the belt of Trade Winds, the islands have heavy precipitation on the sides facing these moist winds. But their mountainous nature shields many parts of the large islands from these wind effects. The areas receiving lesser rainfall generally have savanna climates.

Local climates. A description of the general climate influences on a large land mass such as North America cannot be complete without a discussion of local climate controls. At any place on the continent, certain major climate influences are present. But there are also minor climate influences at the location that help to determine its climate. As a result, there exists within all major climate areas a great number of smaller local climates.

The most common example of local influence on climate is altitude. Temperature decreases rapidly with increasing elevation, giving highlands generally lower average temperatures than surrounding lowlands. In addition, the thin air at higher altitudes retains less heat at night. Thus the difference between day and night temperatures is greater at elevated locations. Mountains create local climate effects because of their various exposures to winds. Mountain slopes facing the wind receive more precipitation than the opposite slopes or the surrounding lowlands. However, mountains do not create a special type of highland climate. A mountainous region is likely to have a variety of local climates as altitude and direction of slope change. As a result, highlands do not fit into any scheme of general climate classification.

Local climates may also be produced by lakes. Lakes can keep temperatures from rising very high or falling very low, and often cause higher precipitation. This is especially true for places on the shore of a lake away from the common wind direction. For example, places on the eastern slope of Lake Michigan generally have more moderate temperatures and higher precipitation than locations on the western shore.

Forests affect local climates by reducing wind speed. They also increase humidity, as compared with nearby open land. Even cities create their own climates as large buildings reduce wind speed and produce haze and smog. Haze and smog reduce the loss of heat from the ground, resulting in locally higher temperatures.

VOCABULARY REVIEW

Match the word or words in the column on the right with the correct phrase in the column on the left. *Do not write in this book.*

1. Lines on a map that connect locations having the same average temperature.
2. Warm winds blowing down the eastern slopes of the Rocky Mountains.
3. A type of tropical climate in which summers are very wet and winters are dry.
4. A climate that causes the land to be covered with heavy forests of cone-bearing trees.
5. The climate found in the coastal region of southern California.
6. A climate whose rainfall supports a heavy growth of grasses.
7. The climate typical of northeastern North America.
8. A climate having very cold winters and short, warm summers.
9. The climate found in southeastern United States.
10. The climate found in the most northern part of Alaska.

a. Mediterranean
b. humid continental
c. foehn
d. subarctic
e. isotherms
f. savanna
g. steppes
h. chinook
i. tundra
j. marine west coast
k. humid subtropical
l. rain forest

QUESTIONS

Group A

Select the best term to complete the following statements. *Do not write in this book.*

1. The word "climate" comes from a Greek word meaning (a) weather (b) wind (c) slope (d) atmosphere.

2. Probably the most important single influence on climate is (a) the general movement of air over the entire planet (b) altitude (c) longitude (d) local conditions.

3. The amount of solar energy received at a particular location is governed mainly by its (a) latitude (b) longitude (c) altitude (d) distance from the prime meridian.

4. The polar regions of the earth are cold due to very long nights and (a) the altitude (b) the fact that they receive slanted rays of the sun (c) absence of moisture (d) very long days.

5. Examination of the isotherms of the earth, as in Figure 23 – 1, shows that temperature (a) depends entirely on the latitude (b) depends entirely on the longitude (c) increases to a high near the poles (d) is determined by other factors as well as latitude.

6. Bodies of water change temperature more slowly than land mainly because water (a) reflects most of the heat (b) has a smooth surface (c) has a high specific heat (d) has a low specific heat.

7. Because of the greenhouse effect, the average temperature of the atmosphere decreases with (a) latitude (b) distance from the equator (c) longitude (d) altitude.

8. Air flowing down mountain slopes, as in a chinook wind, is usually (a) cold and dry (b) warm and dry (c) cold and moist (d) warm and moist.

9. The Gulf Stream usually does not increase the temperature along the east coast of the United States because (a) it is nearly the same temperature as the land (b) it flows too far out at sea (c) it flows too fast (d) winds generally blow off shore.

10. In southern Asia, the monsoon winds during the winter bring (a) heavy rainfall (b) dry weather (c) changeable weather (d) heavy snowfall.

11. The most abundant rainfall of the earth occurs in (a) the polar regions (b) mountainous regions (c) the middle latitudes (d) the equatorial regions.

12. At the horse latitudes of 20° – 30°, rainfall is rare because (a) air is sinking and thus warming up (b) air is rising and thus cooling down (c) there are no mountains at these latitudes (d) there are no land masses at these latitudes.

13. The average temperature of northeastern Europe is usually high as a result of (a) its high latitude (b) its low altitude (c) the North Atlantic Drive and prevailing westerlies (d) the Gulf Stream and prevailing westerlies.

14. The most important influence on the climates of North America is (a) the high latitudes (b) the eastern mountain ranges (c) the longitudes (d) the position of the continent.

15. The summers in Alaska are surprisingly warm due to (a) its high latitude (b) its high altitudes (c) the long periods of daylight (d) the short periods of daylight.

16. Middle latitude climates have weather that is best described as (a) consistently warm (b) consistently cool (c) changeable day to day (d) quite dry.

17. Marine west coast climates generally have considerable precipitation caused by (a) northwest trade winds (b) sinking of air masses (c) westerlies blowing inland (d) easterlies blowing to sea.

18. Middle latitude deserts are unlike the tropical deserts since they (a) are dry only part of the year (b) have higher temperatures (c) have less precipitation (d) have a definite winter season.

19. Steppes are found at the edge of middle latitude deserts. They are also found around (a) subarctic climates (b) marine west coast climates (c) tropical deserts (d) tropical rain forests.

20. In the United States, the Mediterranean climates are found in (a) California (b) Texas (c) Louisiana (d) Florida.

21. Most of the northeastern part of North America has a climate that is (a) Mediterranean (b) humid subtropical (c) humid continental (d) steppe.

22. Most of the precipitation falling on the North American continent comes from (a) sinking air masses (b) cyclonic storms (c) prevailing easterlies (d) hurricanes.

23. Alaska has two types of polar climates: (a) subarctic and tundra (b) humid continental and subarctic (c) tundra and humid continental (d) steppes and tundra.

24. The Hawaiian Islands have both (a) humid continental and rain forest climates (b) rain forest and savanna climates (c) humid continental and savanna climates (d) Mediterranean and rain forest climates.

25. Ocean currents flowing near continental shores modify the climate by making the temperature (a) less extreme (b) more extreme (c) colder (d) warmer.

26. The most common example of local influence on climate is (a) altitude (b) latitude (c) nearness of large bodies of water (d) presence of forests.

27. The climates of highlands (a) are readily classified into the general climate scheme (b) do not fit into any scheme of climate classification (c) are always cold and dry (d) depend entirely on altitude.

28. Places on the eastern shore of Lake Michigan, when compared to places on the western shore, have (a) higher temperatures and lower precipitation (b) lower temperatures and lower precipitation (c) more moderate temperatures and higher precipitation (d) more moderate temperatures and lower precipitation.

29. Forests affect local climates by (a) increasing the temperature (b) reducing wind speed (c) decreasing humidity (d) decreasing precipitation.

30. Cities create their own climates, which generally include (a) haze, smog, and higher temperatures (b) haze, smog, and lower temperatures (c) greater wind speeds and lower temperatures (d) greater wind speeds, haze, and smog.

Group B

1. What factors other than latitude determine the main effect of local wind patterns on climate?

2. What is the chief influence affecting the average temperature of a given place?

3. Give three reasons why the surface of water heats or cools more slowly than the surface of land.

4. What causes the temperature to decrease as altitude increases?

5. Why is the average temperature of northwestern Europe unusually high for its latitude?

6. Describe the kind of weather that is typical of monsoons.

7. What effect do the changing seasons have on the global belts of precipitation?

8. Describe the three general kinds of tropical climate, including the type of air mass associated with each and the annual rainfall of each.

9. What is the principal cause of the tropical deserts?

10. Explain why even though the soils of the tundra climate regions have an abundant supply of water, plant life is restricted to mosses and a few scattered small shrubs.

11. Why do the middle latitude climates have such changeable weather?

12. Why does the west coast of the United States receive most of its precipitation in the winter months, while the middle and east coast portions receive more of their precipitation in the summer?

13. Explain why classification of climates often fails when you consider a specific location.

14. Prepare a table of the various middle latitude climates of North America giving (a) general position (b) type of air involved (c) annual precipitation (d) typical plant growth (e) areas having this climate.

15. Erecting any sizeable buildings, such as radar stations, in regions having tundra climate is quite an engineering challenge. Can you suggest why this is so?

16. Describe the average rainfall expected in specific regions as you go from the equator to the north pole, and explain why these amounts occur.

17. Describe the two different kinds of climate found on the west coast of the United States, and explain why this difference occurs.

18. Explain why the same winds that cause the middle latitude deserts in Nevada, Arizona, and California also give abundant rainfall in the midwestern region of the United States.

19. Give two reasons why highlands have a greater difference in day-and-night temperatures than most other places.

20. How do haze and smog around cities cause an increase in the local temperature?

the cascade volcanoes

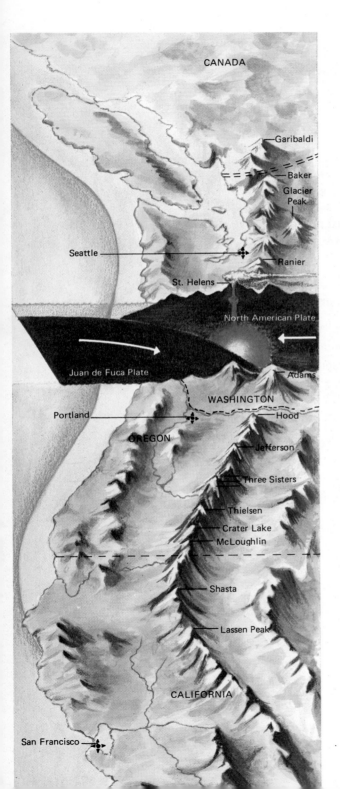

Labels on map: CANADA, Garibaldi, Baker, Glacier Peak, Seattle, Ranier, St. Helens, North American Plate, Juan de Fuca Plate, Adams, WASHINGTON, Portland, Hood, OREGON, Jefferson, Three Sisters, Thielsen, Crater Lake, McLoughlin, Shasta, Lassen Peak, CALIFORNIA, San Francisco

Ice and Fire

Stretching in a single line from California to Canada along the Pacific Coast of the United States are some of the world's largest volcanoes. These majestic peaks lie about 75 km (about 50 mi) apart and form the backbone of the Cascade Mountain Range. The Cascades make up a rugged, heavily forested highlands region that runs north-south about 160 km (about 100 mi) inland from the coast. Glaciers and snowfields are common in the Cascades. Volcanic fire is also a part of these glacier-carved mountains. At least eight of the Cascade volcanoes have erupted during the last 150 years.

It is no accident that these volcanoes are lined up in a single rank along part of the western edge of the North American continent. They are evidence of a collision between two of the earth's crustal plates. Part of the Pacific Ocean floor, the Juan de Fuca plate, is believed to be thrust beneath the edge of the North American plate. See Figure A. The descent of the plate causes deep rocks to become partly melted. Magma rising from this deep source creates the volcanic eruptions as it reaches the surface.

There is evidence that the part of North America where the Cascade Range is located has experienced widespread volcanic activity for about 50 million years. Lava from countless eruptions has covered the entire Pacific Northwest. The greatest outpouring took place about 15 million years ago, when huge amounts of liquid lava flooded much of central Washington and Oregon. About seven million years ago, the region began to be lifted into the mountain range in which today's volcanoes have grown. Much of the old volcanic deposits in the Pacific Northwest have weathered into fertile soil. Farms and ranches now cover the region, and several large cities, along with many smaller communities, today dot this ancient volcanic area. However, the volcanoes are not dead.

FIG. A

A Message from Mt. St. Helens

At 8:32 a.m. on May 18, 1980, one of the younger Cascade volcanoes provided a violent reminder of the power that still lurks beneath its surface. Mt. St. Helens has a history of fairly recent eruptions. Prior to 1980, it had been active between 1831 and 1857. Scientists who studied the mountain concluded that it had erupted at least twenty times during the past 4,500 years, with an average of 225 years between eruptions. In a scientific report issued in 1978, it was predicted that Mt. St. Helens was the one Cascade volcano most likely to erupt within the next 100 years. But there was no way to determine when that eruption would occur. Then, after about two months of earthquakes and very small eruptions, one side of the mountain exploded with a bang heard up to 500 km (about 300 mi) away. A column of volcanic ash climbed to an altitude over 20 km (more than 12 mi) above the volcano. See Figure B. A great, black, debris-filled cloud of hot gases roared out from the blast at speeds between 100 to 400 km/hr (about 60 to 250 mi/hr), destroying everything in its path. See Figure C.

FIG. B

FIG. C

Floods and mudflows created by the eruption were another major cause of destruction. Water was suddenly released from melting snow and ice, as well as from a nearby lake and inside the volcano itself. Ash and fine rock particles mixed with the water to produce a liquid mud that raced down valleys and river channels. A series of mudflows and floods swept away homes, bridges, and logging equipment situated along river valleys between the volcano and the sea. Altogether, the Mt. St. Helens eruption cost the lives of sixty-two persons and more than one billion dollars in damaged property and lost timber. Future eruptions are an ever-present threat to the people of the Pacific Northwest, who now see the silent Cascade volcanoes as an unknown danger to their lives and property.

24
people and the planet

objectives

- ☐ Distinguish between renewable and nonrenewable resources.

- ☐ Describe how fossil fuels are formed.

- ☐ List and explain the ways that atomic, geothermal, and solar energy may be obtained.

- ☐ Identify the most important future energy problems.

- ☐ Describe several ways an ore can be formed.

- ☐ List and explain the uses of some nonmetallic mineral resources.

An aluminum can comes from the earth. Any handful of soil or piece of rock that you might pick up will probably contain some aluminum. It is an element found in many common minerals. However, the aluminum in the earth's crust is rather different than the aluminum in a can. The aluminum that occurs naturally in the earth is extracted and processed in various ways to produce the aluminum we use commercially. Aluminum or any other substance removed from the earth's crust can only be obtained by changing the natural state of the earth's surface. People are the only form of life on earth with the ability to locate, separate, and use materials from the earth in ways that affect the environment.

Any naturally occurring material found in the earth that we can use is a *resource*. In addition to metals such as aluminum, resources include oil, coal, soil, water, air, plants, and animal life. Some resources are said to be renewable because they are never actually used up and they tend to change back again to their natural form. Water is an example of a renewable resource. On the other hand, some resources are nonrenewable. Either they cannot be replaced, or else they take a very long time to replace. Aluminum is an example of a nonrenewable resource. However, aluminum can be *recycled* like many other nonrenewable resources. You may have taken aluminum cans to a site where they are collected for recycling. See Figure 24–1.

Before the metal in any object you turn in for recycling can appear again as a new can or some other useful object, energy must be used. The metal must be melted, purified, and shaped. These processes require a supply of energy. Unlike the metal, the energy used can never be

FIG. 24–1. Recycling of aluminum cans and other objects begins when they are brought to collection centers like the one shown in this photograph. (Andrew Rakoczy)

recycled. The heat energy used to melt the aluminum, for example, is never recaptured. Thus solutions to the problems of obtaining and using the earth's resources first require an answer to the problem of securing a plentiful supply of energy. For this reason, we will first discuss the sources of energy available on our planet.

EARTH'S ENERGY SUPPLY

Fossil fuel resources. Heat and light produced by a piece of burning organic matter, such as wood or charcoal, represent stored solar energy. Plants, and a few animals, are able to absorb radiant energy from the sun and convert it into the organic matter in their cells. Almost all of the remaining solar energy that reaches the earth is held for a time by the atmosphere and oceans before it escapes back into space. The organic remains that have accumulated in the earth's crust are the most important source of trapped solar energy. These organic remains consist mainly of the chemical elements carbon and hydrogen, along with some oxygen. Compounds containing these elements are generally called *hydrocarbons*. Coal, petroleum, and natural gas are made up of hydrocarbons and are known as *fossil fuels*. Fossil fuels, the source of almost all of the energy we use, are found in the sedimentary rocks in the earth's crust.

FIG. 24–2. (Top) A coal strip-mine in Ohio. (Bottom) The same area four years later, after the soil removed to make the strip-mine has been replaced. (American Electric Power Company; David Brock)

Among the fossil fuels, only coal can be considered rock-like. Coal is found in areas that millions of years ago were covered by swamps. Plant remains accumulated in these ancient swamps, but, because of a lack of oxygen in the swamp waters, they did not decay completely. In time, the plant residue formed a brownish-black material called *peat*. Layers of sediment were deposited on top of the peat. The weight of the sediment squeezed the water out of the peat and changed it into a more dense material called *lignite*, or "brown coal." As more sediments accumulated, the pressure increased. The lignite then became *bituminous*, or "soft coal." Higher pressures, created when rocks folded during mountain building, produced a very hard form of coal called *anthracite*. Millions of years are needed to change peat into bituminous or anthracite coal. These forms of coal contain about 80–90 percent pure carbon and produce a large amount of heat when they burn.

Coal is found underground in layers between beds of shale, limestone, and sandstone. These "seams" of coal range from a few centimeters to several meters in thickness. Much of the coal is found in seams that are buried far beneath the surface. Deep mines are needed to reach these deposits, and horizontal tunnels that follow the coal seams are built. Where layers of coal are exposed along a mountain slope, the mine can be driven directly into the coal seam. When coal deposits are close to the surface, *strip-mining* is used. In this method, soil and rock lying over the coal seam are removed with heavy machinery. See Figure 24–2. The coal can then be obtained. In the past, the soil and rock removed during strip-mining were left in large piles. This destroyed the natural environment of the area. As the natural plant cover was removed, a barren scar of sterile sub-soil was left on the landscape. Rapid erosion of the exposed sub-soil clogs streams and reservoirs with silt and debris, which often cover productive land below the strip-mined area. Today when the strip-mining method is used, the soil and rock are replaced, although the land can never be exactly as it was before.

Coal is the most abundant fossil fuel. It is found all over the world. However, only three countries, the United States, the Soviet Union, and China, have almost two thirds of all the known deposits of coal. At the present rate of use, these reserves will last at least 200 years.

There is probably much more coal yet to be discovered. The United States has 28 percent of the world's known coal resources. This is enough to supply the energy needs of the nation for at least 400 years at the present rate of consumption. It might seem that coal is the answer to the energy problems of our country. Unfortunately, the numbers that show a huge supply of coal do not tell the whole story. Coal is expensive and often dangerous to mine. Compared to other fossil fuels, it is difficult to handle and transport. Mining and burning of coal produce waste materials and pollutants that damage the environment. Production and use of large amounts of coal will first require solutions to these problems. Continuing research and wise management by private industry and government will likely provide the answers.

Petroleum is a liquid mixture of hydrocarbons. It is formed by organic matter in sediments that were once part of the floor of shallow seas within the continents or along the continental margins. The organic materials came from sea life or were washed off the land into the sea by rivers. When the sediments became deeply buried, the organic matter was changed by heat flowing upward from the earth's interior. In the same way a roast is cooked in an oven, the organic matter in the rocks was cooked by the heat deep within the earth's crust. A roast that is cooked at the correct temperature turns brown and produces an oily liquid. Organic matter in rocks also turns brown and yields liquid petroleum if the temperature is not too high. Organic matter that is buried very deep for a long time will turn black and yield only natural gas. Some of the sediments containing the organic matter never become buried deep enough for petroleum to be formed. These rocks are found near the surface and are known as *oil shales*. Petroleum can be obtained from oil shales if the rocks are artificially heated.

The sedimentary rocks in which petroleum forms are filled with many small openings, or pores. These pores are filled with water. As the sediments are buried deeper, the fluids are squeezed out of the pores and tend to move upward into neighboring rocks that are porous. The petroleum and water move into more shallow rock layers until they meet some nonporous layer that forms a trap. Because the petroleum is less dense than water, it floats on the top of the trapped liquids, forming an oil pool. An oil pool is actually an oil-bearing rock formation with the

activity

Heating wood in the absence of air causes it to decompose into a number of useful materials. To identify them, perform the following steps:

A. Partially fill a 10 cm test tube with wood splints or dry sawdust. Connect the apparatus as shown below and heat with a burner.

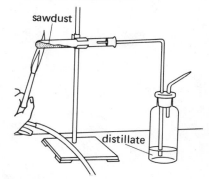

sawdust

distillate

B. Test the gas at the jet with moist, neutral litmus paper.

  1. What happens to the litmus paper?

C. The gas being produced is a mixture of CO, CO_2, CH_4, and some very volatile liquids.

  2. Does the gas burn?

D. Continue heating until no more liquid distillate collects in the bottle. Remove the stopper from the bottle and then stop heating.

  3. Describe the color and odor of the liquid distillate.

The distillate usually forms two liquid layers. The top layer is called pyroligneous acid; the bottom is wood tar.

  4. What is the solid residue left in the tube called?

  5. List the products you have made from wood and use reference books to help you describe the uses of each.

FIG. 24–3. Two common types of oil traps. In A the oil pool is formed in porous rock beneath a nonporous dome. In B the oil is trapped where a nonporous layer is displaced by a fault.

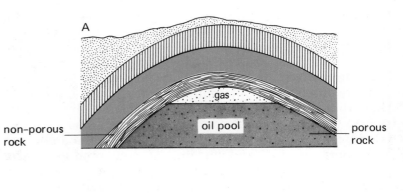

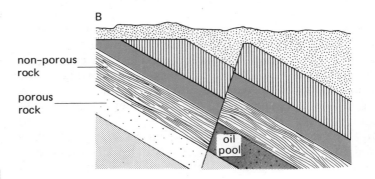

FIG. 24–4. Petroleum and gas are under pressure beneath the surface. Drilling a well into the deposit may release the gas and oil suddenly, spilling the oil and causing the threat of fire. (Olivier Rebbot/ Woodfin Camp)

petroleum trapped in many small pores. Salt beds often form the nonporous cover over oil pools, as shown in Figure 24–3A. Another common type of trap that creates oil pools results from a change in the rock structure, such as a fault, which causes the petroleum to meet a nonporous layer. See Figure 24–3B.

When a well is drilled into an oil pool, the petroleum will often flow out onto the surface. See Figure 24–4. Natural gas trapped with the petroleum also usually flows out. The liquid petroleum rises because water tends to move upward through the oil-bearing rock. Thus the petroleum is pushed up. However, if the petroleum is taken from the well too rapidly, the water cannot rise fast enough to maintain the upward pressure. The petroleum stops flowing. Water or gas must be pumped back into the well to increase the pressure. Too rapid production from an oil pool also causes the oil to separate into small pockets that are bypassed by the water. About 60 to 70 percent of the petroleum usually cannot be recovered from most oil pools. But almost all natural gas in an oil trap can be obtained.

In general, the quantity of natural gas in an oil-bearing formation increases with depth. Higher temperatures in deep formations change more of the petroleum into natural gas. Scientists believe that there are very large quantities of natural gas yet to be discovered in deposits located 4,600 m (15,000 ft) below the surface. The cost of finding such deep traps and drilling into them will be far greater than the present cost of producing petroleum and gas from shallow oil pools.

Suppose that you could follow scientists, like those shown in Figure 24–5, who search the earth for petroleum and natural gas. You would see that their work involves exploring the earth's crust to discover the kinds of rock structures that are likely to contain oil traps. Many years of exploring for oil have shown that the sedimentary rocks that might contain petroleum are not rare. In the United States, for example, thirty-one of the fifty states have commercial oil and gas wells. The United States has been explored for petroleum more completely than any other part of the world. Yet the actual amount of reserves that still exist in this country is not known. Scientists estimate that at least 75 percent of the total supply of petroleum in the entire United States has already been discovered. Much of the undiscovered supply is believed to lie under the sea along the continental shelves. It is almost certain that the production of petroleum in the United States has already reached its peak and will slowly decline in the future. In the rest of the world, however, more than 90 percent of the petroleum is still in the ground. This means that there is probably enough petroleum and natural gas left to supply the world's needs for at least a century, and perhaps much longer. But it takes millions of years for petroleum to accumulate in the earth's crust. Thus the supply is being used at a far faster rate than it is being replaced. Petroleum and natural gas are resources that will certainly be permanently exhausted at some time in the future.

FIG. 24–5. An oil exploration crew at work. The search for new sources of petroleum on land and beneath off-shore waters continues. (James Foote)

Nuclear energy. At the center of every atom is a tiny *nucleus.* Compared to the size of the whole atom, the nucleus takes up about the same space as a gumdrop in a football stadium. However, all of the atom's protons and neutrons are contained in this small nucleus. The protons and neutrons are held in the nucleus by a very strong binding force. When the nucleus undergoes a change, some of the binding force is given off as a large amount of energy. An example of such a nuclear change is fission. Nuclear fission occurs when the nucleus of a large atom splits into two or more smaller nuclei. When fission, or splitting, takes place, an enormous amount of energy is released. Most of this energy is in the form of heat. Nuclear fission can thus be used as an energy source in the same way as fossil fuels.

Only one kind of atom found in nature can be used as a source of nuclear energy. This kind of atom makes up a very rare form of uranium called uranium 235, or U-235. To use U-235 as a nuclear fuel, it is first necessary to extract this scarce metal from deposits of natural uranium in the earth's crust. The U-235 is then shaped into rods. These fuel rods are put together in a bundle. When one U-235 nucleus in these fuel rods splits, it releases energy along with neutrons. These neutrons strike neighboring U-235 atoms and cause their nuclei to undergo fission. This process is called a *chain reaction*. As the chain reaction continues, it causes the bundle of fuel rods to become very hot. A liquid, usually water, is pumped around the fuel rods to carry away the heat. The hot liquid or steam is used as a source of energy for running electric generators. This chain reaction can be controlled. Rods are pushed into the nuclear fuel assembly to absorb the neutrons. Without these neutrons, fission can no longer continue. The equipment in which a controlled nuclear fission reaction is carried out is called a *nuclear reactor*. See Figure 24–6.

One gram of U-235 is a piece about the size of a grain of

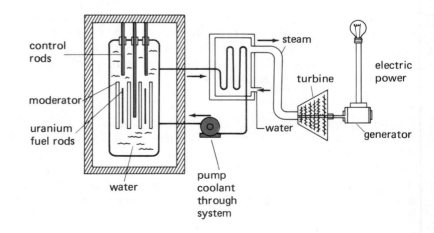

FIG. 24–6. In a nuclear power station, heat is produced by fission of uranium atoms in the fuel rods. Coolant pumped around the fuel rods transfers the heat to a heat exchanger, where steam is produced. The steam drives a turbine and generator.

rice. A piece this size yields an amount of energy equal to the burning of 3 metric tons of coal or about 14 barrels of petroleum. No one knows how much uranium is available in the earth's crust. Very few parts of the world have been as thoroughly explored for the presence of this valuable nuclear fuel as the United States. The findings seem to indicate that the amount of uranium reserves will yield much more energy than the petroleum and natural gas that remain in this country. However, the coal reserves of the United States are a more abundant source of energy for the future than the naturally occurring nuclear fuel. It may be possible to build a nuclear reactor that produces more nuclear fuel than it consumes. This type is called a *breeder reactor*. If breeder reactors can be successfully built and operated, the nuclear fuel available will provide much more energy than any supply of fossil fuel.

At some time in the future, it may also be possible to obtain large amounts of energy from a nuclear process called *fusion*. Nuclear fusion is the joining of the nuclei of atoms to form other nuclei. Like fission, fusion releases large amounts of heat energy. This heat energy can be used to run electric generators. The sun and all other stars get their energy from nuclear fusion. If fusion reactions become a source of energy in the future, hydrogen taken from ocean water will be used as fuel. Thus the amount of energy available from this source would be almost limitless.

Geothermal energy. The earth's interior is very hot. This heat represents a huge supply of energy called *geothermal energy*. Some of this energy appears on the earth's surface naturally in the form of volcanoes and hot springs. Earthquakes may also get their energy from the earth's interior heat. In some places on the earth, it is possible to use geothermal energy to supply a part of our energy needs. Wells drilled in a few special locations yield a steady flow of steam or very hot water. See Figure 24–7. The steam can be used directly, or the hot water can be used to make steam, to run electric power generators. The city of San Francisco gets part of its electricity from geothermal steam produced by wells in nearby mountains. Additional geothermal steam fields presently producing electricity are located in Italy, Mexico, Japan, Iceland, New Zealand, and the Soviet Union. Many other locations able to supply geothermal energy have been discovered

FIG. 24–7. These steam wells in northern California have been drilled into an underground reservoir of geothermally heated water. The temperature and pressure on the water allow it to boil so that it reaches the surface as steam. This is a rare type of geothermal energy resource and is the easiest and least expensive source of geothermal heat used to generate electricity. (J. R. Eyerman/Uniphoto)

and can be developed as additional sources of energy.

Geothermal energy sources already in use rely on steam or hot water that comes from the earth's interior. Another method for making energy would be to drill deep wells into the body of hot rock within the crust. Water pumped down these wells would be changed into steam and returned to the surface. The steam would then be used to operate power stations. However, this use of geothermal energy requires that the deep, heated rocks have a network of channels through which the water can flow. No natural formations with these channels have as yet been discovered. Research is under way to find methods of making artificial fractures in the rock.

In many places in the United States, especially in the western states, there are underground reservoirs of hot water. In several locations wells have been drilled into these geothermally heated water supplies to provide the energy for experimental power plants.

Some deep sedimentary layers along the Gulf Coast of the United States contain hot salt water under great pressure from the weight of the rock above. This water is thought to contain a large amount of dissolved natural gas. In the future, it may be practical to drill wells into these deep deposits in order to use both the heated water and the gas.

The total supply of geothermal energy beneath the United States is so great that it has the potential for providing a large part of our future energy needs. However, there are many serious technical and economic problems associated with its development. These problems will prevent geothermal energy from being a major energy source until well into the next century, if ever.

Solar energy. The term "solar energy" includes a variety of methods for converting the sun's radiation into heat, electric power, and fuels. All of the processes for using the sun's radiant energy can be classified into two groups. The first group consists of *direct* methods for changing solar energy into heat or electricity. For example, a house can be designed with windows that take advantage of the sun's rays to help heat the inside space. This is a passive method for using solar energy because there is no method for transferring the heat. An active system might use a solar heat collector to produce hot water, which is moved by pumps to where the heat is needed. Many houses and commercial buildings now use active and passive methods for space heating and providing hot water. Solar space cooling can also be done, but it is not economical at the present time. Solar energy also provides heat for some industrial uses such as commercial laundering, food processing, and drying of crops.

Another direct use of solar energy is in the generation of electricity. One approach to production of solar electricity uses solar cells. A solar cell is an electronic device that generates a small electric current when sunlight falls on its surface. Many solar cells can be wired together to produce large amounts of electric power. At the present time, however, solar cells are too expensive to be used as a common means of generating electricity. They are now practical for use only in remote locations, such as spacecraft and isolated places on the earth.

Solar electricity can also be generated by concentrating sunshine to produce high temperatures. A large field of mirrors, which move to follow the sun, reflect heat to a central tower. There a boiler produces steam that is used

FIG. 24–8. Concentrated sunlight from 1775 mirror faces strikes the steel target mounted about 35 m (about 114 ft) up on the "power tower" at the U.S. Department of Energy's Solar Thermal Test Facility. In the first major test, a 61 × 91-cm (2 × 3-ft) hole was burned in the .6 cm (.25 in.) thick steel in less than two minutes. (Jim Nachtwey/Black Star)

to run electric generators. See Figure 24–8. Several electric generating plants of this type have been operating for many years in Europe. An American experimental plant has also been built in the California desert. Because solar power plants are shut down when the sun does not shine, they are not able to replace electric generating stations powered by fossil fuels or nuclear energy. They can only help meet the peak demands in an electric power system.

Solar heating and electricity generation use the direct rays of the sun. A second general way to use solar energy involves the sun indirectly. For example, unequal solar heating of the atmosphere causes winds. The energy of moving air can be used to turn various devices to produce electricity or perform some other work, such as pumping water. Windmills have been used to provide energy for many centuries. The energy taken from wind is an example of *indirect* solar energy. Wind energy is now being used worldwide to produce electricity. Small wind generators are used to meet some energy needs of individual homes. Much larger wind devices produce enough power to help with the load in the total electric system. See Figure 24–9. But there are only a few places where the wind blows steadily and with enough speed to make wind generators practical. In the United States, areas such as the New England coast, the western Great Plains, the Pacific Northwest, and Hawaii have winds best suited for generating power. Even in the most favorable locations, the wind does not always blow. Thus wind generators work best if some method for storing electric energy is available. Because wind generators can be built only in limited areas of suitable winds and cannot operate continuously, they will be able to supply only a small part of our total energy needs.

The oceans are also a source of indirect solar energy. About 75 percent of the solar energy reaching the earth's surface is absorbed by the sea. Much of that energy is held in the warm waters near the sea surface. One method of obtaining solar energy from the sea uses the very warm surface water in tropical regions to vaporize a low-boiling-point liquid. The vapor then runs an electric generator in the same way as steam runs most other power plants. Deep, cold water is pumped up to cool the vapor and change it back to a liquid. A small experimental power plant using this principle has worked in the sea off Hawaii. See Figure 24-10. It is the only method for using solar energy that can produce a continuous supply of electricity. Future electric generators using energy from the sea could send the output to shore through a cable or serve to power a floating factory.

Another indirect source of solar energy is the energy trapped in plant matter that has not been buried and altered, as have the fossil fuels. When the plant matter is used as a fuel, it becomes an indirect source of solar energy. All of the plant materials that can be used as fuels are called the *biomass*. For example, wood, when used as a fuel, is part of the biomass. Some of the biomass, such as

FIG. 24–9. The power of wind can be harnessed to provide an additional source of energy. In locations where it blows steadily enough, wind is a dependable energy source. (George Dodge/DPI)

FIG. 24–10. Sun-warmed water near the sea surface is used by this experimental OTEC (Ocean Thermal Energy Conversion) system to generate electricity. (U.S. Department of Energy)

trees used for firewood, consists of plants grown especially for fuel. In some cases, the energy crop is not used directly as a fuel, but is changed into another fuel substance. For example, grain can be changed into alcohol, which can be used as a motor fuel alone or mixed with gasoline to make "gasohol." But the most important source of biomass energy is agricultural and municipal wastes. Agricultural wastes include parts of trees harvested for lumber or paper and unused portions of food crops. These wastes can be used as fuels. It is estimated that the energy available from agricultural wastes is about equal to the amount of energy used to grow the crops. With careful management, the agricultural system could provide its own energy supply in the future. However, crops grown only to supply energy can never make more than a very small contribution to the total energy needs of the United States.

The use of municipal wastes to fuel power plants is a rapidly growing source of energy. In many cities, non-burnable wastes are separated from the trash. The remainder of the waste is either used directly as fuel or converted into gas or petroleum-like substances that are burned. The problem of trash disposal is solved, valuable metals and other materials are recovered, and an energy supply is provided.

Energy and the future. You will probably spend most of your life in a world that has serious problems with its energy supplies. These problems will not come from an overall shortage of energy. There are several energy sources, such as coal, that can supply world energy needs for several centuries. The problems will come from the need to make an efficient and orderly change from the use of decreasing supplies of petroleum and natural gas to the use of alternate sources of energy.

For example, the United States faces a growing problem with the supply of liquid fuels, including gasoline. The production of petroleum, which is the present source of liquid fuels, is likely to drop after the year 2000. Replacing the decreasing supply of domestic petroleum with foreign fuels will cause serious economic and political problems. It will be necessary for this country to reduce its demand for petroleum. This can be done by conserving the available supply and developing other sources of liquid fuels. Oil shale, for example, is available from huge deposits in

FIG. 24–11. Temporary shortages of gasoline are likely to be a part of the general problem of decreasing supplies of petroleum in the United States. (Larry Lee/The Image Bank)

the western United States. Petroleum can be obtained by heating oil shale according to methods that have already been developed. However, petroleum obtained from oil shale is expensive. In addition, large-scale production of petroleum will cause environmental problems of waste, pollutants, and the need for water in arid parts of the country. Although it is difficult to predict its maximum potential for supplying petroleum, oil shale can be a valuable future source of liquid fuels.

Coal can be converted into a type of synthetic petroleum. Like the development of oil shale, production of synthetic fuels from coal is costly and the cause of environmental problems. The process by which coal is changed into liquid fuel releases by-products that could cause serious air and water pollution. No commercial plants now exist in the United States for the production of synthetic fuels from coal. But liquid fuels derived from coal will become more important in the future as the cost of natural petroleum rises and further research enables efficient and safe conversion plants to be constructed.

One of the most efficient uses of coal is as a fuel in electric generating plants. Petroleum used in the generation of electric power could be replaced by coal from the huge reserves in the United States. But coal is much more expensive to burn cleanly. Coal smoke contains large amounts of oxides of sulfur and nitrogen. Moisture in the atmosphere can combine with these oxides to produce acid rain. As the acid rain falls, it can harm plant and animal life in large areas near coal-burning power plants. In some lakes of the northeastern United States, fish have already been killed by the runoff from acid rain. Problems that are caused by pollutants in coal, such as acid rain, can be prevented only by installing equipment that will remove the harmful gases from the coal exhaust. Before large amounts of coal can be used to generate electricity, more research must be done to clearly establish the environmental costs.

Generation of electricity by nuclear energy also presents serious problems. One of the most important is the safe disposal of radioactive nuclear wastes. These wastes must be stored for several centuries until their radioactivity falls to safe levels. To date, there is no known method for disposing of these wastes that does not involve some risk. Nevertheless, there is reason to believe that the total electricity needs of the country cannot be met by the 1990s unless nuclear power plants now under construction or on the drawing boards are completed.

As you can see, there can be no single solution to the energy problems of the world. It is certain that petroleum and natural gas will become less abundant and more expensive. The future will bring a gradual shift to energy sources whose potential usefulness and costs are not completely known at present. Solving the energy problems of the future will require careful planning and constant effort to conserve energy. Without energy, other resources are useless.

THE CRUST AS A SOURCE OF MATERIALS

Metals from the earth's crust. A few metals, such as gold, silver, platinum, and copper, can be found in the earth's crust as native metals. A native metal is one that exists in the pure form. See Figure 24–12. Most metals, however, exist as compounds or as parts of different metals. These metal-bearing minerals are scattered throughout the crust. To obtain the metals at reasonable cost, we must find places where the metal-bearing minerals are concentrated. An accumulation of minerals containing a substance that can be recovered for profit is called an *ore*. There are several ways in which ores are formed within the earth's crust.

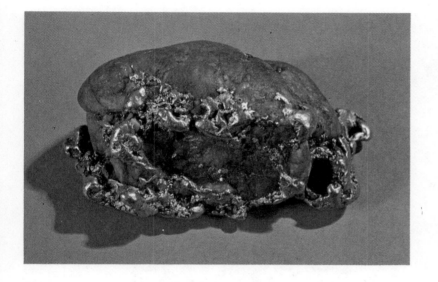

FIG. 24–12. This natural lump, or nugget, of nearly pure gold is an example of a native metal. (Smithsonian Institution)

One common type of ore is formed within a body of magma as it cools. As the magma becomes solid, dense crystals form. These crystals sink through the remaining liquid magma. Layers or bodies of certain minerals accumulate within the mass of cooled magma and form deposits of ore. The ores of chromium, nickel, and iron are commonly formed by separation from magma.

The high temperature of magma can also cause ores to be formed when the magma comes in contact with surrounding rock. Hot water carrying dissolved minerals flows from the magma and soaks into the surrounding rock. The original minerals in the surrounding rock are replaced by those carried in the hot magma fluids. This process often forms a band of ore minerals around a body of cooled magma. The ores of lead, copper, and zinc are formed in this way.

In later stages of cooling, a body of magma will often produce large amounts of hot solutions that contain dissolved minerals. These hot solutions can enter cracks in surrounding rocks and deposit their minerals in narrow bands called *veins*. Veins often contain valuable minerals like gold and silver. A large number of thick veins filled with mineral deposits is called a *lode*. Sometimes the mineral solution can spread through many small cracks in a large mass of rock. Some of the largest deposits of copper ore were formed in this way. See Figure 24–13.

Ores can also be deposited by the action of streams or by

FIG. 24–13. A lode of copper ore is being dug out of this open pit mine in Utah. (Kennecoh Copper Co.)

waves along the shore. Small fragments of native metals such as gold or platinum are released from their parent rock by weathering. These metal fragments are carried away by streams. Along the stream bed, or sometimes on beaches, the density of the metal causes it to become concentrated in layers of gravel. These layers are called *placer deposits*. The placer deposits are likely sites in which to find native gold and platinum. Ores of tin, as well as diamonds and other precious gems, are sometimes found in placer deposits.

Water moving downward through the ground can also produce ore deposits. As rainwater moves down from the surface through the cracks in the rocks, it dissolves minerals. These dissolved minerals can accumulate in some places and become ore deposits. Ores of iron, lead, copper, zinc, and aluminum are often formed in this way.

Nonmetallic mineral resources. In addition to metals, large amounts of other substances are taken from the earth. Two of the most important of these nonmetallic earth resources are coal and petroleum. Other nonmetals taken from the crust are used in building. Some of the building materials include sand, gravel, crushed rock, and blocks of stone. Cement is made from limestone and clay. Gypsum is used to make plaster and wallboard. Almost all the nonmetallic materials used in building come directly from the earth.

Other minerals are used to manufacture many products we use in everyday life. One of these minerals is sulfur. Huge amounts of sulfur are used in the production of iron and steel, chemicals, fertilizers, and rubber. Some minerals, like phosphate rock, are used directly as fertilizers. Natural rock salt, or halite, is also a valuable mineral. It is used in the manufacture of chemicals. The list of minerals used in industry and agriculture is very long. The minerals mentioned earlier provide only an example of these important nonmetallic earth resources.

Mineral resources and the future. You probably have no trouble remembering that the world's supply of petroleum is limited. Every visit to a gas station is a reminder of this fact. The cost of gasoline is increasing, and there may be occasional shortages. Some day we could have the same kind of problem with mineral resources. The rate at which various minerals are being used is increasing each year. For example, the amount of aluminum consumed by each person in the United States has doubled during the past 50 years. Other mineral resources also show similar, but smaller, increases in the rate of consumption. Each person in the world's growing population represents a need for additional mineral resources.

The natural processes that concentrate minerals into usable ores take millions of years. Deposits of ore that are used up cannot be replaced. A new supply must be discovered in order to meet the continuing demand. At the present rate of use, the known supplies of most high-grade mineral ores will soon be exhausted. For example, it is estimated that the supplies of aluminum ore already discovered will last for about the next 30 years. Although the known reserve of each mineral resource is different, it appears that none can last more than 100 years into the future.

New scientific knowledge, however, may help to discover important mineral deposits still hidden in the earth's crust. Scientists now understand that the location of mineral resources is controlled by plate tectonics. The most important kinds of ore deposits are formed when minerals in rock dissolve in hot water. When the water cools, the minerals separate from the water and are deposited as solids. This process is continuously taking place along the mid-ocean ridges where huge amounts of sea water come into contact with magma. Hot springs develop along the ridges as fountains of mineral-rich hot water rise from the lava. See Figure 24–14. Mineral deposits build up on the sea floor around these submarine hot springs. Thus the places where the crustal plates are spreading apart are known to be probable locations of mineral deposits. These deposits are carried on the spreading sea floors to plate boundaries where volcanic activity can cause the minerals to separate into distinct zones. See Figure 24–15. Scientists now understand that mineral deposits are most likely to be found on the land where the plates are now, or once were, colliding and where there is evidence that the continents have split apart in the past. This knowledge, combined with new methods of surveying the earth's surface, such as the images from satellites, may be helpful in discovering new supplies of minerals. At the same time, conservation of our existing mineral resources by recycling will still be necessary to insure that the earth's resources can continue to support our planet's human population.

FIG. 24–14. A submarine hot spring photographed on the deep sea floor near a mid-ocean ridge. The dissolved minerals in the hot water give it a dark color. (Woods Hole Oceanographic Institution)

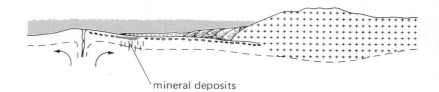

mineral deposits

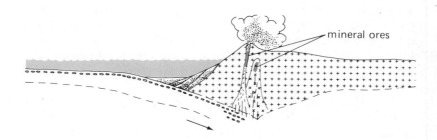

mineral ores

FIG. 24–15. (Top) Mineral deposits formed along the mid-ocean ridges are carried away as the sea floor spreads. (Bottom) At a plate boundary where the plate carrying the sea floor descends and is heated, the mineral deposits are separated into bodies of ore and move toward the surface.

VOCABULARY REVIEW

Match the words in the column on the right with the correct phrase in the column on the left. *Do not write in this book.*

1. Any natural material provided by the earth.
2. An accumulation of minerals containing a substance that can be obtained for a profit.
3. A large number of thick mineral veins.
4. Reusing certain metals such as iron and aluminum.
5. Organic remains consisting mostly of carbon and hydrogen.
6. Commonly called brown coal.
7. Known as soft coal.
8. A very hard form of coal.
9. Method of mining coal deposits close to the surface.
10. A sedimentary rock containing a heavy, waxy form of hydrocarbons.
11. Splitting of a large nucleus into two or more smaller nuclei.
12. Process by which energy is released during fission.
13. Produces more nuclear fuel than it uses.
14. Nuclear process by which the sun releases energy.

a. anthracite
b. bituminous
c. breeder reactor
d. chain reaction
e. fission
f. fusion
g. geothermal energy
h. hydrocarbons
i. lignite
j. lode
k. oil shale
l. ore
m. recycle
n. resource
o. strip-mining

QUESTIONS

Group A

Select the best term to complete the following statements. *Do not write in this book.*

1. The fuel not considered a fossil fuel is (a) coal (b) petroleum (c) uranium-235 (d) natural gas.
2. Fossil fuels are organic remains consisting mostly of (a) hydrocarbons (b) carbon (c) uranium (d) carbon and uranium.
3. Of the following, which is the hardest and most pure form of coal? (a) peat (b) anthracite (c) bituminous (d) lignite.
4. Coal is now found as a sedimentary rock (a) only on a few continents (b) only deep within the crust (c) scattered equally throughout the continents (d) in layers between beds of shale, limestone, and sandstone.
5. At the present rate of use, coal could supply the energy needs of the world for (a) several decades (b) several centuries (c) many centuries (d) several years.
6. A method of extracting coal that is close to the surface is (a) strip-mining (b) deep mining (c) scam-mining (d) pit-mining.
7. A deposit of sedimentary rocks containing many small openings filled with petroleum makes up a(n) (a) seam (b) lode (c) oil pool (d) placer.

8. The supply of coal and petroleum is limited on earth because (a) they are not renewable resources (b) they are destroyed when they are used (c) earth conditions are such that more coal and petroleum are not being produced in nature at a rapid rate (d) all of these ideas are involved in limiting coal and petroleum.

9. Nuclear fission happens when (a) a nucleus of a large atom splits into two or more smaller nuclei (b) an atom is heated (c) two or more nuclei form a larger nucleus (d) the sun gives off energy.

10. During the fission of an atom, the usable energy is given off as (a) light (b) heat (c) electricity (d) radioactivity.

11. The chain reaction in a nuclear generator is controlled by (a) releasing excess steam to the atmosphere (b) limiting the amount of fuel in the rods (c) pushing rods into the fuel bundles to absorb neutrons (d) pumping more liquid around the fuel rods.

12. It is possible that the supply of nuclear energy may be made far greater than any fossil fuel if (a) more atoms can be made to undergo fission (b) an artificial nuclear source is found (c) nuclear reactions are sped up (d) breeder reactors can be used.

13. The nuclear reaction that occurs on the sun is (a) fission (b) fusion (c) radioactivity (d) burning.

14. The source of geothermal energy is (a) earthquakes (b) volcanoes (c) the earth's hot interior (d) hot springs.

15. The hot water or steam produced from geothermal energy is generally used to (a) run electric generators (b) heat homes (c) power machines (d) produce fresh water.

16. Only a small fraction of the energy requirements of the world is expected to come from (a) geothermal energy (b) wind (c) tides (d) all of these.

17. An example of a direct method of changing solar energy into heat or electricity is (a) windows designed to let the sun's rays into the house (b) using a windmill (c) using warm ocean water to vaporize a low-boiling-point liquid (d) burning wood.

18. An example of an indirect method for changing solar energy into heat or electricity is (a) windows designed to let the sun's rays into the house (b) a solar heat collector used to heat water (c) using a part of the biomass as a fuel (d) using solar cells to produce electricity.

19. Passive and active methods of using solar energy can be used for each of the following, but are *not* economical for (a) producing hot water (b) space heating (c) space cooling (d) producing electricity.

20. Using a portion of the biomass for energy is demonstrated by (a) burning wood (b) burning agricultural wastes (c) using alcohol to produce gasohol (d) all three of these examples.

21. A rapidly growing use of biomass for energy is (a) converting coal to synthetic fuels (b) burning coal to produce electricity (c) burning municipal wastes (d) extracting petroleum from oil shale.

22. Acid rain, which is threatening the plant and animal life in a number of areas, would be greatly increased by using more (a) nuclear reactors (b) natural gas (c) fuel oil (d) coal.

23. One of the most important unsolved problems related to use of nuclear energy is (a) a reliable source of uranium-235 (b) the safe disposal of radioactive wastes (c) the cooling of the reactor (d) keeping a safe level of radiation in the area around the generating plant.

24. An example of a native metal is (a) gold (b) iron (c) chromium (d) nickel.

25. A native metal or an ore may be transported by a stream and concentrated in layers of gravel called (a) veins (b) a lode (c) placer deposits (d) magma liquid.

26. A nonrenewable resource is (a) air (b) water (c) soil (d) gold.

27. When natural processes concentrate metal-bearing minerals, the deposit is called (a) native metals (b) an ore (c) magma (d) hydrocarbons.

28. The most important nonmetallic earth resources are (a) sand and gravel (b) crushed rock and blocks of stone (c) coal and petroleum (d) limestone and clay.

29. A nonmetallic mineral resource that is used in almost every product available is (a) phosphate rock (b) halite (c) sulfur (d) gypsum.

Group B

1. Describe three ways in which magma is involved in forming ores.

2. What is meant by a placer deposit? List two native metals found in placer deposits.

3. Describe two ways in which water is involved in forming ores.

4. List one or more nonmetallic mineral resource used for each of the following:
  a. building materials
  b. production of iron and steel
  c. production of fertilizer
  d. production of chemicals

5. Explain why the outlook for the future supply of metals seems to be an encouraging one.

6. Describe the processes involved in the natural formation of coal.

7. Explain how petroleum and natural gas are thought to have been formed.

8. How is petroleum produced from oil shale?

9. Describe what happens in the nuclear fission process.

10. Give a brief description of how uranium-235 is obtained and processed to produce fuel.

11. Describe how energy is produced and used in a nuclear reactor.

12. Why would the fusion reaction give us an almost limitless energy source?

13. List several natural ways geothermal energy is released at the earth's surface.

14. Most geothermal energy comes from deep wells that contain hot water or steam. Describe another future method for obtaining geothermal energy.

15. Explain how tidal power could be used as a source of energy.

16. List four *direct* methods for changing solar energy into heat or electricity.

17. List four *indirect* methods of using solar energy.

18. Describe how a farm could provide all of its own required energy.

Prospecting for Wind

Like the deposits of oil, coal, and natural gas hidden beneath the earth's surface, unseen supplies of energy are present in the sky. That energy is contained in the winds. Winds can be made to turn electric generators. In locations where the wind blows strongly and steadily enough, it is practical to build wind generators. Power from the wind is proportional to the cube of the wind speed. Thus a wind blowing at 10 km/hr has eight times as much power as a wind blowing at 5 km/hr. The potential ability of a wind generator to produce electricity is greatly increased if the generator is at a site where the average wind speed is high.

In the past, wind speeds have usually been measured at airports and weather stations. These measurements do not necessarily point to locations with strong winds. Scientists have had to develop methods for finding the best locations for wind generators. One method involves studying the way in which trees and shrubs grow. Strong winds blowing steadily from one direction cause "flagging" of trees and shrubs. The trunk of a flagged tree, for example, is permanently bent away from the wind. Growth rings in trees are also usually thicker on the side of the trunk away from the wind. Wood from trees already cut down can be examined to give information about wind conditions where the trees grew. Photographs made from orbiting satellites and from aircraft at lower altitudes are also being studied. They provide evidence of flagging in plants, as well as general surface features, such as sand dunes, that might indicate strong winds.

When favorable locations for wind generators have been discovered, the strength of actual winds at that site can be measured with special kites. The kite is flown at certain altitudes while attached to a device that measures its pull. Measurements are repeated at higher altitudes until a profile of the winds up to about 100 m (about 350 ft) is obtained. This information is very useful in deciding the location and design of large wind generators. Wind prospecting will play an important role in the future development of the earth's moving air.

Salt Plus Sun Equals Electricity

Salt water, or brine, is more dense than fresh water. This means that a salt-water pond can have a layer of fresher, lighter water floating on top of the brine. Sunlight passes through the top layer and heats the brine below, sometimes to temperatures of 90° C (about 200° F) or more. A pond with warm brine covered by cooler, fresher water is called a solar salt pond, and represents an important potential source of electric energy.

A generating plant using energy from a solar salt pond is built beside a shallow, diked-off section of a salt lake. Floating windbreaks prevent waves that would tend to mix the water. Electricity is generated by using the hot brine to evaporate a low-boiling-point liquid. The vapor turns a turbine wheel along with an electric generator. After leaving the turbine, the vapor is condensed by the cooler water from the upper layers of the pond. A solar-pond electric plant can run continuously because the brine holds heat overnight and through periods of cloudy weather.

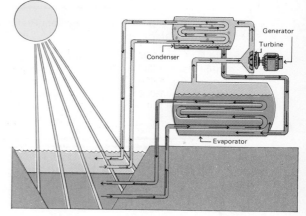

Energy Specialist

The earth is rich in energy resources. But effective use of these supplies requires the skills of people who are specially trained to find and develop new sources of energy; meteorologists, oceanographers, engineers, and technicians are but a few.

appendix

A. International System of Measurement
B. Geology in the National Parks
C. Geologic Calendar
D. Stones
E. Key for Identification of Minerals
F. Temperature Scales

A. International System of Measurement

METRIC UNITS

(UNIT SYMBOLS ARE IN PARENTHESES THE FIRST TIME ONLY.)

Metric System Prefixes	
Greater than 1	Less than 1
tera (T) = 1,000,000,000,000	deci (d) = 1/10
giga (G) = 1,000,000,000	centi (c) = 1/100
mega (M) = 1,000,000	milli (m) = 1/1,000
kilo (k) = 1,000	micro (μ) = 1/1,000,000
hecto (h) = 100	nano (n) = 1/1,000,000,000
deka (da) = 10	pico (p) = 1/1,000,000,000,000

Metric-English Equivalents

1 inch = 2.54 cm
1 foot (ft) = 30.48 cm
1 yard (yd) = 91.44 cm
1 mi = 1609.3 m
1 mi = 1.61 km
1 cm = 0.3937 in
1 cm = 0.0109 yd
1 m = 39.37 in
1 km = .621 mi
1 lb = 453.592 g
1 lb = 0.4536 kg
1 g = 0.002205 lb
1 g = 0.03527 oz
1 oz = 28.35 g
1 kg = 2.2046 lb

Commonly Used Metric Unit	Examples		English System Equivalents
Length: Meter (m)			39.37 inches (in)
	1 *kilo*meter (km)	= 1000m	0.621 mile (mi)
	1 *centi*meter (cm)	= 1/100m	0.394 in
	1 m	= 100cm	
	1 *milli*meter (mm)	= 1/1000m	0.0394 in
	1 m	= 1000mm	
	1 micron (μ)	= 1/1,000,000m	
	(this is one exception to the use of the prefixes)		
Mass (Weight): gram (g)			0.0353 ounce (oz)
	1 *kilo*gram (kg)	= 1000g	2.205 pounds (lb)
	1 *milli*gram (mg)	= 1/1000g	.0000022 lb
	1 g	= 1000mg	
Volume: Liter (L)			1.06 quarts (qt)
	1 *milli*liter (mL)	= 1/1000 L	0.00106 qt
	1 L	= 1000 mL	

B. Geology in the National Parks

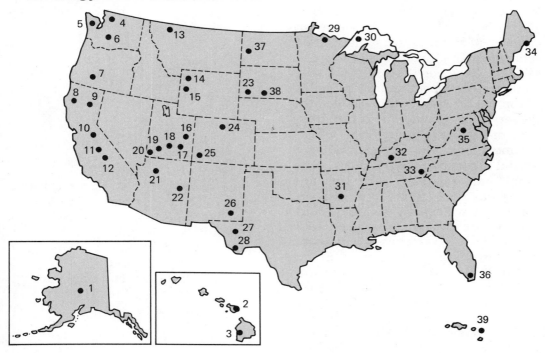

Key to National Park Locations

1. Mount McKinley
2. Haleakala
3. Hawaii Volcanoes
4. North Cascades
5. Olympic
6. Mount Rainier
7. Crater Lake
8. Redwood
9. Lassen Volcanic
10. Yosemite
11. Kings Canyon
12. Sequoia
13. Glacier
14. Yellowstone
15. Grand Teton
16. Arches
17. Canyonlands
18. Capitol Reef
19. Bryce Canyon
20. Zion

21. Grand Canyon
22. Petrified Forest
23. Wind Cave
24. Rocky Mountain
25. Mesa Verde
26. Carlsbad Caverns
27. Guadalupe Mountains
28. Big Bend
29. Voyageurs
30. Isle Royale
31. Hot Springs
32. Mammoth Cave
33. Great Smoky Mountains
34. Acadia
35. Shenandoah
36. Everglades
37. Theodore Roosevelt
38. Badlands
39. Virgin Islands

America's national parks are natural outdoor geological laboratories. There are no better places to see the results of geological processes. These areas, set aside as national parks, have been selected because they contain spectacular landscapes. An understanding of the natural forces that created this scenery can add much to the enjoyment of a visit to these special places.

Almost all national parks have special facilities to help visitors understand and appreciate how the landscape was created. Marked trails and roads that will guide visitors to main features are found in every park. Many of the parks also provide conducted tours, evening campfire programs, nature hikes, and other activities designed to provide information about things to do and see. Each park usually has a visitor center where information can be obtained. The visitor centers frequently have exhibits and displays calling attention to the most interesting features of the park. A visit to one of the national parks is also an opportunity to enjoy outdoor recreational activities, such as camping, picnicking, fishing, boating, swimming, and horseback riding.

Twenty of the national parks that have special geologic features are listed in the sections that follow. Each listing gives a brief description of the principal geologic features that can be seen. The state in which each park is located is given after its name. The numbered location of each park is shown on the map on p. 523. The number corresponding to the park's location on the map is given along with its name. An address that can be used for requests and questions is included. Below the description of each park is a section that gives refer-

ences in the text to features found in the park, as well as text references to the park.

Volcanic Activity

2. Haleakala (Hawaii)

One of the world's largest volcanic craters located at the summit of a huge volcanic mountain; rare plants and animals. (Open all year)
Address: Superintendent
P. O. Box 456
Haleakala National Park
Kahului, Maui, Hawaii
96732

3. Hawaii Volcanoes (Hawaii)

Two active volcanoes that often produce flows of lava; many examples of land forms created by lava flows; wilderness areas with unusual plants and animals. (Open all year)
Address: Superintendent
Hawaii Volcanoes
National Park
Hawaii, Hawaii 96718
References in text:
Hawaii volcanoes, 196 – 198
lava flows, 196, Fig. 9 – 17; 197, Fig. 9 – 18

6. Mount Rainier (Washington)

Main feature is Mount Rainier, which is a composite volcanic cone of great height in the Cascade Mountain

Range formed by volcanic activity; active glaciers cover much of the mountain and have created many glacial features; dense forests surround the base of the mountain and contain much wildlife. (Open from May to October, depending upon weather)

Address: Superintendent
Mount Rainier
National Park
Longmire, Washington
98397

References in text:
composite volcanic cone (Mount Rainier), 198
glacial features, Chapter 13

7. Crater Lake (Oregon)

An unusual, deep lake of great beauty lies in the caldera of an ancient volcano; walls of the crater around the lake show a variety of colors; volcanic features are common in the area; located in a region of forests and mountains. (Open all year)

Address: Superintendent
Crater Lake National Park
Crater Lake, Oregon 97604

References in text:
Crater Lake, 199, Fig. 9–23

9. Lassen Volcanic (California)

Site of Lassen Peak, one of the active volcanoes in the continental United States; examples of volcanic features include cinder cones, thermal springs, and lava flows; very rugged mountainous wilderness area. (Open approximately June 1–October 31)

Address: Superintendent
Lassen Volcanic
National Park
Mineral, California 96063

References in text:
volcanic block (Mt. Lassen), 198, Fig. 9–20
cinder cone, 198, Fig. 9–22
lava flows, 197, Fig. 9–18
hot spring (thermal), 269, Fig. 12–21

Stream Activity and Weathering

17. Canyonlands (Utah)

Large and spectacular eroded region containing numerous canyons with natural arches, standing rocks, pinnacles, and other features carved from brightly colored rock. (Open all year)

Address: Superintendent
Canyonlands National Park
Post Office Building
Moab, Utah 84532

References in text:
natural bridges, arches, spires (photo), 232

19. Bryce Canyon (Utah)

A horseshoe-shaped depression containing a large number of erosion features in the shapes of spires, arches, natural bridges, and windows carved from brightly colored rock. (Open all year)

Address: Superintendent
Bryce Canyon National Park
Utah 84717

References in text:
Bryce Canyon, 221, Fig. 10–24
spires, arches, natural bridges (photo), 232

20. Zion (Utah)

A plateau that has been carved into steep-sided gorges and sheer cliffs by water and wind erosion; exposed rocks reveal a history going back more than 200 million years. (Open all year)

Address: Superintendent
         Zion National Park
         Springdale, Utah 84767

References in text:
   Zion Canyon, 221, Fig. 10–24

21. Grand Canyon (Arizona)

Spectacular gorge cut by the Colorado River exposing a giant slice through layers of sedimentary rocks; rock layers form a record of a large part of the earth's history. (South Rim open all year; North Rim open mid-May to mid-October)

Address: Superintendent
         Grand Canyon National Park
         Grand Canyon, Arizona
         86023

References in text:
   Grand Canyon, 221, Fig. 10–24; 398–400; 398, Figs. 19–3, 19–4; 399, Fig. 19–5

33. Great Smoky Mountains (Tennessee)

A section of the southern end of the very old, folded Appalachian Mountain Range; name comes from haze that normally covers the valleys; covered by extensive hardwood forests. (Open all year)

Address: Superintendent
         Great Smoky Mountains
          National Park
         Gatlinburg, Tennessee
         37738

References in text:
   folded mountains, 209, Figs. 10–3, 10–4
   Appalachian Mountains, 224

35. Shenandoah (Virginia)

A region along the crest of the very old Blue Ridge Mountains made of folded and fractured sedimentary rocks; extensive hardwood forest.

Address: Superintendent
         Shenandoah National Park
         Luray, Virginia 22835

Groundwater Action

14. Yellowstone (Wyoming)

Largest hydrothermal region in the world with about three thousand geysers, hot springs, mud pots, and volcanic vents; evidence of ancient volcanic activity in the form of many kinds of igneous rocks and petrified forests; a miniature version of the Grand Canyon with a waterfall twice the height of Niagara; mountain scenery and wilderness areas with abundant wildlife. (Open May 1 – Oct. 31)

Address: Superintendent
         Yellowstone National Park
         Wyoming 83020

References in text:
   geyser, 269, Fig. 12–22; 270, Fig. 12–23
   hot springs, 269, Fig. 12–21; 270, Fig. 12–23
   Yellowstone Park, 269, Figs. 12–21, 12–22; 193, Fig. 9–27
   waterfalls, 259
   volcanic vents, 197

26. Carlsbad Caverns (New Mexico)

Largest and most beautiful limestone caverns yet discovered; many large rooms with examples of stalagmites, stalactites, and similar cave formations. (Open all year)

Address: Superintendent
         Box 1598
         Carlsbad Caverns
          National Park
         Carlsbad, New Mexico 88220

References in text:
  Carlsbad Caverns, 271
  limestone caverns, 271, Fig. 12–26

32. Mammoth Cave (Kentucky)

Large limestone cavern with many chambers and cave formations made of gypsum and travertine; unusual historical significance. (Open all year)

Address: Superintendent
  Mammoth Cave
  National Park
  Mammoth Cave,
  Kentucky 42259

References in text:
  limestone caverns, 271, Fig. 12–26
  Mammoth Cave, 271

36. Everglades (Florida)

Area of subtropical wilderness with extensive fresh-water and salt-water regions; mangrove forests and open prairies; many unusual kinds of wildlife. (Open all year)

Address: Superintendent
  Box 279
  Everglades National Park
  Homestead, Florida 33030

Glacial Features

13. Mount McKinley (Alaska)

Highest mountain in North America; large area of the complex Alaska Mountain Range; active glaciers with many examples of glacial features; animals and plants of the subarctic climate zone. (Open June 1 – September 19)

Address: Superintendent
  Mount McKinley
  National Park
  Alaska 99755

References in text:
  glacial features, Chapter 13
  subarctic climates, 488–489

10. Yosemite (California)

Some of the world's most impressive and easily reached glaciated scenery; Yosemite Valley is probably the most spectacular ice-carved valley that exists, with its sheer granite walls and waterfalls; many examples of rock domes formed by exfoliation; giant sequoia trees and mountain wilderness. (Open all year)

Address: Superintendent
  Box 577
  Yosemite National Park
  California 95389

References in text:
  Yosemite Valley, 247, Fig. 11–22A
  exfoliation domes, 237, Fig. 11–7
  waterfalls, 259

24. Rocky Mountain (Colorado)

A section of the Rocky Mountains with lofty snow-capped peaks, glacier-carved valleys, lakes, and meadows; higher elevations with examples of tundra vegetation. (Open all year)

Address: Superintendent
  Rocky Mountain
  National Park
  Estes Park, Colorado 80517

References in text:
  glacier-carved valley, 283, Fig. 13–11
  glacial lakes, 287–288

13. Glacier (Montana)

Large and unusually rugged section of the Rocky Mountains, showing much of the history of this complex mountain range; many active glaciers, ice fields, and examples of glaciated landscapes; mountain plants and wild-life. (Open approximately June 15–September 15)

Address: Superintendent
        Glacier National Park
        West Glacier, Montana
        59936

References in text:
glacial features, Chapter 13
ice field, 276, Fig. 13–1

Mountain Building

15. Grand Teton (Wyoming)

A series of rugged peaks formed by erosion of a large fault-block mountain range; mountain landscape with glaciers, hot springs, and glacial features; wilderness areas include large herd of American elk. (Open all year)

Address: Superintendent
        Box 67
        Grand Teton National Park
        Moose, Wyoming 83012

References in text:
fault-block mountains, 220, Fig. 10–22
glacial features, Chapter 13
hot springs, 268–269

The following nineteen national parks offer many opportunities for outdoor recreation and observation of various geologic features. Other points of interest are also listed. Information about any particular park can be obtained by writing to:

National Park Service
U.S. Department of the Interior
Washington, D.C. 20240

4. North Cascades (Washington)
Wilderness mountain region.

5. Olympic (Washington)
Mountains, rainforest, and seashore.

8. Redwood (California)
World's tallest trees.

11. Kings Canyon (California)
Rugged mountains and deep canyons.

12. Sequoia (California)
World's largest trees; highest peak in United States (Mt. Whitney).

16. Arches (Utah)
Natural bridges and eroded rock pedestals.

18. Capitol Reef (Utah)
Large cliffs and rock domes of colorful rock.

22. Petrified Forest (Arizona)
Trees turned to stone; includes a part of the Painted Desert.

23. Wind Cave (South Dakota)
Cavern with rare dripstone formations.

25. Mesa Verde (Colorado)
Ancient ruins of cliff-dwelling Indians.

27. Guadalupe Mountains (Texas)
Ancient fossilized coral reef.

28. Big Bend (Texas)
Rugged gorge of the Rio Grande River.

29. **Voyageurs** (Minnesota)
    Glacial lakes in heavy forest.

30. **Isle Royale** (Michigan)
    Wilderness island.

31. **Hot Springs** (Arkansas)
    Thermal springs.

34. **Acadia** (Maine)
    Rocky coast with many submergent features.

37. **Theodore Roosevelt** (North Dakota)
    Wilderness along the Little Missouri River.

38. **Badlands** (South Dakota)
    Eroded gullies and buttes.

39. **Virgin Islands** (Virgin Islands)
    Tropical island.

C. Geologic Calendar

ERAS	PERIODS	EPOCHS	DURATION	Years before present
CENOZOIC	QUATERNARY	RECENT	25,000	
		PLEISTOCENE	975,000	
				1,000,000
	TERTIARY	PLIOCENE	11,000,000	
				12,000,000
		MIOCENE	13,000,000	
				25,000,000
		OLIGOCENE	10,000,000	
				35,000,000
		EOCENE	25,000,000	
				60,000,000
		PALEOCENE	10,000,000	
				70,000,000
MESOZOIC	CRETACEOUS		65,000,000	
				135,000,000
	JURASSIC		45,000,000	
				180,000,000
	TRIASSIC		45,000,000	
				225,000,000
PALEOZOIC	PERMIAN		45,000,000	270,000,000
	PENNSYLVANIAN		60,000,000	
	MISSISSIPPIAN		20,000,000	330,000,000
	DEVONIAN		50,000,000	350,000,000
	SILURIAN		40,000,000	400,000,000
	ORDOVICIAN		60,000,000	440,000,000
				500,000,000
	CAMBRIAN		100,000,000	
				600,000,000
PRECAMBRIAN	LATE PRECAMBRIAN		1,100,000,000	
		KEEWEENAWAN		1,700,000,000
		HURONIAN		
	EARLY PRECAMBRIAN	TIMISKAMING	2,800,000,000	
		ONTARIAN		4,500,000,000

Geologic Calendar

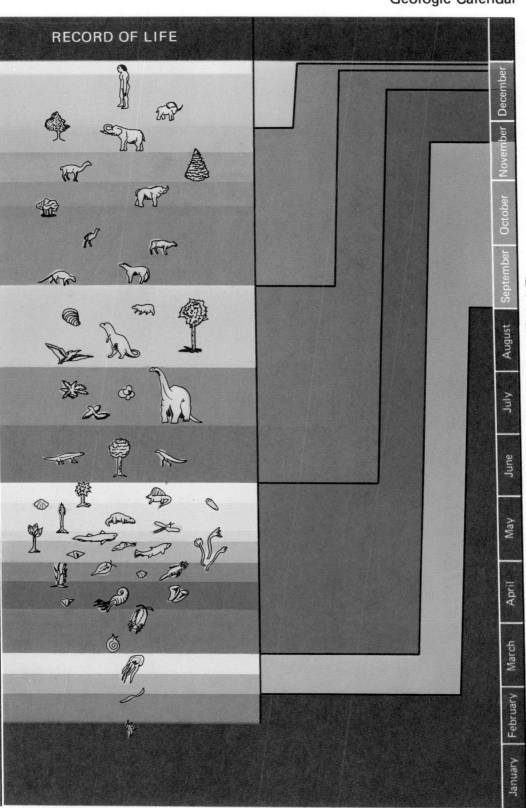

RECORD OF LIFE

January | February | March | April | May | June | July | August | September | October | November | December

Calendar Date if ONE YEAR represented all of Geologic Time

D. Stones

What is a stone?

Stones are the loose rock and mineral fragments larger than a sand grain that have been broken off the continuous solid rock of the earth's crust. Some natural process must have separated the stone from its parent bedrock. Stones are found loose on the ground or directly beneath the surface. Each stone has a history, beginning with its parent rock and continuing with the weathering processes that separate it from bedrock and erosion processes that transport the fragments.

Everyone has picked up stones with unusual shapes or colors and perhaps wondered about their history. This section will show you how to apply your knowledge of earth science to interpret some of the histories of stones.

What to look for in stones

1. *Shape*

The history of a stone begins when it is separated from bedrock. If it remains close to its source, its shape is usually the result of weathering processes. However, most stones are moved by one or more of the agents of erosion. Movement during erosion usually gives a stone its characteristic shape and surface texture. Some of the characteristics given to stones by the common agents of erosion are listed in Table 1.

Table 1

Example	Agent of erosion	Characteristics
	Water (streams and wave action)	Rounded shape with no sharp corners or edges; dull surface

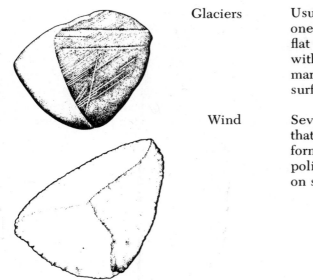

Glaciers	Usually have one or more flat surfaces with scratch marks; dull surface
Wind	Several faces that meet to form ridges; polished surface on some sides

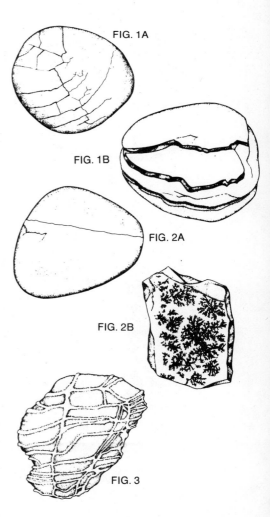

FIG. 1A

FIG. 1B

FIG. 2A

FIG. 2B

FIG. 3

2. *Evidence of weathering*

Physical weathering often causes stones to have cracks or to separate into layers. This can be caused by water entering small openings, then freezing and expanding, thus breaking the stone. Various minerals in stones may also respond to heating and cooling in different ways, causing cracks to develop. Figures 1A and 1B show common ways in which stones may give evidence of physical weathering.

Chemical weathering frequently causes color changes along cracks when water is absorbed. See Figure 2A. Mineral solutions that enter fractures in some stones may deposit a tree-like pattern on the rock surface. See Figure 2B. These growths, called *dendrites,* are often mistaken for fossils.

Stones that are affected mainly by physical weathering processes tend to have irregular shapes with sharp edges. Those affected mostly by chemical weathering are usually more rounded. Some stones may develop strange shapes and very rough surfaces when different minerals within them weather at unequal rates. See Figure 3.

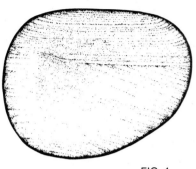

FIG. 4

3. *Structure*

The structure of a stone may be determined by the arrangement of the materials from which its parent bedrock was formed. For example, a stone shows layers at angles to each other if it comes from sedimentary rock whose layers were cross-bedded. See Figure 4. Other structures seen in stones are the result of changes in the parent rock after it was formed. For example, a stone may show part of a fault. See Figure 5. Metamorphic rocks frequently have folds like those shown in Figure 6. Some stones coming from igneous rocks may have dikes where magma has entered open spaces. See Figure 7. Small fractures may show veins where minerals have been deposited from solutions.

4. *Origin of parent rock*

Almost all stones begin their histories as part of a body of igneous, sedimentary, or metamorphic rock. With some practice, you can identify the most common kinds of rock from which stones are made by comparison with the photographs of different kinds of rock on pp. 173–175. However, you will also probably find many stones whose parent rock is difficult to identify. Use of the mineral key on pp. 536–549 may be helpful in determining the kinds of minerals present in the unknown stones.

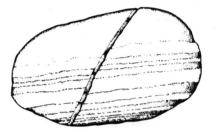

FIG. 5

Collecting Stones

Some stones can be found in the same place or near where they were separated from their parent rock. However, the best collecting areas are in the deposits made by the agents of erosion that have moved the stones. Good places to collect a variety of stones are:

1. *Beaches.* The edges of the sea and larger lakes are often lined with deposits of stones of many sizes and kinds. The spring season is the best time to look for stones along a beach. Winter storms bring stones to the shore and remove the sand that may conceal them.

2. *Streams.* Almost all stream channels are lined with stones. Larger stones are generally found near the head of the stream, with the size decreasing toward the mouth. Stones found in stream beds often show evidence of the tumbling action that has changed their size, shape, and surfaces.

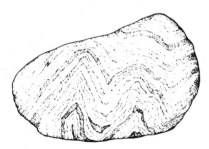

FIG. 6

3. *Glacial deposits.* Any location that has been eroded by glaciers or is near an existing glacier is likely to be covered with stones transported by the ice. There will be a few stones that have been changed by the movement of the glacier. Most stones, however, will show little evidence of being carried and dropped by the ice. In glaciated regions, good places to look for stones are in road or railroad cuts through glacial deposits, farm fields, and sand and gravel pits.

When collecting stones, you might concentrate on those with shapes resulting from the different erosion and weathering processes. Another kind of collection could consist of stones made up of certain rocks or minerals.

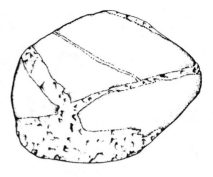

FIG. 7

Hints for collecting stones

1. Always collect in safe locations. Avoid cliffs or other locations where there is danger of falling.
2. Colors of stones are best seen if they are coated with wax, shellac, clear fingernail polish, or clear plastic. You can see how a coated stone will look by wetting it.
3. Permission must be obtained to collect stones on private property. Collecting of stones is not permitted in national parks nor in many state and local parks.

E. Key for Identification of Minerals

The identification of particular mineral specimens is made simpler if a series of tables, or *key*, is used. The key in this book of 11 tables in which a number of the more important minerals are arranged systematically according to their characteristics.

Before attempting to use the key for identification of minerals, the student should thoroughly review the text material in Chapter 8 dealing with the characteristic properties of minerals (pages 160–169). In the key, the first step in arrangement of minerals is a division into two groups: those having metallic luster and those without metallic luster. The metallic luster group is further divided according to color; the non-metallic luster group is divided according to the color of the streak. The mineral species which fall into these groups are then arranged according to hardness.

The procedure in using the tables is a step-by-step determination of the properties of the specimen to be identified. First examine the mineral and determine its luster, color and streak, and hardness. Turn to the general classification list, page 538, and find the number of the table which applies. Then read carefully across the appropriate column of the table, and examine the mineral, making any required tests, to determine its properties as described there.

To illustrate the use of the tables, assume that we have a specimen of galenite. By examination, we see that it has metallic luster, which places it in the left column of the classification list. Next, we see that it is gray, which places it in the last group of the column. Now we determine the hardness and find that it is very soft. Turning to Table 2 as directed in the list, we find a detailed analysis of four minerals in the section labeled *gray, very soft*. In the righthand section of the *Key* column, distinctive features of each mineral are given as a help in distinguishing among the members of a small group. We determine that our sample does not have the low specific gravity of molybdenite and pyrolusite, nor is it sectile, like argentite. Galenite remains, and by elimination, we find that our sample is most likely to be this mineral. A check of the other characteristics listed, including testing for fusibility, helps to make the identification more certain.

Some additional information on the arrangement and terminology used in the tables will be helpful.

Luster. The presence of metallic luster is the first basis of classification in the key. Therefore it is of great importance to determine the luster of a mineral specimen. It is helpful to note that minerals with metallic luster are opaque; that is, they do not allow light to pass through them. However, some dark, non-metallic minerals appear opaque, and so some of these have been included in both the metallic and non-metallic tables in the key. Metallic luster is common to all true metals. In some minerals, the luster is indistinctly metallic, ranging between metallic and nonmetallic: this is called *submetallic*. The luster of minerals composed mostly of nonmetallic elements may be described as follows: *adamantine,* exceedingly brilliant; *vitreous,* glassy; *resinous,* with appearance of resin; *pearly,* with appearance of mother-of-pearl; *greasy or waxy,* with appearance of oiled or waxed surface; *silky,* similar to pearly, found in finely fibrous minerals; *dull,* without bright or shiny surface, earthy. A particularly bright luster may be called *splendent.*

Color, Streak, Hardness and Specific Gravity. The methods and scales used in determining these properties as given in Chapter 11 are referred to in the tables.

Cleavage and Fracture. Cleavage, the breaking of a mineral along the faces of its crystal planes, may be described in geometrical terms. A few of the more common terms used in the table are *cubic,* in the form of a cube; *rhombohedral,* in a form having six faces intersecting at angles other than 90°; *octahedral,* in a form having eight faces; *basal,* cleaving parallel to the base of the crystal; *clinopinacoidal,* in monoclinic crystals, cleaving parallel to the oblique and the vertical axes; *prismatic,* cleaving in forms with faces parallel to the vertical axis.

Most of the terms describing fracture are self-explanatory. Many minerals show *conchoidal* fracture: they break in smooth, curving surfaces resembling the inside of a shell; in *hackly*

fracture, the mineral breaks with thin points that catch the skin as one scrapes a finger across the surface.

Fusibility. The fusibility of a mineral is an important aid to identification. The temperature of fusion is not accurately known except for a few species. The approximate fusing points of different minerals, that is, their relative fusing points, can be determined by comparison. A small fragment of the mineral is heated with the oxidizing flame in the forceps or on a charcoal block. Seven species of minerals have been chosen as a scale of fusibility. Beginning with the most easily melted, they are stibnite—1; natrolite—2; garnet—3; actinolite—4; orthoclase—5; bronzite—6; quartz—7. If the specimen to be identified fuses as readily as stibnite, it has a fusibility of 1; if it cannot be rounded on the thinnest edges, like quartz, it is 7, and so on.

Other Characteristics. The table uses standard terms to indicate the ability of a mineral to transmit light: *opaque,* no light transmitted even through thin edges or layers; *translucent,* some light transmitted but objects appear indistinctly through the substance; *transparent,* objects distinctly seen through the substance. The **tenacity** of a mineral, that is, its behavior when an attempt is made to break, cut, hammer, or similarly alter it, is described as *brittle,* breaks under distortion; *sectile,* can be cut without crumbling; *malleable,* can be hammered into thin sheets; *flexible,* can be bent visibly without breaking; *elastic,* thin layers can be bent and spring back when released. Some brittle minerals are very rigid and strong and break with great difficulty. There are said to be *tough.*

The external forms and internal structures of minerals are described in the tables by a variety of terms. Some types of crystalline or amorphous masses are *acicular*, needlelike crystals; *botryoidal*, closely joined, small rounded masses, like grapes; *fibrous*, very thin crystals or filaments; *foliated*, plates or leaves which are easily separated; *granular*, grains, closely packed; *nodular*, round masses of irregular shape; *pisolitic*, small masses about the size of peas; *scaly*, thin scales or plates; *sheaflike*, crystals resembling a sheaf of wheat. Many similar terms are in general use in describing minerals. Further discussion of this subject may be found in any standard textbook of mineralogy.

The last column of each table in the key includes, as well as the name of the mineral species, its chemical formula and a roman numeral indicating the system in which the mineral crystallizes. The crystal systems are I, Isometric; II, Tetragonal; III, Hexagonal; IV, Orthorhombic; V, Monoclinic; and VI, Triclinic. See page 164. If no system number is given, the mineral is amorphous.

Classification of Minerals

Minerals with metallic luster

Color red or brown	table
Very soft	1
Hard	1

Color yellow	
Very soft	1
Soft	1
Hard	1

Color white or silver	
Very soft	1
Hard	1

Color gray	
Very soft	2
Hard	2

Color gray to black	
Very soft	2
Soft	2
Hard	2
Very hard	2

Minerals without metallic luster

Streak reddish	table
Very soft	3
Soft	3
Hard	3

Streak brown	
Very soft	3
Soft	3,4
Hard	4
Very Hard	4

Streak yellow	
Very soft	4
Soft	4
Hard	4,5
Very hard	5

Streak blue or green	
Very soft	5
Soft	5
Hard	5

Streak black	
Very soft	5
Soft	5
Hard	5

Streak gray or white	
Very soft	6,7
Soft	6,7
Hard	6
Very hard	6,7

Streak white	
Soft	8
Hard	9
Very Hard	9,10,11
Adamantine	10,11

MINERALS WITH METALLIC LUSTER

TABLE 1

Color Group	Hardness	KEY	LUSTER	COLOR	STREAK	HARDNESS, S.G., AND CLEAVAGE	CHARACTERISTICS	SPECIES
COLOR RED OR BROWN	VERY SOFT	STREAK WHITE	Pearly to metallic on cleavage	Light to dark brown	White	2.5 to 3 / 2.75 / Basal, perfect	Infusible. Yields a little water in closed tube. Decomposes in strong sulfuric acid. Transparent to opaque. Sectile, elastic. Foliated	MICA Phlogopite complex silicate V
COLOR RED OR BROWN		STREAK RED MALLEABLE	Metallic	Copper red, tarnishes black	Copper red	2.5 to 3 / 8.8 to 8.9 / None Fracture hackly	Fuses at 780 C. Gives green solution in nitric acid, then blue in ammonia (DANGER). Malleable. Sometimes cubic; rounded branches	NATIVE COPPER Cu I
COLOR RED OR BROWN	HARD	STREAK DEEP RED	Submetallic to dull	Reddish black	Red	4 to 6 / 4.3 to 4.7 / None Fracture uneven	Infusible; becomes magnetic on charcoal. Yields water in closed tube. Soluble in hydrochloric acid. Opaque. Brittle. Massive, botryoidal	TURGITE $2Fe_2O_3 + H_2O$ III
COLOR RED OR BROWN	HARD	COMPARE S.G.	Metallic	Reddish brown	Reddish brown	4 to 6 / 4.9 to 5.3 / Basal Fracture uneven	Infusible. Yields no water in a closed tube. Sometimes magnetic. Opaque. Brittle. Massive, botryoidal	HEMATITE Red Ocher Fe_2O_3 III
COLOR RED OR BROWN	HARD	S.G. 6.4 TO 9.7	Submetallic to greasy	Gray, green, brown to black	Brownish black to olive green or grayish	5.5 to 6 / 6.4 to 9.7 / None. Fracture uneven or conchoidal	Infusible. Gives coating of lead oxide with soda on charcoal. Soluble in acids. Nitric acid leaves fluorescent spot. Opaque. Brittle. Massive, botryoidal	URANINITE UO_2 I
COLOR YELLOW	VERY SOFT	DOES NOT TARNISH MALLEABLE	Metallic	Rich yellow to silvery yellow	Gold yellow	2.5 to 3 / 19.3 / None	Fuses readily. Soluble only in aqua regia. Does not tarnish. Very malleable. Grains or nuggets	NATIVE GOLD Au I
COLOR YELLOW	SOFT	IRIDESCENT TARNISH	Metallic	Bronze yellow (brass)	Dark greenish black	3.5 to 4 / 4.1 to 4.3 / One poor Fracture uneven	Gives copper on charcoal. Yields sulfur in test tube. Produces green solution with nitric acid. Often tarnished; iridescent	CHALCOPYRITE $CuFeS_2$ II
COLOR YELLOW	SOFT	BROWN TARNISH	Metallic	Bronze yellow	Dark grayish black	3.5 to 4.5 / 4.4 to 4.7 / Basal Fracture uneven	Fuses to black, magnetic mass. Dissolves in hydrochloric acid to form H_2S. Often has dark brown tarnish. Opaque. Massive, granular	PYRRHOTITE $Fe_{1-x}S$ or Fe_7S_8 III
COLOR YELLOW	HARD	BRASS YELLOW COLOR	Metallic	Pale brass yellow	Brownish black to greenish black	6 to 6.5 / 4.8 to 5.2 / Indistinct Fracture uneven	Fuses easily, leaving magnetic residue. Yields sulfur in a closed tube. Opaque. Massive or in cubes	PYRITE FeS_2 I
COLOR YELLOW	HARD	GRAYISH YELLOW COLOR	Metallic	Pale grayish yellow	Dark brownish black	6 to 6.5 / 4.9 to 5 / Indistinct. Fracture uneven	Fuses easily, leaving magnetic residue. Yields sulfur in a closed tube. Brittle. Massive, fibrous, or crystalline	MARCASITE FeS_2 IV
WHITE OR SILVER	VERY SOFT TO HARD	COMPARE S.G.	Metallic (mirrorlike)	Silvery or steel gray	Red	1 to 6 / 4.9 to 5.3 / None Fracture uneven	Infusible. No cracking when heated. Soluble in hydrochloric acid. Opaque. Brittle	HEMATITE Specular Fe_2O_3 III
WHITE OR SILVER	VERY SOFT TO HARD	COMPARE S.G.	Metallic	Silver white Bright white Black if tarnished	White	2.5 to 3 / 10 to 11 / None Fracture hackly	Fuses readily. Dissolves in nitric acid. Opaque. Malleable. Cubic crystals	NATIVE SILVER Ag I
WHITE OR SILVER	HARD	S.G. 6 OR ABOVE	Metallic	Silver white to steel gray	Dark gray to black	5.5 to 6 / 6 to 6.4 / Prismatic, distinct Fracture uneven	Fuses easily, leaving magnetic residue and arsenic coating. Tarnishes yellowish. Gives garlic smell after hammer blow. Opaque. Massive, wedgelike crystals	ARSENOPYRITE FeAsS IV

MINERALS WITH METALLIC LUSTER

TABLE 2

Species	Luster	Color	Streak	Hardness	S.G.	Cleavage / Fracture	Characteristics	Key
MOLYBDENITE MoS_2 III	Metallic	Lead gray	Blackish lead gray with bluish tinge	1 to 1.5	4.7 to 4.8	Basal, perfect	Infusible. Yields sulfur fumes in oxidizing flame. Opaque. Sectile. Usually foliated. Hexagonal crystals; greasy, flexible plates	SECTILE
PYROLUSITE MnO_2 II	Metallic	Iron black to steel gray	Black	2.5 *	4.8	Two cleavages. Prismatic. Fracture uneven, splintery	Infusible. Yields chlorine with hydrochloric acid. Gives amethyst bead with borax. Opaque. Brittle. Radiating fibers or massive	BRITTLE
ARGENTITE Ag_2S I	Metallic	Blackish lead gray	Blackish lead gray	2.5	7.3	Poor cubic and dodecahedral Fracture uneven or subconchoidal	Fuses to form silver button in oxidizing flame. Reacts to test for sulfur. Yields silver with soda. Tarnishes dull black. Sectile. Cubic crystals	SECTILE
GALENITE PbS I	Metallic (Dull if coated)	Dark lead gray	Dark lead gray to black	2.5 to 2.75	7.4 to 7.6	Cubic, perfect	Fuses readily, producing lead and sulfur fumes. Opaque. Brittle. Cubic crystals	NOT SECTILE
MAGNETITE Fe_3O_4 I	Metallic	Dark gray to iron black	Black	5.5	5	Octahedral parting Fracture conchoidal, uneven	Infusible. Soluble in hydrochloric acid. Magnetic. Striated. Granular	MAGNETIC
GRAPHITE C III	Metallic	Black	Black	1 to 2	2 to 2.3	Basal, perfect Flexible scales	Infusible. Insoluble. Greasy feel. Opaque. Sectile. Hexagonal crystals, fibrous	GREASY
MANGANITE $MnO(OH)$ IV	Submetallic	Iron black to steel gray	Reddish brown and black	4	4.2 to 4.4	Prismatic and basal poor; perfect side. Fracture uneven	Infusible. Gives amethyst bead with borax. Yields water in closed tube. Yields chlorine with hydrochloric acid. Translucent. Brittle. Striated prisms	NOT MAGNETIC
ILMENITE Menaccanite $FeTiO_3$ III	Submetallic to metallic	Iron black to brownish black	Brownish red and black	5 to 6	4.1 to 5	None Fracture conchoidal	Infusible. Pale yellow when held in hot oxidizing flame, turning to colorless or white when cooled. Weakly magnetic. Opaque. Brittle. Granular	STREAK BROWNISH RED AND BLACK WEAKLY MAGNETIC
CHROMITE $FeCr_2O_4$ I	Submetallic	Iron black	Light grayish brown	5.5	4.3 to 4.6	None Fracture uneven	Infusible. Becomes magnetic when heated. Gives green bead in borax after cooling. May be slightly magnetic. Opaque. Brittle. Massive or octahedral crystals	STREAK LIGHT GRAYISH BROWN SLIGHTLY MAGNETIC
URANINITE Pitchblende UO_2 I	Metallic	Steel black	Gray to brown to black	5.5	9 to 9.5 †	None. Fracture uneven or conchoidal	Infusible. Gives coating of lead oxide with soda on charcoal. Soluble in acids. Opaque. Brittle. Massive, fibrous, botryoidal	S.G. 9 TO 9.5 (IF PURE)
RUTILE TiO_2 II	Metallic to adamantine	Black, reddish brown or red	Gray to light yellow brown	6 to 6.5	4.2	Prismatic Fracture uneven	Infusible. Insoluble. Opaque. Brittle. Crystals or massive	S.G. 4.2

Left-margin classification:

COLOR GRAY	VERY SOFT	S.G. BELOW 5
	VERY SOFT	S.G. ABOVE 7
COLOR GRAY TO BLACK	HARD / VERY SOFT	S.G. 4 TO 5
	SOFT / HARD	
	VERY HARD	

* Crystal hardness 6 to 6.5   * Crystal hardness 6 to 6.5   † If pure

TABLE 3

MINERALS WITHOUT METALLIC LUSTER

KEY	SPECIES	CHARACTERISTICS	HARDNESS, S.G., AND CLEAVAGE	STREAK	COLOR	LUSTER
S.G. LESS THAN 3 (VERY SOFT, STREAK REDDISH)	BAUXITE $Al_2O_3 + 2H_2O$	Infusible. Insoluble. Colors blue in cobalt nitrate test. Opaque. Brittle. Usually massive like hard clay; pisolitic	1 to 3 — 2.0 to 2.6; None; Fracture earthy	Red	Red	Dull
BRITTLE (SOFT, S.G. 3 OR MORE)	TURGITE Red Ocher $2Fe_2O_3 + H_2O$	Infusible. Becomes black and magnetic on charcoal. Yields water in closed tube. Soluble in hydrochloric acid. Opaque. Brittle. Massive or botryoidal	1 to 2.5 — 3 to 4; None; Fracture earthy	Red	Red	Dull
SECTILE	REALGAR AsS V	Volatile, combustible—burns with blue flame. Gives arsenic (garlic) odor. Transparent to translucent. Often has yellow spots. Sectile	1.5 to 2 — 3.5; Basal and clinopinacoidal; Fracture conchoidal	Red to orange	Red and orange	Resinous
COMPARE S.G.	CUPRITE CuO I	Fuses readily, yielding copper. Colors flame green; blue with hydrochloric acid. Translucent to opaque. Brittle. Often has green or blue spots	3.5 to 4 — 5.8 to 6.1; Octahedral; Fracture uneven	Red to brown	Red to brown	Adamantine to dull
COMPARE S.G.	CINNABAR HgS III	Volatile. Yields sulfur fumes and mercury droplets. Transparent to opaque. Brittle to sectile. Hexagonal crystals, massive, granular, acicular	4 — 8 to 8.2; Prismatic, perfect; Fracture uneven	Scarlet	Red to brownish red	Adamantine to dull
STREAK ORANGE, S.G. ABOVE 5 (HARD)	ZINCITE ZnO III	Infusible—turns black. Soluble in acid. Translucent to opaque. Brittle. Pyramidal crystals rare, granular, foliated	4 to 4.5 — 5.4 to 5.7; Prismatic. Basal, perfect; Fracture conchoidal	Orange	Red to orange	Subadamantine
YIELDS WATER IN CLOSED TUBE	TURGITE $2Fe_2O_3 + H_2O$	Infusible. Becomes black and magnetic on charcoal. Yields water in closed tube. Soluble in hydrochloric acid. Opaque. Brittle. Massive or botryoidal	4 to 6 — 4.3 to 4.7; None; Fracture uneven	Red	Reddish black	Submetallic to dull
FLUORESCES GREEN	WILLEMITE Zn_2SiO_4 III	Fuses with difficulty. Yields zinc oxide with soda on charcoal. Gelatinizes with HCl. Fluoresces green. Translucent to opaque. Brittle. Massive	5.5 — 3.9 to 4.2; Imperfect; Fracture uneven	Reddish and brownish	Yellowish, greenish, brownish	Vitreous-resinous
S.G. 2.6 (VERY SOFT, STREAK BROWN)	KAOLINITE $Al_2Si_2O_6(OH)_4$ IV	Infusible. Yields water in closed tube. Insoluble. Colors blue in cobalt nitrate test. Opaque. Brittle (clay)	1 — 2.6; Basal; Fracture earthy	Brown	Yellow	Dull
S.G. 3.6 TO 4	LIMONITE Yellow Ocher $2Fe_2O_3 \cdot 3H_2O$	Infusible. Turns black and magnetic on charcoal. Yields water in a closed tube. Opaque	1 to 2.5 — 3.6 to 4; None; Fracture earthy	Brown	Brown	Dull
S.G. 2 TO 2.5	BAUXITE $Al_2O_3 + 2H_2O$	Infusible. Insoluble. Colors blue in cobalt nitrate test. Opaque. Brittle. Usually massive like hard clay; pisolitic	1 to 3 — 2.0 to 2.5; None; Fracture earthy	Brown	Brown	Dull
BECOMES MAGNETIC ON CHARCOAL (SOFT, STREAK BROWN)	SIDERITE $FeCO_3$ III	Infusible. Turns black and magnetic on charcoal. Effervesces in hot acid. Translucent to near opaque. Brittle. Curved crystals	3.5 to 4 — 3.8; Rhombohedral, perfect; Fracture conchoidal	Brown	Dark brown to gray	Dull to vitreous
REMAINS NONMAGNETIC AFTER HEATING	SPHALERITE ZnS I	Infusible. Reacts to test for sulfur. Remains nonmagnetic after heating. Effervesces in hot acid, giving H_2S	3.5 to 4 — 3.9 to 4.1; Perfect. Six cleavages; Fracture uneven	Brown	Yellow to brown	Resinous

TABLE 4 — MINERALS WITHOUT METALLIC LUSTER

Group	Key	Luster	Color	Streak	Hardness, S.G., and Cleavage	Characteristics	Species
STREAK BROWN — SOFT	COLORS FLAME GREEN. FUSES READILY	Dull to adamantine	Red to brown	Brown	3.5 to 4 — 5.8 to 6.1; Octahedral; Fracture uneven	Fuses readily, yielding copper. Colors flame green; blue with hydrochloric acid. Translucent to opaque. Brittle. Often has green or blue spots	CUPRITE Cu_2O — I
STREAK BROWN — HARD	BECOMES MAGNETIC WHEN HEATED	Submetallic	Iron black	Brown	5.5 — 4.3 to 4.6; None; Fracture uneven	Infusible. Becomes magnetic when heated. Gives green bead in borax after cooling. May be slightly magnetic. Opaque. Brittle. Massive or octahedral crystals	CHROMITE $FeCr_2O_4$ — I
STREAK BROWN — VERY HARD	NOT MAGNETIC S.G. 4.2	Adamantine	Reddish brown	Light brown	6 to 6.5 — 4.2; Prismatic; Fracture uneven	Infusible. Insoluble. Opaque. Brittle. Crystals or massive	RUTILE TiO_2 — II
STREAK BROWN — VERY HARD	NOT MAGNETIC S.G. 6.8	Adamantine to greasy	Brown to black	Light brown	6.5 — 6.8; Imperfect; Fracture uneven	Infusible. Yields metallic tin with soda on charcoal. Not magnetic. Translucent to opaque. Brittle. Prismatic crystals; granular; fibrous	CASSITERITE SnO_2 — II
STREAK YELLOW — VERY SOFT		Dull to vitreous	White to yellow or gray	Yellow	0.5 to 1.5 — 1.5; Fracture conchoidal	Infusible. Yields water in closed tube. Insoluble in acid. Transparent to opaque. Brittle. Massive, botryoidal. Earthy variety, diatomite	OPAL $SiO_2 + H_2O$ — IV
STREAK YELLOW — VERY SOFT		Dull	Yellow	Yellow	1 to 2.5 — 2.6; Basal; Fracture earthy	Infusible. Yields water in closed tube. Insoluble. Colors blue in cobalt nitrate test. Opaque. Brittle (yellow clay)	KAOLINITE $Al_2Si_2O_5(OH)_4$ — IV
STREAK YELLOW — VERY SOFT		Dull	Yellow to brown	Yellow to brown	1 to 2.5 — 3.6 to 4; None; Fracture earthy	Infusible. Turns black and magnetic on charcoal. Yields water in a closed tube. Opaque	LIMONITE Yellow Ocher $2Fe_2O_3 \cdot 3H_2O$ — IV
STREAK YELLOW — VERY SOFT	COMPARE S.G.	Dull	Bright yellow	Yellow	1 to 2 — 4.1 to 5.0; Basal, perfect; Fracture earthy	Infusible. Cold borax bead is fluorescent green. Powder turns red brown in boiling nitric acid and dissolves to green solution. Opaque. Sectile. Scaly powder	CARNOTITE $K_2(UO_2)_2(VO_4)_2 \cdot 3H_2O$ — IV
STREAK YELLOW — VERY SOFT		Resinous to pearly	Yellow often with orange spots	Yellow	1.5 to 2 — 3.5; Perfect	Reacts to tests for arsenic and sulfur. Translucent. Slightly sectile; flexible plates. Massive, foliated, botryoidal	ORPIMENT As_2S_3 — IV
STREAK YELLOW — VERY SOFT		Resinous	Light yellow to brown	Yellow to white	1.5 to 2.5 — 2.0 to 2.1; Prismatic, basal, imperfect; Fracture conchoidal	Burns with blue flame and odor of sulfur. Translucent. Brittle. Orthorhombic crystals, massive	NATIVE SULFUR S — IV
STREAK YELLOW — SOFT	COMPARE S.G.	Dull to vitreous	Gray to dark brown	Gray to yellow and brown	3.5 to 4 — 3.8; Rhombohedral, perfect; Fracture conchoidal	Infusible. Turns black and magnetic on charcoal. Effervesces in hot acid. Translucent to near opaque. Brittle. Curved crystals	SIDERITE $FeCO_3$ — III
STREAK YELLOW — SOFT	COMPARE S.G.	Resinous	Yellow to brown	Light yellow to brown	3.5 to 4 — 3.9 to 4.1; Perfect. Six cleavages	Infusible. Reacts to test for sulfur. Remains nonmagnetic after heating. Effervesces in hot acid, giving H_2S	SPHALERITE ZnS — I
STREAK YELLOW — HARD	COLOR BROWN TO BLACK	Dull to adamantine	Brown to black	Yellow	5.0 to 5.5 — 4 to 4.4; Prismatic, perfect; Fracture conchoidal	Infusible. Turns black and magnetic on charcoal. Yields water in a closed tube. Opaque. Dark and light bands. Stalactitic; radiating plates	LIMONITE Goethite $2Fe_2O_3 \cdot 3H_2O$ — IV

TABLE 5

MINERALS WITHOUT METALLIC LUSTER

		KEY	LUSTER	COLOR	STREAK	HARDNESS, S.G., AND CLEAVAGE	CHARACTERISTICS	SPECIES
STREAK YELLOW	HARD	COLOR OCHER YELLOW	Submetallic	Ocher yellow	Ocher yellow	5 to 6 — 4.1 to 5 — None — Fracture conchoidal	Infusible. Pale yellow when held in hot oxidizing flame, turning to colorless or white when cooled. Weakly magnetic. Opaque. Brittle. Granular	ILMENITE Menaccanite $FeTiO_3$ III
	VERY HARD	S.G. 2.6	Nearly dull	Brown to yellow	Light yellow	7 — 2.6 — None — Fracture conchoidal	Infusible. Insoluble. Translucent to opaque. Brittle to tough. Crystals six-sided with striations. Often with inclusions	QUARTZ Jasper $SiO_2 + Fe_2O_3$ III
STREAK BLUE OR GREEN	SOFT	COLOR BLUE	Dull to vitreous	Blue	Pale blue	3.5 to 4 — 3.8 — Perfect. Fracture conchoidal	Fuses readily. Colors flame green. Yields copper. Yields water in closed tube. Effervesces in acid. Translucent to opaque. Brittle, often velvety	AZURITE $Cu_3(OH)_2(CO_3)_2$ V
	VERY SOFT	S.G. 2.6 TO 3	Pearly to dull	White to dark green, black, brown, rose, yellow	Lighter green	1 to 2.5 — 2.6 to 3.0 — Basal, perfect. Fracture scaly, earthy	Fuses with difficulty. Yields water in closed tube. Translucent to opaque. Tough to brittle. Somewhat sectile. Usually finely foliated to massive	CHLORITE $MgFeAlSi_3 \cdot H_2O$ V
	SOFT	S.G. 3.9 TO 4	Adamantine, silky or dull to vitreous	Green	Pale green	3.5 to 4 — 3.9 to 4 — Basal, perfect. Fracture conchoidal	Fuses readily, coloring flame green. Leaves copper on charcoal. Yields water in closed tube. Effervesces in acid. Translucent to opaque; often banded. Brittle	MALACHITE $CuCO_3 + Cu(OH)_2$ V
	HARD	COMPARE S.G.	Vitreous	Greenish black	Grayish green	5 to 6 — 3.2 to 3.6 — Prismatic, perfect; two cleavages, often one parting	Fusible at 2.5–5. Insoluble. Translucent to opaque. Brittle. Massive, granular	AUGITE Pyroxene complex silicate V
STREAK BLACK	VERY SOFT	S.G. 2.6	Submetallic to greasy	Gray, green, brown, black	Olive green	5.5 to 6 — 6.4 to 9.7 — None — Fracture uneven or conchoidal	Infusible. Gives coating of lead oxide with soda on charcoal. Soluble in acids. Opaque. Brittle. Massive, fibrous, botryoidal	URANINITE Pitchblende UO_2 I
			Dull	Red or reddish brown to black or white	Black	1 to 2.5 — 2.6 — Micaceous. Fracture earthy	Infusible. Yields water in closed tube. Insoluble. Colors blue in cobalt nitrate test. Opaque. Brittle	KAOLINITE $Al_2Si_2O_5(OH)_4$ III
	SOFT	S.G. 4.2 TO 4.4	Submetallic to metallic	Iron black	Black	4 — 4.2 to 4.4 — Prismatic and basal poor, side perfect. Fracture uneven	Infusible. Gives amethyst bead with borax. Yields water in closed tube. Yields chlorine with hydrochloric acid. Translucent. Brittle. Striated prisms	MANGANITE $MnO(OH)$ IV
	HARD	COMPARE S.G.	Submetallic to metallic	Iron black to brownish black	Black	5 to 6 — 4.1 to 5 — None — Fracture conchoidal	Infusible. Pale yellow when held in hot oxidizing flame, turning to colorless or white when cooled. Weakly magnetic. Opaque. Brittle. Granular	ILMENITE Menaccanite $FeTiO_3$ III
			Submetallic to greasy	Gray, green, brown to black	Brownish black to grayish	5.5 to 6 — 6.4 to 9.7 — None — Fracture uneven or conchoidal	Infusible. Gives coating of lead oxide with soda on charcoal. Soluble in acids. Opaque. Brittle. Massive, fibrous, botryoidal	URANINITE Pitchblende UO_2 I

MINERALS WITHOUT METALLIC LUSTER
STREAK GRAY OR WHITE

TABLE 6

Hardness	KEY	SPECIES	CHARACTERISTICS	HARDNESS, S.G., AND CLEAVAGE	STREAK	COLOR	LUSTER
VERY SOFT	S.G. 2 TO 2.6 COLOR GRAY	BAUXITE $Al_2O_3 + 2H_2O$	Infusible. Insoluble. Colors blue in cobalt nitrate test. Opaque. Brittle. Usually massive like hard clay; pisolitic	1 to 3; 2.0 to 2.6; None; Fracture earthy	Gray	Gray, red, white, brown	Dull
HARD	STREAK GRAYISH BROWN	CHROMITE $FeCr_2O_4$ I	Infusible. Becomes magnetic when heated. (Gives green bead in borax after cooling. May be slightly magnetic. Opaque. Brittle. Octahedral crystals, massive	5.5; 4.3 to 4.6; None; Fracture uneven	Light grayish brown	Iron-black	Submetallic
VERY SOFT	COLOR DARK GRAY	KAOLINITE $Al_2Si_2O_5(OH)_4$	Infusible. Yields water in closed tube. Insoluble. Colors blue in cobalt nitrate test. Opaque. Brittle (clay)	1; 2.6; Basal; Fracture earthy	Dark gray	Gray or brown to black	Dull
HARD TO VERY HARD	COMPARE HARDNESS	CASSITERITE SnO_2 II	Infusible. Yields metallic tin with soda on charcoal. Not magnetic. Translucent to opaque. Brittle. Prismatic crystals; granular; fibrods	6.5; 6.8; Imperfect; Fracture uneven	Gray	Brown to black	Adamantine
HARD TO VERY HARD	COMPARE HARDNESS	URANINITE UO_2 I	Infusible. Gives coating of lead oxide with soda on charcoal. Soluble in acids. Opaque. Brittle. Massive, fibrous, botryoidal	5.5 to 6; 6.4 to 9.7; None; Fracture uneven or conchoidal	Grayish	Gray, green, brown to black	Submetallic to greasy
SOFT	COMPARE S.G.	EPIDOTE $HCa_2(Al,Fe)_3Si_3O_{13}$ V	Fuses at 3–3.5 to a magnetic mass. Yields water when strongly heated. Opaque. Brittle. Crystals darker with parallel striations	6 to 7; 3.25 to 3.5; One basal; Fracture uneven	Gray	Yellow green	Vitreous
SOFT	COMPARE S.G.	SIDERITE $FeCO_3$ III	Infusible. Turns black and magnetic on charcoal. Effervesces in hot acid. Translucent to near opaque. Brittle. Curved crystals	3.5 to 4; 3.8; Perfect, rhombohedral; Fracture conchoidal	Gray	Gray to dark brown	Vitreous to dull
VERY SOFT	S.G. 1.5	OPAL $SiO_2 + H_2O$	Infusible. Yields water in closed tube. Insoluble in acid. Transparent to opaque. Brittle. Massive, botryoidal. Earthy variety, diatomite	.05 to 1.5; 1.5; Fracture conchoidal	Gray	Gray	Dull
VERY SOFT	STRONG DOUBLE REFRACTION	CALCITE Iceland Spar $CaCO_3$ III	Infusible. Effervesces in dilute, cold hydrochloric acid. Fluoresces red, pink, yellow. Strong double refraction. Transparent. Brittle	2 to 3; 2.7; Rhombohedral, perfect; Fracture conchoidal	Gray to white	White when pure	Vitreous
VERY SOFT	OFTEN TINTED OR DARKENED	CALCITE $CaCO_3$ III	Infusible. Effervesces in dilute, cold hydrochloric acid. Transparent to nearly opaque. Often tinted or darkened. Brittle	2 to 3; 2.7; Rhombohedral, perfect; Fracture conchoidal	Gray to white	White when pure	Vitreous to dull or pearly
HARD	CONCHOIDAL FRACTURE	OBSIDIAN complex silicate	Fuses at 3.5–4 with intumescence. Insoluble in acids. Brittle, with sharp edges	6; 2.2 to 2.8; Fracture conchoidal	Gray to white	Black to dark gray, red	Vitreous
HARD	PRISMATIC CRYSTALS YIELDS WATER IN CLOSED TUBE	AMPHIBOLE Hornblende complex silicate V	Fusible with difficulty. Yields water in a closed tube. Translucent to opaque. Brittle. Prismatic crystals, fibrous, granular, massive	5 to 6; 2 to 3.4; Cleavage angle 56° and 124°; Fracture subconchoidal to uneven	Gray to white	Black	Vitreous
VERY HARD	S.G. 3.15 TO 4.3	GARNET Pyrope complex silicate I	Fuses at 3–3.5. Gelatinizes with HCl after fusing. Insoluble. Translucent to opaque. Brittle. Crystals 4, 6 or 8-sided, granular, massive	6.5 to 7.5; 3.15 to 4.3; None; Fracture uneven	Gray to white	Red, brown, yellow, green, black, white	Vitreous to resinous

MINERALS WITHOUT METALLIC LUSTER
STREAK GRAY OR WHITE

TABLE 7

KEY	SPECIES	CHARACTERISTICS	HARDNESS, S.G., AND CLEAVAGE	STREAK	COLOR	LUSTER
VERY HARD — USUALLY MORE BRITTLE WITH IMPURITIES	QUARTZ Chert SiO_2 III	Infusible. Insoluble. Flake held in flame breaks up. Translucent. Brittle. Massive, botryoidal, nodular	7 — 2.6 to 2.7; None; Conchoidal fracture	Gray to white	Gray, brown, black	Waxy to dull
USUALLY DARKER IN COLOR THAN CHERT	QUARTZ Flint SiO_2 III	See above.	7 — 2.6 to 2.7; None; Conchoidal fracture	Gray to white	Gray, brown, black	Waxy to dull
S.G. 2.9 TO 3.2	TOURMALINE complex silicate III	Mostly infusible. Gelatinizes with HCl after fusion. Insoluble. Transparent to opaque. Often zoned with bands. Very brittle. Prismatic crystals, often striated	7 to 7.5 — 2.9 to 3.2; None; Fracture uneven or conchoidal	Gray to white	Black, red, green, pink	Vitreous
VERY SOFT	TALC $Mg_3(OH)_2Si_4O_{10}$ V	Infusible. Insoluble. Swells, turns white, and gives violet color in cobalt nitrate test. Opaque to transparent. Sectile, plates flexible.	1 to 1.5 — 2.7 to 2.8; Basal, perfect (micaceous); Fracture uneven	White	Apple green to white, gray, etc.	Pearly to dull or greasy
COMPARE S.G.	CALCITE Chalk $CaCO_3$ III	Infusible. Effervesces in dilute, cold hydrochloric acid. Transparent to nearly opaque. Brittle	0.5 to 1.5 — 2.6; Fracture earthy	White	White	Dull
	GYPSUM $CaSO_4 + H_2O$ V	Fuses at 3. Yields water in closed tube. Phosphorescent. Fluorescent yellow. Transparent to opaque. Brittle, plates flexible. Varied crystal forms, fibrous	1.5 to 2 — 2.3; Clinopinacoidal, perfect, two; Fracture conchoidal	White	White, gray, brown	Pearly, silky, dull, glassy
COLOR YELLOW	NATIVE SULFUR S IV	Burns with blue flame and odor of sulfur. Translucent. Brittle. Orthorhombic crystals, massive	1.5 to 2.5 — 2.0 to 2.1; Prismatic, basal, imperfect; Conchoidal fracture	White	Yellow	Resinous
COMPARE S.G.	BAUXITE $Al_2O_3 + 2H_2O$	Infusible. Insoluble. Colors blue in cobalt nitrate test. Opaque. Brittle. Usually massive like hard clay; pisolitic	1 to 3 — 2.0 to 2.6; None; Fracture earthy	White	White, gray, red, brown	Dull
VERY SOFT TO SOFT	MICA Muscovite complex silicate V	Infusible. Insoluble. Gives little water in closed tube. Transparent to translucent. Flexible, elastic plates or sheets	2 to 2.5 — 2.75 to 3; Basal, perfect	White	White, light yellow, brown, pale green	Vitreous to pearly
TRANSPARENT SHOWS STRONG DOUBLE REFRACTION	CALCITE Iceland Spar $CaCO_3$ III	Infusible. Effervesces in dilute, cold hydrochloric acid. Fluoresces red, pink, yellow. Strong double refraction. Transparent. Brittle	2 to 3 — 2.7; Rhombohedral, perfect; Fracture conchoidal	White	White when pure	Vitreous
OFTEN TINTED OR DARKENED	CALCITE $CaCO_3$ III	Infusible. Effervesces in dilute, cold hydrochloric acid. Transparent to nearly opaque. Often tinted or darkened. Brittle	2 to 3 — 2.7; Rhombohedral, perfect; Fracture conchoidal	White	White when pure	Vitreous to dull or pearly
SOLUBLE IN WATER TASTES SALTY	HALITE NaCl I	Fuses readily. Colors flame deep yellow. Soluble in water. Tastes salty. Transparent to translucent. Brittle	2.5 — 2.1 to 2.6; Cubic, perfect; Fracture conchoidal	White	White to gray or brown	Vitreous, dull when moist
FUSES READILY COLORS FLAME YELLOW	CRYOLITE Na_3AlF_6 V	Fuses very readily, coloring flame yellow. Transparent to translucent. Almost invisible in water. Brittle	2.5 — 3; Imperfect; Fracture uneven	White	White to brown	Vitreous or greasy

MINERALS WITHOUT METALLIC LUSTER
STREAK WHITE   SOFT

TABLE 8	KEY	LUSTER	COLOR	STREAK	HARDNESS, S.G., AND CLEAVAGE	CHARACTERISTICS	SPECIES
FLEXIBLE, ELASTIC SHEETS	DARKER COLOR	Vitreous, pearly, splendent	Dark brown to black	White	2.5 to 3 2.7 to 3.4; Basal, perfect	Fuses on thin edges. Decomposes in hot, strong sulfuric acid. Transparent to translucent. Flexible, elastic plates or sheets	MICA Biotite complex silicate V
		Pearly to metallic on cleavage surface	Light to dark brown	White	2.5 to 3 2.75; Basal, perfect	Fuses with difficulty. Decomposes in hot, strong sulfuric acid. Transparent to translucent. Flexible, elastic plates or sheets	MICA Phlogopite complex silicate V
	TRANSPARENT	Vitreous	White to bluish or brownish or reddish	White	2.5 to 3.5 4.3 to 4.6; Basal, prismatic, perfect (diamond shaped cleavage). Fracture uneven	Colors flame yellow green. Insoluble in acid. Reacts to sulfur test with soda. May fluoresce orange after heating. Translucent. Brittle	BARITE $BaSO_4$ IV
	COLORS FLAME YELLOWISH GREEN INSOLUBLE IN ACIDS / GREENISH COLOR FIBROUS	Silky or greasy	Green to yellow green	White	2.5 to 4 2.2 to 2.6; None Fracture fibrous	Yields water in closed tube. Decomposed by hydrochloric acid. Yellow variety fluoresces cream yellow. Opaque to translucent. Flexible to sectile	ASBESTOS Chrysotile $Mg_3Si_2O_5(OH)_4$ V
	COLORS FLAME PURPLE RED	Pearly	Gray green, rose red and violet to white or pale yellow	White	2.5 to 4 2.7 to 3.3; Basal, perfect (micaceous) Fracture scaly	Fuses and expands. Colors flame purple red. Yields little water in closed tube. Translucent to transparent. Tough	LEPIDOLITE complex silicate V
	REACTS LIKE BORAX UNDER BLOWPIPE HARDNESS 3 S.G. 1.9	Vitreous to dull	Colorless or white upon exposure to air	White	3 1.9; Basal, perfect. Produces long splinters. Fracture conchoidal	Reacts like borax under blowpipe but with less swelling. Surface often chalky. Transparent to translucent. Brittle	KERNITE $Na_2B_4O_7 + 4H_2O$
	COLORS FLAME RED, CRACKLES WHEN HEATED	Vitreous	White to bluish	White	3 to 3.5 4; Basal, prismatic, perfect (diamond shaped cleavage). Fracture uneven	Fuses with difficulty. Colors flame red. Cracks when heated. After heating fluoresces bright green. Transparent to translucent. Brittle	CELESTITE $SrSO_4$ IV
	FUSES, YIELDING METALLIC LEAD. BUBBLES IN ACID	Vitreous to adamantine	White to gray	White	3 to 3.5 6.5; Prismatic, imperfect Fracture conchoidal	Fuses readily yielding metallic lead with soda on charcoal. Bubbles in acid. Transparent to translucent. Brittle	CERUSSITE $PbCO_3$ IV
	FALLS TO PIECES WHEN HEATED WITH A BLOWPIPE ON CHARCOAL	Vitreous to dull	White to gray	White	3 to 4 2.9 to 3; Prismatic, imperfect Fracture conchoidal	Falls to pieces when heated with a blowpipe on charcoal. Bubbles in acid. Transparent to translucent. Brittle. Fibrous	ARAGONITE $CaCO_3$ IV
	FUSES EASILY YIELDS WATER IN CLOSED TUBE	Vitreous or pearly on cleavage surface	White to yellow and red	White	3.5 to 4 2.1 to 2.2; Perfect, one Fracture uneven	Fuses at 2.5. Yields water in closed tube. Swells, writhes in wormlike manner when heated. Transparent to translucent. Brittle. Sheaflike crystals	STILBITE Zeolite complex silicate V
	DOES NOT EFFERVESCE IN COLD, DILUTE ACID	Vitreous to dull or pearly	White to gray	White	3.5 to 4 *2.8 to 2.9; Rhombohedral, perfect Fracture conchoidal	Infusible. Does not effervesce in cold dilute acid. Not fluorescent. Transparent to translucent. Curved surfaces when broken. Brittle	DOLOMITE $MgCa(CO_3)_2$ III
	DECREPITATES PHOSPHORESCENT WHEN GENTLY HEATED	Vitreous	White, green, violet, blue, brown, yellow	White	4 3 to 3.25; Octahedral, perfect	Fuses at 3. Phosphorescent when gently heated. Decrepitates. Transparent to nearly opaque. Brittle	FLUORITE CaF_2 I

* When pure

MINERALS WITHOUT METALLIC LUSTER

STREAK WHITE   VERY HARD TO HARD

TABLE 9	KEY	LUSTER	COLOR	STREAK	HARDNESS, S.G., AND CLEAVAGE	CHARACTERISTICS	SPECIES
	TURNS FLAME RED YELLOW (Ca)	Vitreous to almost resinous	Green, brown, yellow, white	White	5 3.2 Basal, imperfect Fracture conchoidal	Infusible. Soluble in acid. Turns flame reddish yellow. Fluoresces orange when heated. Opaque. Brittle. Prismatic crystals, granular, massive	APATITE $Ca_5(Cl,F)\ (PO_4)_3$ III
	FUSES IN CANDLE FLAME	Vitreous	White	White	5 to 5.5 2.2 Prismatic, perfect, two cleavages Fracture uneven across the prism	Fuses in candle flame. Yields water in a closed tube. May fluoresce orange or blue to greenish white when heated. Transparent to translucent. Brittle	NATROLITE $Na_2Al_2Si_3O_{10} + 2H_2O$ IV
	FUSES WITH DIFFICULTY CLEAVAGE ANGLES 56° AND 124°	Vitreous	Black	White	5 to 6 2 to 3.4 Prismatic, 56° and 124° Fracture subconchoidal to uneven	Fuses with difficulty. Yields water in closed tube. Translucent to opaque. Brittle. Prismatic crystals, fibrous, granular, massive	HORNBLENDE Amphibole complex silicate V
	S.G. 2.71 TWIN STRIATIONS	Vitreous to pearly	Gray to greenish and reddish	White	5 to 6 2.71 Basal, perfect Fracture uneven	Fuses at 3.5. Insoluble. Translucent to near opaque. Often has bluish iridescence. Brittle	LABRADORITE Feldspar $(Na,Ca)Al_2Si_3O_{10}$ VI
COMPARE S.G.	YIELDS WATER IN CLOSED TUBE	Vitreous	White, light green to dark green	White	5 to 6 2.9 to 3.4 Prismatic, perfect Two cleavages	Fuses to black or white glass. Yields water in closed tube. Insoluble in acid. Transparent to opaque. Massive. One of two jade minerals	NEPHRITE Actinolite, Amphibole complex silicate V
COMPARE S.G.	POWDER GELATINIZES WITH HCl	Vitreous, resinous	Yellowish greenish, brownish	White	5.5 3.9 to 4.2 Imperfect Fracture uneven	Fuses with difficulty. Yields zinc oxide with soda on charcoal. Gelatinizes with hydrochloric acid. Fluoresces green. Translucent to opaque. Brittle. Massive	WILLEMITE Zn_2SiO_4 III
COMPARE S.G.	FUSES AT 2.5 AND BLACKENS	Vitreous	Red and brown to gray	White	5.5 to 6.5 3.4 to 3.7 Prismatic, perfect with two cleavages. Fracture uneven	Fuses at 2.5. Gives amethyst bead with borax. Insoluble. Blackens on exposure. Transparent to opaque. Tough, crystals brittle. Often with brown or black spots	RHODONITE $MnSiO_3$ VI
	FUSES WITH INTUMESCENCE, GLASSY	Vitreous	Black to dark gray or red	White	6 2.2 to 2.8 Fracture conchoidal	Fuses at 3.5 to 4 with intumescence. Insoluble in acids. Brittle, with sharp edges	OBSIDIAN complex silicate
FUSES WITH DIFFICULTY	LACKS TWIN STRIATIONS ON CRYSTAL FACES	Vitreous	White, red, green, flesh color	White	6 to 6.5 2.55 Basal, perfect. Blocky Fracture conchoidal	Fuses at 5. Not affected by acids. Translucent to transparent. Brittle. Often very large crystals	MICROCLINE Feldspar $KAlSi_3O_8$ VII
FUSES WITH DIFFICULTY	WHEN PURE, S.G. 2.63, TWIN STRIATIONS	Vitreous to pearly	White to gray, red, green	White	6 to 6.5 * 2.63 Basal, perfect Fracture conchoidal	Fuses with difficulty (4–5). Insoluble. Transparent to translucent. Brittle. Sometimes in thin, flat crystals	ALBITE Feldspar $Na_2Al_2Si_6O_{16}$ VI
FUSES WITH DIFFICULTY	BLOCKY CLEAVAGES TWO AT 90°, ONE GOOD	Vitreous to pearly	White to gray, red, green, etc.	White	6 to 6.5 2.5 to 2.65 Basal, perfect. Blocky at 90° Fracture conchoidal	Fuses with difficulty. Insoluble. Not fluorescent after blowpiping. Transparent to opaque. Brittle. Short prismatic crystals	ORTHOCLASE Feldspar $KAlSi_3O_8$ V
FUSES WITH DIFFICULTY	S.G. 2.76 TWIN STRIATIONS	Vitreous	White, gray, red, brown, green	White	6 to 6.5 * 2.76 Basal, perfect, two at right angles. Fracture conchoidal	Fuses at 5. Insoluble. Transparent to translucent. Brittle. Rare	ANORTHITE Feldspar $CaAl_2Si_2O_8$ VI
	S.G. ABOVE 3 PURPLISH RED FLAME	Vitreous to pearly	White to gray and buff, green	White	6.5 to 7 3.1 to 3.2 Perfect, prismatic 87° and 93° Splinters. Fracture uneven * When pure	Fuses at 3.5 with red purple flame. Insoluble. Fluoresces orange. Transparent to opaque. Brittle. Long, striated crystals	SPODUMENE Pyroxene $LiAlSi_2O_6$ V

MINERALS WITHOUT METALLIC LUSTER
STREAK WHITE   VERY HARD TO ADAMANTINE

TABLE 10 — KEY	LUSTER	COLOR	STREAK	HARDNESS, S.G., AND CLEAVAGE	CHARACTERISTICS	SPECIES
FUSES, NOT AFFECTED BY ACIDS	Subvitreous to silky	Green to white; often spotty or patterned	White	6.5 to 7 3.3 to 3.5 Prismatic. Two cleavages Fracture splintery	Fuses readily. Not affected by acids. Colors sodium flame yellow. Tough. Massive, fibrous, granular	JADEITE Jade $NaAlSi_2O_6$
INFUSIBLE GREEN CRYSTALLINE GRAINS	Vitreous	Green	White	6.5 to 7.5 3.3 to 3.4 One fair, one poor Fracture conchoidal	Infusible. Gelatinizes with hydrochloric acid. May be magnetic. Transparent to translucent. Brittle	OLIVINE (Chrysolite) $(MgFe)_2SiO_4$ IV
FUSES GELATINIZES WITH HCl AFTER FUSION	Vitreous to resinous	Red, brown, yellow, green, black, white	White	6.5 to 7.5 3.15 to 4.3 None Fracture uneven	Fuses at 3–3.5. Gelatinizes with HCl after fusing. Insoluble. Translucent to opaque. Brittle. Crystals 4, 6, or 8-sided, granular, massive	GARNET Pyrope complex silicate I
S.G. ABOVE 4, SQUARE PRISMS AND PYRAMIDS RARELY IRREGULAR GRAINS	Adamantine	Colorless, brown, gray, blue, violet, reddish, green	Colorless	6.5 to 7.5 4.2 to 4.8 Two cleavages poor Fracture conchoidal to uneven	Insoluble. Colors whiten when heated and glow briefly only once. Fluorescent yellow orange. Transparent to opaque. Brittle	ZIRCON SiO_2 II
MICROSCOPIC CRYSTALS ARRANGED IN SLENDER, BANDED, PARALLEL FIBERS	Waxy	White, red, green, blue, brown, gray, black	White	7 2.6 to 2.65 None Fracture conchoidal	Infusible. Insoluble. Flake held in flame breaks up. Dissolves with effervescence with soda on platinum wire. Translucent. Brittle	QUARTZ Chalcedony SiO_2 III
WITH WAVY COLOR BANDS	Waxy or vitreous	Red, brown, white, blue, yellow, gray, green	White	7 2.6 to 2.65 None Fracture conchoidal	See above	QUARTZ Agate SiO_2 III
WITH STRAIGHT COLOR BANDS	Waxy or vitreous	White, gray, brown, red, green	White	7 2.6 to 2.65 None Fracture conchoidal	See above	QUARTZ Onyx SiO_2 III
TRANSPARENT SIX-SIDED CRYSTALS	Vitreous	Colorless	White	7 2.65 to 2.7 None Fracture conchoidal	See above	QUARTZ Rock Crystal SiO_2 III
PURPLE OR VIOLET COLOR	Vitreous	Purple color due to traces of manganese or iron	White	7 2.65 to 2.7 None Fracture conchoidal	See above	QUARTZ Amethyst SiO_2 III

HARDNESS NOT BELOW 7

S.G., NOT ABOVE 2.7

MINERALS WITHOUT METALLIC LUSTER
STREAK WHITE

TABLE 11 — KEY	LUSTER	COLOR	STREAK	HARDNESS, S.G., AND CLEAVAGE	CHARACTERISTICS	SPECIES
LIKE GLASS	Vitreous	Gray, black, brownish	White	7 — 2.65 to 2.7 — None, Fracture conchoidal	Infusible. Insoluble. Flake held in flame breaks up. Dissolves with effervescence with soda on platinum wire. Translucent. Brittle	QUARTZ Smoky SiO_2 III
MORE BRITTLE, OFTEN WITH IMPURITIES	Waxy to dull	Gray, black, brown	White	7 — 2.6 to 2.7 — None, Fracture conchoidal	See above	QUARTZ Chert SiO_2 III
USUALLY DARKER IN COLOR THAN CHERT, COMPACT	Waxy to dull	Gray, black, brown	White	7 — 2.6 to 2.7 — None, Fracture conchoidal	See above	QUARTZ Flint SiO_2 III
STRIATED PRISMS, PYROELECTRIC. SHOWS DICHROISM, S.G. 2.9–3.2	Vitreous	Black, red, pink, green, brown	White	7 to 7.5 — 2.9 to 3.2 — None, Fracture uneven or conchoidal	Mostly infusible. Insoluble. Gelatinizes with HCl after fusion. Transparent to opaque. Often zoned with bands. Very brittle. Prismatic crystals	TOURMALINE Complex silicate III
S.G. BELOW 3	Vitreous	Greenish to bluish and yellow green	White	7.5 to 8 — 2.6 to 2.8 — Basal, poor, Fracture conchoidal	Infusible. Does not decrepitate violently. Insoluble. Transparent to translucent. Brittle. Large, long, prismatic crystals, striated	BERYL $Be_3Al_2(SiO_3)_6$ III
CRYSTALS PRISMS, CLEAVAGE BASAL, PERFECT	Vitreous	Yellow, white, blue, red, green	White	8 — 3.4 to 3.65 — Basal, perfect, one cleavage, Fracture conchoidal, uneven	Infusible. Insoluble. Gives blue color in cobalt nitrate test. Transparent to opaque. Brittle	TOPAZ $Al_2SiO_4F_2$ IV
HARDNESS 9, S.G. ABOUT 4	Vitreous to adamantine	Gray, brown, red, yellow, blue, black, pink	White	9 — 3.95 to 4.1 — Basal, rhombic parting, Fracture conchoidal	Infusible. Insoluble. Powder gives blue color in cobalt nitrate test. Transparent to opaque. Brittle to tough	CORUNDUM Al_2O_3 III
HARDNESS 10—SCRATCHES CORUNDUM	Adamantine	White, colorless, gray, blue, pink, black	White	10 — 3.5 — Octahedral, perfect, Fracture conchoidal	Infusible. Insoluble. Often fluorescent. Translucent to opaque. Brittle. Crystals, grains, pebbles	DIAMOND C I

VERY HARD, S.G. NOT ABOVE 2.7 — WAXY TO DULL

HARDNESS ADAMANTINE

F. Temperature Scales

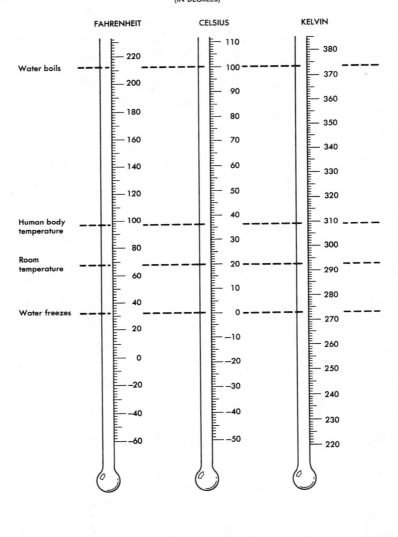

Temperature Scales
(IN DEGREES)

FAHRENHEIT   CELSIUS   KELVIN

Water boils — Human body temperature — Room temperature — Water freezes

---

Temperature Equivalents / Conversion Formulas

$1°F \cong 0.56°C$ $°F = 9/5°C+32$

$1°C \cong 1.8°F$ $°C = 5/9(°F-32)$

$1°K \cong 1°C$ $°K = °C+273$

glossary

aa. Black lava.

absolute humidity. The weight of water vapor actually contained in a given quantity of air.

absolute magnitude. Brightness of a star viewed from a fixed distance.

abyssal plains. Extremely level regions that cover half of all the ocean floor.

adiabatic change. A temperature change that takes place without the addition or loss of heat to or from the surroundings.

advection fog. A fog produced by the horizontal movement of air over a cool surface.

agonic line. A line connecting locations of zero magnetic declination.

air mass. A large body of air that has about the same temperature and humidity throughout.

albedo. Average reflectivity of a body in space, such as the moon or earth.

alluvial fan. A delta-shaped deposit formed when a stream deposits its load upon reaching a level surface.

alpha particle. An electrically charged particle produced by many radioactive substances and consisting of two protons and two neutrons.

alpine (or valley) **glacier.** A glacier formed from ice moving out from a mountain ice field.

altimeter. An instrument used to measure altitude.

amorphous. A non-crystalline solid having no orderly arrangement of molecules.

anemometer. An instrument to measure wind velocity.

aneroid barometer. A type of barometer that operates without the use of liquids.

annular eclipse. A type of partial eclipse of the sun in which a ring of light is seen around the edge of the moon.

anthracite. Black hard coal.

anticline. The crest or upturn of a rock fold.

anticyclone. A large mass of high-pressure air, with winds moving out from its center in a clockwise whirl; a high.

aphelion. The point on the earth's orbit farthest from the sun.

apogee. The point in a satellite's orbit farthest from the object around which it revolves.

apparent magnitude. The observed brightness of a star as it appears from the earth.

apparent solar time. Time as determined by the apparent position of the sun.

aquaculture. The farming of living plants or animals in a body of water.

aquifer. A water-bearing layer of porous rock.

Arctic Circle. An imaginary line at $66\frac{1}{2}°$ north latitude, north of which the sun never rises on the winter solstice and never sets on the summer solstice.

arête. A sharp, narrow ridge separating two cirques or glacial valleys.

artesian well. A well located between two permeable rock layers, in which water rises under its own pressure.

ash. Smallest of the solid materials produced during a volcanic eruption.

asteroid. One of the many tiny planets between the orbits of Mars and Jupiter.

asthenosphere. A softer region of the lithosphere that consists of rock able to flow slowly.

atmosphere. The layer of gases that surrounds an entire planet.

atoll. A roughly circular coral reef that encloses a lagoon.

atom. The smallest subdivision of an element that has all the properties of that element.

aurora. Streamers of glowing ionized gases in the earth's atmosphere.

axes of a crystal. Imaginary lines running from the center of a crystal face to the center of the opposite face.

axis of fold. An imaginary line running along the top of an anticline or along the bottom of a syncline.

bar. Underwater ridge just offshore.

barchan. A crescent-shaped dune.

barograph. A recording barometer.

barometer. An instrument to measure air pressure.

barrier island. Offshore bar that has been built up by wave action to a height above mean sea level.

barrier reef. A long narrow coral reef some distance from the shore.

batholith. A large mass of intrusive igneous rock; its lower limit is at an unknown depth.

bead test. A color test in which a borax bead formed on a platinum wire loop is used to determine the chemical makeup of minerals.

bedrock. The unweathered solid rock of the earth's crust.

bench mark. Points of accurately known location and elevation that are usually identified by metal marker disks.

berm. Inner raised portion of a beach.

beta particle. A negative radioactive particle consisting of one electron.

biomass. All plant materials that can be used to supply energy.

bituminous. A dark brown to black form of soft coal.

blowout. A depression whose origin can be traced to removal of material by wind. Also called *deflation hollow*.

body waves. Earthquake waves that travel through the body of the earth; primary and secondary waves.

breeder reactor. A kind of nuclear reactor that produces more nuclear fuel than it consumes.

butte. A flat-topped, steep-walled hill, usually a remnant of horizontal beds, smaller and narrower than a mesa.

caldera. A basin-like depression formed by the destruction of a volcanic cone.

calorie. Amount of heat required to raise the temperature of 1 gram of water to 1°C at near freezing temperatures.

capillary fringe. A zone in the rocks just above the water table in which water is drawn upward into the spaces by capillary tension.

capillary tension. Tendency for water to be drawn upward in very narrow spaces.

carbonation. The chemical combination of substances with carbon dioxide.

carbon dioxide cycle. A series of processes by which the balance of carbon dioxide and oxygen in the atmosphere is maintained.

carbonization. A process in which remains of living things are partly decomposed, leaving a residue of carbon.

casting. A trail of fossilized sand or mud left by worms.

cavern. An opening in rock formed when the dissolving action of ground water enlarges and connects cracks.

celestial north pole. Point in the sky directly over the earth's North Pole.

Cepheid variable. A variable star whose bright and dim phases are related to its true brightness.

chemical compound. A substance formed by the combination of two or more different elements in a definite weight relationship.

chemical weathering. Natural chemical processes that alter the mineral composition of rock while reducing it to small fragments.

chromosphere. The sun's surface atmosphere, made up of glowing gases.

cinder cone. Cone formed by the explosive type of volcanic eruption; it has a narrow base and steep, symmetrical slopes of interlocking, angular cinders.

circumpolar whirl. High-altitude winds moving around the earth from west to east.

cirque. A bowl-shaped depression in a mountainside, formed at the head of a valley glacier.

cirrus clouds. A family of high-altitude clouds.

clay. Fine grains of silicate minerals also containing aluminum and water.

cleavage. The tendency of a mineral to break so that it yields definite flat surfaces.

climate. Average weather over a number of years at a particular location.

cloud seeding. Methods of artificially producing precipitation from clouds.

coalescence. Process by which small water droplets suspended in the air combine and grow into larger droplets.

cold front. A front at which a colder air mass thrusts up under a warmer air mass.

column. A structure formed when a stalactite and a stalagmite meet.

comet. A member of the solar system composed of rocks and frozen gases, usually moving in a very eccentric orbit.

composite cone. Cone formed by intermediate type of volcanic eruption; consists of alternating layers of cinders and lava.

conchoidal. Shell-like surface produced by the fracture of certain minerals.

concretion. Rounded accumulation of mineral matter, often resembling petrified eggs.

condensation nucleus. The center around which a cloud droplet is formed.

conduction. Transfer of heat by direct contact.

conformity. Boundary between two rock layers deposited in succession.

constellation. An apparent grouping of stars in a recognizable pattern.

continental drift. The theory that the continents were not stationary, but slowly drifted along the earth's surface.

continental glacier. A glacier that covers an entire land surface with a continuous sheet of ice. It is found only in the polar regions.

continental rise. A bulging accumulation of sediments at the base of the continental slope.

continental shelf. Relatively shallow ocean floor bordering a continental land mass.

continental slope. A zone of rapidly increasing depth in the sea between the edge of the continental shelf and the deeper ocean basin.

contour interval. The difference in elevation between two successive contour lines.

contour lines. Lines on a map joining points on the earth of equal elevation.

convection. Movement of a liquid or gas brought about by uneven heating.

coprolites. Fossil excrement.

core. The very dense innermost region of the earth.

Coriolis effect. The tendency of moving objects to veer to the right in the Northern Hemisphere and to the left in the Southern Hemisphere.

corona. A halo of faintly glowing gases outside the chromosphere of the sun, usually visible only during a solar eclipse.

crater. The funnel-shaped pit at the top of a volcanic cone.

cratons. Precambrian rocks that form the centers of nearly all continents.

crest. The ridge of a wave, elevated above the surrounding water.

crevasse. A deep crack in a glacier.

cross bedding. Angled layers in sedimentary rocks.

crust. The outer layer of the solid earth.

crystal. A mineral form resulting from the arrangement of atoms, ions, or molecules in definite geometric patterns.

cumulus clouds. A family of clouds that resemble balls of cotton, usually having flat bases.

cyclone. A large mass of low-pressure air with winds moving toward the center in a counterclockwise whirl; a low.

deflation. The effect of wind blowing away loose soil and rock particles.

delta. A triangular deposit at the mouth of a stream.

density. The weight per unit volume of any material.

desert pavement. A layer of close-fitting larger stones left on the desert floor after the lighter material has been blown away.

dew point. The temperature at which a given quantity of air reaches 100 percent relative humidity.

diatoms. Microscopic plants whose silica skeletons are often found in sea floor sediments.

dike. Solidified magma in vertical cracks or fissures.

dip. The angle of inclination of a rock bed or a fault from a horizontal surface, measured in degrees.

disconformity. Parallel horizontal rock layers separated by an erosional surface.

distillation. A method used to extract fresh water from sea water. The sea water is changed into vapor, then condensed into pure fresh water.

divide. A ridge or region of high ground that separates the drainage system of a stream.

doldrums. Belt of calm air under lower pressure near the equator.

dome mountains. Mountains that form when sedimentary beds are uplifted into broad, circular domes.

Doppler effect. The apparent effect on a train of waves if there is relative motion between the source and the receiver.

double stars. Two stars that move around one another, held together by mutual gravity.

drift. A general term for the material deposited by a glacier or by glacial meltwater.

drizzle. Precipitation caused when cloud or fog droplets fall to earth.

drumlin. A long, low mound of glacial till standing parallel to the direction of ice movement. They are generally steepest near the rounded end and gradually descend toward the narrow side.

dune. A hill or ridge of wind-blown sand.

earthshine. Illumination of the darker portion of the moon by reflection of light from the earth.

eclipse. The cutting off of light from one celestial body by another.

elastic rebound theory. The theory that earthquakes take place when the strain of many years of slow movement causes rocks to stretch and bend until they break and rebound (fault).

electromagnetic radiation. Energy that travels through space in the form of waves.

electron. An atomic particle having a negative electrical charge.

element. A substance that cannot be decomposed to a simpler substance by ordinary means.

ellipse. Oval figure produced when a cone is cut in a diagonal direction.

emergent. Term used to describe shoreline features resulting mainly from a drop in sea level or a rise of the land surface.

epicenter. The point or line on the earth's surface directly above the center or focus of an earthquake's origin.

epoch. Subdivision of a period in the geologic time scale.

equinox. A time when the sun is directly overhead at noon on the equator.

era. Largest unit of geologic time.

erosion. The removal of soil and rock fragments by natural agents.

erratic. A large boulder, deposited by glacial action, whose composition is different from that of the native bedrock.

escape velocity. The speed a rocket must reach in order to leave the earth's immediate gravitational field.

esker. A winding ridge of stratified drift deposited by streams running under or through glacial ice.

estuary. A shallow bay extending inland; a drowned river valley.

evaporites. Chemical sedimentary rocks formed by evaporation of water.

exfoliation. The splitting off of scales or flakes of rock as a result of weathering.

exosphere. The layer of the atmosphere, above the ionosphere, that merges with outer space.

extrusive rocks. Igneous rocks formed when magma flows out over the earth's surface.

fault. A fracture in a rock surface, along which there is displacement of the broken surfaces.

fault-block mountains. Mountains formed by faults that break part of the crust into large blocks which are then tilted.

faulting. The movement of rock layers along a break.

fetch. The extent of open water across which a wind blows.

fiord. A drowned glacial valley.

firn. The grainy ice in an accumulation of snow that later becomes glacial ice, also called *névé.*

fissure. An open fracture in the rock surface.

floe. A drifting section of arctic ice.

flood plain. The part of a stream valley that is covered with water during flood stage.

fluorescence. A luminescence given off by substances that are irradiated by ultraviolet or X-rays.

focus. The location below the earth's surface at which an earthquake originates; the source and center of the earthquake.

folded mountains. Mountains that result from the folding of rocks.

folding. The bending and breaking of horizontal rocks due to lateral pressures.

Foraminifera. Microscopic animals whose calcium carbonate skeletons often form sea-floor sediments.

fossil. Preserved remains or traces of prehistoric life on earth, buried in the rocks.

fracture. The way that a mineral breaks when it does not yield along a cleavage or parting surface.

fringing reef. A coral reef closely attached to the shore.

front. The boundary between two air masses.

frost action. Prying off of rock fragments by the expansion of freezing water in crevices.

fumaroles. Fissures or holes in volcanic regions from which steam and other volcanic gases are emitted.

fusion. The union or blending of materials by chemical or nuclear reaction.

galaxy. An astronomical system composed of billions of stars.

gastroliths. Fossilized gizzard stones of reptiles.

Geologic Column. The total sequence of rock layers from oldest to youngest.

geology. Scientific study of the solid part of the earth.

geomagnetic poles. The points on the earth's surface directly above the magnetic axis.

geosyncline. Very broad downfold in the earth's crust, extending for hundreds of miles.

geothermal energy. Energy supplied from heat in the earth's interior.

geyser. A hot spring that periodically erupts steam and hot water.

geyserite. A silicate mineral deposited around the opening of a geyser.

gibbous. The phase of the moon that is more than half full but less than full.

glaze ice. Formed when rain freezes upon striking the ground.

graben. A trough that develops when parallel faults allow the blocks between them to sink, forming broad valleys flanked on each side by steep cliffs. Also called *rift valley*.

gradient. The difference in elevation between the head and mouth of a stream.

gravity. A force that tends to pull every particle of matter toward every other particle.

great circle. Any circle drawn on the earth's surface that divides the sphere in half.

greenhouse effect. The process by which the atmosphere traps the sun's radiations.

Greenwich time. Time as measured on the prime meridian; used to determine longitude.

ground water. Water that penetrates into spaces within the rocks of the earth's crust.

gullying. A type of erosion in which deep channels form as water flows down a slope.

guyot. A flat-topped seamount.

gyre. Circular pattern of movement of surface water in the oceans.

hachures. A method of indicating a depression on a topographic map; short straight lines represent the direction that water would take in flowing down the slopes.

hail. Ice pellets formed by the successive freezing of layers of water.

half-life. The time interval during which half of a given amount of radioactive material disintegrates.

hanging valley. The valley of a tributary that enters the main valley from a considerable height above the main stream bed.

hardness (H). The resistance of a mineral to scratching.

headward erosion. Extension of the gullies that form the beginning or head of a stream.

highlands. Mountainous regions on the surface of the moon created by volcanic activity. They reflect more light than do the smoother parts.

horn. Sharp peak formed where several arêtes join.

hot spots. Areas on the earth's surface where magma rises from the asthenosphere.

humidity. Water vapor in the air.

humus. Organic matter in the soil, produced by the decomposition of plant and animal materials.

hurricane. A destructive tropical storm over the Atlantic Ocean; such a storm over the Pacific Ocean is called a *typhoon*.

hydration. The chemical combination of substances with water.

hydrocarbons. Organic compounds containing carbon, hydrogen, and oxygen.

hydrologic cycle. The continuous process of water being evaporated from the sea, precipitated over the land, and eventually returned to the sea.

hydrosphere. The earth's envelope of water.

hygrometer. An instrument used to measure humidity.

icebergs. Large blocks of ice broken off from the leading front of a continental glacier ice shelf.

ice field. Formed where the amount of snow and ice that accumulates during a year is greater than the amount that melts during the warm season.

ice front. The advancing front wall of a glacier.

igneous rocks. Rocks formed by the solidification of magma.

index contours. Heavier and darker contour lines, usually every fifth line.

index fossils. Fossils that are found only in rocks of one particular period of geologic time; guide fossils.

International Date Line. An imaginary line at about 180° longitude; when it is crossed, standard time moves back or ahead by 24 hours.

intrusive rocks. Rocks formed by magma solidifying among other rocks below the earth's surface; also called *plutonic rocks.*

invertebrates. Animals without backbones.

ion. An atom or group of atoms that carries an electric charge.

ionosphere. The layer of the atmosphere above the stratosphere containing molecules of ionized gases.

island arc. A chain of islands lying close to the edge of a continent and adjacent to a deep submarine trench.

isobar. A line drawn on a weather map connecting locations of the same atmospheric pressure.

isostasy. Principle describing the state of balance of the earth's crust.

isotherm. A line on a map joining locations that record the same temperature.

isotopes. Atoms of the same element having different atomic weights.

jet streams. Swift, high-altitude winds at the edges of the circumpolar whirls.

joint. A fracture in a rock surface.

karst plain. A type of landscape characteristic of some limestone regions; drainage is mostly through underground streams in caverns.

kettle. A depression left after the melting of large blocks of ice buried in glacial drift.

knob. Large rock projection smoothed by glacial action.

laccolith. A domed mass of igneous rock formed by intrusion beneath other rocks.

lagoon. The body of water between an offshore bar and the mainland.

latent energy. Heat released when water vapor changes back to liquid water.

latitude. Location measured in degrees north and south of the equator.

lava. Liquid rock material that flows out on the surface of the earth from underground sources.

leaching. Removal of minerals from soil and rock in solution as water seeps down from the surface.

levee. The raised bank of a stream.

light year. The distance that light travels in one year.

lignite. A brownish-black, low-grade coal.

lithosphere. The solid part of the earth.

load. The total amount of material carried by a stream.

local group. A cluster of galaxies containing the Milky Way Galaxy.

lode. A large number of thick ore veins.

loess. An extensive deposit of wind-blown silt.

longitude. Location measured in degrees east or west of the prime meridian.

longshore current. Ocean current that moves close to and parallel with the shore.

luster (L). The appearance of the surface of a mineral in reflected light.

magma. Molten rock materials below the earth's surface.

magnetic declination. The difference, measured in degrees, between true north and magnetic north; also called *variation*.

magnetic storm. Electrical effects in the upper atmosphere usually caused by solar flares.

mantle. The thicker, more dense part of the earth beneath the crust.

map projection. A representation of the globe, or a portion of it, on a flat surface.

maria. Dark regions seen on the moon's surface.

massive. Rock masses without definite form (unstratified).

mass-wasting. Erosional processes chiefly caused by gravity.

meanders. Wide curves typical of well-developed streams.

mean sea level. The point midway between the highest and lowest tide.

mean solar time. Time as measured by the average position of the sun.

mechanical weathering. Natural processes that reduce rock to small fragments without changing its mineral composition.

meridian. An imaginary line extend-ing from pole to pole that crosses the equator at a right angle.

mesa. A large, wide, flat-topped hill, usually a remnant of horizontal rock strata.

mesopause. Boundary between the mesosphere and thermosphere.

mesosphere. Layer of the atmosphere immediately above the stratosphere.

metamorphic rocks. Rocks that have been changed from their original form by great heat and pressure.

meteor. A rock-like particle in space; one of the smallest members of the solar system.

meteorite. A meteor that actually strikes the earth's surface.

meteorology. The science of the atmosphere.

mid-ocean ridges. A series of underwater mountain ranges in the ocean basins.

millibar. A unit for measuring air pressure; used in international weather observations.

mineral. Chemical compounds or uncombined elements found in rocks.

mirage. An optical illusion caused by the refraction of light as it passes through air layers of varying densities.

Mohorovicic discontinuity (Moho). The zone of contact between the crust and the mantle.

Mohs' scale. A numerical scale of hardness represented by ten minerals, ranging in hardness from 1 (talc) to 10 (diamond).

mold. The empty cavity of a fossil type left after the original organism has disappeared.

molecule. The smallest particle of a substance that can exist and still retain the properties of the substance.

monadnocks. Isolated hills or mountains of resistant rock rising above the general level of a peneplain.

monsoon. Seasonal wind that is more pronounced over large continental areas near the equator.

moraine. A ridge or mound of boulders, gravel, sand, and clay carried by and deposited at the leading edge of a glacier.

mudflow. The movement down a valley of a large mass of mud and rock debris during a flood.

native metals. Metals found naturally in the pure form.

nautical mile. A distance equal to one minute of latitude; about 1.15 statute miles.

neap tides. Lower than normal tides produced when the sun and moon are at right angles to the earth.

nebula. A cloud of gases or dust in space.

nebular theory. The theory that the solar system formed from the condensation of a large cloud of dust and gases.

neutron star. A collapsed star whose atoms have been stripped down to the neutrons.

neutrons. The electrically neutral parts of atoms.

nitrogen cycle. The complex series of actions by which the nitrogen content of the atmosphere is maintained.

nodules. Black lumps of material on the sea floor, many of which contain valuable minerals.

nova. A star whose brightness suddenly increases many times the normal.

nuclear fission. A nuclear reaction in which heavy atoms split into lighter nuclei.

nuclear fusion. A nuclear reaction in which lighter-weight particles combine to form heavier nuclei.

nuclear reactor. Equipment used to control nuclear fission.

nucleus (of an atom). The central part of the atom, containing most of its total mass.

oblate spheroid. The shape produced if a sphere is slightly flattened; the true shape of the earth.

occluded front. A type of weather front created by an advancing cold front overtaking a warm front.

oceanography. The scientific study of the sea; also called oceanology.

offshore bar. A sand bar extending more or less parallel to the mainland.

oil pool. An underground accumulation of oil and gas.

oil shale. A fine-grained sedimentary rock containing large amounts of hydrocarbons.

ooze. An extremely fine layer of sediment on the sea floor.

ore. A mixture of minerals containing at least one substance that can profitably be recovered.

outcrop. An exposure of crustal rock at the surface.

outwash plains. Plains formed by the deposition of materials washed out from the leading edge of a glacier.

overturn. Sinking of chilled surface water in the sea or in a lake.

oxbow lake. A lake formed by the isolation of a meander from the main stream.

oxidation. The chemical combination of substances with oxygen.

ozone layer. A concentration of the gas ozone, lying largely in the region from 19 to 34 km (stratosphere) above the earth.

pack ice. A floating layer of ice completely covering the sea surface.

pahoehoe lava. Lava that has solidified, having a smooth, ropy, or billowy appearance.

paleontology. The science that deals with the study of fossils.

Pangaea. A supercontinent that broke apart to form the present continents.

parallax. The apparent movement of

an object due to a change in the observer's position.

parallel fault. A fault in which there is evidence of both lateral and vertical motion.

parallel of latitude. An imaginary line drawn on the earth's surface parallel to the equator.

parasitic cones. Volcanic cones that develop at openings some distance below the main vent.

peat. Soft brown or black plant material formed in a decay process lacking oxygen.

peneplain. The flat surface left after landforms have been worn down by the agents of weathering and erosion.

penumbra. The lighter, outer part of a shadow.

perigee. The point of a satellite's orbit nearest the object around which it revolves.

perihelion. The point on the earth's orbit nearest the sun.

period. Subdivision of an era in the geologic time scale.

permafrost. A permanently frozen area in the deeper soil of the tundra.

permeability. The ability of water to penetrate and pass through rock.

Perseids. A meteor shower seen each year in mid-August.

petrifaction. A process in which the original substance of a fossil is replaced by mineral matter.

phosphorescence. The continued emission of light rays from certain minerals after exposure to ultraviolet light.

photosphere. A layer of brilliantly glowing gases beneath the chromosphere; the source of most of the radiant energy from the sun.

phytoplankton. Microscopic marine plants.

pitching folds. Folds whose axes slant downward at each end.

placer deposit. Mineral deposits in the gravel of a stream bed.

planetary circulation. The movement of the atmosphere over the entire earth.

plankton. Microscopic free-floating plants and animals in water.

plate tectonics. The theory that the earth's crust is made up of a number of separate rigid plates exposed to forces that cause them to move.

pluton. A body of igneous rock pushed up into the crust of the earth.

Polaris. The star very near the celestial north pole (North Star).

polyp. Body form of a coral animal.

pothole. A rounded depression in the rock of a stream bed.

precession. A slight wobbling motion of the earth as it turns on its axis.

precipitation. Moisture that condenses and falls from the air. Also, separation of an insoluble substance from a solution.

pressure ridge. Formations of ice near the surface of ice due to uneven surface movements of an alpine glacier.

primary wave. A kind of seismic wave formed by back and forth vibrations.

prime meridian. The 0° longitude line passing through Greenwich, England.

prominence. A bright arch-shaped structure appearing above the solar surface.

protons. An atomic particle having a positive electrical charge found in the nucleus.

pseudofossils. Rock structures superficially resembling fossils but not actually made by prehistoric organisms.

psychrometer. A hygrometer that uses a wet-and-dry bulb thermometer.

pulsars. Stars that give off radio signals in regular pulses.

pyroclastics. Volcanic rock formed from material ejected during an eruption.

quadrangle. Detailed topographic maps commonly 7.5 minutes of latitude and 7.5 minutes of longitude. They are available from the Geological Survey, U.S. Department of Interior.

quasars (quasi-stellar radio sources). Stars that produce more energy than can be accounted for by any known process.

radar. A short-wave radio device that can be used to locate and track storms.

radiation. The process by which energy is given off in the form of rays.

radiation fog. A fog formed as a result of radiation heat lost by the earth. Also called *ground fog.*

radioactivity. The spontaneous breakdown of uranium and certain other radioactive elements into invisible radiations.

radiolarians. Protozoans that secrete siliceous shells of intricate design.

radiosonde. A set of electronic instruments that record and broadcast temperature, pressure, and humidity conditions at high altitudes.

radio telescope. An electronic device that focuses radio waves from outer space.

rays. Light streaks spreading out from some of the moon's craters.

red shift. A shift in the spectral lines of light from distant galaxies, caused by their receding motion.

reflector. A type of telescope in which a curved mirror produces an enlarged image.

refraction. The bending of a water wave when it approaches the shore at an angle, caused by the slowing down of the part that reaches shallow water first.

refractor. A type of telescope that uses only lenses to enlarge the image of the object viewed.

rejuvenation. Any action that tends to increase the gradient of a stream, thereby renewing youthful features of topography.

relative humidity. The ratio between the actual amount of water vapor in the air and the maximum amount it could hold at a given temperature.

relief. The irregularity in elevation of parts of the earth's surface; also the difference in elevation between the high and low points of a land surface.

residual boulders. Large rock fragments in their original position of the weathered bedrock.

resource. A natural material found in the earth that we can use to our advantage.

revolution. The movement of a celestial body in its orbit.

Richter scale. A scale that measures earthquake magnitude as determined by seismograph records.

rift. A narrow depression running along the center of the mid-ocean ridges.

rift valley. See *graben.*

rip currents. Swift currents moving away from the shore through a line of breakers.

rocks. A combination of different minerals found in the earth's crust.

rotation. The turning of a celestial body on its axis.

salinity. The total amount of dissolved solids in sea water.

satellite. Any celestial object that revolves around a larger object; a natural or artificial moon.

saturation value. The amount of water vapor a quantity of air can hold at any particular temperature.

savanna. A type of tropical climate found between rain forests and tropical deserts.

scale. The mathematical comparison between actual distance and distance on a particular map.

SCUBA (self-contained underwater breathing apparatus). An apparatus used by an individual diver as an air supply.

sea-floor spreading. A process that creates new sea floor when plates separate and molten material fills in from below.

seamount. An underwater volcanic peak.

secondary wave. A kind of seismic wave formed by up-and-down vibrations.

section. A square region of land measuring one mile on each side.

sedimentary rocks. Rocks formed in layers from materials deposited by water, wind, ice, or other erosional agents.

sediments. Rock fragments produced by the weathering of solid rock.

seismic waves. A form of energy moving through the solid earth.

seismograph. An instrument for detecting and measuring movements of the earth's crust.

shadow zone. A region on the earth's surface where a particular earthquake wave cannot easily be detected.

sheet erosion. A type of erosion in which water on a slope strips away exposed topsoil slowly and evenly.

shield. An exposure of Precambrian rocks over a wide area.

shield cone. A broadly based cone with a gentle slope, formed by the quiet type of volcanic eruption; composed of successive layers of lava. Also called *lava dome.*

sidereal year. The time required for the earth to complete one revolution in relation to the stars.

silicates. Minerals whose basic structure is a product of the joining of silicon and oxygen.

sill. Solidified magma in horizontal rock formations.

silt. Soil particles in an intermediate size range between clay particles and sand grains.

sink. A depression in the earth's surface formed by the collapse of the roof of an underground cavern.

sleet. Frozen raindrops.

slumping. A minor landslide involving only a small amount of loose material.

snow line. The elevation above sea level where snow remains all year.

soil. Mixture of loose rock fragments and materials produced by weathering.

soil creep. Slow movement of a mass of soil down a slope, caused by the soil's own weight.

soil horizon. One of the layers in a soil profile.

soil profile. Arrangement of layers in a mature soil.

solar flare. An unusually bright cloud of gases erupting from the sun, lasting for a few hours.

solar system. The sun and the celestial objects that revolve around it.

solar wind. A stream of high-speed atomic particles moving away from the sun.

solstice. A time when the sun seems to reverse its apparent movement north or south of the equator.

spatter cones. Small cones that form in lava fields away from the main vent. Lava is spattered out through holes in a thin crust.

specific gravity (**sp. gr.**). A number that relates the density of a substance to that of water.

spectrograph. A type of spectroscope that makes a photographic record of the spectrum of an object viewed through a telescope.

spectroscope. An instrument that separates a beam of light into its various components.

spectrum. The band of colors seen

when light is separated by a prism.

spit. A sand bar extending outward from the shore.

spring. Ground water that comes naturally to any exposed surface between the water table and the ground.

spring tides. Higher than usual tides produced when the moon and sun are lined up with each other.

squall line. A long line of heavy thunderstorms often advancing just ahead of a fast-moving cold front.

stack. An isolated column of rock left standing after waves eroded a shoreline.

stalactite. A stony projection of minerals deposited from dripping water on the roof of a cavern.

stalagmite. A raised deposit of minerals deposited from dripping water on the floor of a cavern.

standard time. Time determined by division of the earth's surface into 24 time zones, the centers of which are approximately 15 degrees apart in longitude.

star cluster. A close grouping of stars, probably of a common origin.

stationary front. Either a cold or warm front that comes to a halt for a few days.

station model. A group of symbols on a weather map describing weather conditions at a particular location.

steppes. Regions bordering the tropical deserts, often extending into the middle latitudes.

stock. A small batholith.

strata. Rock layers or beds.

stratified rocks. Rocks that are found in parallel layers.

stratopause. The top boundary of the stratosphere.

stratosphere. The layer of the atmosphere above the troposphere where the temperature remains fairly constant.

stratovolcanoes. Volcanoes with cones of alternate layers of lava as well as solid fragments.

stratus clouds. A family of clouds that form in layers.

streak. The color of a thin layer of any finely powdered mineral.

stream piracy. Capture of one stream by another as a result of erosion of the divide separating the stream valleys.

strike. The direction of a line along the edge of an inclined bed where it meets the horizontal plane. The strike is always at right angles to the dip.

strike-slip fault. A fault in which horizontal movement of one or both sides has occurred.

strip-mining. The method of mining in which layers of ore are exposed at the surface.

subduction zone. The region along plate boundaries where one plate is forced under another.

submarine canyon. An unusually deep-cut valley in the continental slope.

submergent. Term used to describe shoreline features resulting mainly from a rising sea level or from a drop in the land surface.

sub-tropical high. A belt of calm air under increased pressure; located about 30° north and south of the equator.

sunspots. Large dark areas of cooler gases on the sun's surface.

superposition. A principle that states that the overlying rock layer is younger than the layers below it.

supersaturation. A condition in which air reaches better than 100 percent relative humidity.

surface (long) waves. Earthquake waves that move over the surface of the crust.

swells. Groups of large ocean waves alike in size.

synchronous orbit. An orbit that allows an earth satellite to remain at the same point above the earth's surface.

syncline. A trough or downturn of a rock fold.

talus. A mass of rock debris at the base of a steep mountain or cliff.

tarn. A lake formed in a glacial cirque.

temperature inversion. An unusual condition in which the layer of air at the surface is colder than the layer above it.

terraces. Step-like formations along the sides of a rejuvenated stream valley.

tetrahedron. Four-sided pyramid.

thermocline. A zone of rapid temperature change usually found near the surface of a body of water.

thermonuclear reaction. A nuclear *fusion* reaction taking place at high temperature.

thermosphere. Layer of the atmosphere above the mesopause.

tidal bore. A wave that passes upstream in a river from the sea as the tide rises.

tidal bulge. A bulge produced in the part of the sea facing the moon or on the opposite side of the earth.

tidal oscillations. Very slow rocking motions in parts of the oceans, occurring in response to the tidal bulges.

till. Glacial deposits that have not been stratified or sorted by water action.

tombolo. Island connected to the shore by a ridge of sand.

topographic map. A map showing surface features of a portion of the earth.

topography. The physical features of a region.

tornado. A relatively small, destructive middle-latitude storm, usually originating over the land.

township. A square region of land measuring six miles on each side.

trade winds. Planetary winds in tropical areas blowing generally toward the equator.

transit. A crossing between the earth and sun by Mercury or Venus.

transpiration. The process by which plants release water vapor into the air.

travertine. A form of calcite deposited around the opening of a hot spring.

trenches. Deep fissures on the ocean floor caused by faulting.

tributary. A stream that flows into a larger one.

trilobite. A common invertebrate animal that became extinct during the Permian period.

Tropic of Cancer. An imaginary line found parallel to the equator at $23\frac{1}{2}°$ north latitude; marks the most northern point where the sun is ever directly overhead.

Tropic of Capricorn. An imaginary line found parallel to the equator at $23\frac{1}{2}°$ south latitude; marks the most southern point where the sun is ever directly overhead.

tropical year. The time from one vernal equinox to the next.

tropopause. The upper boundary of the troposphere.

troposphere. The layer of the atmosphere closest to the earth's surface in which most weather activities occur.

trough. The depression on either side of a wave crest.

tsunami. A giant wave produced by disturbances on the sea floor; incorrectly called a "tidal wave."

tundra. A type of climate in the zone of transition between the subarctic regions and the icecaps.

turbidity currents. Strong currents produced by material sliding down the continental slopes.

umbra. The dark, inner part of a shadow.

unconformity. An eroded bedrock surface that separates younger strata from older rocks.

undertow. A current that runs constantly beneath a line of breakers as water from the breaking waves is pulled back to deeper water.

uniformitarianism. The principle that the same forces that changed rocks in the past are still operating today and causing the same kinds of changes.

upslope fog. A fog produced by the adiabatic cooling of air as it moves up a slope.

upwelling. Movement of deeper sea water to the surface.

variable star. Any star whose brightness appears to change from time to time.

varves. Annual, double layers of fine clay-like sediments.

vein. Ore deposits that as hot solutions have entered cracks in rock and then cooled.

ventifact. A stone smoothed by wind abrasion.

vertebrates. Animals with backbones.

vertical faults. Several kinds of faults in which movement of the rocks is mainly vertical in direction.

volcanic bomb. Rounded rock fragments thrown out during a volcanic eruption.

volcanism. A general term including all types of activity due to movement of magma.

volcano. The vent from which molten rock materials move out onto the surface; includes the accumulation of volcanic materials deposited around the vent.

warm front. A front at which a warmer air mass overrides a colder air mass.

water budget. A comparison between the amount of water received from precipitation and the amount lost as water vapor passes back again into the atmosphere.

water gap. A valley with an existing stream that cuts across a mountain ridge.

watershed. A series of slopes that drain into a river system.

waterspouts. Tornadoes over the sea.

water table. The upper boundary of the ground water below which all spaces within the rock are completely filled with water.

wave-built terrace. A seaward extension of a wave-cut terrace, produced by the accumulation of debris from wave action.

wave-cut cliff. A cliff formed from wave erosion in which waves strike directly against the rock of the land.

wave-cut terrace. A level surface of rock under the water along the shore, formed as waves cut back the shoreline.

wavelength. The distance from one wave crest to the next, or from one wave trough to the next.

wave period. Time taken for two successive crests of a wave to pass a given point.

weathering. The natural disintegration and decomposition of rocks and minerals.

wind gap. An abandoned stream valley cutting across a ridge.

zone of aeration. A region beneath the ground surface where rocks contain both water and air.

zone of saturation. Region beneath the surface in which all spaces between rocks are filled with water.

zooplankton. Microscopic marine animals.

index

aa lava, 197
absolute humidity, **441**
absolute magnitude, *29–30*
abyssal plains, 325–326
acid rain, 512
adiabatic changes, *443*
adiabatic cooling, 443–444
Adirondack Mountains, 221–222
advection fog, 448
aeration, zone of, 265, **266**
Agassiz, Lake, **288**
agonic lines, *189*
air, composition of, 420–421; gases in (Table), 421; pollution of, 426–**427**; pressure of, 421–424, 466; supersaturated, 444, 446; *see also* atmosphere
air masses, classification of, 458; cold, 458, **460**; North American, 458–**459**; origin of, 458; warm, 458, **460**
air pollutants, 426
Alabama Hills, **248**
Alaskan glaciers, **284**, 285, 290
albedo, 429
Aleutian Islands, 202–203
alluvial fan, **263**
alpha particle, 381
alpine glacier, 277, **278, 280, 283,** 290
altimeter, 423–424
altitude, temperature and, 482

altocumulus clouds, **445,** 446, 447
altostratus clouds, 446, **447**
aluminum, recycling of, *498–499*
amber, 390
Andes Mountains, 223, 290
Andromeda, **16,** 17
anemometer, **467**
aneroid barometer, 422–**423**
animals, microscopic, in sea water, 329–**330,** 343, 344, **345;** and nitrogen cycle, **424;** weathering and, 235
annular eclipse, **83**
Antarctic Current, 356
Antarctic Ocean, **317,** 318
Antarctica, **277**–278, 281, 487
antennas, radio telescope, **14**–15
anthracite coal, 500
anticline, **208**–*209,* **209**
anticyclones (highs), **465**
aphelion, *7–8*
apogee, *79*
Appalachian Mountains, 209, 213, 224, 406, 408, 409
apparent magnitude, of stars, **28, 29, 30**
apparent solar time, *63–64*
applications satellites, 102–107
aquaculture, 347–348
aquifer, **268**
Arctic Circle, **61**
Arctic Ocean, **317,** 318; ice layer in, 292, 337
Arcturus, spectra of, **33**

aretes, *281,* **282**
argon, 342, 421
artesian wells, **268**
asteroids, 47–48
asthenosphere, *188*
astrology, 5
astronomical observatories, ancient, *2–3*
Atlantic Ocean, **317**–318, 326
atmosphere, 420; carbon dioxide in, **425,** 429; chemical balance of, 424–425; composition of, 420–421; continuous motion of, **353;** gases of, 421; layers of, 429–**431,** 432–433; moisture in, measurement of, 440, **441–442;** solar energy and, 427–**428,** 429; water vapor in, 439–442; *see also* air
atoll, **309**
atomic mass, 141–142; of hydrogen, 142; of oxygen, 142
atomic number(s), 141; Table of, 141
atomic theory, 138
atoms, electron arrangement in, 143–144; kinds of, 140–142; nucleus of, 22, 140, 513; structure of, 138–140
auroras, **27**
autumnal equinox, **61,** *62*
axis, of earth, 60–63, **61;** geographic, 190, 191; of rock fold, 210; of rotation, 190

bacteria, nitrogen-fixing,

(Note: Page numbers in **boldface** type indicate illustrations and those in *italic* type indicate definitions.)

424
banner clouds, **444**
bar (shoreline), **300,** 301
barchans, **243,** *244*
barograph, **423**
barometer, aneroid, 422–**423;** mercurial, 422, **423;** Torricellian, **422**
barrier islands, *307,* **307–308**
barrier reef, **309**
Barringer Crater, **48**
basalt, 171, **173**
basaltic rocks, 171, **178**
basement complex, 381
basins, drainage, **257,** 258; lake, 287, 288; ocean, 323
batholiths, *199,* 213, 221
beach materials (Table), 299–300
beaches, *299,* **299, 300**–301; balance of, **302**
bead tests for minerals, 167, **168,** 169
bedrock, 235–236
bends, 321
berm, *300,* **300**
beta particle, 382
Big Bang Theory of universe, 39–**40**
biomass, *509*–510
bituminous coal, *500*
Black Hills, 221–222
black holes, **39**
block lava, *197*
blowout, *242*
body waves (earthquake), 215
bog, 288, **289**
borax, 289
breaker, wave, 362, **363**
breeder reactor, 505
breezes, land, 472; mountain, 472; sea, 472; valley, 472
bright-line spectrum, 31–32

brown clay, 331
"brown" coal, 500
building materials, 514
Bunsen burner, 167
buttes, **249,** *250*

Caesar, Julius, 87
calcite, 271
calcium, 157
calcium carbonate, 329, 330
caldera, *199,* **199**
calendar, Gregorian, **87**–88; Julian, **87;** Roman, 87; Universal, 88; World, **88**
calorie, *440*
Cambrian period, **407,** 407–408, **530–531**
Canadian Rockies, **276, 283,** 285
Canadian Shield, *404,* **404–405**
Cape Cod, 297
capillary fringe, **266**
capillary movement, zone of, **266**
capillary tension, 266
carbon dioxide, in air (Table), 421; in atmosphere, **425,** 429
carbon dioxide cycle, **425**
carbon-14 dating, 383
carbonation, *235,* 237, 270
carbonic acid, 234–235
Carboniferous period, **409**
carbonization, *389*
Carlsbad Caverns (New Mexico), 270–271
Cascade Mountains, 221
Catskill Mountains, **248,** 250
caverns, 270–**271**
celestial north pole, **4**
Celsius scale, 466
cement, 514
Cenozoic era, 401–**402, 412–413**

cepheid variables, *30*
chain reaction, nuclear, *504*
chalcedony, **155**
charcoal block tests, **168,** 169
chemical bonds, 144–145
chemical deposits in ocean, **330**
chemical formula, 144
chemical properties, of ocean, 319–320; of sea water, 335, 340–346
chemical rocks, 172
chemical weathering, 233–235, 297
chinook winds, **482**
chromosphere of sun, **25**–26
cinder cones, **198**
circumpolar whirl, **465**–466
cirques, **281–282**
cirrocumulus clouds, **445,** 446
cirrostratus clouds, **445,** 446
cirrus clouds, **445,** 446
clay, 238; brown, 331; in cement, 514; red, 330–331
cleavage of minerals, 164–**165**
cliffs, formation of, 177, **178**
climate(s), *479–496;* control of, 479–484; humid continental, **488;** humid subtropical, **488;** local, 489–490; maps of, 125; marine west coast, **487, 488;** Mediterranean, 487, **488;** middle latitude, 484, **485;** North American, 492–496; polar, 484, **485;** savanna, 486; steppe, 487; subarctic, 488–489; and temperature, 479–484; tropical, 484, **485;** tundra, 489; and weath-

ering, 237–238; *see also* weather

clouds, banner, **444;** and cold front, **461;** formation of, **444, 445,** 446; high, 446; low, 446–448; middle, 446; seeding of, **450**–451; types of, 446, **447**–448; with vertical development, 448; and weather fronts, **461, 462**

coal, 172, 389, 499, **500**–501; anthracite, 500; bituminous, 500; "brown," 500; formation of, 500; pollution by, 511–512; stripmining of, **501;** supply of, 501, 505, 510

coalescence of raindrops, **449**–450

coast, *303*

coastal resources, 309–310

coastlines, conservation of, 309–310; creation of, 304–**305, 306;** emergent, **306**–307; and sea level, **303**–304; submerged, 304–**305, 306;** uses of, 309–**310**

color of minerals, 163

Colorado Plateau, **398**

Columbia ice field, **276**

Columbia Plateau, **196**

comets, 49– 50, **50–51**

composite cones, 198

compound, *143*

concretion, 391

condensation nuclei, 446

condensation of water vapor, 442–443

conduction, heat transfer by, 433

conformity of rocks, 378

conservation, soil erosion and, **245–246,** 247; of

water, 256

constellations, *4–5*

contact metamorphism, *380, 380*

continental collisions, 224

continental drift, theory of, *182*–183

continental glaciers, 277, 281, **283**

continental margins, **324**

continental rise, **325**

continental shelf, 324–**325**

continental slope, 325

continents, history of, 402–**403, 404;** movements of, 402–**403, 404;** North America, growth of, 404–414

continuous spectrum, **31**–32

contour interval, **128**–129

contour lines, **127,** *128,* **128, 129**

contour loop, 130

contour map, interpretation of, 129–130

contour plowing, **246**

convection, heat transfer by, 433, 434

Copernicus, Nicolaus, **6**–7

copper ore, 513, **514**

coprolites, *390*

coquina, 172

coral reefs, **309–310**

core, of earth, **184–186;** of sun, 24, **25**

core samples, 320, **330**

Coriolis effect, 354–356, 358, **434,** 435, 465

corona of sun, **25**–26

Crab nebula, **38**

Crater Lake (Oregon), **199**

craters, formation of, **73;** meteorite, **48;** on moon, **72–73;** volcanic, *199*

cratons, *404*

crescent moon, **80**–81

Cretaceous period, **411**–412

crevasses, glacial, **280, 281**

crop rotation, 247

cross bedding of rocks, **378**

crust of earth, **184;** metals in, 512–**513,** 514

crustal plates, 186–192; *see also* plate tectonics

crystals, ice, 148, **450**–451; mineral, 163–**164;** sodium chloride, **146**

cumulonimbus clouds, **447,** 448

cumulus clouds, **444, 447,** 448

current(s), 320, 352, 353, **354**–359; Antarctic, 356; causes of, 353–354; deep, **359;** equatorial, **354,** 355; Japan, 356; Labrador, 356, 358; longshore, *301*–**302;** North Pacific, 356; rip, **363;** and temperature, 483; tidal, 366; turbidity, 354; undertow, 362–363

cyclone, wave, 461–462, **463–464,** 465

dark-line spectrum, **31**–32

day, measurement of, 66–67

daylight, length of, 63–65

Daylight Saving Time, 65

Dead Sea, 212

deflation, *242*

deflation hollow, 242

deltas, **263**

depression contour, **130**

desert pavement, **242**

deserts, middle latitude, 489, **490;** tropical, 486

Devonian period, **408**

dew, **443**
dew point, *442*–443
diatoms, **330,** 331
dikes, volcanic, *202*
dinosaur tracks, **389**
dinosaurs, **411**
dip of rock bed, **210**
direct tide, *84*
direction, points of, 120–121
disconformity of rocks, **379**
divides, **257,** 258
doldrums, 435
domed mountains, 221–**222**
Doppler effect, *33, 33,* **34,** 41
double stars, **37**–38
drainage basins, **257,** 258
drizzle, 451
drumlins, 285, **286**
dunes, sand, **243**–244
dust storms, **245,** 246

earth, 54; asthenosphere of, **188;** core of, **184**–**186;** crust of, **184;** crustal plates (*see* plate tectonics); directions on, 120–122; distance from moon, 78–**79;** distance from sun, 8; elements in crust of, 141, 153, **154;** Eratosthenes' measurement of, 119–**120;** exploration of from space, **102**–**103, 104;** formation of, **184**–**185, 186;** gravity of, 10–11; interior of, **184**–186; locations on, 120–122; magnetism of, 190–192; mantle of, **184;** maps of surface of, 122–125; models of, 116–117, 184–186; motions of, 60–66; orbit

of, **4, 8,** 60–**61,** 62–63; physical history of, 396–400; rotation of, 63–65, 85; satellites of, 100–107; shape of, 117–118, 120; size of, 119–120; solar energy on, 23–24
earthquakes, 213–220; landslides and, 241; on sea floor, 354
earthshine, *80*
eclipse, annular, **83;** of moon, **82**–83, 85; of sun, **24,** 82–83; total, **24,** 82–83
Einstein, Albert, 22–23
elastic rebound theory of earthquakes, **214**
electrical charge, 139, **140,** 145
electrical currents, magnetism and, 191
electromagnetic radiation, *11,* **12**–**13**
electromagnetic spectrum, *13*
electromagnetic waves, 428
electronic weather instruments, 467–**468**
electrons, 140, 141; arrangement in atoms, 143–144; loss of, 144–**145;** repulsion of, **140;** sharing of, 144–145, 149; transfer of, 144–**145**
element(s), *141;* in earth's crust (Table), 141; percent by weight in earth's crust (Graph), **154**
ellipse, **7**
elliptical galaxy, **17**
elliptical orbit, 7, 78–**79**
emergent coastline, **306**–307
energy, earth's supply of,

499–512; geothermal, 505–507; heat, 505; latent, 440; nuclear, 503–505; of sun (*see* solar energy); from tides, 506–507; wave, 296; wind, 508–**509**
epicenter of earthquake, **214, 216, 218**
epochs, of earth's history, *401*–402
equatorial currents, **354,** 355
equinox(es), 62
eras, of earth's history, *400*–**402**
Eratosthenes, 119–120; and earth measurement, **120**
erosion, *232;* and geologic time, 383–**384;** glacial, 281–283, 287–288; gravity and, 239, **240**–**241,** 242; headward, **257**–**258;** by running water, **256**–263; and sand, **242**–**243;** sheet, 245–246, 261; soil, and conservation, **245**–**246,** 247; water and, 253; wave, **296**–**299;** wind, **242**–**243**
erratics (boulders), **284**
eskers, *286*
estuary, *305*
evaporation, patterns of, 254–255; of sea water, 253, 338–339, 341, **342, 440**
evaporites, **172**
exfoliation, *236,* **237**
exosphere, 431
extrusive igneous rocks, 171, **173**

Fahrenheit scale, 466
fault-block mountain, **220**

faults, 210–*211,* **211,** 212; and earthquakes, 214; in mid-ocean ridges, **327;** San Andreas, 218–**219;** strike-slip, **210**
feldspars, 154, **156**
fertilizers, 507
fetch, *363*
Finger Lakes (New York), 287
fiords, *305*–**306**
firn, 275–276
first quarter moon, **80**–81
fish, 347–348
fissures, *200*
flame tests for minerals, 167, **168**
flood plain, **264**
flood stage, 263–264
floods, 263–**264,** 265; control of, 265
fluorescent minerals, 166, **167**
focus of earthquake, 214, 216
foehn, **482**
fog, advection, 448; radiation, 448; upslope, 449
folded mountains, 220
folding of rocks, *209*–**210**
food from sea water, 347–348
food chain in sea water, 344, **345**
Foraminifera, **330,** 331
fossil fuels, *449*–503; supplies of, 501–503; 510–512; underwater deposits of, 347; *see also* individual fossil fuels
fossils, Cambrian, **407–408;** finding of, 390–391; formation of, 387, **388–390;** in Grand Canyon, 398–**399;** index, **386**–387; indirect evidence

of, **389–390;** molds of, **390;** record of, 385–391
fractional scale, *127*
fracture of minerals, **165**
Franklin, Benjamin, 139, 357
fringing reef, 308–**309**
frost, 443
frost action, **233**
fuels (*see* fossil fuels)
full phases, of moon, **80**–81
Fundy, Bay of, **366,** 367

galaxies, *15*–**17;** Andromeda, **16**–17; Milky Way, 15–**16;** types of, **17**
Galileo, 117
gases, in air (Table), 421; dissolved, in sea water, 340, 342–343; molecular motion in, **147;** volcanic, 197, 340
gasoline, supplies of, 510–**511**
gastroliths, *390*
Geiger counter, 167
gem minerals, 158, **162**
geographic axis, **190**
geographic poles, **190**–192
Geologic Column, **399**–*400,* **400**
geologic eras, *400*–**402**
geologic maps, 125
geology, *207*
geomagnetic poles, *190,* **190**–192
geosyncline, 213, 223
geothermal energy, 505–507
geyser basins, **270**
geyserite, 269
geysers, **269–270**
gibbous phase of moon, **80**–81
glacial drift, 283–284
glacial erratic boulders, **284**

glacial meltwater, 286
glacial periods, 289–290, **291**
glacial till, **282, 284**
glaciated rocks, **282**–283
glaciers, *213,* 275–292; Alaskan, **284,** 285, 290; alpine, 277, **278, 280, 283,** 290; continental, 277, 281, **283;** crevasses in, **280, 281;** cycle of formation of, **282;** deposits left by, 283–285; erosion by 281–283, 287–288; formation of, 275–278; of Greenland, 278, 281; ice front of, 278; lakes created by, 287–288; movement of, **278–280,** 281; pressure ridges in, **280, 338;** relative speeds of flowage in, **278;** valley, 277
glaze ice, **451**–452
Glomar Challenger, 361–**317**
gneiss, **175**
gold, **513;** in sea water, 347
graben, 211–**212**
gradient, stream, 258
Grand Canyon, **221, 398**–399
granite, **171,** 238
granitic rock, 171
graphic scale, *127*
graphite, 176, 401
gravitation, Newton's Law of, 9–10
gravitational force, **9**–10, 93–95
gravity, 10; barrier of, 93–95; of earth, 10–11; and erosion, 239, **240–241,** 242; of moon, 84; pit of, 94–95; and space travel, 93–95; specific, of minerals, 165–166; and

stress in rocks, 213; and weight, 11; *see also* gravitation
great circle, **122**
Great Lakes, formation of, **287**–288
Great Red Spot of Jupiter, **56**–57
greenhouse effect, **429,** 430, 439
Greenland glaciers, 278, 281
Gregorian calendar, **87**–88
Gregory, Pope, 88
ground moraine, 285, 287
ground water, 254, **266,** 267
Gulf Stream, 355–356, **357**–359, 483; movement of water in, **358**
gullies, **257**
gullying, **245**
guyots, *326*
gypsum, 172
gyres, *352*

hachure lines, *130,* **130**
hailstones, **451,** 452
hair hygrometer, 442
half-life of uranium, 382, **383**
halite, 172
Halley's Comet, 50
hanging valleys, **282,** 283
hardness of minerals, 165
Hawaiian Islands, 326–**327;** climate of, 489
headward erosion, **257**–**258**
heat energy, 505
heating, air movements and, **433**–434
helium, 46; in air (Table), 421
highs (anticyclones), **465**
Himalaya Mountains, 224
horizons of soil, **239**
horn peaks, 281, **282**
hornblende, **157**
Horse Head nebula, 37

hot spot, *195, 326*
hot springs, 268–**269, 270**
humid continental climate, **488**
humid subtropical climate, **488**
humidity, *440;* absolute, **441;** relative, 441, 442; saturation value of, 441
humus, 238–239
hurricanes, 472–**473, 474**
Hutton, James, 207
hydration, *235*
hydrocarbons, *499; see also* fossil fuels
hydrogen, 46; in air (Table), 421; atomic mass of, 142; isotopes of, **142**
hydrographic maps, 125
hydrologic cycle, 253–**255,** 256
hydrosphere, 420

Ice Age, 413
ice ages, 289–290, 324; causes of, 290–292
ice crystals, 148, **450**–451
ice field, **276**
ice floes, 337–338
ice front of glacier, 278
icebergs, 281, 328
igneous rocks, 169–**171, 173;** extrusive, 171, **173;** intrusive, 170–**171;** massive, **176,** 177
index contours, 129, **130**
index fossils, **386**–387
Indian Ocean, **317**
inferior mirage, **432**–433
International Date Line, **66**
intrusive igneous rocks, 170–**171**
invertebrates, *407*–408
ion(s), *145,* 149; sodium chloride, 145–**146**

ionic bond, 147
ionosphere, 431
iron, 76, 191
irregular galaxies, **17**
island arcs, *195,* 223
isobars, **470**
isogonic lines, *189*
isostasy, *212*–213
isotherms, *480,* **481**
isotopes, 142–143; of hydrogen, **142**

Japan Current, 356
jet streams, 465–**466**
joints of rocks, 177, 210–*211*
Julian calendar, **87**
Jupiter, **56**–57
Jurassic period, **411**

karst plain, **271**
Kepler, Johannes, **6,** 117; his laws of planetary motion, **6**–9
kettles, *286*
knob, glaciated rock, 282–**283**

Labrador Current, 356, 358
La Brea tar pits, **388**
laccoliths, *201*
lagoon, *307*–308
lakes, basins of, 287, 288; formation of, **267;** life history of, 288–289; origin of, 287–288; oxbow; **262;** salt, 288–289
land breezes, 472
landforms, guide to study of, 127–128; locating of, 130–131; minor, origin of, **247**–249; structure of, 248
LANDSAT, **102**–**103**

landslides, 214, 241, 288; underwater, 328
latent energy, 440
lateral moraines, **284**
latitude, *121;* parallels of, *121,* **121;** and temperature (Table), 480
launch windows, 98
lava, 170, 171, *194,* aa, 197; block, 197; pahoehoe, **197;** plateaus of, **197**
leaching, *235,* 239
leap year, 87–88
Lembert Dome (Yosemite National Park), **283**
levees, 264
light rays, absorption of by sea water, **336**
light waves, **12,** 33
light-year, *15*
lightning, 474–**475**
lignite, 500
limestone, 172, **174,** 176, 514
limestone column, **271**
limestone rock, 270, **271**
limestone sink, **270,** 271
liquids, molecular arrangement in, 147; molecular motion in, **147;** *see also* water
lithosphere, 187–*188,* 420
local climates, 489–490
lode, *513*–**514**
lodestone, 166
loess, **244,** *245*
longitude, 121; meridians of, 121–**122**
longshore currents, *301*–**302**
looming mirage, 433
lunar eclipse, **82**–83
luster of minerals, 163

magma, *170*–171, *194,* 513
magnetic declination, *190*

magnetic disturbances of sun, 26–27
magnetic field, **190–191, 192**
magnetic poles, 190
magnetic storm, 26–**27**
magnetism, of earth, 188, **189–190;** of minerals, 166
magnetite, 166
magnitude, absolute, 29–30, **35;** apparent, **30;** of earthquakes, 217; scale of, 30
main sequence stars, 34–**35**
Mammoth Cave (Kentucky), 270–271
mantle, of earth, *184*
map projections, 123–125; of plane surface, **125**
maps, climate, 125; contour, interpretation of, 129–130; of earth's surface, 122–125; geologic, 125; hydrographic, 125; political, 125; relief, 125; star, **5;** topographic, 127–131, **132–133;** use of, 125, 127; weather, 125, 422, **470**–471
marble, **175,** 176
maria of moon, *72*
marine cliff, 297
marine west coast climate, **487, 488**
Mars, 54–**55;** exploration of, 100; rocks of, 138; travel to, 96–97
mass-wasting, *241*
massive rocks, 176, 177
mature mountains, **248,** 249
mature soil, **239**
mean sea level, *128*
mean solar time, *65*–66
meanders, 261–**262**
mechanical weathering, 233, 237, 238, 296–297

median moraines, **284**–285
Mediterranean climate, 487, **488**
mercurial barometer, 422, **423**
Mercury, 49–**50,** 51; transit of, **52**
meridians, 121–**122**
mesa, **249,** 250
mesopause, 430, **431**
mesosphere, 430, **431**
Mesozoic era, 401–**402,** **410–411, 412–413**
metal ore minerals, 158, **160–161,** 512–514
metals, 512–514; native, *512;* recycling of, 498–**499**
metamorphic rock, 170, 172, **175,** 176, **177**
meteorites, *48, 48–49,* 331
meteorology, *420*
meteors, 49
mica, **156,** 157
middle latitude climates, 484, **485**
mid-ocean ridges, *188*–189, **324;** faults in, **327**
Milky Way, 15–16, **45**
mineral resources, future supplies, 515–516
minerals, *153*–169; cleavage of, 164–**165;** color of, 163; crystals of, 163–**164;** fluorescent, 166, **167;** fracture of, **165;** gem, 158, **162,** 506; hardness of, 165; identification of, 158, 163–169; luster of, 163; magnetism of, 166; metal ore, 158, **160–161,** 512–514; nonmetallic, 514–515; optical properties of, **167;** phosphorescent, 166; radioactive,

167; rock-forming, 158, **159;** in sea water, 346–347; on sea floor, 515–**516;** specific gravity of, 165–166; streak of, 163; structure of, **154;** tests for identification of, 167, **168,** 169; types of, 153–158

minimum energy orbit, **96**

minutes, 121

mirages, **432**–433

Mississippian period, 409

mixing in sea water, 345

Mohorovicic discontinuity (Moho), *186*

Moh's scale, 165, 166

molds, fossil, **390**

molecule(s), *143*–149; chemical bonds in, 144–145; forces holding, 145–149; relative motion of, **147;** water, **144,** 147–**148**

monadnocks, **248,** 249

monsoons, **483**

moon, 71–88; albedo of, 429; and calendar, 86–88; craters of, 72–73; diameter of, 78; distance from earth, 78–79; eclipse of, 82–83; history of, 76–77; maria of, 72; movement of, 4; orbit of, 77–**78, 79;** phases of, 80–81; rotation of, 74; surface of. 72–76; temperature of, 73–74; and tidal bulges, 84, 85–86; travel to, 95–96

moon rocks, 75–76, 138

moraines, **284–285,** 287

Mount Everest, **207**

Mount St. Helens, **203, 501**

mountain breezes, 472

mountains, domed, 221–**222;** fault-block, **220;** folded 220; formation of, 220–224; life cycle of, **248,** 249; mature, 248, 249; peneplain of, **248,** 249; plate tectonics and building of, 222–224; synclinal, **209–210;** volcanic, 221, 326; youthful, **248,** 249

mudflow, 241

muds in ocean sediment, 331

mutual gravity, 37

native metal, *512*

natural gas, 499, 502–503; underwater deposits of, 347

nautical mile, *121*

navigation charts, 125

neap tides, **86**

nebula, *37, 37,* **37, 38,** 48; Crab, **38;** Horse Head, **37**

nebular theory of solar system, 45, 46

Neptune, 60

neutron stars, **38**–39

neutrons, *22,* 140–142, 503

névé, 275–276

new phase of moon, **80**

Newton, Isaac, **9,** 120; his law of gravitation, **9**–10; and spectra, 32

Niagara Falls, **259,** 383–**384**

nickel, 189

Nile River delta, **263**

nimbostratus clouds, **445,** 448

NIMBUS (weather satellite), 468

nitrogen, 148; in air (Table), 421

nitrogen cycle, **424**–425

nitrogen-fixing bacteria, 424

North American air masses, 458–**459**

North American climates, 486–**487, 488**–490

North American continent, building of, 404–414; cross-section of, **413**

North Atlantic Drift, 356, 358, 483

North Atlantic Ocean, **317**–318; floor of, **324**

North Equatorial Current, 355

North Pacific Current, 356

North Pacific Ocean, **317**

North Pole, 61–**63**

North Star (Polaris), 4, **118**

northern lights, **27**

nuclear energy, 503–505

nuclear fission, 503–504

nuclear fusion, *22*–23, *505*

nuclear reactions, 21–*22,* **22**

nuclear reactor, **504;** breeder, 505

nuclear wastes, disposal of, 512

nucleus, atomic, 140, 503

oblate spheroid, *120*

obsidian, 171, **172, 173**

ocean basin floor, **323**–327; *see also* sea floor

ocean farming, 347–348

ocean waves, 359–365; breakers, 362, **363;** crest of, 360, **361;** energy of, 296; erosion by, 296–300; giant, 363–365; height of, 360, **361;** longshore, 301–**302,** 303; period of, 360–**361;** refraction of, 361–**362,** 363; speed of, 360–361; swells, *360,* **360;**

tidal, 354, 363; trough of, 360, **361;** tsunamis, 363-**365;** wavelength of, 360-**361;** wind and, 359-360

oceanography, 316

oceans, 316-331; basins of, **323**-327; chemical deposits in, **330;** chemical properties of, 319; currents in (*see* currents); depths of, 318-**319;** floor of (*see* sea floor); and light rays, **336;** pollution of, 342; solar energy from, **509;** surface temperature of, 337-339; temperatures of, 319; underwater exploration of, 316, **321**-322, **323;** water movements of, 320; of the world, **317**-318; *see also* sea water

oil pool, 501-**502**

oil shale, 501

oil traps, **502**

olivine, 157

ooze on sea floor, 331

opposite tide, *84*

optical window, **12**-*13*

orbit, of earth, **4, 8,** 60-**61,** 62-63; elliptical, 78-**79;** minimum energy, **90;** of moon, 77-**78, 79;** planetary, **51;** polar, *101;* of rocket, 96-97; of satellite, 100-102; of sun, 26; synchronous, *101;* of Venus, 96-**97**

orbiters, space, **105-106, 107**

Ordovician period, **408**

ores, 158, 513-**514**

organic rocks, 172

organic sediments, 329-**330**

outcrops, *397*

outer core of earth, **184-185**

outwash plain, 286

overturn in sea water, 345

oxbow lake, **262**

oxidation, *235*

oxygen, 148; in air (Table), 421; atomic mass of (Table), 141; atomic number of (Table), 141

ozone, 421

ozone layer, 430

Pacific Ocean, **317,** 318, 326

Pacific Standard Time, **65**

pack ice, 337, **338**

pahoehoe lava, 197

paleontology, *385*

Paleozoic era, 401-**402, 406**-407, **412-413**

Pangaea, 182, 403, **410**

parallax, 28-**29**

parallels of latitude, *121,* **121**

parasitic cones, *199*

peat, 500

peneplain, **248,** 249

Pennsylvanian period, 409

penumbra, *82*

perigee, 78-79

perihelion, 7-8

permafrost, 489

permeability of rock, **266**

Permian period, 409-410

petrification, 388, **389**

petroleum 501-503, supplies of, 503, 510-511; underwater deposits of, 347

phosphorescent minerals, 166

photosphere of sun, **25-26**

photosynthesis, 424

phyllite, **175,** 176

physical properties of sea water, 335-339

phytoplankton, 344

pitching fold, **210**

placer deposits, *514*

plains, abyssal, 325-326; life cycle of, 249-250; outwash, 286

planets, 11, 12; birth of, **46**-47; circulation system of, 433; Earth (*see* earth); inner, 46; Kepler's laws of motion of, 7-9; knowledge gained from exploration of, **99**-100; miniature, 56-**57;** orbits of, 5, 7-8, 60-**61;** outer, 46-47; protoplanets, *46;* travel to, 95-96, 97-98; *see also* specific planets

plankton, 344, **345**

plants, and photosynthesis, 424; as energy source 509-510; in sea water, 329, **330,** 343-344; weathering and, **235**

plate tectonics (theory), *183;* and volcanism, 192-198; *see also* crustal plates

plateaus, lava, **196;** life cycle of, **249**-250

Pleistocene epoch, 414

Pluto, 60

plutonic rocks, **171**

polar Atlantic air masses, **459**

polar Canadian air masses, **459**

polar climates, 484, **485,**

polar easterlies, 435

polar front, 465-466

polar orbit, *101*

polar Pacific air masses, **459**

Polaris (North Star), 4, **118**

political maps, 125

pollution, air, 426-**427;** of

sea water, 342
polyconic projection, *125*
polyps, coral, 304
pothole, **261**
Precambrian era, 401–**402,** 404–405
precession, *62–63*
precipitation, 254–255; causes of, 449–451; measurement of, **452;** rainfall patterns, **484, 494;** and temperature, 483–**484;** types of, **451**–452; *see also* rain
pressure, air, 421–424, 466; of sea water, 320–321
pressure ridges, **280, 338**
primary earthquake wave, 215
prime meridian, *122,* **122**
prominences of sun, *26–27*
proton-proton nuclear reaction, **22–23**
protons, 22, 140–142
protoplanets, **46**–47
pseudofossils, *391*
psychrometer, 441–**442**
Ptolemy, **6**–7
pulsars, *38–39*
pumice, 171
pyroclastics, *197,* 198
pyroxene, **157**
pyrrhotite, 166

quadrangles, 127
Quaternary period, 414
quartz, 237, 238; structure of, **154;** types of, **155**
quartzite, **175,** 176
quasars, *39*

radar, 468

radiation, electromagnetic, *11*–**12, 13;** heating by, 433
radiation fog, 448
radio telescope antennas, **14**
radio telescopes, **14**–15
radio waves, **12**
radio window, **12,** 14
radioactive decay, rate of, 381–**382**
radioactive minerals, 167
radioactivity, geologic time and, 381–383
radiolaria, 329–**330,** 331
radiosonde, 467–**468,** 470
rain, 457; causes of, 449–451; coalescence of drops of, **449**–450; and ice crystals, **450**–451; *see also* precipitation
rain forest climate, **485**
rain gauges, **452**
rainbow, **431,** 432
rainfall, pattern of, **484**
rapids, 259, 260
recycling of metals, 498–**499**
red clay, 330–331
red giants (stars), 34–**35**
reefs, barrier, 309; coral, **309–310;** fringing, 308–**309**
reflector telescope, **13**–*14*
refraction of waves, 361–**362,** 363
refractor telescope, **13**–*14*
relative humidity, 441, 442
relief, *129*
relief maps, 125
residual boulders, **236**
resources, *498;* recycling of, 498–**499**
revolutions, rock layers and, 377
Richter scale, 217

rift valley, 211–**212**
rip currents, **363**
ripple marks in rocks, **378**
rivers, formation of, 256–**257,** 258–260
rocket engine, **94**–95
rocket lift-off, 93, **94**–95
rockets, 94–97; fuel-oxidizer type, 95–96; orbit of, 96–**97**
rocks, 153, 169–178; basaltic, 171, **178;** carbonation and, 270; chemical, 172; conformity of, 378; cross bedding of, **378;** deforming of, **208;** dip of, **210;** disconformity of, **379;** and early history, 379–381; extrusive igneous, 171, **173;** faulting in (*see* faults); folding of, **208–209, 210;** glaciated, **282**–283; granitic, 171; igneous, 169–**171, 173, 176;** intrusive igneous, 170–**171;** joints of, 177, **210**–211; layers of, **397–398,** 399–400; limestone, 270, **271;** Martian, 138; massive, **176,** 177; melting temperature of, 192; metamorphic, 170, 172, **175,** 176, **177;** moon, 76, 138; organic, 172; permeability of, **266;** plutonic, **171;** residual boulders, **236;** and revolutions, 377; ripple marks on, **378;** sedimentary (*see* sedimentary rocks); strata of, 176–**177;** stress in, 212–213; strike of, 209–**210;** types of, 169–172; unconformity

of, 378, **379;** weathering of, 235, **236–237**

Rocky Mountains, 290, 413; Canadian, **276, 283,** 285

Roman calendar, 87

Sahara desert, **486**

salinity, *341;* graph of, 341

salt(s), dissolved in sea water, 337–**342;** for Table (*see* sodium chloride)

salt lakes, 288–**289**

San Andreas fault, 218–**219**

San Francisco earthquake (1906), 219

sand, as building material, 514; grains of, 299–300; wind erosion and, **242–243**

sand bars, **300**

sand dunes, **243**–244

sand spit, *302*

sandstones, 171, **174,** 385–386

sandy soils, 238

Sargasso Sea, 356

sargassum, 356

satellites, communications, **104;** earth, **100–101,** 102; LANDSAT, 102–103; orbit of, 96–**97;** weather, **104**

saturation, zone of, **266**

saturation value, *441*

Saturn, **57, 58, 59**

savanna climates, 486

scale of map, 127

schist, **175,** 176

scientific laws, *7*

Scorpius, *5*

SCUBA, **320**–321

sea arches, **298**

sea breezes, 472

sea caves, 298

sea cliffs, **297, 298, 300**

sea floor, earthquakes on, 354; features of, **323–324;** mineral deposits on, 515–**516;** nodules on, 347; sediments on, 319, 320, 325–**330,** 331; spreading of, 188–190, 327; volcanic eruptions on, 221, 326, 328, 354

sea-floor spreading, *189*

sea level, changing of, **303**–304; mean, *128*

sea stack, **298**

sea water, 335–348; chemical properties of, 335, 340–346; density of, **339;** desalting of, 346; dissolved gases in, 340, 342–343; dissolved salts in, 337, 339–**342;** distillation of, 346; evaporation of, 253, 338–339, 341, **342, 440;** food from, 347–348; food chain in, 344, **345;** life in, 343–346; light rays absorbed by, **336;** microscopic organisms in, 329–**330,** 343, 344, **345;** minerals in, 346–347; mixing in, 345; movements of, 320, 352–354; overturn in, 345; physical properties of, 335–339; plants in, 329, **330,** 343, 344; pollution of, 342; pressure of, 321–322; salinity of, *341;* solar energy and, 253, 335–336, 352–353, **440;** substances dissolved in (Table), 341; temperature and depth of, **339;** thermo-cline of, **339;** upwelling in, 345, **346;** *see also* oceans

seamounts, 326

seasons, passage of, 3

secondary earthquake wave, 215

seconds, 121

sedimentary rocks, 169, **170**–172, 397; conglomerate, 171, **174;** fragmental, 171; stratified, **176;** types of, 171–172

sediments, 169, 213, 219, 221; deposition of, **384**–385; microscopic organisms and, 329–**330;** organic, 329–**330;** on sea floor, 319, 320, 325–**330,** 331; stream, 288, **289**

Seismic Sea Wave Warning System, 365

seismic waves, *183*–184

seismograph, 214–**215**

shadow zone, **184**

shale, 172, **174,** 176

Sheep Mountain (Wyoming), **210**

sheet erosion, 245–246, 261

shield cones, *198*

Shiprock volcano, **200**

shoreline, *296;* emergent, **306**–307; features of, 304–**305, 306;** submerged, 304–305

sidereal year, *63*

Sierra Nevada Mountains, 220–**221, 283,** 412

silica, 329–330

silicates, *153*

sills, volcanic, 201–*202*

silt, 238, 331

sky, night **4–5**

slate, **175,** 176
sleet, 451
slumping, **241**
smog, **427**
snow, 451, 457
snow line, *276*
sodium chloride, crystals of, **146;** formation of, 144–**145;** ions of, 145–**146**
soil(s), *232;* formation of, 238–239; horizons of, **239;** lunar, **79;** mature, **239;** profile of, **239;** sandy, 238
soil creep, *240*
soil erosion and conservation, **245–246,** 247
solar cell, **24,** 507
solar eclipse, **82**–83
solar energy, 21–24, 507–510; and atmosphere, 427–**428,** 429; changes in amount of, 290–291; reflection of (Table), 428; and sea water, 253, 335–336, 352–353, **440;** trapping of, 429
solar flares, 27
solar nebula, **46**
solar prominence, *26–27*
solar system, 44–66; dust cloud theory of, 44–47; formation of, 44–49; *see also* planets, sun, universe
solar telescope, **24**
solar time, apparent, 63–64; mean, 65–66
solar wind, 25–26
solids, molecular arrangement in, 146–147; molecular motion in, **147**
sonic depth recorder, **318**
sound waves, 318–319
South Atlantic Ocean, **317**
South Equatorial Current, 355

South Pacific Ocean, **317**
southern lights, 27
space shuttle, 104, **105–106,** 107
space telescopes, 15
space travel, 95–100
space lab, **106**
spatter cones, *199*
specific gravity of minerals, 165–166
specific heat of water, 482
spectra, **31–33,** 34
spectrograph, 32
spectroscopy, 30–34
spectrum, bright-line, **31**–32; continuous, **31**–32; dark-line, **31–32;** electromagnetic, *13*
spiral galaxy, **17**
spit, **301**–302
spring tides, **85**
springs, **268;** hot, 268–**269, 270**
squall line, 460
stacks, **298**
stalactite, **271**
stalagmite, **271**
standard time, 65–66; zones of, **65**
star(s), absolute magnitude of, *29*–30; apparent magnitude of, *28–**29, 30;** birth of, 35–39; brightness of, 4, 28–30; cepheid variables, 30; death of, 38; distance from earth, measurement of, 28; double, 37; kinds of, 34–35; life cycle of, 35–39; magnitude of, 28–30; main sequence, 34; movement of, **4–5;** neutron, 38; nova, 38; parallax angle of, 28–**29;** pul-

sars, 38; red giants, 34; supergiant, 34–35; temperature of, **35;** white dwarfs, 34–**35**
star charts, **5**
star tracks, **4**
starlight, measurement of, 30–34
station model, symbols used in, **469,** 470
steam wells, **506**
stellar parallax, measurement of, 28–29
steppes climate, 487
Stonehenge, 2–**3**
strata of rocks, 176–**177**
stratified drift, 284
stratocumulus clouds, **447,** 448
stratopause, 430, **431**
stratosphere, 430, **431**
stratovolcanoes, *198*
stratus clouds, **445,** 448
streak of minerals, 163
stream piracy, **258**
streams, load of, 262–263; rejuvenated, 260; sediment in, 288, **289;** terraces of, **260;** youthful, 259
stress in rocks, 212–213
strike of rock layer, *210,* **210**
strike-slip fault, *210,* **210**
strip cropping, **246**
strip-mining, **500**
Stromboli volcano, 195
subarctic climate, **488**–489
subduction zone, *189*
submarine canyon, **325**
submarine vehicles, 321–**322**
submerged coastline, 304–**305**
subpolar lows, 435
subtropical highs, **435**
summer solstice, *61,* **61,** 62

sun, absolute magnitude of, 30; atmosphere of, 24–**25;** body of, 24, **25;** chromosphere of, **25;** core of, 24, **25;** corona of, **25;** distance from earth, 8; eclipse of, **82**–83; energy of (*see* solar energy); magnetic disturbances of, 26–27; movement of, **3;** nuclear reactions in, 21–22; orbit of, 26; perihelion and aphelion of, 7–8; photosphere of, **25;** prominences of, **26–27;** proton-proton reaction of, 22–23; structure of, 24–27; temperature of, 24; and tides, **85**

sundial, **63**

sunspots, **26**

supergiant stars, 34–**35**

superior mirage, **432,** 433

supernovas, 38

superposition, principle of, 377, 378

supersaturated air, 444, 446

surfacelong earthquake wave, 215

surfing, **296**

swamps, formation of, **267**

swells, *360,* **360**

synchronous orbit, *101*

synclinal mountains, **208–209**

syncline, **208**–*209*

talus slope, **240**

tarn, **281,** 287

telescopes, optical, *13–***14,** 43; radio, **14**–15, 43; reflector, **13**–*14;* refractor, **13**–*14;* solar, **24;**

space, 15

temperature, and air pollution, 426–**427;** and altitude, 482; Celsius scale, 466; and climate, 479–484; Fahrenheit scale, 466; latitude and (Table), 480; of moon, 73–74; ocean currents and, 483; of oceans, 319; precipitation and, 483–**484;** of stars, **35;** of sun, 24; wind and, 483

temperature inversion, *427,* **427**

terminal moraines, **285,** 287

terracing, **246**–247

Tertiary period, **412**–413

tetrahedron, **154**

thermocline of sea water, **339**

thermometers, 466, **467**

thermonuclear reaction, 21–22, **22**

thermosphere, 430–**431**

thrust of rocket engine, *94,* **94**

thunderstorms, 457, 460, **474–475**

tidal bore, 366, **367**

tidal bulges, moon and, **84**–85

tidal currents, 366

tidal flats, *308*

tidal oscillations, *366*

tidal waves, 354, 363

tides, 84–86, 352, 365–**366,** 367; energy from, 506–507; power of, 366–367

time zones, **65**–66

TIROS, **468**

tombolos, *302*

topographic conditions, ice ages and, 291–292; and

weathering, 238

topographic maps, 127–131, **132–133**

topographic sheets, 127

topography, *127*

tornadoes, **475**–476

Torricelli, Evangelista, 422

trade winds, 435

transit, *52*

transpiration, **254,** 440

travertine, 269

trenches, ocean, *194, 325*

Triassic period, **411**

tributaries, **257**

trilobites, **408**

Tropic of Cancer, **61**

Tropic of Capricorn, **61,** 62

tropical Atlantic air masses, **459**

tropical climates, 484, **485**

tropical continental air masses, **459**

tropical Gulf air masses, **459**

tropical Pacific air masses, **459**

tropopause, 430, **431**

troposphere, 430, **431**

tsunamis, 363–**365**

tundra climates, 489

turbidity currents, 354

Tycho, **72**–73

umbra, *82*

unconformity of rock, 378, **379**

undertow, 362–363

uniformitarianism, *208*

United States, major climatic regions of, **488**

Universal Calendar, 88

universe, Egyptian concept of, **117;** Hindu concept of, **117;** middle-ages concept of, **117;** models